1250 MC
01

Critical Problems in the History of Science

CRITICAL PROBLEMS

IN THE HISTORY OF SCIENCE

Proceedings of the INSTITUTE FOR THE HISTORY OF SCIENCE

at the University of Wisconsin, September 1–11, 1957

Edited by **Marshall Clagett**

THE UNIVERSITY OF WISCONSIN PRESS

Madison

1962

Published by the University of Wisconsin Press
430 Sterling Court, Madison 6, Wisconsin

Copyright © 1959 by the Regents of the University
of Wisconsin. Copyright, Canada, 1959

Second Printing, 1962

Printed in the United States of America

Library of Congress Catalog Card Number 59-5304

The preparation of a preface to this volume of distinguished papers gives me considerable personal satisfaction. For they answer so well the hopes of those of us who helped plan the Institute of the History of Science held at the University of Wisconsin, September 1–11, 1957. As director of that Institute I think it not inappropriate for me to say something about how the Institute took form.

In the Spring of 1956, the Joint Committee for the History and Philosophy of Science of the Social Science Research Council and the National Research Council—a committee under the chairmanship of Professor Richard Shryock—discussed various ways and means of stimulating the study of the history of science. One of the means suggested was to have a meeting of somewhat longer duration and at the same time more intensive than the usual meetings of the History of Science Society. The Wisconsin member of the Joint Committee, Dean Mark Ingraham, was asked to explore the possibilities of holding such a meeting at the University of Wisconsin. Upon my return from Europe in July, 1956, Dean Ingraham turned over to me—as Chairman of the Department of the History of Science—the task of preparing a program and making the other necessary arrangements for the proposed conference. I in turn met with the members of the History of Science group at the University of Wisconsin and discussed various topics that might lead to fruitful papers and discussions. The members of this group other than myself were Aaron Ihde, Professor of Chemistry and the History of Science, Robert Stauffer, Assistant Professor of the History of Science, Erwin Ackerknecht, Professor of the History of Medicine, and Glenn Sonnedecker, Associate Professor of the History of Pharmacy and the History of Science. Our group was later depleted by the resignation of Professor Ackerknecht from the University of Wisconsin and by the departure on European leave of Professor Stauffer, but was augmented by the addition of Erwin Hiebert, recently appointed as Assistant Professor of the History of Science.

On the basis of the discussions of this group I drafted a proposed program, which after consultation with Dean Ingraham I submitted to the National Science Foundation as the basis for a request of funds. At the same time valuable suggestions were made regarding the program by members of the Joint Committee of the SSRC and the NRC. The most important suggestion of the Joint Committee was that I make every effort to invite in addition to professional historians of science such scientists, philosophers, and sociologists as might be genuinely interested in the history of science.

While negotiations for financial support from the NSF were under way a supplementary fund of $5,000 was requested and received from the Research Committee of the University of Wisconsin. I hardly need say without this grant and the final grant of $24,270 from the NSF our highly stimulating meeting could not have been held. I might further add that the NSF grant specifically included funds allocated for the publication of the various papers of the Institute. Hence it is the National Science Foundation (and particularly its scientific personnel and educational program representatives, Harry Kelly and Donald Anderson) that makes possible the publication of this volume.

When our group here at Wisconsin discussed the possible program in detail, we of course realized that we could concentrate on only a very few of the important problems now under investigation by historians of science. Hence we were forced to be somewhat arbitrary in our selection of topics, and thus it might seem at first glance that we put too little emphasis on developments of the last century. The Committee would certainly agree that this is so. But I would stress that so few historians are doing serious and professional historical work in the history of science of the last few decades, that the presentation of a critical discussion of such problems would be most difficult. It might also seem that we slighted the biological developments in favor of those in the physical sciences. This was not our original intention. But our preliminary efforts to line up a notable group of people in the discussion of nineteenth-century biology was only partially successful. The field of those engaged in active research in biological history is so narrow that when we received some advance refusals, we had as a result to eliminate an additional day we had hoped to devote to biology.

Our choice of problems was also dictated to some extent by the necessity of taking problems which were of significance to the development of science and would at the same time serve as the basis of a general discussion from the floor. This latter consideration was particularly in our minds when we devoted three whole days to the so-called "Scientific Revolution" of the seventeenth century, i.e., the revolution in astronomy and mechanics. The interpretation and evaluation of this revolution is a central point of focus for all who teach the history of science and for many who are engaged in detailed research on some phase of medieval or early modern science. As the papers in the first part of this volume indicate, a great diversity of opinion is held as to the significance and originality of the antecedents to seventeenth-century science. On the first day Messrs. Hall and de Santillana delineated the participants in the revolution and the main lines that were crucial to the revolution, and particularly they discussed the parts the scholar, artisan, and artist played in the revolution. The four succeeding speakers attempted a detailed exploration in what the Revolution consisted, first in the general area of method (Crombie and Clark), and then in each of the two most important scientific areas in the seventeenth century: mechanics (Dijksterhuis) and astronomy (Price).

Of equally wide interest as the problem of the scientific revolution was that of the teaching of the history of science itself. We felt that the varying difficulties encountered in teaching art students as well as students of engineering and the sciences ought to be treated with some discussion of the ways in which various teachers had attempted to solve these difficulties. Miss Stimson examined the importance of the subject in a Liberal Arts college, particularly in relationship to the course she gave for so many years to girls of Goucher College. On the other hand, Henry Guerlac reported his experiences in a course for engineering students. The commentators naturally stressed their own experiences in still different academic environments.

One of the areas of research in the history of science which is being actively explored is the history of science in the French Revolution. The first of our contributors (Gillispie) examined the philosophy of science of some of the principal Jacobin figures in the Revolution, connecting that philosophy with the *Encyclopédie*. The next speaker (Pierce Williams) examined other phases of science during the Revolution, using primarily educational sources, and indicated other lines than those suggested by Gillispie. The contrasting views of these two papers makes for extremely interesting reading.

The committee felt that in addition to the previously noted problems, others, relating to specific sciences, should be explored. In nineteenth-century physical science the doctrine of the conservation of energy played an important role. Hence its manifold origins seemed of interest and Thomas Kuhn's careful examination of the contribution and relationships of the various "discoveries" of energy conservation will constitute a necessary starting point for further investigation of the part played by the concept of energy conservation in modern science. That the energy concept was fruitful for other areas of science other than the investigation of heat and mechanics was shown by I. B. Cohen, who particularly concerned himself with the origin and development of the concept of conservation of charge in electrical investigations.

In the area of biology, J. Walter Wilson showed that evolution was only one of the important factors in the attainment of maturity in the nineteenth century, citing, for example, the cellular hypothesis and the idea that all organisms proceed from organisms. That biological concepts were most fruitful in areas of nineteenth-century social thought is a truism, but their precise use by individual authors is always worthy of investigation and John Greene's careful analysis of the ideas of Comte and Spencer constitute just such investigations.

A last problem that seemed important to us was to see how chemical developments affected concepts of the structure of matter. Cyril Smith turned his attention to seeing how a technical study, namely metallurgical investigations, affected ideas of material structure, while Miss Boas occupied herself with the concepts of matter of the general chemists of the seventeenth and eighteenth centuries.

The commentators have, the reader will readily see, often disagreed radically with the authors of the principal papers, and needless to say the editor—in those areas where he feels he has a right to his opinion—finds himself in occasional disagreement with the conclusions and even in some cases the method followed by authors, but rarely have the various points of view on these fundamental problems been so well presented. It will be observed that on a few occasions the commentators quote statements that cannot be found in so many words in the papers on which they are commenting. This results from the fact that the authors have in some cases altered their original papers either as the result of the commentaries and the general discussion or simply for literary purposes. I believe, however, that the reader will have no difficulty in following the points being made by the commentators even when the original text has been slightly altered. Furthermore, leaving the commentaries much as they were when delivered will give the reader some hint of the effectiveness of the discussions. There can be little doubt that these essays will prove of considerable value to the historian of science as he pursues his teaching and research. I can personally testify that listening to and reading these papers has brought to focus or emended a number of my own ideas. But it was, of course, not only the formal papers themselves that proved so useful and stimulating, but the continuous and sometimes peripheral discussion of these and many other problems that went on for the ten days of the Institute, days of close association—both during the daytime sessions at the Hillel Foundation and during the evening at the Villa Maria where all seventy-two invitees were staying. In fact, it is probably this close association that those of us who participated in the Institute will longest remember. It is thus regrettable that the excitement and content of these informal discussions cannot somehow be included here. But still we are certain that much of the wisdom of those who spoke and discussed critical problems in the history of science at the Institute is layed before the readers of this volume.

Madison, Wisconsin MARSHALL CLAGETT
January, 1958

CONTENTS

Contents

Contents

ILLUSTRATIONS

🜓 Critical Problems in the History of Science

THE SCHOLAR AND THE CRAFTSMAN

IN THE SCIENTIFIC REVOLUTION

Rupert Hall

Never has there been such a time as that during the later sixteenth and the seventeenth centuries for the great diversity of men in the forefront of scientific achievement.[1] A proportion of those who contributed to the swelling literature of science were in a broad sense professionals: indeed, a sizable proportion, since many minor figures enlarge this group. Among these professionals were university teachers, professors of mathematics, anatomy, and medicine; teachers of these subjects, especially applied mathematics, outside the universities; and their various practitioners —physicians, surveyors, mariners, engineers and so on; and lastly the instrument-makers, opticians, apothecaries, surgeons, and other tradesmen, though their great period in science is to be found rather in the eighteenth century than in the seventeenth.[2] These men, widely divergent as they were in social origins and intellectual attainments, at least occupied positions in a recognizable scientific hierarchy. Some had won them through academic study, others through private education and research, others again by apprenticeship and pursuit of an occupation closely related to scientific inquiry. All were trained men in some way, whether in mathematics, physic and dissection, or the exercise of a manual craft. Now it is surprising enough, whether we make comparison with the scientific world of recent times, or with that of the later Middle Ages, to find such disparity in the professional group, that is, to find that the definition of scientific professionalism must be so loosely drawn; yet it is still more astonishing that many minor figures in the history of seventeenth-century science, and not a few notable ones, constitute an even more heterogeneous collection. Among these true "amateurs" of science (the distinction has really little meaning), some, it is true, had been exposed to scientific influences of a kind in college or university; yet the creation of a permanent interest thus, in an ordinary passing student, must have been as rare then as the acquisition of a taste for Latin verse is now. A few also, no doubt,

3

were quietly encouraged by discerning fathers or by private patrons. The rest remain as "sports"; diffusionist and environmental principles hardly suffice to explain their appearance on the scene. One thinks of such men as William Petty, son of a clothier, Otto von Guericke, Mayor of Magdeburg, John Flamsteed, an independent gentleman of modest means, or, most extraordinary of all, Antony van Leeuwenhoek, an unschooled borough official.

Thus one can never predict the social circumstances or personal history of a seventeenth-century scientist. Given the taste, the ability, and freedom from the immediate necessities of the struggle for subsistence, any man who could read and write might become such. Latin was no longer essential, nor mathematics, nor wide knowledge of books, nor a professorial chair. Publication in journals, even membership in scientific societies, was open to all; no man's work needed the stamp of academic approval. This was the free age between the medieval M.A. and the modern Ph.D. In the virtual absence of systematic scientific training, when far more was learned from books than from lectures, the wholly self-educated man was hardly at a disadvantage as compared with his more fortunate colleague who had attended the seats of learning, except perhaps in such special fields as theoretical astronomy or human anatomy. There were no important barriers blocking entry into the newer areas of exploration, such as chemistry, microscopy, qualitative astronomy, where all types of ability, manual and intellectual, were almost equally required. Obviously it was statistically more probable that a scientist would spring from the gentry class (if I may use this disputed term) than any other, and that he would be a university man rather than not. But the considerations determining the probability were sociological rather than scientific; if the texture of science was almost infinitely receptive of first-rate ability of any kind, the texture of society was such that it was more likely to emerge from some quarters than from others.

It is needful to traverse this familiar ground in order to set in perspective the dichotomy to which I shall turn—that of craftsman and scholar. It is a quadruple dichotomy—social, intellectual, teleological, and educational. It marks off, broadly, men of one class in society from another—those who earn their bread from scientific trades of one kind or another from those who do not. It distinguishes likewise those achievements in science which are in the main practical or operational from those which are cerebral or conceptual. Thirdly, it draws attention to the different objects of those who seek mainly practical success through science, and those who seek mainly understanding. And finally, if we consider only the group whom I have previously called professional, we may discern on the one hand the "scholars" who have been introduced to science by university or similar studies, and on the other the "craftsmen" who have learnt something of practical science in a trade. But we must be cautious in detecting polar opposites where there is in reality a spectrum. The scientific movement of the seventeenth century was infinitely varied, its successes demanded an infinite range of different qualities, and it is against this background of wide inclusion that we must set any attempt at analysis in particular terms.

By far the most closely-knit, homogeneous, and intellectually influential of the groups I have described was that of the university men, including both those who remained as teachers and those who departed to other walks of life. Some of the harshest critics of the contemporary "schools," like Bacon, Descartes, or Webster, were nevertheless their products. The opponents of the Aristotelian "forms and qualities" had been firmly grounded in that doctrine; many future scientists found stimulus in the universally required mathematical studies. To exemplify this point, one may consider the earliest membership of the Royal Society in 1663. Of the 115 names listed,[3] I find that 65 had definitely attended a university, while only 16 were certainly non-academic. The remaining 34 are doubtful, but at any rate the university men had the majority. It is still more telling to single out the names which have a definite association-value on inspection; I rate 38 on this test, of whom 32 are "U" and only 6 "non-U." Whether or not we term such men "scholars" is largely a rather unimportant question of definition: at any rate they had in common a knowledge of Latin, some training in mathematics, and an introduction at least to logic and natural philosophy; quite a proportion would also have had such experience of the biological and medical sciences as was available at the time.

It appears then that the medieval association of scientific activity with the universities was weakened, but not disrupted, in the seventeenth century, though the association certainly became less strong as the century advanced. It was weakened not only by the importance in science of men who were not academically trained at all, but by the shift in the locus of scientific activity from the universities, where it had remained securely fixed throughout the Middle Ages,[4] to new institutions like Gresham College, to the scientific societies meeting in capital cities, and to the circles basking in the patronage of a Montmor or a Medici.[5] If a majority of creative scientists had been at the university, they were so no longer in their mature age. Moreover, while in the medieval university there had been little disparity between the instruction given to the student, and the advanced researches of the master, this was no longer the case in the seventeenth century. In the schools of the fourteenth century the master who remained to teach pushed forward his knowledge, in the main, within the framework of ideas, and through study of authorities, with which he had become familiar at a more elementary level. The seventeenth-century university, on the other hand, almost ignored observational and experimental science. The unprecedented advances in scientific technique occurring in physics, astronomy, botany and zoology, and chemistry were not made widely available to students: there was a fairly good grounding only in mathematics and human medicine. The potential investigator had to learn the techniques he required from practice, by the aid of books, and through personal contact with an experienced scientist, often only obtainable elsewhere. Perhaps even more serious was the absence from university courses of the leading principles of the scientific revolution and of the ideas of the new natural philosophy. In the last quarter of the seventeenth century Cartesian science was indeed expounded in some of the colleges of France, and less widely elsewhere,[6] but dissemination of the thought of Galileo, of Bacon, and of the exponents of the mechanical philosophy owed little to university courses. Occasional examples of a university teacher having a decided influence upon a circle of pupils—as was the case with John Wilkins at Wadham College, Oxford, and Isaac Barrow at Trinity, Cambridge—hardly vitiate the general conclusion that the activities of various societies, books, and journals were far more potent vehicles of proselytization, which is supported by many personal biographies. However stimulating the exceptional teacher, formal courses were commonly conservative and pedestrian: it is curious to note that the two greatest scientists of the age who were also professors, Galileo and Newton, seem to have been singularly unremarkable in their public instruction. If the universities could produce scholars, they were ill-adapted to turning out scientists; the scientist had to train himself. Many who accomplished this transition regarded it, indeed, as a revulsion from the ordinary conception of scholarship. The learning they genuinely prized, in their own scientific disciplines, they had hardly won for themselves. It would surely be absurd to argue that Newton was less a self-made scientist than Huyghens, or Malpighi than Leeuwenhoek, because the former had attended a university and the latter not.

It lies outside my brief to discuss the fossilization of the universities, which, from what I can learn, the Renaissance did little to diminish so far as science was concerned, nor the rise of the new science as a rejection of academic dogma. Recent investigations would, I believe, tend to make one hesitant in concluding that the innovations and criticisms in the academic sciences—astronomy, physics, anatomy—which we call the scientific revolution, were the product solely, or even chiefly, of forces and changes operating outside the universities. Rather it would seem that, in relation to these subjects, it was a case of internal strife, one party of academic innovators trying to wrest the field from a more numerous one of academic conservatives. Certainly this was the case with Vesalius and his fellow-anatomists, with Copernicus, with Galileo. It was the academic and professional world that was passionately divided on the question of the inviolability of the Galenic, Aristotelian, or Ptolemaic doctrines; these quarrels of learned men had as little to do with capitalism as with the protestant ethic. Only towards the middle of the seventeenth century were they extended through the wider range of the educated class.

In the long run—that is to say within a century or so in each instance —the innovators won. In the short run they were defeated; academic conservatism prevented the recognition and implementation of the victories of the revolution in each science until long after they were universally applauded by thoughtful men outside. Whereas in the thirteenth century the schools had swung over to the Greeks and Muslims, despite their paganism and their often unorthodox philosophy, whereas in the fourteenth century the development of mechanics, of astronomy theoretical and practical, of anatomical and other medical studies, had been centered upon them, in the later sixteenth and seventeenth centuries teaching failed to adapt itself to the pace with which philosophy and science were moving. In the mid-sixteenth century the universities could still have formed the spear-head of this astonishing intellectual advance; in Galileo's life-time the opportunity was lost, and despite the invaluable efforts of individual teachers, as institutions the universities figured only in the army of occupation, a fantastic position not reversed until the nineteenth century. The innovators really failed, at the critical period, to capture the universities and bring them over to their side as centers of teaching and research in the new scientific manner. There were, for instance, many schemes in the seventeenth century for organizing scientific research, and for the provision of observatories, museums, laboratories and so on: yet no one, I think, thought of basing such new institutions on a university. That would have seemed, during the last century, a natural course to follow; and it would presumably have seemed equally natural in the Middle Ages.[7]

Hence it happened that the academic type, the scholar, book-learned in Aristotle or Galen, the Simplicius, the professor who could see the holes in the septum of the heart but was blind to the spots on the face of the sun, became the butt of the scientific revolutionaries.

> Oxford and Cambridge are our laughter,
> Their learning is but pedantry,

as the ballad has it. The passage in the *Discourse on Method* may be recalled, in which Descartes reviews critically the content of education and learning as ordinarily understood:

> Of philosophy I will say nothing, except that when I saw it had been cultivated for many ages by the most distinguished men, and that yet there is not a single matter within its sphere which is not still in dispute, and nothing therefore which is above doubt, I did not presume to anticipate that my success would be greater in it than that of others; and further, when I considered the number of conflicting opinions touching a single matter that may be upheld by learned men, while there can be but one true, I reckoned as well-nigh false all that was only probable.

After observing that the other sciences derived their principles from philosophy, which was itself infirm, so that "neither the honour nor the gain held out by them was sufficient to determine one to their cultivation," Descartes abandoned the study of letters "and resolved no longer to seek any other science than the knowledge of myself, or of the great book of the world." With this one may compare Bacon's "surprise, that among so many illustrious colleges in Europe, all the foundations are engrossed by the professions, none being left for the free cultivation of the arts and sciences." This restriction, he declares, "has not only dwarfed the growth of the sciences, but been prejudicial to states and governments themselves." The candid appraisal of the first chapter of the *Advancement of Learning* could have been applied to many academic institutions more than two centuries after it was penned.

8

Admittedly the period when Bacon and Descartes formed such adverse opinions was one early in the scientific revolution; but there is little evidence to show that academic reform progressed rapidly thereafter, and it would not be difficult to quote parallel judgments from a later time. It was not the case, of course, that learned conservatives could see no merit in the study of science. This was no science *versus* humanities wrangle, for the conservatives were themselves teachers of science, of Renaissance science in fact. Their science was Aristotelian and formal; it denounced both Copernicanism and the mechanical philosophy, and distrusted the new instruments and experiments. An analogous situation existed in medicine, where the modernists who were experimenting with chemical preparations and new drugs such as a guaiacum and Jesuits' bark, who followed Harvey and attempted the transfusion of blood, were opposed by the entrenched faculties of so-called Galenists, enemies of every innovation. Nevertheless, the effect was much the same: The "new philosophy" and science were forced to take root outside the academic garden where they should have found most fertile soil.

The effect of the development of new scientific ideas and methods in diminishing the role of the universities as creative centers reinforced rather than initiated the decline of their intellectual prestige, which had begun with the Renaissance. Then, too, new movements in learning and scholarship were at least as much associated with the activities of private scholars, as with those of university teachers. Private patrons had been as energetic in encouraging neo-classical modes of writing and sculpture, as they were to be in promoting science in the seventeenth century. Already in the Renaissance academic learning was reproached for its inelegance in the classical tongues, its imported Arabisms, its lingering attachment to imperfect texts, its barren philosophy. No one was more scathing of academic pedantry than Erasmus, not to say Paracelsus. "Then grew the learning of the schoolmen to be utterly despised as barbarous," says Bacon, so that when he himself attacked the fine philosophic web of scholasticism—too many words spun out of too little matter—he was but repeating an old canard. This revulsion of the Renaissance scholar from the "barbarousness" of still-medieval universities was, as is well known, a linguistic and textual one in the main; it did not touch so much the content of thought as its expression, nor did it, in particular, greatly disturb the pre-eminent position of the ancient masters of science. This aspect of the Renaissance can most clearly be seen in the history of medicine during the first half of the sixteenth century. Some of the lost ground the universities recovered; they began to teach Greek and Ciceronian Latin; more attention was paid to history and literature and less to disputative philosophy. But they could not recover their medieval pre-eminence as cultural centers—particularly perhaps in northern Europe—and the scientists of the seventeenth century had only, in a sense, to follow the path which Renaissance humanists had trodden, in rejecting it.

The object of the preceding remarks is to justify my conception of the scientific scholar of the sixteenth and seventeenth centuries, as a man learned not merely in recent scientific activities and methods, but in the thought of the past. It seems superfluous to argue that the majority of the scientists of the time were of this type, neither technicians nor ignorant empiricists. Certainly the learning of Galileo, or Mersenne, or Huyghens, or Newton, was not quite like learning in the medieval or Renaissance conception; they may have been as deficient in the subtleties of Thomist philosophy as in the niceties of Greek syntax; but to deny that they were learned scholars in their field and in their outlook, would be to deny that any scientist is entitled to be called learned.

I have tried also to trace in outline the way in which, at this time, scientific learning diverged from other branches of scholarship, without wholly severing its affiliations with academic institutions. One might also ask the question: how far was the new scientific spirit of the seventeenth century brought into being by activities of a purely scholarly kind—for example, through the evolution of certain principles of logic during the Middle Ages, or through the activities of the persistent students of Greek science in the Renaissance?

The latter especially furnished the core of an interpretation of the scientific revolution which held favor until recent times. To put it crudely, the scientific revolution was seen, according to this view, as the terminal stage of a scientific renaissance beginning about the mid-fifteenth century, and characterized chiefly by its full exploration of classical scientific texts, which was aided particularly by the invention of printing; the scientific renaissance was itself regarded as a classical reaction against the gothic barbarity of the Middle Ages.[8] This interpretation is in effect an extension of Bacon's, to which I referred earlier; an extension which Bacon himself was unable to make because he did not know that the revolution he sought was going on around him. Clearly, if such a view is accepted, it attaches a very great importance indeed to the activity of the scholar-scientists of the Renaissance, who besides polishing and extending the works of the most authoritative ancient authors, shed a full light on others, such as Lucretius, Celsus, and Archimedes, whose writings had not previously been widely studied.

The merits of this hypothesis of the origin of the scientific revolution are as obvious as its defects. It draws attention to the weight of the contribution of sheer scholarship, and of the amazing Hellenophile instinct of the Renaissance, to the change in science which occurred between 1550 and 1700. No one would deny the connection between the mechanical, corpuscular philosophy of the seventeenth century, and *De natura rerum;* nor the significance for anatomy of the intensive study of Galen; nor would he dispute that the virtual rediscovery of Archimedes transformed geometry, and ultimately algebra. Equally, however, it is clear that this is far from being the whole story: the instances I have quoted are not universally typical ones. The history of mechanics before Galileo, which has been so elaborately worked out in the present half-century, proves the point. Medieval science was not abruptly cut short by a classical revival called a renaissance: it had much—how much must be the subject of continuing research—to contribute to the formation of modern science. Very important threads in the scientific revolution are not really traceable to antiquity at all, at least not through the channels of scholarship; here the chemical sciences furnish examples. Above all, the renaissance-scholarship interpretation fails to account for the *change* in science. If anything is fairly certain, it is that the intention of the Renaissance was the imitation of antiquity, and there is evidence that this ideal extended to the scholar-scientists. Yet the pursuit of this ideal seems to have endured least long in science, of all the learned subjects; it had ceased to have force long before the end of the sixteenth century. There never was a true Palladian age in science, and the limitations that had bound the Greeks themselves were relatively soon transcended in Europe. Why this was so is really the whole point at issue, and the Renaissance-scholarship interpretation does not squarely face it.

Nevertheless, if that view is not completely adequate, it must serve as an element in any more complete interpretation. The different view of the importance of scholarly activities, this time in the Middle Ages, that I mentioned previously, has won ground much more recently. It is one that the non-medievalist has some difficulty in evaluating, and it would be inappropriate for him to criticize it. I had better state my conception of its tenets at the risk of oversimplification: that medieval philosophers evolved a theory for the investigation of natural phenomena which was essentially that applied with success in the scientific revolution. It is not claimed for those who elaborated this theory that they were themselves as eminent in experiment, or observation, or the use of mathematics as their successors; its applications—other than to optical phenomena and the discussion of impetus—seem to have been few and sporadic. It was a scholar's method of science, vindicated by some successes, which only awaited general application to transform the whole exploration of nature, and this the method found in the late sixteenth and seventeenth centuries. Again, then, great importance is attached to the role of the scholar; the scientific revolution, it might be said, is the direct consequence of a philosophic revolution.

At the same time, it is evident that there is a measure of incompatibility between these two alternative appraisals of the supposed contribution of scholars to the genesis of modern science. One lays emphasis on content, the other on method; if the medieval ideas on method are pre-eminently important, then the Renaissance revival of classical science is irrelevant, and vice versa. One view, if allowed to fill the whole picture, tends to obscure the other. We are not forced to an exclusive choice, however, and I think it may be granted that a compromise which allows room for both views is possible. It would seem to be the case that while one theory best accounts for certain aspects of the development of science in the sixteenth and seventeenth centuries, the other best explains other aspects. Nor should it be forgotten that changes of emphasis within the scope of the classical tradition may be attributable, in part at least, to changed ideas of method derived from the Middle Ages. I mentioned earlier the rediscovery of Lucretius in 1417, and the connection of this with the mechanical philosophy of two and a half centuries later; it may well be that new ideas on the form and structure of a scientific theory had much to do with the preference for the atomistic tradition in Greek thought over the Aristotelian thus evinced. The fuller acquaintance that textual scholarship conferred with those relics of Greek science which best exemplified the newer medieval notions of scientific procedure might have built up a greater pressure for the further application of those notions. For example, if Galileo, unknown to himself, inherited a method of scientific enquiry from medieval philosophers, he thought of himself (on occasion) as practicing a method used with success by Archimedes.

The medievalist view, if I may so term it, raises in a peculiarly acute form the question which seems central to my problem. Is the effective and creative impulse, which urged men to abandon not merely the philosophy and doctrines of medieval science but even their Greek foundations, to be found in the dissatisfaction of learned men with established modes of inquiry, and the theories and practices to which they gave rise? In short, was the scientific revolution in the main the product of a sense of intellectual frustration and sterility? If we think this was the case, and that it was the same philosophers, scholars and intellectuals who suffered this frustration, who found a way of breaking through it to a more rewarding kind of inquiry and a more satisfying mode of scientific explanation, then our historical seeking is at an end. We might of course go on to enquire where this frustration originated and what brought it into being, and we might also ask what factors enabled the learned men of science to break through it, but at least we should have established their crucial role in the actual break-through, and all else would be ancillary.

That such frustration was experienced hardly requires demonstration. It is expressed by Vesalius, when he laments—with whatever element of exaggeration and ingratitude—the wasted effort into which too uncritical a confidence in the exactitude of Galen had led him; by Copernicus, when he speaks of the disagreement of mathematicians, and their ineptitude: "Rem quoque praecipuam, hoc est mundi formam, ac partium eius certam symmetriam non potuerunt invenire, vel ex illis colligere"; and surely conspicuously enough by Galileo. The latter's is the attitude of one who has broken out of the dead circle of ancient thought, and who can, from reliance on his own new knowledge, pity as well as condemn those still bound by the chains of authority:

Oh, the inexpressible baseness of abject minds! To make themselves slaves willingly; to accept decrees as inviolable; to place themselves under obligation and to call themselves persuaded and convinced by arguments that are so "powerful" and "clearly conclusive" that they themselves cannot tell the purpose for which they were written, or what conclusion they serve to prove! . . . Now what is this but to make an oracle out of a log of wood, and run to it for answers; to fear it, revere it, and adore it? [9]

Now what the medievalists contend for is, I take it, that such an attitude to authority was already nascent in the Middle Ages, and that it was not merely negative but creative. I quote Dr. Crombie's very plain statement, from the first page of his book on *Grosseteste and Experimental Science* (Oxford, 1953):

Modern science owes most of its success to the use of these inductive and experimental procedures, constituting what is often called "the experimental method." The thesis of this book is that the modern, systematic understanding of at least the qualitative aspects of this method was created by the philosophers of the West in the thirteenth century. It was they who transformed the Greek geometrical method into the experimental science of the modern world.

Why was it necessary to devise new inductive and experimental procedures at all at this point? Dr. Crombie finds the answer to this question in the problem presented to Western natural philosophers by the scientific texts recently made available to them: "How is it possible to reach with the greatest possible certainty true premisses for demonstrated knowledge of the world of experience, as for example the conclusions of Euclid's theorems are demonstrated?"

This view places the genesis of the scientific revolution at a very high level of intellectual achievement, which is still maintained if we transfer our attention to a somewhat different field from Crombie's, namely the history of theories of mechanics from the Middle Ages down to the time of Galileo. Here again we may note, not merely striking dissatisfaction with the Aristotelian explanation of continued motion founded on the total separation of the moving force from the moved inanimate body, as well as with certain other features of mechanics of supposedly Aristotelian formulation, but definite and partially successful steps towards more satisfactory concepts.[10] When we come to the critical point, with Galileo himself, we contemplate an intellectual struggle of the most sublime kind, which Professor Koyré has analyzed for us. If the ultimate victory here is not the result of prolonged and arduous cerebration, then it is difficult to see what successes could be attributable to thought and reason in science. Just as the medieval criticism of Aristotle had come from scholars, so also it was in the minds of scholars that the battle between old and new in science had to be fought. I should find it difficult to cite an exponent of the "new philosophy" who did not visualize its fate in those terms.

There is a point here, however, that deserves fuller consideration, and allows the craftsman to enter on the scene. For while we recognize science as a scholarly activity, and the reform of science as an act of learned men, it may plausibly be asked whether the impulse to reform was spontaneously generated among the learned. Was it perhaps stimulated elsewhere? Some support for this suspicion might seem to spring from the emphasis that has been laid on empiricism, not merely in the scientific revolution itself, but among its philosophical precursors. Thus, to quote Dr. Crombie again: "The outstanding scientific event of the twelfth and thirteenth centuries was the confrontation of the empiricism long present in the West in the practical arts, with the conception of rational explanation contained in scientific texts recently translated from Greek and Arabic." It is unnecessary to dwell on the well-known interest of at least a few learned men, during the Middle Ages, in such fruits of empirical invention as the magnetic compass, the grinding of lenses, and above all, the important advances in the chemical and metallurgical arts.[11] Similarly, everyone is familiar with the arguments of the Baconian school: that true command—and therefore real if unwitting knowledge—of natural processes had been won by the arts rather than by sciences, and that the scholar would often become more learned if he would consent to apprentice himself to the craftsman. All this might suggest that the increasingly spectacular achievements of empirical technology arrested the attention of scholarly scientists, enforcing some doubt of the rectitude of their own procedures, and still more, leading them to accept as an ideal of science itself that subjection of the natural environment to human purposes which had formerly seemed to belong only to the arts and crafts.

There are two issues here. One is the fact of technological progress, which some philosophical critics contrasted with the stagnation of science.[12] The other is the reaction of learned men to the state of technology, and this is more properly our concern. Technological progress was not simply a feature of the Middle Ages and Renaissance: it occurred in the ancient empires, in the Greek world, under the Roman dominion, and even in the so-called "Dark Ages." It would be difficult to think of a long period of complete technical stagnation in European history, though individual arts suffered temporary periods of decline. Some craftsmen at some places seem always to have been making their way forward by trial and error. In short, a philosopher of antiquity had as great an opportunity of appreciating the inventiveness of craftsmen as his successors of the sixteenth and seventeenth centuries, and of drawing the same lessons as were drawn then. Indeed, ancient writers were aware of the importance of the crafts in creating the means of civilized existence, and praised works of ingenuity and dexterity; where they differed from the moderns was in their preservation of the distinction between *understanding* and *doing*. They did not conclude that the progressive success of the crafts set up any model of empiricism for philosophy to emulate. They would not have written, as Francis Bacon did, in the opening lines of the *Novum Organum:* "Man, as the minister and interpreter of nature, does and understands as much as his observations on the order of nature, either with regard to things or the mind, permit him, and neither knows nor is capable of more. The unassisted hand and the understanding left to itself possess but little power. . . . Knowledge and human power are synonymous."

It is the philosopher who has modified his attitude, not the craftsman, and the change is essentially subjective. The success of craft empiricism was nothing new in late medieval and early modern times, and if the philosopher became conscious of its significance for science it was not because such success was more dramatic now than in the past. It was always there to be seen, by those who had eyes to see it, and the change was in the eye of the beholder. It is absurd, for instance, to suppose that the introduction of gunpowder and cannon into warfare was in any serious sense the cause of a revival of interest in dynamics, and especially in the theory of the motion of projectiles, during the sixteenth and early seventeenth century. The ancient torsion artillery provided equally dramatic machines in its day, not to mention the crossbow, mangonel and trebuchet of the Middle Ages. The simplest methods of hurling projectiles —the human arm, the sling, the bow—pose problems of motion no less emphatically than more complex or powerful devices, and as everyone knows, appeal to practical experience of this primitive kind was the basis for the development of the concept of impetus. The earliest "scientific" writers on explosive artillery, such as Tartaglia, did no more than transfer this concept to the operation of a different device.

Such an example reminds us that it may be naive to assume that even major technological advances suggested, contemporaneously, such questions worthy of scientific enquiry as would, indeed, immediately spring to our own minds. The scientific examination of the three useful forms of iron—cast-iron, wrought iron, and steel—did not begin until the early eighteenth century; the geometrical theory of gear-wheels was initiated about fifty years earlier; the serious study of the chemistry of the ceramics industry was undertaken a little later. I choose deliberately examples of practical science each associated with notable developments in late-medieval craftsmanship: the introduction, respectively, of the effective blast-furnace; of the gear-train in the windmill, water-mill, mechanical clock, and other devices; and of fine, brightly pigmented, tin-glazed earthenware.[13] The time-lag in each instance between the establishment of a new craft-skill, and the effective appearance of scientific interest in it, is of the order of 250 years, and in each of these examples it appears *after* the scientific revolution was well under way. If there is some truth in the view that interest in crafts promoted a change in scientific procedures, it is also true that, at a later date, the very success of the new scientific knowledge and methods opened up the possibility of examining craft procedures systematically, which had not existed before.

It would be a *non sequitur* to argue that, because an important measure of technological progress occurred in the Middle Ages (as *we* are aware), medieval scholars recognized the fact and appreciated its significance. Clearly in many instances they did not—that is why the history of medieval technology is so difficult to reconstruct. Our literary records of the Middle Ages were in large part compiled by scholars; the paucity in them of technological documentation—concerning not merely the use of tools like the carpenter's brace and lathe, but major industries such as paper-making and iron-working—is very conspicuous. The historian of medieval technology is notably better served by the artist than by the scribe. This could hardly have happened, had more than a very few scholars been impressed by the empiricism which brought in the mindmill, the magnetic compass, the mechanical clock, and so on.

In any case, I hesitate to conclude that the behavior of an empirical scientist—that is, I take it, one who observes and experiments, both to discover new information and to confirm his statements and ideas—is derivable by virtually direct imitation from the trial-and-error, haphazard, and fortuitous progress of the crafts. This seems to me to be the defect of the view that sees the new scientist of the seventeenth century as a sort of hybrid between the older natural philosopher and the craftsman. It is easy enough to say that the philosopher thought much and did little, while the craftsman did much but had no ideas, and to see the scientist as one who both thinks and does. But is such a gross simplification really very helpful in describing or explaining a complex historical transition? Neither Copernicus, nor Vesalius, nor Descartes, to name only three, were more craftsmanlike than Ptolemy, Galen, or Aristotle. Surely scientific empiricism is itself a philosophical artefact, or at least the creation of learned men—here I believe Dr. Crombie has a very strong point—and it stands in about the same relation to craftsmanship as the theory of evolution does to the practices of pigeon-fanciers. It is a highly sophisticated way of finding out about the world in which we live; on the other hand, the notion that direct immersion in the lore of tradesmen was the essential baptism preceding scientific discovery was one of the sterile by-paths from which the scientists of the seventeenth century fortunately emerged after a short time. Modern studies combine in revealing that the empirical element in the scientific revolution, taking the word in its crudest, least philosophical and most craftsmanlike sense, has been greatly exaggerated; correspondingly we are learning to attach more and more significance to its conceptual and intellectual aspects.

This is not to deny that the processes of artisans constituted an important part of the natural environment. If, by an internal displacement, the attention of the natural philosopher was more closely directed to this, and less to his own consciousness and limited academic horizon, he could learn much of what the world is like. As the Middle Ages verged on the Renaissance, an increasingly rich technological experience offered ample problems for enquiry, and besides, much knowledge of facts and techniques. This, apart from their direct technological importance, was the significance for science of the great works of craft-description and invention by Cellini, Agricola, Biringuccio, Palissy, Ercker, Ramelli and others that appeared in the sixteenth century, for while their own scientific content was slight, these authors provided materials and methods for the use of others more philosophically equipped than themselves. Science indeed owes much to technology: but we must remember that the debt was itself created by natural philosophers and other men of learning.

There is no straightforward answer to any question about the whole nature of the scientific revolution. Here it may again be useful to recall the deep distinction between the academic sciences (astronomy, anatomy, mechanics, medicine) and the nonacademic (experimental physics, chemistry, botany and zoology, metallurgy)—the latter group being so described because it had no regular place in university studies. Comparing paradigm cases from the two groups, say, astronomy and chemistry, we note that the former was already highly organized, with an elaborate theoretical structure, in the Middle Ages; it used relatively sophisticated techniques, both instrumental and mathematical; searching criticism of one of its fundamental axioms, that is, the stability of the Earth, occurred in the fourteenth century (and indeed long before) while dissatisfaction with its existing condition was vocal and definite before the end of the fifteenth. A fundamental change in ideas came early—in 1543—and was followed, not preceded, by great activity in the acquisition of new factual material, which in turn prompted fresh essays in theory. All this was the work of learned men, and there was little possibility of craft-influence; even if the pivotal invention of the telescope were a craft invention, its scientific potentialities were perceived by scholars.[14] Chemistry reveals a very different historical pattern, in which almost everything said of astronomy is negated. There was no organized chemical science before a comparatively late stage in the scientific revolution; there was no coherent theory of chemical change and reaction; there was no clearly definable classical and medieval tradition to challenge; the conception of chemistry as a branch of natural philosophy was late in establishing itself, and involved a lengthy fact-gathering stage that preceded the formulation of general theories; and in all these developments the influence of craft-empiricism was strong. It can hardly be doubted that the range of chemical phenomena known to craftsmen about 1550 was much greater than that known to scholars, and that, as Professor C. S. Smith has pointed out, craftsmen had developed both qualitative and quantitative techniques of vital necessity to the growth of chemistry as an exact science.[15]

Sometimes, when one turns from considering the history of such a science as mechanics or astronomy to that of, say, chemistry or a biological subject, it seems as though the transition is from one discipline to another completely alien to the first. Nor is it enough simply to admit that some sciences developed more slowly than others; the situations are really different, so that Lavoisier's work in chemistry cannot be made strictly analogous, point by point, to that of Newton in celestial mechanics or optics. Hence all generalizations concerning the scientific revolution require qualification when the attempt is made to apply them to a particular science.

Perhaps I may illustrate this in the following way. The contributions of craftsmanship to the development of scientific knowledge in the sixteenth and seventeenth centuries seem to be analyzable under five heads:

1 the presentation of striking problems worthy of rational and systematic enquiry;

2 the accumulation of technological information susceptible to scientific study;

3 the exemplification of techniques and apparatus adaptable from the purposes of manufacture to those of scientific research;

4 the realization of the scientific need for instruments and apparatus;

5 the development of topics not embraced in the organization of science proper.

The incidence of these contributions is highly variable among the individual sciences. None are strongly relevant in anatomy, medicine, or indeed any biological science, except that 4 would apply to microscopy. All the sciences demonstrate an increasing dependence on the instrument-maker's craft. Again, 4 is relevant to astronomy, while mechanics draws very slightly upon 1 and 2. Chemistry, on the other hand, exemplifies all these possible contributions, and most forms of applied science—other than mathematical sciences—owe much to the fifth contribution. All we can conclude, therefore, is an obvious truism: that those sciences in whose development empiricism played the greatest part are those in which elements derived from craftsmanship have the most effect. It does not follow, however, that the empirical sciences are those that best exhibit the profundity or the nature of the change in scientific thought and work, nor that the theoretical function of scholars is insignificant even in these sciences. Rather the converse would seem to be true, namely that some of those scientists, like Robert Boyle, who at first sight seem to be highly empirical in their scientific work and attitude, were in fact deeply engaged in the search for general theories and laws.[16] The academic and above all the mathematical sciences were not only those that advanced fastest, but they were already regarded as the models for the structure of other sciences, when these should have reached a sufficiently mature stage. In an ascending scale of sophistication, it was regarded as desirable to render all physical science of the same pattern as mechanics and astronomy, and to interpret all other phenomena in terms of the basic physical laws. The first great step towards the attainment of such an ambition was Newton's *Principia,* a work soon regarded by many as the ultimate manifestation of man's capacity for scientific knowledge. I believe it would be wrong to suppose that the scientists of the late seventeenth century, with such rich examples before them, were content to remain indefinitely at the level of empiricism or sublimated craftsmanship, though indeed in many branches of enquiry it was not yet possible to soar far above it. They were aware that the more abstruse and theoretical sciences, where the contributions of learned men had been greatest, were of a higher type than this.

Perhaps I may now summarize the position I have sought to delineate and justify in the following six propositions, in which it is assumed as an axiom that a science is distinguished by its coherent structure of theory and explanation from a mass of information about the way the world is, however carefully arranged.

1 The scientific revolution appears primarily as a revolution in theory and explanation, whether we view it in the most general fashion, considering the methods and philosophy of the new scientists, or whether we consider the critical points of evolution in any single science.

2 There is a tradition of logical (or, more broadly, philosophical) preoccupation with the problem of understanding natural phenomena of which the later stages, from the thirteenth to the seventeenth century, have at the lowest estimate some bearing on the attitudes to this problem of seventeenth century scientists.

3 Some of the most splendid successes of the scientific revolution sprang from its novel treatment of questions much discussed by medieval scholars.

4 These may be distinguished from the "contrary instances" of success (or an approximation to it) in handling types of natural phenomena previously ignored by philosophers, though familiar in technological experience.

5 While "scholars" showed increasing readiness to make use of the information acquired by craftsmen, and their special techniques for criticizing established ideas and exploring phenomena afresh, it is far less clear that craftsmen were apt or equipped to criticize the theories and procedures of science.

6 Though the early exploitation of observation and experiment as methods of scientific enquiry drew heavily on straightforward workshop practice, the initiative for this borrowing seems to be with scholars rather than craftsmen.

I dislike dichotomies: of two propositions, so often neither *a* nor *b* by itself can be wholly true. The roles of the scholar and the craftsman in the scientific revolution are complementary ones, and if the former holds the prime place in its story, the plot would lack many rich overtones had the latter also not played his part. The scholar's function was active, to transform science; the craftsman's was passive, to provide some of the raw material with which the transformation was to be effected. If science is not constructed from pure empiricism, neither can it be created by pure thought. I do not believe that the scientific revolution was enforced by a necessity for technological progress, but equally in a more backward technological setting it could not have occurred at all. If the genesis of the scientific revolution is in the mind, with its need and capacity for explanation, as I believe, it is also true that the nascent movement would have proved nugatory, had it not occurred in a world which offered the means and incentive for its success.

References

1 Robert K. Merton, "Science, Technology and Society in Seventeenth-Century England," *Osiris*, IV (1938), 360–632. This is the major study of the sociology of science in a single country.

2 The most useful work on the instrument-making craft generally is Maurice Daumas, *Les instruments scientifiques aux XVIIe et XVIIIe siècles* (Paris, 1953). Cf. also the two chapters by Derek J. Price in Singer, Holmyard, Hall, and Williams, *A History of Technology* (Oxford, 1957), Volume III.

3 *The Record of the Royal Society of London* (3rd ed.; London, 1912), pp. 309–11.

4 The more "practical" departments of science, such as alchemy, metallurgy, and cartography, admittedly had little direct dependence on the universities; but it should be remembered that knowledge of fundamental texts in these as well as other topics was derived from them. The universities played an important role in the development of the mathematical and astronomical techniques required for practical ends.

5 On the origins of scientific societies: Martha Ornstein, *The Role of Scientific Societies in the Seventeenth Century* (Chicago, 1938); Harcourt Brown, *Scientific Organization in Seventeenth Century France* (Baltimore, 1934).

6 Paul Mouy, *Le Developpement de la physique cartésienne, 1646–1712* (Paris, 1934).

7 Universities had their anatomy theatres and libraries (often of medieval foundation) and, later, museums and laboratories; the latter were, however, private creations (as at Bologna and Oxford) and failed to become living and growing features of academic life.

8 For bibliographical details on the scientific activities of Renaissance scholars, cf. George Sarton, *The Appreciation of Ancient and Medieval Science during the Renaissance (1450–1600)* (Philadelphia, 1955).

9 Galileo Galilei, *Dialogues Concerning the Two Chief World Systems—Ptolemaic & Copernican*, tr. Stillman Drake (Berkeley and Los Angeles, 1953), p. 112.

10 Besides the extensive studies of the history of medieval mechanics by Pierre Duhem, Anneliese Maier and others, convenient short discussions (with further bibliographical details) are in A. C. Crombie, *Augustine to Galileo* (London, 1952); René Dugas, *Histoire de la mécanique* (Neuchâtel, 1950). On Galileo and impetus, cf. Alexandre Koyré, *Etudes Galiléennes* (Paris, 1939), *Actualités scientifiques et industrielles*, 852–54.

11 The extent to which the technological progress of Europe during the Middle

Ages was due to transmission rather than indigenous invention is immaterial here, since such transmission seems to have occurred at the level of craftsmanship rather than scholarship. Cf. Thomas Francis Carter, *The Invention of Printing in China and its Spread Westward,* ed. L. Carrington Goodrich (New York, 1955); Joseph Needham, *Science and Civilization in China* (Cambridge, Eng., 1954—).

12 The idea that ancient technological secrets had been lost (in the same way as scientific knowledge and artistic skill) was, however, voiced during the Renaissance.

13 Here again it does not affect the argument that three or more of the inventions mentioned were made outside Europe; each became significant for European technology in the period *c.* 1300–1500.

14 If the evidence for the invention of the telescope in Italy *c.* 1590 is accepted (as seems reasonable), it would appear that this invention was connected with the experiments of Giovanbaptista Porta and other scholars shortly before this date. It is thus difficult to believe that the discovery was completely accidental.

15 *Lazarus Ercker's Treatise on Ores and Assaying,* tr. Anneliese Grünhaldt Sisco and Cyril Stanley Smith (Chicago, 1951), pp. xv–xix. Charles Singer *et al., A History of Technology,* III, 27–68.

16 On Boyle, I am much indebted to the work of Marie Boas, "Boyle as a Theoretical Scientist," *Isis,* XLI (1950); "The Establishment of the Mechanical Philosophy," *Osiris,* X (1952); "La Méthode scientifique de Robert Boyle," *Rev. d'Hist. des Sciences,* IX (1956). Cf. also her *Robert Boyle and Seventeenth Century Chemistry* (Cambridge, 1958).

ROBERT K. MERTON

on the Paper of Rupert Hall

At the close of his paper, Mr. Hall sets out the essentials of his discussion in six succinct propositions. It would therefore seem that my work is largely done. Indeed, it would be, if I were to accept Mr. Hall's synopsis as adequate. But I do not. Excellent as the synopsis is in so many ways, it does not begin to do justice to the paper, to its richness of content, its subtlety of analysis, and its variety of implications. (I do not hesitate to differ with Mr. Hall on this score; after all, he only wrote the paper; I have studied it.)

I shall try my hand at another summary which, though it overlaps Mr. Hall's, tells what I have learned from his paper in the form of its major themes, for space will not allow me to review the subsidiary themes as well.

Mr. Hall has given us five major themes about the origins and nature of the revolution in science which occurred in the sixteenth and seventeenth centuries. With each of these, he has proceeded in much the same fashion: He has first sketched out an interpretative idea; secondly, he has analyzed the idea into its components, so that the particular issues requiring interpretation shall be clear; thirdly, he has drawn upon apposite historical examples; fourthly, he has examined interpretations alternative to his; and finally, he has set down his far-reaching though still tentative conclusions. In the course of this orderly exposition, Mr. Hall has also brought us to see historical problems which he left unresolved, problems which deserve and will probably receive systematic inquiry, now that he has formulated them so clearly.

Mr. Hall begins with the seemingly evident but often neglected thought that if we are to examine the distinctive roles of the scholar and the craftsman in the scientific revolution, it might be a good idea to know what we mean by the two kinds of men. He suggests a fourfold basis for comparison: The craftsman, having been drawn from the scientific trades of one kind or another, was concerned with practical achievements, aimed at practical successes, and was knowledgeable in empiric lore. The scientist, in comparison, was largely drawn from university circles, though he typically did not remain there; was concerned with intellectual rather than practical formulations; aimed at understanding, and was introduced to science by university or similar studies.

And here Mr. Hall reminds us that the scientists of that time could have diverse social origins largely because the complexities of science were not yet so great that prolonged and systematic training was required to approach them. The professionalization of science was not yet developed; the interested, self-taught man of science could make significant contributions.

This theme, briefly developed, is a prelude to the second, dealing with what might be called the social ecology of seventeenth-century science. (Here, fortunately for me, Mr. Hall confines himself largely to science in England, so that I can follow him with less temerity than if he had ventured substantially onto the continent. He would be the first to say, no doubt, that a comparable ecological inquiry is needed there.) He finds that some two in every three members of the Royal Society had attended a university. Nevertheless, this was the age of the autodidact in science, for even the men who had been educated in one or the other of the universities and had remained there for a time gained only the rudiments of science and, for the most part, had to teach themselves if they were to do any serious work in science. It may not be too much to say that it was in the coffee-house rather than the classroom, to say nothing of the teaching laboratory, that the mutual education of scientists took place.

Mr. Hall does not merely repeat the time-worn and probably true thesis that the universities lay outside the main stream of scientific work during the seventeenth century, and increasingly so, as the century moved toward its greatest accomplishments; he clarifies and extends that thesis. I found particularly suggestive his observation that no one thought of basing research laboratories in a university; an institutional invention which seems simple enough, once the idea of a laboratory occurred at all, and yet one which, despite Salomon's House, was remote from the prevailing conception of a university. (We are now witnessing a comparable difficulty in arriving at the idea that universities should house research laboratories and organizations devoted to the social sciences; these were largely developed as independent institutions before coming to be slowly accepted, during the last ten or twenty years, by universities.)

In developing the theme of the role of universities in the scientific revolution, Mr. Hall presents us with an unresolved problem for further inquiry; he describes rather than attempts to account for the conservatism of the seventeenth-century universities. This observation has been repeatedly made in the histories of universities and scientific societies. But the fact remains a puzzle. What were the sources of this conservatism? What was there about the structure of the universities, their recruitment of staff, the policies governing instruction and their conception of purpose which together resulted in the conspicuous incapacity of the universities to advance the sciences greatly? To describe this as the persistence of tradition or to call it academic conservatism is only to affix a label to our ignorance.

Mr. Hall's third theme introduces the concept of the scholar-scientist, the man learned "in the thought of the past" and not merely in the more recent activities and methods of science. And this gives rise to the question: How far was the new scientific outlook brought into being by the scholarship which led the new men of science to continue with the rediscovery of ancient science, both directly and as mediated by medieval scholarship? Were the attitudes toward authority and the image of empirical science simply an extension of those generated centuries before? This theme, and its associated problematics, I simply announce, but do not venture to discuss. I dare not rush in where Mr. Crombie need not fear to tread; I can safely leave this part of the paper to him. But the importance of the issues raised in this connexion by Mr. Hall must be evident to all of us, even those of us who are very far from being medievalists.

I do want to dwell, however, on the fourth theme which treats the question whether the impulse to reform science was generated spontaneously among the scholar-scientists or was stimulated by craftsmen and by empirically developed technology. It is Mr. Hall's view—at least, it is his view at first—that, at the most, technological developments afforded only the *occasion* for a merging of the empirical orientation of the practical arts and the rationality of the scholar-scientist; technological change was not its cause. The basic change came about, not in the objective realities of technology, but in the eye of the scientific beholder; the Greek just as the medieval scholars could have taken their technology as a model, but did not; the early modern scientists did.

I find myself ending in agreement with Mr. Hall, but only after having become nervous and apprehensive about the road which he travels to get him (and us) to the right destination. Let me try to re-trace that road.

Mr. Hall questions the image "of the new scientist of the seventeenth century as a sort of hybrid between the older natural philosopher and the craftsman"; he rejects the view that "the behavior of an empirical scientist . . . is derivable by virtually direct imitation from the trial-and-error, haphazard, and fortuitous progress of the crafts." But who holds this image, and who adopts this notion of direct imitation?

He says further that "it is absurd . . . to suppose [as an example] that the introduction of gunpowder and cannon into warfare was in any serious sense the cause of a revival of interest in dynamics, and especially in the theory of the motion of projectiles, during the sixteenth and early seventeenth century." But who has argued that this was "the" cause, that this alone was enough to bring about the observed result?

He suggests, finally, that "it may be naive to assume that even major technological advances suggested, contemporaneously, such questions worthy of scientific enquiry as would, indeed, immediately spring to our own minds." And it can truly be said that such anachronistic thinking would have little to recommend it.

In short, doctrines which claim that technology was the *single* cause of the empirical component in science, that the behavior of scientists was *only* imitative of craft behavior and that present-day implications can be anachronistically read into earlier technological developments are all properly rejected; but history has anticipated us here. And indeed, the rejection of these exaggerated views is only the beginning to Mr. Hall's own analysis.

The crafts and technological development played quite another part, it would seem, in the genesis of modern science. The availability of tools for scientific experimentation coincided with the new interest in experimenting. However little Bacon may have understood the nature of experimentation, he had already warned that "neither the naked hand or the understanding left to itself can effect much." The needs of mining had focussed the attention of craftsmen upon the improvement of pumps; correlatively, these attracted the notice and advanced the work of scholar-scientists, who needed such equipment not only to get on with experiments previously conceived but to think of experiments which the apparatus made possible. Consider only Boyle's repeated practice of placing all manner of objects into the receiver of his air pump to see what would happen when he pumped out the air. (This, incidentally, is more than a little reminiscent of present-day practices in psychology, anthropology, and sociology in which a new instrument—such as projective tests or attitude-scales—will be used almost at random in various groups, just to see what will turn up.)

Galileo's acknowledgement of the assistance given him in the Florentine arsenal does not mean, of course, that his work on the trajectory of a projectile was "caused" by possible military applications of his work any more than was the case for Boyle's proposal to the Royal Society "that it might be examined what is really the expansion of gunpowder, when fired." Nor when Halley indicated that his rule for the trajectory "may be of good use to all Bombardiers and Gunners, not only that they may use no more Powder than is necessary to cast their Bombs into the place assigned," can we conclude that military technology alone invited the new experimental spirit of science, any more than present-day work on atomic explosives can be regarded as the exclusive source of scientific inquiry into atomic physics. But if these craft and technological considerations were not all, neither were they nothing.

When the Boyles and Huygens of the time drew upon the skills of craftsmen to help build their apparatus, or when that craftsman-scholar-scientist, Hooke, served as curator of experiments, they exhibited the indispensable role of the crafts and technology, but not, of course, in any of the senses properly repudiated by Mr. Hall. Now all this must have a ring of familiarity, for Mr. Hall has said it all; but only at the close of this theme. Before then, he leads us repeatedly to the edge of profound disbelief, to the verge of wanting to quarrel with what seems manifestly untrue or a truth exaggerated into error. But just as we totter and threaten to fall over the edge, he takes us quietly by the hand and directs us to ground which seems more solid than ever, once he had shown us the dangers of exaggerating the place of technology in the advancement of early modern science.

When he comes to list five kinds of contributions which craftsmanship and technology made to the development of science in the sixteenth and seventeenth centuries, he presents the most thorough anatomizing of these contributions that I have seen. This list will bear repetition in the course of our discussion.

I have dwelt upon this one theme at such length that I have little space for Mr. Hall's fifth and final theme. This one, too, is a product of Mr. Hall's superb capacity for clarifying historical problems by first anatomizing them into their components. In this theme, he reminds us that the relations between the crafts and science not only differ at different times and places, but greatly differ also among the various sciences at any one time and place. He says, in effect, that science is not all of a piece at any time in history; that, to adopt the phrase of an historian of art, we must recognize the "noncontemporaneity of the contemporaneous." By distinguishing between the "academic sciences" (astronomy, anatomy, mechanics, medicine) and the others, such as experimental physics, chemistry, botany, zoology, and metallurgy, which had no regular place in the university, he enables us to anticipate that these were diversely connected with technology and the crafts. This *analysis* of the historical materials prepares the way for solution of the historical problem which must remain unclear, as long as it is left undifferentiated by dealing with the relations between "the crafts" and "science" in the large.

Mr. Hall's propensity for anatomizing the problem confers many benefits upon us. It avoids the all-or-none formulations which hold that technology precipitated the scientific revolution of the time, or that technology was wholly inconsequential for it. At the same time, his analysis avoids the trite pseudo-solution of saying that it's all very complex, that technology was somehow engaged in the revolution but in complicated ways which are beyond comprehension. Most of all, he provides a model, in both the emulative and interpretative senses of the word "model," which joins historical scholarship with the practice of systematic analysis to effect a distinct step forward in interpreting an absorbing problem in the history and sociology of science.

FRANCIS R. JOHNSON

on the Paper of Rupert Hall

The most important feature of Mr. Hall's paper is his consistent emphasis upon the complexity of the problem of weighing the relative roles of the scholar and of the craftsman in the scientific revolution of the sixteenth and seventeenth centuries. I quite agree with his warning against over-simplification. The best motto for the historian of science to follow is, "Nothing is ever quite as simple as you would like to believe."

Mr. Hall has indicated the difficulties of determining which men one should classify as "craftsmen" and which as "scholars." In fact, the rough "rule of thumb" of labeling among the "scholars" any man who at some time during his career held an official position in a university, breaks down the moment one realizes how many among the great names of the traditional pioneers of experimental science—Galileo, Vesalius, Boyle, Kepler, Harvey—would have to be ranked on the side of the "scholars." In fact, in the biographies of most of the eminent scientists involved in what we term the scientific revolution, the effort to separate the strands of influence that were due to the "scholar" and the "craftsman" poses a problem that defies precise analysis—even by the man himself. I would suggest that it might be more helpful to assign to the "scholar" all factors derived from the knowledge of foreign languages—classical and modern—that in the sixteenth century were the keys to accurate information about the experimentally acquired knowledge of past generations of scientists.

Much as I agree with most of the things that Mr. Hall has said, I have an uneasy feeling that some of the implications of his presentation, stressing as it does the lamentable failure of the universities in this period to assume the position of leadership in the scientific revolution, often does less than justice to the "scholar" and to the institution that represented him—the university. My further remarks will be aimed primarily to remind you of a few facts that should be given adequate weight and which will help in some measure to redress the precarious balance.

Certainly it is regrettable that the universities in the late Middle Ages and the early Renaissance failed to take advantage of the opportunity to become the centers and the spearhead of the new attitudes in science. The new interest in the classical languages that was the mark of the humanistic scholar of the Renaissance, resulted in clear gains for the scientist. At last he could read what his ancient predecessors had actually said, the data they had recorded in their works, and the inferences they had drawn from it. At last he was in the position to analyze accurately and to appraise what Hippocrates, Aristotle, and Galen had done. The criticism of the deficiencies of ancient authors could at last rest on a basis of sound texts. The alert and forward-looking scientist took full advantage of the new opportunities conferred upon him by the humanistic scholar who produced, and disseminated by means of the printing press, more accurate texts and translations of the works of ancient scientists—Euclid, Ptolemy, Archimedes, Hero of Alexandria, Hippocrates, Galen, and I should include Aristotle himself, for his work in experimental biology. The scientist of the time thus had placed at his disposal a great body of new material to assimilate and to test. If he blindly accepted this material on the authority of the ancients, without testing it, the fault was his alone. Many, as we know, were inspired to test this new material. Therefore the recovery of the *corpus* of ancient science remains a vital factor—even though it does not account fully for everything, in explaining the scientific revolution of the seventeenth century; it can be minimized or neglected only at the expense of serious distortion.

Nevertheless, by an unfortunate coincidence, the Renaissance humanist was attacked at this critical juncture by a pernicious disease usually labeled as "Ciceronianism"—the notion that Cicero's practice must be accepted as the only standard of correct Latinity. Every locution for which a passage in Cicero could not be cited as authority was branded as "barbarous" Latin. The best scholars among the humanists—witness Erasmus—fortunately escaped this pernicious disease, and satirized their less rugged colleagues who succumbed to it. Unfortunately this disease was to kill Latin as a living language; today we observe the paradox of our live and vigorous sciences being dependent for their vocabulary upon a dead language.

Again, we should remember, as the prefaces of countless scientific books written in the vernacular continually remind us, that in the sixteenth century there was a deep sense of the obligation that rested upon any scholar who was a master of Latin or Greek, to make scientific masterpieces available in the vernacular to the scientist who knew no other language than his native tongue. Here again are many examples of where the contribution of the "scholar" was to make possible the later contribution of the "craftsman."

Also, I would like to interject at this point the observation that the people living in the seventeenth century were fully aware that the older English universities were, to use a familiar colloquial expression, "missing the boat," and that newer foundations were stepping in to fill the places they had abdicated. For example, observe how frequently in the early seventeenth century one finds Gresham College described as the "third university in England."

Furthermore, I would like to raise now the question of the extent to which the opposition to the new science in the universities was occasioned by the flamboyant materialistic emphasis that was flaunted in the propaganda of its principal supporters. Plato's *Dialogues* taught that the first goal of man was to "Know Thyself." Socrates' injunction was, in the Renaissance, given a moral interpretation, not the materialistic interpretation that came into vogue after Francis Bacon. The goal of self-knowledge was virtuous living, not the material enhancement of man's control of nature (the physical world) for his own somewhat dubious ends—potentially dangerous when not controlled. Perhaps we are more conscious today than at any time since the seventeenth century of the moral dilemma posed by perils inherent in the uncontrolled and irresponsible increase in man's material knowledge. In the sentence that Mr. Hall quoted from Descartes, in which Descartes equated knowledge of himself with "knowledge of the great book of the world," we find aptly epitomized the materialistic attitude of the period which the moral teachers of the time were seeking to combat.

I disagree with Mr. Hall when he states that the contemporary opposition to science displayed by many to the rampant materialism of Descartes and Hobbes was no "science *vs.* humanities wrangle."

In its fundamentals *this* was precisely the real issue at stake. Much was to be said on both sides and little is to be gained for man and for the world by further inflaming the conflict. But the seventeenth century, by failing to resolve in a sensible manner has left it for the twentieth to work out a solution in the face of much greater handicaps—handicaps that have been multiplied many times during the intervening centuries so that now the issue has come to involve the very survival of the human race.

32

THE ROLE OF ART
IN THE SCIENTIFIC RENAISSANCE

Giorgio de Santillana

Let me start straight at the heart of the matter, and try to pinpoint the positive technical achievements of art over the early scientific time-scale. We have then the direct contributions of art in the rendering of observed reality. There are here cases of true symbiosis, as in the team of Vesalius and Calcar, or of Brunfels the botanist and Weiditz his illustrator. In this last case indeed, art runs ahead of science, for it can draw only what it sees, and Weiditz's German plants were in ludicrous disagreement with Dioscorides' Eastern Flora. Once the gap is bridged, we have the wonderful flowering of natural history, from Leonhard Fuchs' *Historia stirpium* (1542) all the way to Gesner and Aldrovandi, where the artist often achieves the greater scientific contribution. But where does art cease, and technology begin? What then of copperplate reproduction, which makes Dürer in a way more important than Leonardo?

Erwin Panofsky has analyzed with great learning and insight the contribution of art to science in many fields. He has rightly insisted that above all the discovery of perspective, and the related methods of drawing three-dimensional objects to scale, were as necessary for the development of the "descriptive" sciences in a pre-Galilean period as were the telescope and the microscope in the next centuries, and as is photography today. This was particularly the case, in the Renaissance, with anatomy. In this field, it was really the painters, beginning with Pollaiolo, and not the doctors, who practiced the thing in person and for purposes of exploration rather than demonstration. In all this I cannot but accept Panofsky's verdict: Art, from a certain point on, provides the means of transmitting observations which no amount of learned words could achieve in many fields. For myself, I am reminded of a letter of Ludovico Cigoli, the painter, apropos of Father Clavius and his bizarre ideas of the surface of the Moon being made of something like mother-of-pearl: "This

33

proves to me again," he writes, "that a mathematician however great who does not understand drawing is not only half a mathematician, but indeed a man deprived of sight." In such words there lives intact the spirit of Leonardo, although they were written a century later. And so there we are back at Leonardo. But I would rather avoid, at this point, being drawn into the Vincian jungle, where the historian of art meets the historian of science, at best, as Stanley met Dr. Livingston—to get him out.

What I intend to do, in this essay, is to concentrate on the early period, which centers around Filippo Brunelleschi (1377–1446).[1] Its importance, I think, has not been sufficiently brought to light. We are here at the initial point where the historian has to unscramble ideas. Brunelleschi around 1400 should be considered the most creative scientist as well as the most creative artist of his time, since there was nothing much else then that could go by the name of creative science.

We must not superimpose our own image of science as a criterion for the past. It is Brunelleschi who seems to define the way of science for his generation. Who was it that gave him his initial scientific ideas? Was it Manetti or, as others suggest, Toscanelli? Or someone else, still? Under what form? A historian of art, Pierre Francastel, has some veiled reproaches against historians of science. "So little is known," he says, "about history of science in that period, that we have no bearings."[2] He goes even further; he quotes from Henry Guerlac's report to the Ninth Historical Congress in Paris, describing the as yet embryonic and fragmentary state of our discipline, to show why the historian of art is left helpless, or has to help himself out as best he may, as Focillon did concerning the relations of mathematics and art in the twelfth century.[3]

We are, then, on the spot, and we had better accept the reminder. We have not yet created our own techniques of analysis in that no man's land between art and science; we have all too few of the facts and none of the critical tools, if we except some penetrating remarks by philosophers like Brunschvicg, who indicated in this period a moving from "Mechanicism" to "Mathematicism." And even so, what do such words really mean, and in what context? My master, Federigo Enriques, used to say in a light vein that good intellectual history is made *a priori*. I would follow his advice so far as to suggest that we be allowed to characterize *a priori* the crucial issues as we see them now, and then try to see where they were effectively tackled.

Two great issues come to mind when we think of that epoch-making change. One is the rise of the modern concept of natural law over and against the medieval one which applies to society, another the intellectual "change of axes" which allowed an explanation of nature no longer in terms of form but in terms of mathematical function.

On these issues, history of science has not much to say before the time of Galileo and Descartes, except to point up the well-known scholastic attempts. Prophetic utterances are quoted, which, scattered over three centuries, fall short of conclusiveness. We are still looking for where the thing actually *took place* in its first form.

Let us then try to make a landfall at a point where art and science, undeniably, join. Brunelleschi created his theory of perspective by experimental means. He built the earliest optical instrument after the eyeglasses. We have Manetti's description of the device, a wooden tablet of about half an ell, in which he had painted "with such diligence and excellence and care of color, that it seemed the work of a miniaturist," the square of the Cathedral in Florence, seen from a point three feet inside the main door of the cathedral. What there was of open sky within the painting he had filled in with a plate of burnished silver, "so that the air and sky should be reflected in it as they are, and so the clouds, which are seen moving on that silver as they are borne by the winds." In the front, at the point where the perpendicular of vision met the portrayed scene, he had bored a hole not much bigger than the pupil of the eye, which funnelled out to the other side. Opposite the tablet, at arm's length, he had mounted a mirror. If you looked then, through the hole from the back of the tablet at its reflection in the mounted mirror, you saw the painting exactly from its perspective point, "so that you thought you saw the proper truth and not an image."

The next step, as we see it now, is to invert the device and let in the light through the pinhole, to portray by itself the exterior scene on an oiled paper screen. This is the *camera obscura,* not quite the one that Alberti described for the first time in 1430, but its next of kin, and it took time to be properly understood. But the whole train of ideas originated with Brunelleschi, between 1390 and 1420.

We have thus not *one* device, but a set of experimental devices of enormous import, comparable in importance to that next device which came two centuries later, namely, Galileo's telescope. Galileo, it will be remembered, announced his instrument as derived "from the more recondite laws of perspective," and in their very inappositeness (at least in modern usage), the words are revealing. The new thing could be understood by opinion only as one more "perspective instrument" and indeed its earliest Latin name was *perspicillum,* whence the English "perspective glass," a name which might apply just as well to Brunelleschi's devices.

We may note that Copernicus himself, untroubled by any modern thought of dynamics, had candidly proposed his system as a proper perspective construction: "Where should the Lamp of the Universe be rightly placed except in the center?" Conversely, there is a "Copernican" spirit about perspective. People will be portrayed as small as they have to be, if they are that distant. There is nothing wrong in looking small if the mind knows the law of proportion; whereas the medieval artists had felt that change of size had to be held within the limits of the symbolic relationship and importance of the figures to the whole.

What the devices of Brunelleschi have done is in every way as significant as the achievement of the telescope. If the telescope established the Copernican system as a physical reality, and gave men an idea of true astronomical space, Brunelleschi's devices went a long way towards establishing in natural philosophy a new idea concerning the nature of light. On these effects of the *camera obscura* I need not dwell, as they have become by now a fairly well-established item in the history of science. By showing the passive character of vision, it cut the ground from under a vast set of theories, primitive and also Platonic in origin, which assume vision to be an "active function," a reaching out, as it were, of the soul. After the discovery of the formation of an image on a screen, the theorizing may well go on for some time, since it is an acquired cultural capital, but the best minds cannot participate in it wholeheartedly, and therefore it is sentenced to eventual decay and fall.

This, then, is set; I mean, all set to be worked out properly. To see the change in the period which concerns us here, we need only compare the consistency and fertility of the new outlook with the optical theories of Ghiberti, Brunelleschi's older friend and contemporary. They are really confused and uncertain footnotes to Alhazen. He sees dimly the new thing that Brunelleschi has established, but he cannot grasp it, he flounders between old and new. Here, too, he is as Frankl has described him stylistically, the last medieval. Fifty years later, with Piero della Francesca and Leonardo, we are almost wholly within the new ideas concerning light. It may take until Benedetto Castelli in 1620 to give them uncontested scientific citizenship, but that is because in philosophical theory the old is so much slower to die, the present a coexistence of many epochs.

When it comes, however, to Brunelleschi's first optical apparatus, the perspective mirror device, understanding becomes much more complex and difficult. For if our *camera obscura* is a passive instrument like the telescope, this one is meant to project correctly a work of man, the painted tablet; yet it is undoubtedly an instrument, based on the geometrical construction later expounded by Alberti, the "visual pyramid" axed on the "centric ray." It is strictly speaking a perspicil, meant to show reality as is, and the portrayed reality even meshes with the natural in the reflection of sky and clouds in the burnished silver. The sky, we realize, is not to be portrayed. (Nor is it in the second model, where it was used directly.) Why, we shall see.

Surely, here we have the beginnings of a science—a science of visive rays, as Leonardo calls it, so scientific that there is even an apparatus designed for it. But what is it for? To help us portray rightly what we see around us—essentially, to give the illusion of it. Illusionism is a strong motive, inasmuch as the most direct application is to scenographic design. This in itself is no mean thing, nor merely a way to amuse the rich. It has been proved that those monumental town perspectives—like that of the main square of Urbino by a pupil of Piero—are no mere exercises in drawing, they are actually projects for an architecture that is not yet there, and the first sketches of town planning. This is in the true line of development, for Brunelleschi himself had devised his instrument as an aid to architectural planning. But where does it leave us as far as science is concerned? George Sarton appraised it crisply in *Six Wings,* his last book on the Renaissance. After giving a brief historical summary on perspective, he concludes: "Linear perspective implied a certain amount of mathematical knowledge, but not a great deal. The best mathematical work was done in other fields, such as trigonometry and algebra." And that's that; or rather, let it stand, even if we are tempted to qualify it. The treatment that Kepler gives in his *Optics* to just such constructions, involving catoptrics, shows the thing to require no inconsiderable mathematical skill. The point remains that Sarton considers the subject not particularly important.

Yet we all agree that if the history of science is to be understood as George Sarton himself wanted it understood, a dominant cultural factor, we cannot push it off again at such a critical juncture to a couple of algebraists working in a corner. At a time when the main avenue to reality was through art, it is inconceivable that the artist should be dismissed. But Sarton also knew that beyond his curt appraisal, one would have to enter the uncertain region of interpretation, where the art critic may feel at ease, but the scientific mind finds itself desperately uncomfortable. He restrained himself according to the by-laws of his craft. Yet he knew, better than most, that it is these by-laws which kept so much of the history of science in a condition like that of Gerontion, having neither youth nor age,

> but as it were an after-dinner sleep
> dreaming of both,

caught between a blurred past and a misapprehended present. The Scientific Revolution should be seen and studied as a major intellectual mutation, which obviously started about the time when both interest and relevance seemed to have slipped out from under the scholastic system without anyone being able to say why. But the old political axiom still holds: "On ne détruit que ce qu'on remplace." Beyond the mass of heterogeneous social factors, the rise of this and the decline of that, which are strung together in the common histories, what do we know about the actual "replacing" element? The shift of opinion can be traced to the early Renaissance. It cannot be accounted for by still non-existent experimental results, nor by "industrial growth," nor, even less, by any advances in trigonometry. That is why research seems to center at present on the new interest among scholars for the mathematical "method." I submit that, valuable as it is, it is not the answer, and I hope later on to show that indeed it could provide no "replacement."

May I be allowed, then, to ignore the by-laws for once, and to enter at my own risk the region of discomfort.

Much, of course, has been written about perspective by cultural historians. It has been said, authoritatively, that Quattrocento painting places man inside a world, the real world, exactly portrayed according to physical laws, that it asserts naturalism against medieval symbolism, and thus creates the natural presupposition for science. This may be very rightly said, but one goes on feeling uncomfortable. Such a piling up of obvious factors until science is then produced out of a hat looks suspicious. Panofsky, as we quoted him, insisted on art as a tool of science, which is methodically strict. I tried to stay on the same ground with the perspective device. There are risks in going beyond. We remember Whitehead saying that the subtle logic of the scholastics was another of those necessary presuppositions for science, and it remains a most unverifiable statement. There would be, indeed, strong reason for saying the opposite, that scholastic thought was and remained a retarding factor. The ultimate proof of the pudding lies in the Galilean consummation; behind it one might see very different factors at work: the engineering and Archimedean line of thought for example. And there again we must see how that factor works. Invocations are not enough. Such a qualified scholar as Mr. Martin Johnson has insisted on the Archimedean factor in Leonardo's development, and it can be proved—Sarton has done it—that he is wrong. Archimedes is still for Leonardo a legend, a name to conjure with, he is not a set of concepts: Leonardo's thought remains utterly non-Archimedean, committed to the concepts of a biological physics of a wholly different style. Even more does it behoove us to be careful in dealing with large words like "naturalism." There is surely full-blown naturalism in Leonardo, but he expresses it mainly by being a natural philosopher: There the ambiguity is lifted. For he certainly *is* a natural philosopher.

Then what is naturalism in art? The art critic may legitimately use this term in a way which has nothing to do with natural science. Not as realism, for instance: Leonardo, the wizard of trompe-l'oeil, insisted that here could never be truth, but a *simulacro*. The new power, or shall we say the new freedom, of a Masaccio or Masolino applies to the way people *stand* inside the setting, it indicates a new canon that is plastic and visual, a new choice perhaps—not naturalistic objectivity. And certainly not a new discovery of space. That way of placing people has been maturing since the time of Giotto. I think it is safe to say that up to Leonardo, pictorial space will remain tied up with the human action in it, and that again is continuous with Giotto. If there is an evolution, we might call it a critical revision, with the help of perspective, of the "social" conception of art which belongs to the fourteenth century.

But there is a striking change nonetheless.

If we try to define what is meant, we are led to expressions like "the presence of a charged reality," a feeling of "vital autonomy" of these beings—in short, a feeling of *immanence* as against the former sacramental transcendence. These are metaphysical terms. Since we have to enter the metaphysician's den, strewn with the bones of former explorers, let me suggest a password. It will mean here simply that our attention is held, not so much by what we see, as by what we do *not* see, by what might be called, with apologies to Maxwell, a field of force.

The invisible element up to then in the paintings has been the living dispensation of God, his mysterious presence in all things: It was symbolized by the golden background which had no depth and yet suggested infinite depth. The image was strictly an intellectual symbol in the Plotinian sense. The world of things was, as it should be, an emanation of the hypostasis of Soul, it had no depth of its own. In the new representation, imagined matter becomes truly spatial, it is bodily on its own, and not only creaturely, as it were. It is surrounded by an invisible field which does not, however, escape comprehension like the other, for it is an intelligible reality, it is space itself. One, of course, may speak in terms of plastic or tactile values. But in this way, I hope to have avoided all danger of doubletalk.

What I am saying about the new painting is still interpretation. But when we come to architecture, it turns out to be the central fact itself. Brunelleschi's perspective machine is designed to place objects correctly in space, the nearest thing to a stereopticon, each thing being defined, as Alberti says, by having *uno luogo,* one place. And the reason why the sky is not represented but used directly is highly systematic: The sky does not have *uno luogo.* It cannot be brought back to measure, proportion and comparison, hence it cannot be represented in the proportional system which defines the form. It is space itself, it is of the essence of the field. It is what represents but cannot be represented.

This is not to say that from that moment on, art becomes based on field theory. It is rather of the nature of science to be led from step to step by the firm consequentiality of abstract propositions, which as soon as discovered are developed in their fullest consequences. In art many interests have to coexist which are certainly not abstract. If Brunelleschi himself had been asked to explain what he was doing, about the same time, in his submission for the famous portals of the Baptistery, he would have stated it no differently from his competitor Ghiberti, whose painstaking report shows the humility of the medieval craftsman. Those are the Bible stories, he insists, just as they are told, carefully represented, and to show it he goes through the detail of objective particulars specifying the materials used, stone and bronze and gold each in its place.

Even the man who obviously got the theoretical illumination, he who is said to have become obsessed by perspective, Paolo Uccello, can be extremely unsystematic about it in his work. His case is clearly that of a divided personality, which accounts for his being, as Schlosser put it a bit drastically, more than half a failure. The extraordinary novelty and power in his portrayal of individual figures goes with a strange helplessness in composition, a crankiness of interests, which caused him to sink into an obscure old age. And yet, when we look at those descriptive geometry studies of absurd complication, like the *mazzocchi,* we realize that here a true engineer was lost, a simple man who would have been happy designing some new radio tower or steel pylon for power lines, who would have revelled in industrial design and become a master for Freyssinet and Nervi—but who lived at a time when there was no use for such talents.

As with Uccello, so—in a measure—with Piero della Francesca and with Leonardo himself. The breakthrough of any genuine line of scientific pursuit works to the detriment of the artist's true productivity. It is not in Piero's theorems on solid geometry that his philosophical contribution lies, nor in his *Perspectiva pingendi,* it is in his sense of the mystery and power of pure geometric form and volume which shine through his composition and heighten its metaphysical import. There is a variety of meanings you can put in such a simple thing as a circle. The circles of Piero are an unmoved reminder of perfection. Those of Paul Klee look as if they were spinning. They are the symbol of a rat in a cage. Yet both are just circles to the unaided eye. And so back we are in the region of the undefinable.

It may have something to do with the elusive nature of the artist's quest, which in turn is certainly tied with the socially long undefined nature of his activity. Let us be properly historical about it for a moment. Let us not forget that the Middle Ages had given music a place in the Quadrivium, among the Liberal Arts, but none to painting. If we follow the scheme down into the seven Mechanical Arts which are supposed to be as it were a pallid reflection of the higher ones, we find *lanificium, armatura* (i.e., building), *navigatio, agricultura, venatio, theatrica.* Woolcarding and hunting are recognized arts, while painting and sculpture are nowhere. The figurative arts are left to a purely servile status. They are not supposed to be a freeman's activity. This is not just forgetfulness. Hrabanus Maurus had taught long since, at the onset of the Middle Ages, the primacy of the Word. Dante himself, who admires Giotto, shows by his idealized use of sculptures in Purgatory that he considers them a kind of teaching aid, heightened maybe by divine touch into some kind of 3-D cineramic animation effect, but still subsidiary and illustrative, whereas poetry, we know, is for him a true philosophical medium, a transfiguration of the object to cosmic significance. And even Cennini, who wrote in 1390, in time for young Brunelleschi to read him, who insists on giving art a professional status beside poetry, provides for it only the old proportion rules of antique convention, plus the new and important caution that the rules do not apply concerning woman, who has no symmetry or proportions, being irrational and a pure thing of nature.

This is the intellectual state of the question at the bursting of the early Renaissance, and it derives little clarification from the first helpless and arbitrary theorizings, or even from Vitruvius whom Brunelleschi, it seems, never troubled to study. As for "science," it was held in trust by professors who cared for none of these things. How then, can these men effect a *prise de conscience?* How does the new intellectual constellation arise?

The problem may become clearer if we think of the second stage, so much more familiar to us. Two centuries later, at the time of Galileo, the struggle between the old and the new form of knowledge is out in the open. Galileo is an acknowledged master of acknowledged science, in mathematics and in astronomy, yet he has to contend all his life with the prejudice which denies him philosophical status. He is treated as a technician and denied any capacity to deal with the true causes of things, which are of the domain of cosmology, as taught in the schools. We all remember how much persuasion he had to spend to show that when we have found a "necessary," i.e., a mathematical cause, we have as true a cause as the mind can encompass and that we need not look farther afield. We know that it took the combined effort of Galileo, Descartes, and Newton to establish the new idea of a natural philosophy.

If now we come back to 1400 instead of 1600, when the medieval frame of ideas was still an overarching and unchallenged presence, when what we call science was still a hole-and-corner affair without a character of its own, if we think of this group of men who had no connection with the universities, little access to books, who hardly even dressed as burghers and went around girt in the leather apron of their trade—then, I say, it is no use vaguely talking about genius; we have to show some concrete possibilities underlying this sudden creation of a new world of theoretical conceptions.

The period I am referring to is fairly well defined. It is the fifteenth century itself; it starts with Brunelleschi's early work; it has reached a conclusion by 1500 with Leonardo and Luca Pacioli.

The first thing was for these men to have some conception of their social role which allowed them to think legitimately. Here we can see that they found an anchoring point in the old theory, reinterpreted. For just as medicine claimed the "physics" of the schools as its patron science (hence the names "physic" and "physician"), those craftsmen had two of the mechanical arts, architecture and *theatrica,* and above those, music and mathematics. By the time of Pollaiolo, it is commonly understood that architecture is a straight liberal art.

But mark the difference. The painter remained with only a mechanical art, *theatrica,* back of him. He did in fact tie up with it, in a measure of which we would scarcely be aware if G. R. Kernodle[4] had not collected all the evidence of the connection there was, back and forth, between scenographic art with its repertory of props and the settings of the paintings. This went one way as far as town-planning, as we have seen previously, but even when it remained pure decoration, or play scenes, it shifted the accent from the *liber pictus* of devotional art to the productions and the magnificence of court and city pageants. Ghirlandaio, Benozzo Gozzoli, are names that come to mind: But even in the production of Raphael and Titian, the new "spectacular" element is stronger than the devotional occasion. The architect, instead, has something unequivocally solid to handle. He builds reality.

This leads up to the *third* factor, the new type of patronage from a new ruling class. It is significant that just about the time when Brunelleschi was called in as a consultant to the Opera del Duomo (the Works Committee for the Cathedral) its last ecclesiastical members were dropped and it became an entirely lay body: Brunelleschi on his side, once he is the executive, shows a new technocratic high-handedness in firing the master builders *en masse* for obstructionism and re-hiring them on his own terms. But he had not risen to the top without a decisive fight with the Guild of the Masters of Stone and Wood, who resisted his plans and even managed to get him, an outsider, arrested for non-payment of dues. He appealed to the Works Committee and had them clapped in jail. This was the end of the medieval power of those he called scornfully the "Grand masters of the trowel," whom the softer Ghiberti had joined believing no one could ever lick them.

43

Here we have, then, for the first time the Master Engineer of a new type, backed by the prestige of mathematics and of the "recondite secrets of perspective" (Galileo's tongue-in-cheek description of his own achievements is certainly valid here), the man whose capacity is not supposed to depend only on long experience and trade secrets, but on strength of intellect and theoretical boldness, who derides and sidesteps the usual thinking-by-committee, who can speak his mind in the councils of the city and is granted patents for his engineering devices, such as that one of 1421 which describes in the somewhat casual Latinity of those circles a machine "pro trahendo et conducendo super muris Cupolae lapides, macignos et alia opportuna," or the other, of the same year, for a river lighter equipped with cranes. He is, in fact, the first professional engineer as opposed to the old and tradition-bound figure of the "master builder": He is the first man to be consulted by the Signoria as a professional military engineer and to design the fortifications of several towns: His work is the precondition of the first text-book, that of Francesco di Giorgio Martini. But still and withal, he is acknowledged to the end of his life as the great designer and artist; not only that, but as the man who masters the philosophical implications of what he is doing. Donatello may be acquainted with the Latin classics; he is still considered a craftsman. Brunelleschi is not; he stands as an intellectual. It is only a century later that the fateful distinction emerges between pure and applied art. By that time, the pure artist himself is hardly an intellectual.

Finally, we have the fact that this complex of achievements by a well-led group of great talents—Manetti, Ghiberti, Donatello, Masaccio, Uccello, Luca della Robbia, with Brunelleschi as leader—found a literary expounder of comparable talent in the person of Leon Battista Alberti (d. 1472) who gave their ideas full citizenship in the robed world of letters and humanism—something that only Galileo was able later to achieve by himself. It will have been a fragile and fleeting conjunction, no doubt. By insisting on a "science of beauty" Alberti perpetuated the rigidity of medieval disciplines with their ancient idea of "method," and their dictation of what is right. It will end up in mere academism, about the time when science breaks forth with its own idea of method and of truth. But as long as it lasted, in the period of creation, it has been a true conjunction, both in one. Alberti only paraphrases Filippo's words (we know that) when he says of the new art of architecture: "If it ever was written in the past, we have dug it up, and if it was not, we have drawn it from heaven." That "social breakthrough" that the new science of Galileo effected through the telescope, we find here in its early counterpart or rather first rehearsal. Everyone in 1450 was aware that a boldly speculative theory had preceded the complex of achievements, until the Dome of the Cathedral rose unsupported in its greatness, "ample enough," says Alberti, "to hold in its shade all the peoples of Tuscany."

We have lined up, one, two, three, four, the preconditions, or what our dialectical colleagues would call the objective possibilities for a scientific renascence, as well as, I daresay, the proper revolutionary consciousness; yet, even if we go and comb the patent records in approved dialectical style, we shall not be able to register the birth of modern empirical science. We are left with the question we started out with: "What was it? In the name of what?"

A revolution, surely. But what does that mean? Are we not reading upheavals into the past, when it is very possible that we should find, below this commotion, a sequence of slow and unnoticed alluvial effects? I suggest that it is best to be coldly phenomenological about it. Revolutionary is as revolutionarily does. Uprisings, *jacqueries,* justified as they may be, are revolts of the slighted or the oppressed; they are not revolutions. It is only when a group of individuals arises in whom the community recognizes in some way the right to think legitimately in universal terms, that a revolution is on its way. "What is the Third Estate?" said Sieyès. "Nothing. What could it be? Everything. What is it asking to be? Something." This is fair and reasonable, but that "something" has not been granted or taught from above, it is dictated to them by an inner reasoned certainty, and it is that no longer disputable certainty which makes all the difference. Here in 1789 are men whose philosophy has grown to impose itself, as it does in the calm utterances of our Declaration of Independence, men who know they can assume responsibility for the whole body social, not only in the running of its affairs, but in its decisions about first and last things. When these decisions sweep even the entrenched opposition off its feet, and move it to yield freely its privileges in the historic night of August 4th, then we know that a revolution has taken place. It is the resolute assumption of responsibility which forms the criterion.

May I insist upon this line of discrimination, as it might enable us to cut across the large amounts of indecisive data and shaded pros and cons which encumber our research.

To be sure, there had been in our case, over more than a century, a discernible trend among scholars towards the mathematical knowledge of nature; but in what spirit? Alistair Crombie's contribution to this volume, even if conducted *ex altera parte,* would provide by itself enough material for my contention, which stands as a tribute to his fairness. We read that Hugh of St. Victor and Gundisalvi pleaded for the admission of technical concerns to educational theory, because the *architector et ingeniator* is a useful figure indeed. "The science of the engines," says Gundisalvi, "teaches us how to imagine and to find a way of arranging natural bodies etc." Why surely. But what was Gundisalvi conceding from his book-lined study that a man like Villard de Honnecourt did not know much better by himself? What contributions did the patronizing scholars bring to the still primitive efforts at writing of men like Walter of Henley, Guido da Vigevano, or Conrad Kyeser? J. Ackerman[5] has published a case which is very much to our point, since it deals with the predicament of the Lombard masters, Brunelleschi's rivals, when it came to establishing the plans for Milan Cathedral in 1386. On one side there is the local guild commission, suggesting that geometry might be of use in some way for designing the buttresses, on the other the consultant Pierre Mignot, described as "maximus inzignierus" (or Big Injuneer as our students would call him). As he is unable to give any positive advice, he takes evasive action in a fog of doubletalk about Aristotle's principle that mathematics should not be used in the science of locomotion, and since statics as a part of mechanics belongs to that science, let's just make the buttresses doubly thick to be on the safe side. The Masters are not impressed but bow to authority. So much for the collaboration of the scholars with technology.

As the scholars cannot see beyond the immediate utility, so they cannot overcome their own school-bred respect for the raw data of observation and the commonsense physics that goes with these. More than once I find in Crombie the term "irresolution" which acutely characterizes their attitude either in equipping themselves with the proper knowledge of mathematics, or in their way of attempting experiment. Buridan does not really hope that nature will provide the conditions for mathematical laws to be fulfilled, although "it could happen that they should be realized by the Omnipotence of God." In other words, wouldn't it be wonderful, but only a miracle of divine benevolence could free us from the Aristotelian bondage. This is exactly the attitude of the traditionalist vs. the revolutionary. Yet Buridan is no timid spirit. His words show him to be a true rationalist, but he has to defer to long-established authority, whose dictates become akin in his mind to the deposit of the Faith or to the divine right of kings. For two centuries this kind of speculation has been going on with nothing much happening.

Here and there, the scholar can risk bold theorizing, he can intimate, adumbrate, and prophesy; but he must be prepared for an intervention of authority which tells him to drop his playthings and come back to a correct attitude. This intervention did come at last, decisively, with the anti-Copernican decree of 1616. By this time, however, even if alone and abandoned by the scholars in retreat, there was Galileo. He stated in no uncertain terms that in such grave matters, his authority was fully equal to that of the Church Fathers themselves. This is what I call the assumption of responsibility. Galileo does not hesitate to denounce the authorities for playing irresponsibly with reactionary subversion. The freedoms granted by tradition, he insists, are his protection, the reason that God gave us to understand His laws is on his side. He uses all the proper language of submission, but he makes it clear by his attitude that he will not compromise, that he will not retreat, and that he will be heard.

As we know, Galileo could have gone on establishing the pure and simple science of dynamics without all this fuss. He actually did—by the time he could do nothing else. This would be enough to prove that he did not consider his thought the empiricistic outcome of industrial division of labor or advanced technology or book-keeping or whatever gadget it is that amateur sociologists have devised for his rationale. He felt he had to face the central issue: To the well-worked-out cosmos of his predecessors he opposed another cosmology, another way of knowledge whereby man has to go ahead forever in discovery, trusting Providence that it will not lead him to perdition. This Galileo maintained even when told by the Successor to Peter, the Vicar of Christ, that his doctrine was "pernicious in the most extreme degree." He alone, with very few men of his time, perhaps only Kepler and Castelli, could really know what he was doing. Even his trusted Ciàmpoli, who sacrificed his career to get the Dialogue published, romantically thought that this was some novel and wondrous line of Neoplatonic speculation, in the spirit of Robert Grosseteste. Here is what I consider Galileo's assumption of responsibility for the whole body social in first and last things. It stands with us to this day.

Now that we have defined our terms, let us go back to the situation around 1400, and I daresay we can see some reason for the fog of indecisiveness which hangs over the period to the eyes of the historian of science. The gap was too wide between the world of the scholar and that of the technologist to be crossed by anything profoundly significant or fruitful. Encouraging noises were the most of it, and they can be discounted. Even in the extremities of nominalism, the world of the scholar is too well-knit, spiritually and conceptually, not to keep dangerous deviations in check. Nicole d'Autricourt is free to suggest atomism as a natural philosophy, but once the chips are down, he has to back out or become a heretic. The investigator of mathematical methods is given more latitude, but he, too, knows there is an end to his tether, hence his "irresoluteness."

Yet there is agreement among historians, as I note from both Crombie and Hall, about the highly intellectual and theoretical character of the Scientific Renaissance from its inception. No accumulation of progress in the crafts could bring it forth. Now if we search for the point of conceptual start, where the "flashover" could take place from the world of intellect to that of the crafts, we are certainly on better methodical ground than if we juxtapose incongruous factors into a specious syncretism. The idea that I would submit for reflection is that the flashover occurred in the one art capable of receiving a high theoretical content, namely, architecture. It will not lead directly to physical science, that is obvious, but in those early years of the sixteenth century it is, in Whitehead's simile, the laboratory in which the new ideas are inconspicuously taking shape, for it provides the point at which the old dominant ideas of *form* can be resolved effectively into mathematical law. The resoluteness with which Brunelleschi carries through this transformation of intellectual axes makes of him a more decisive figure than Leonardo, who, faced with vaster problems, tends to lapse back into a medieval "irresolution." Not being a branch of philosophy, not even quite a liberal art yet in the estimation of the learned, architecture is free to assume its commitment unchallenged, and to lead it to concrete fruition and victorious achievement. There is nothing like success to establish a certain way of thinking; it is the success that had never attended the efforts of Oresme or Buridan which allows Brunelleschi and Alberti to work out their ideas meaningfully, to their far implications. That is why I have tried here to concentrate on the figure of those Founding Fathers.

Of Brunelleschi there is no question but that he stands there as the

48

scientific-minded Innovator *par excellence*. He is that, if I may say so, from snout to tail. Take only the way he went at studying the Roman monuments in his early years, while earning his living on the side as a clockmaker (this small point, too, has its relevance). He took down "on strips of vellum" the measures of each part, written in terms of one basic measure. It may look simple enough, but it is described as a great new idea. The old master builders had never come upon it. They did not write. They took their measures, we are told, with pieces of string, and left it at that. No collaboration of practical necessity with the wisdom of the learned monks who did advise them, no interplay of scholarly thought with technology had gotten beyond this stage for centuries.

Small wonder that we find in Brunelleschi's theory, through Alberti, an analysis of proportion which leads one back to Greek mathematics at the time, say, of Theodorus. His method shows that it took one mind from start to finish to bring in with the technique of measurement the new idea of proportion, and make it work; just as it took one mind, Faraday's, to bring in from start to finish the electro-magnetic idea. He is the man, too, who could assume a commitment and maintain it in the face of opposition. What we know of him (and I do not mean the watered-down generalities to be found in Vasari) shows him as a daring leader but also a careful strategist. Witness the time when he capably withdrew and allowed his rival to try his hand at building the Cathedral Dome, but as soon as the chain-and-girder system devised by the latter had snapped, he was free of his "sickness" and ready to take on the job. The easy-going and humanistic Ghiberti provides the needed foil for his affirmative and polemic personality. When Machiavelli, in his famous preface, says that it is time to do for politics what the great artists have done in building, he has clearly in mind Filippo's way of dealing with Roman antiquities: not to stop entranced at the grandeur of them as Donatello did, but to study the joints and structure of the buildings, how the stuff was actually put together so as to stand forever. Which is just the way Galileo studied Archimedes. We have there what Vitruvius defines as invention, "the mobile vigor of the mind." In all three cases, it has to face from contemporaries the bitter old reproach, that of the cold and abstract—if not destructive—pride of intellect,[6] while the poetic and tender-minded uphold the virtues of tradition.

There is indeed in the personality of Brunelleschi much that reminds us of the other two "patricians of Florence"; his love of good company and vivid argument, his irony and sarcasm (he was, says the biographer, "most facetious and prompt in repartee"), not to mention his pranks, some of which were incredibly elaborate, and required half the town in connivance, like the one he played on Grasso the carpenter, to make him believe that he was not himself at all, but another man called Matteo, who was being looked for by the police. With the diligent help of Donatello and the police chief who had him booked, Grasso was made to lose his identity so that he went around for two days in a daze asking people "Who, then, am I?" until the friends restored him in sleep to his own bed. This climate of jokesmanship in terms of the ultimate is to me not incidental. In Machiavelli's letters, in Galileo's poems and correspondence, we realize what it stands for: Freedom of spirit, *homo ludens,* the crackling of the sharply critical and relativistic mind; the play that can go with single-minded dedication, and even with flashes of profound metaphysical poetry. If the flavor of that salt be lost, we are bereft of one of our senses in historical perception. We miss the power of strong irony and liberating wit, from Erasmus to Voltaire. There is more than one way of merriment: This one is the straight opposite to that, otherwise charming, of *mansuetudo,* and the much later reaction of a man caught between the two ways might show us a kind of litmus test. Magalotti is one of those men who had been under Galileo's influence, and who had never, like their master, lost their sincere concern with salvation. Late in life he entered a monastery, but had to leave it after two years of fruitless endeavor. As he writes rather shamefacedly to Cosimo III, it was not the ascetic life that drove him out, that was indeed what he had come to seek; "it was rather, I have to admit it, the good Fathers' way of having fun."

Brunelleschi is akin to Galileo in his sense both of fun and of purpose. Like Galileo, he could be bold, clear, and decisive in a way that compelled assent even in matters but dimly understood by the public. One is reminded, too, of Guicciardini's good-natured note to Machiavelli on reading the *Discourses*: "Well, here you go, *ut plurimum* paradoxical and upsetting in your conclusions, but always very notable."

What the revolution was about is less easy to define, because of the intellectually elusive quality of art. Such a theory as the Copernican will put the issue squarely, as between closed world and infinite universe. It will bring violent reactions and counter-actions beyond the bounds of reason. Averroist naturalism—or Pomponazzi's hard logic—will evoke sectarianism, condemnation and refutation without end. Whereas the artist simply projects the thing, something that stands there by itself, he brings in magically a new way of seeing, and then leaves imagination to take its course. What comes out in this case, for instance, is that when the artistic revolution is done, all the old emotions about realism and nominalism are as dead as a door nail. Something new is in the world, but the creators have gone unchallenged.

It could not happen today, granted. No combination of Picasso, Gropius, and Mies van der Rohe could sway our philosophy that much. But that is because we live in a world where modern science has taken over, and will not allow thought to be moved around beyond certain bounds. Five centuries ago, science being conceived as of forms, the appeal of a new science of forms could blot out the old one, if those forms were felt as particularly significant. The proof is such a well-known work as the *Hypnerotomachia Polyphili,* the mystic tale of a quest similar to those of the Rose or of the Grail. The symbols are no longer the ancient and medieval ones, they are drawn from architectural fantasies. Architecture, as I said earlier, seems to be the point where art and knowledge are felt to join.

It is then left to us to decide what the new thing actually *is,* and to evaluate it. This will be the subject of the last part of this paper.

It might be well to see first what they *thought* it was. The literary theorist of the movement is Leon Battista Alberti, himself an artist of sufficient stature to assure us that he understood what his friends were about. The label that he puts on it is the neoplatonic one. Neoplatonism was orthodox enough to be safe, strong and speculative enough to be new. The mathematical bent of the whole school was so marked, that such a label is hardly unexpected. Here again, we are faced with the fact so well marked by Professor Koyré, that the new thought arising against the old logic is of a necessity under the invocation of Plato, and will stay so until and after Galileo. The Galilean way in dynamics, as he points out, is to explain the real case by way of a theoretical one that can never be brought under observation, the concrete by way of the abstract, what *is* by way of what cannot be said to "be."

The theory of linear perspective might seem too simple to lend itself to such weighty thoughts. It can be kept down to a workout in elementary geometry, just as Sarton described it, and one wonders why the medievals who tried found such difficulties in relating the objects portrayed to a fixed position—if it is really a matter of drawing lines. But let us not forget that Alberti, like most theoreticians, is putting his new wine in old bottles, whence the label of Platonism. He can use only then-known, i.e., elementary terms, hence his explanation falls short of what is really the case. What we call "a system of mental representation," and "the ordering of a conventional space in depth," and so on, are things which cannot be expressed in his language but only pointed up to. The system that is being evolved is part mathematical, part symbolic and mythical; it is fully there only in the actual works of the artists. What comes out of those, we shall try to work out later. But it must be extracted from implication, it is not stated in simple words. Unless they be, by rare chance, some few poetic words of Brunelleschi's very own. We have those few words in the scornful sonnet he wrote in reply to Giovanni di Gherardo:

> Ogni falso pensier non vede l'essere
> Che l'arte dà, quando natura invola[7]

"False thinking does not see the being to which art gives birth when it steals [its secrets from] nature."

The visual space of these early theoreticians is seen as containing somehow the secrets of nature. It is not mere analytical relation as in Descartes, nor does it only express the togetherness of beings as in Gothic. The creative mind is always midway in process; so here. Geometry is used instrumentally, for unravelling that space; the business of perspective unravelling goes on for centuries to pose problems still to such a distinguished mathematician as Desargues in 1648.[8] Optical space tends to wriggle away from the geometric straitshirt. What is truly abstract, hence the guiding mystery throughout, is *logos,* is proportion, is number.

Now, as to Brunelleschi; how much was he, whose business after all it was to make things, aware of the intellectual implications? We know that he was very keen on theory, were it only from the passing mention that he considered Alberti a bit too simple in his thoughts. The few poetic words we have quoted of him certainly go deeper than Alberti's theorizing, but whither do they tend? We would have to know more about his background to have an exact answer. The real books of his early education, we are told, were Dante and the Bible (we read, just as in Helmholtz's autobiography, that instead of following Latin classes in school, he would draw houses and geometrical diagrams). Manetti and Toscanelli cannot be properly called his teachers, since they are younger than he. But when he decided to move from sculpture into architecture, they became his advisers and confidantes. There is some documentary evidence of their influence in his planning and thinking. And here, even with insufficient facts, we might come close enough with *a priori* inference to help M. Francastel out of his difficulties. What these men imparted was not so much the elements of geometry which were needed for perspective and graphical statics; it was a certain frame of thought connected with the subject, which had come down to them from the world of scholars. It might be said by some that it was a mantle of formal dignity intended to place them above the teachers of the hornbook and the abacus. It would be truer to describe it as the garment of their own thought, what went with education, and the "gentle soul." I have heard critics wonder how Dante could thank so tenderly and solemnly old Brunetto Latini, the dry encyclopedist and professional backroom politician, whom he finds broiling in Hell, for "having taught him time and again how man makes himself eternal."[9] But after all, if he says it, we have to believe him, who believed so much in the importance of philosophy for the "cuor gentile," hence we must conclude that even the hard-boiled Brunetto knew how to dispense more than factual information. In short, whatever his trade, this type of teacher had to communicate a philosophical understanding, it was part of the medieval style, just as the German professor of the good epoch could not but transmit his emotions about the dignity of *Geist* and *Bildung*.

All this would hardly need to be said, were it not that I fear we tend to lose the feeling of past life in our schemes of "transmission." What did those men, then, actually transmit? The feeling of the high dignity of mathematics, for one, as vouched for by Platonic wisdom; the hope in mathematics, as revived in the preceding century by famous scholars like Bacon and Grosseteste; most of all, the beliefs and wondrous intimations to which Christian intellectual mysticism had clung through the centuries, and which expressed themselves in the "metaphysics of light."

With this body of doctrine we are fully acquainted through many expositions. It starts from the Platonic analogy of God with the sun, that is set forth in the myth of the Cave. It pursues the analogy to suggest that, as God is the life of the soul, so the physical world is held together and animated by the force of light and heat, which should turn out to be, as it were, the ultimate constituent of reality.[10] One could not express it better than Galileo does in the Third Letter to Dini in 1615.

It appears to me that there is in nature a substance most spiritual, tenous and rapid, which in diffusing through the universe, penetrates all its parts without hindrance, warms, vivifies and makes fertile all living things, of which power the body of the sun seems to be a principal receptacle. . . . This should be reasonably supposed to be somewhat more than the light we actually see, since it goes through all bodies however dense, as warmth does, invisibly, yet it is conjoined to light as its spirit and power, and concentrates in the Sun whence it issues fortified and more splendid, circulating through the universe, as blood does through the body from the heart. This ought to be, if I may risk a supposition, "the spirit that hovered on the face of the waters," the light that was created on the First Day, while the Sun was made only on the Fourth. . .

Later, Galileo was to insist that he believed the ultimate substrate of reality to be light itself. There is no need, after that, to weigh his explicit professions and reservations with respect to Platonism. The doctrine of light, "a principal access to the philosophic contemplation of nature," as he calls it, is the true Platonic watermark.

Should I pursue the lineage into the Cambridge Platonists and beyond? I trust the case can rest here. There are minds to whom the geometric virtues of the light ray, its sovereign diffusion and instantaneous transmission, are symbols of its closeness to creative omnipotence, and intimations of its mysterious role as a prime element. There are other minds, call them Aristotelians, to whom all these are fine and poetic thoughts, but cats is cats, and dogs is dogs, and light is best characterized as the entelechy of transparency, that is, the perfection of an attribute, not a substance, hence incapable by itself of bringing forth even the simplest squid.

We might add that there are still other kinds of minds, fascinated by the protean changes of living substance, to whom fire and warmth seem a truer essence than light. These are the Stoics, but they belong to quite another story.

Of these so different lineages, the men that concern us here belong definitely to the first. From them springs the persistent interest in optics, from Alhazen to Witelo the Pole and Johannes Kepler. From them descends the line of speculation which becomes significant in great scholastics like Grosseteste or Oresme, men who have taken to scanning the world of mathematical ideas and entities for pure schemes, like the triangle or the quadrangle, which may be superimposed on the phenomena of change as a grid to reveal their law. Such diagrams are forms still disembodied, floating around inconclusively, keys searching for their lock; nothing will happen until Galileo brings them down to mechanical reality, the only lock that will fit them. It will mean then a deep transformation of the languages and their formal content, as e.g., from "uniformly difform qualities" to "uniformly accelerated motion"; but the schemes are there already, more than pure shapes, trains of thought in mid-air, asking to be brought down to earth. Father Clark gives a striking example in the diagram that Oresme tries vainly to apply to qualities, and will appear later, with only a change of axes, as Galileo's first integral of motion.

Such were the ideas hovering around, say, in 1360 (the time of Oresme's writing). They were disembodied and yet very pregnant ideas. They had superadded themselves to the old traditional wonderings about the cosmic role of the five geometrical solids. They are the inspiration that teachers, from Toscanelli to Pacioli, communicated with passion. At this point, in these years, they hit fertile ground, with the invention of perspective by Brunelleschi and Uccello; here we are back in our story, but we may understand better all that Brunelleschi would see in his invention—if he was the intellectual that history shows him to be. We have not yet investigated Toscanelli as an individual; but we have characterized enough of his ideas to provide an adequate first answer to Francastel's query. We have reconstructed this education as safely as we would Newton's (supposing we knew nothing about it) from the fact that his educators had to be Protestant divines and classical scholars.

The documents will then tell us that Toscanelli was an intensely ascetic and spiritual man, "a physician, philosopher, and astrologer of most holy life" Landucci calls him; a man who shunned meat and held it was wrong to take the life of animals. When we see this unusual doctrine so strongly emphasized by his successor Leonardo, with an explicit and equally unusual Pythagorean connotation, we may conclude that we know something of Toscanelli's line of thought.

We understand better, after this, certain peculiar characteristics of Alberti's avowed Platonism. Ficino is already starting a trend, and it is natural for Alberti to "move in" on it. But what he brings is very far indeed from Ficino's *Schwärmerei*.

Says Alberti: "Great, small, high, low, light, dark, and all such which are called the accidents of things are such that all knowledge of them is by comparison and proportion . . ." Does this remind us of something we heard elsewhere? Why surely it does. "In our speculating we either seek to penetrate the true and internal essence of natural substances, or content ourselves with a knowledge of some of their accidents. . . . But I shall discover that in truth I understand no more about the essences of such familiar things as water, earth or fire than about those of the moon and the sun, for that knowledge is withheld from us, and is not to be obtained until we reach the state of blessedness. But if what we want to grasp is the apprehension of some accidents or affections of things, then it seems to me that we need not despair of acquiring this by means springing from measurement and geometry, and respecting distant bodies as well as those close at hand—perchance in some cases even more precisely in the former than in the latter."[11] This is Galileo writing on the sunspots, and the kinship of thought is unmistakable. Galileo is using here diplomatic restraint, for he is arguing against the authorized system. In the *Dialogue,* he will go further and suggest that what is called change of form might be nothing but a rearrangement of invariant parts; later, Robert Boyle will be forthright about it: "This convention of essential accidents [i.e., the order of constituent particles] being taken together for the specific difference that constitutes the body and discriminates it from all other forms, is by one name, because consider'd as a collective thing, called its form."

The word is out, the essential knowledge is not of forms but of the order and position of parts. Although a very religious man, Boyle will not have to retract his words like Nicole d'Autricourt, he will take his chance with the "Epicurean error"; for it has become what we call science.

Alberti has been writing the manifesto of something that he cannot grasp in its full import, for theoreticians know in part, and they prophesy in part; but he—and Leonardo—call it a science, not an art like music, which is a daring commitment, proper science having been properly defined long ago by the authorities in the Trivium and Quadrivium. The new science of light and space is one of "accidents" or properties,[12] something which to the mind of the scholastic could not possibly hang together, or make sense.

Is it that they are unaware of the official stop sign? Surely not. The current language of their time would be enough to point it up unmistakably; that is why they have to work their way around the language. But they do not have to come into conflict with the philosophic doctrine, because they, the artists, are not compelled, at that point, to choose between "substance" and "accidents."

That grave choice will have to be Galileo's, who is after abstract knowledge. If he decides he cannot know the substance of things, he will have to be explicitly agnostic about it. The artist can be completely confident about the "substance" of his quest, for it is his own artistic creation and nothing else. If it does not exist yet, it is going to exist, of that he is pretty sure, or his life would be bereft of meaning.

Let us restate the thing in Alberti's simple terms. There is a science, perspective, whose aim is correct comparison and proportion, projection in a visual space (we shall see later all that this implies, but Alberti himself need not be aware of it as yet). It allows us, by way of geometric properties, to deal with the object, the substance, architecture itself, or rather *il murare,* as they say so much more concretely, the *act* of raising walls brick by brick. The right walls. If they know what they are doing, that is all they need.

We have come to a point where we can take stock of the situation. We have here a type of knowledge which refers strictly to what "we ourselves are doing," a *conversio veri et facti* which will find its theoretical development with Vico three centuries later. That future development does not concern us here, but its immediate continuity in Leonardo, the student of nature, surely does so. We cannot forget how, after his restless search on all planes for the essential forms and mainsprings of nature, a search which cannot lead to an abstract physics because it presupposes so much more, Leonardo is led at last to set limits to his understanding: "You who speculate on the nature of things, do not expect to know the things that nature according to her own order leads to her own ends, but be glad if so be you know the issue of such things as your mind designs."[13]

This is the end of that trail; the artist has to go back to a knowledge which is not that of nature. Leonardo's attempted system breaks up and dissolves into a magic of form and color that rises to the stratosphere. But no one could deny that the fallout was considerable.

By transforming the concept of substance into something which could be designed and built up through their science of proportion, the mathematical artists have crossed the otherwise unmanageable distance between Substance and Function. Any attempt at bridging it directly by philosophy would have led to an intellectual impasse, worse, to a breakdown.

This ought to be clear at least in one aspect. In the whole of medieval philosophy the "principle of individuation" had to be carefully worked out and structurally established, for we must account for a multiplicity of true individual "beings" (there would be no proper natural foundation otherwise for what we call now the "dignity of the individual"). Aquinas individualizes Aristotle's generic natures, made up of form and matter, by having them endowed with individual "existence" by God's will. Scotus, as befits his nominalistic attitude, has to sharpen the issue; he makes of the individual natures true essences, *haecceitates*.[14] This atomization of "being" may not be very convincing but it is needed. Three centuries later, in the Cartesian system, individuation has vanished as if by a conjuror's trick. This is inherent in mathematical ideation. Even in Galileo's thinking, it is no longer clearly justified. Now all of medieval and Renaissance thought is still strongly based on individual beings, on the *signatura rerum*. That is why the sorry predicament of Grasso the carpenter, who was made to lose his individuality for a while, is far from philosophically irrelevant, and shows the prankster's awareness of ultimates. But it was a prank. To remove the props from under individuation unequivocally and in a responsible manner as is done at present would have meant in that intellectual climate a psychological trauma or a revulsion. The philosophy of the artists, although it rests on such mere "attributes" of being as light and space, does bring forth out of those again a "being," the actual individual creation; it provides thus a gradual relativizing and an acceptable transition.

It took this mediation. Are the ladies in Piero della Francesca's *Queen of Sheba* substance or function? If we affirm either thing of them, or even both at once, it will always be true—on a level from which there is no way down again to scholastic common sense. This is what Tuscany had done for Galileo.

Shall we try to describe what we called the fallout, as it seeps in invisibly, all around, for generations before the birth of Galileo?

The new science around 1430 is, as we have concluded, operational in character; that is, it defines the object of its quest by what it does about it, with the difference that whereas the modern object is the experimental procedure, its object is *il murare,* the building procedure. It remains now to work out its theoretical structure, which is far-ranging.

The peculiar "substance" of that quest is a system of planes and volumes rigorously thought out by way of its properties, known and understood geometrically, physically, functionally, aesthetically, and even symbolically. It imports a fullness of knowledge. The perceptual raw material, as it were, of that knowledge is provided by the past of civilization, for it is in the traditional architecture which is already around, on which judgment and criticism have been able to sharpen themselves. That is the stuff which is now going to be transformed. The huge variety and multiplicity of the pre-existent Gothic pile is a wonderful mass of decorative singularities, where the eye is led without break from the minute particular to the immensity of the whole. Ockhamism concluded, as it were. The Gothic structure, thinks Alberti, is a denial of proportion, a seething cosmos of things streaking off towards heaven; we have to bring into it a constructive law which is dictated by ratio and perspective: "A speciebus ad rationes" as Ficino transcribes the Platonic principle.

Let us make this a little clearer. Form, here, for the artist, has the function of *ratio* or cause of all the species, insofar as there are not really many aspects or forms, but *the* form or the Idea ("concetto"), given by draftsmanship. Perspective shows us the actual size of a thing that looks small in the distance; its position in space determines the truth and invariance of the individual object: its projection the true form abiding in it. The module and reality of the particular are shifted from the thing to geometric space. It is in this space of *ratios* that true construction takes place.

I am trying to paraphrase as best I can the actual ideas of a contemporary who thought he saw what was taking place but could perceive only a dim outline. You cannot expect it to be as clear as a theorem. But if this goes as Platonism, it is certainly not the literary variety with its fashionable uplift. We are treated to diagrams and visual pyramids, to a coördinates net on a screen. We are asked to see longitudinal perspective in terms of that other inverted pyramid ending in the flight-point placed on the infinite circle: that mathematical point at infinity in which all the forms and ratios of reality are absorbed or rather "contracted." No one could deny that we are here on Cusan territory, although there could be no question of direct influence, since Nicholas of Cusa was just having his initial ideas in 1433 when Alberti was writing. What these men have in common can only be the source, the unformulated Pythagorean element just then being transformed. Alberti is fully aware that the longitudinal perspective carries the theme of contemplation; it loses itself as Plotinus suggests in that one flight-point.[15] In fact, it is the equivalent of the medieval golden background: But the imagined ensemble of flight points is, with respect to that sensuous gold, utterly abstract—a true intellectual construction, like that of Nicholas of Cusa, ending up on the "circle of infinite radius."

This metaphysical aspect had been fully grasped by Brunelleschi himself, as is shown by his main line of research, the inflexible endeavor pursued through his life to achieve a synthesis between the longitudinal perspective, implying transcendence, and the central perspective, which implied to him an intrinsic organization of space, or, in philosophical terms, immanence. We can follow the successive stages of this study in his great buildings, San Lorenzo, Santo Spirito, the Ospedale degli Innocenti, until the synthesis is achieved right at the time of his death (1446), with the topping of his great Dome. For the Dome is not only his most outstanding solution in static engineering (what was felt as the miracle of unsupported growth), it is also, in its "rib-and-sail" structure as it was called, the conclusive formal solution of a philosophical issue. The slowly convergent triangles are pure geometric forms leading up to infinity, as no hemispherical dome ever could. The Gothic vertical flight has been transposed into another key; it is concluded and held together by the topping Lantern.

At this moment, as if to mark the scientific inspiration of the whole, astronomy comes in with a significant note. Paolo Toscanelli, Brunelleschi's trusted consultant, and no doubt in accord with his friend's wish, had a dial device built into the Lantern as it was being terminated. It was the third novel instrument in the series: a perforated bronze plate, placed so that the sun's beams struck the pavement below along a graded strip cemented into the floor. It turned the Dome, as Lalande was to write in 1765, into the greatest astronomical instrument ever built. The beam was 240 feet long, and it allowed Toscanelli to effect his solstitial measurements of the inclination of the ecliptic reported by Regiomontanus, which gave 23° 30'. It is correct within 1', which is probably luck, but in any case better for the time than Regiomontanus' own, 23° 28'. (It is at present 23° 26' 40", owing to the yearly precessional variation.)

Have I been trying, then, to read philosophy or science into art, a thing reproved both by the scientist and by the aesthetician? I trust I have not. We have only to read Alberti to realize these men's keen awareness of their intellectual quest. At a time when what *we* mean by science was still beyond the horizon, when the *name* of science was monopolized by scholastic officials, who officially denied to mathematics any link with physical reality, these men had conceived of an original prototype of science based on mathematics, which was to provide them with a creative knowledge of reality, repeat—creative, and could claim the name of true knowledge in that it dealt with first and last things. There should be some proper way of placing this attempt, in its true dimensions, inside the history of science, but it has yet to be made.

The historian of art, left to himself, will help us only to a point, because he is thinking of different terms. It is characteristic that so acute a critic as Heydenreich, who contributed much to the knowledge of Brunelleschi's evolution,[16] left the Dome to one side as a technical stunt—a strictly scientific job in graphic statics, we might say, raising a vault without centering. It took the careful monograph of Sanpaolesi,[17] conducted on the archival material of budgets, notes, and sketches, to show how the remote theoretical problems never left Brunelleschi's mind.

That he was left free to undertake it is curiously tied up with very practical and economic reasons. The original plans for the Cathedral, designed by Arnolfo di Cambio two hundred years earlier, established the measures for the circular base of the dome, conceived as a hemispheric vault. Those measures were found quite proper by his successors. It turned out, however, that by 1400, architects could count no longer on the huge availabilities of timber that a conventional centering frame implied. Even the carpenters able to deal with it were scarce; the task appeared "impossible, nay more than impossible," as the chronicler puts it. This is what moved the *Operai,* a committee of conservative businessmen, to consider any suggestions for a new working solution. But that solution sprang in Brunelleschi's mind from theoretical considerations. It was a triumph of the speculative, and not only of the mechanical, mind over matter, hence we should not lose sight of that speculative element.

It has its source in the Pythagorean tradition implicit in Platonism. It is a bewildering mixture of ancient tradition and daring modernity, as the old Pythagorean mentality itself had been. In his Christian faith, in his canons, in his certainty (he said once: "Our faith should be called certainty"), Brunelleschi is profoundly a medieval. He is even older than that, for Wittkower has shown that the Renaissance canon relied on integers and rational numbers, whereas the medievals had been more open to the use of irrationals. But it is all the more strongly founded on proportion and *logos.*

It is this intellectual element, overarching that whole early Renaissance, which explains many things in the rise of the new architecture (Brunelleschi himself, as we shall see, had no inhibitions about the Golden Section, irrational though it may be). It will drive Uccello and Piero deep into geometrical speculation, it will eventually find its concluding manifesto in Pacioli's *Divina Proportione,* which is expressly dedicated to Piero della Francesca, "the most worthy monarch of the art of our time." In this kind of theory, elementary abstraction has hardened and simplified much of what had been profound and original creation. But the development, we said it earlier, is as linear and rigorous as one could wish.

Whither does it lead us? To what Alberti, as essentially a man of letters, could not see, nor for that matter Pacioli (who was barely above a teacher of the abacus), but seems to be present as a deep intuition in the creative masters. We mean, to an impressive generalization of the Pythagorean system. In the original doctrine, there were only a few entities which embodied geometry, but they were felt to be enough in that they acted by participation. The circle, the cube, the Harmonic Fifth, existed as absolutes, and caused things to behave "in imitation" of them. Of such manifest imitation the cases could not but be very few, and so science was limited to the charismatic regions of music and astronomy. But if we begin to think in projective terms, then we begin to see circularity inherent in ellipses, squareness in rhomboidal shapes, and so on. The "imitation" of the Ideas turns up everywhere, even if the original forms be no longer obvious: The square becomes one of its own perspectives, the circle a conic section. In fact, what used to be "form" becomes a collection of very abstract relations whose mathematical treatment is surely not elementary—and may have been above anyone's resources at the time. What would Uccello be tormenting himself about with the absurd complication of all the perspectives of the square in his *mazzocchi,* why should he be exploring the possibilities of invariance under distortion, if he did not have some such idea at the back of his mind? This is, then, a vision of the Pythagorean system become truly universal and permeating all of reality.

What is suggested here is surely inference *a priori.* I am not aware as yet of any statement documenting it unequivocally within the fifteenth century. A check that comes to mind could be found in Maurolico's treatment of the conic sections as the perspectives of a circle. It would be a definite subject of research to look for more evidence. What I was suggesting there was not even the probability of an explicit theory along the lines mentioned above, but essentially the presence of an intuition, which allows the mind to envision a new generalized Pythagorean approach to the whole of reality. This intuition becomes then, in the image of Galileo, like a sun in which the old ideas concentrate "to issue thence fortified and more splendid" to circulate through the whole culture.

However that may be, we have brought into evidence a *fifth* factor which provides what was certainly missing, the vision, the unifying intellectual element behind the artistic revolution. It is not in direct contrast with current scholastic theory, and in fact we discern no quarrel with it in the whole contemporary Florentine Platonist trend. It merely bypasses it. But it will carry thought far into the future.

I trust it is plain by now that what is involved in this story are the great categories of scientific philosophy. Erwin Panofsky has characterized the space of classical art as "aggregative." There would have been no better way to describe the space of Aristotle himself, which is nothing but an orderly pile of containers. This commonsensical kind of space is, for a modern, utterly irrelevant. The space of the Renaissance Panofsky describes by contrast as "homogeneous." In our language we might call it a metric continuum. We have seen earlier what it can imply. It is a new potential richness that Leonardo expresses with awestruck phrases.

It is not a matter of realizing its three-dimensionality. It had been three-dimensional for everybody all along, even for the most medieval of artists; we do not hear of their bumping their heads into corners from lack of space perception. It was simply that they had not felt three-dimensionality as relevant to their art. Here it has become a subject for relevance and intense abstract imagination. No one will mind making distant figures as small as they have to be: The space which comprises them in its structure will restore and define their meaning. "Grasp firmly with thy mind the far and the near together . . ." These are the words of old Parmenides, and they stood for a great beginning already at that time. It is this firm, this creative grasp which makes all the difference.

The closed space of Aristotle is only the tidy arrangement of a simple multiplicity of things, not unlike, let us say, the shipping department of Sears Roebuck. Whichever way we understand the new space, that of Nicholas of Cusa and Brunelleschi, it is certainly not that. It is for the artist a pure space of diaphanous light articulated throughout by the central design, bringing into action the law of forms from every point of view at once. It is described by Cusanus, in his famous phrase, as "that whose circumference is nowhere and the center everywhere" and the phrase should be enough to show the whole import of the revolution, for Nicholas of Cusa has borrowed it from the description that "Hermes Thrice-Greatest" gives of the soul. Such a transfer to cosmic space of the properties of the soul, with the accent on a "central perspective" of the intellect, is something no medieval would have dared.

Space is for the new imagination a matrix for infinite potential complexities and states and tensions—a matrix awaiting total structure, rather, a manifold of structures. It is on its way to becoming what is for Newton the organ of perception of God, for Malebranche the only intelligible reality, for the theory of central forces the carrier of that incomprehensible property, action at a distance; it is rich enough to bring forth set theory, group transformations, phase spaces, the electromagnetic ether, Riemann's geometry, and the Einsteinian reduction of all reality to properties of the time-space continuum.

To sum up, this investigation seems to suggest that two of the major features of the Scientific Renaissance, namely, the change-over from Form to Function, and the rise of a "natural law" unconnected with the affairs of human society, have their origin in a specific transformation of the arts. They cannot be said to arise out of the scholar's interest in mathematics, which remains wishful, nor out of the development of the crafts per se, nor out of any statistical accumulation of small interactions between the two zones. They are coherently worked out and brought to bear at the time when the representative and building arts form a new idea of themselves, and go through a theoretical elaboration of that new idea, in such a way as to be able to bring it to grips with reality in their creation. This seems to be the moment when the actual shaping of a new operative thought takes place, and it provides some fundamental categories for nascent scientific thought.

I have barely sketched out the outline of the problem. The analytical tools have yet to be forged. The scientist and the historian of art have hardly ever met, and even then under a cloud of misunderstanding. I am only trying to enter a plea for collaboration in a subject which is still tricky and most difficult.

References

1 Filippo di ser Brunellesco di Lippo di Tura is his exact name; if the sopranym from his father had not stuck to him, he would have been Filippo Lapi, for he belonged to the ancient noble family of Lapi-Aldobrandi, and his arms were the same as theirs. His father, Ser Brunellesco, had been a notary and a judge, who had belonged to the Council of Ten for Defense (Dieci di Balia) and had been ambassador for the Republic several times. The family was a wealthy one, and young Filippo had been sent to a humanist school, but his disinclination towards the study of the classics prompted his father to have him apprenticed in the guild of goldsmiths, which included at the time painters and sculptors.

2 Pierre Francastel, *Peinture et Société* (1951).

3 Henri Focillon, *L'art des sculpteurs romans* (Paris, 1932).

4 G. R. Kernodle, *From Art to Theatre, Form and Convention in the Renaissance* (Chicago, 1944).

5 J. Ackerman, "Ars sine scientia nihil est," *Art Bulletin,* XXXI (1949), 84–111. "Mechanics examines bodies at rest, their natural tendency (to a locus), and their locomotion in general, not only assigning causes of natural motion, but devising means of impelling bodies to change their position, contrary to their natures. In this (last) the science of mechanics uses theorems . . ." These are words not even of Aristotle, but of Pappus in his *Math. Collection,* VIII, 1.

6 Giovanni di Gherardo wrote against him in a sonnet which begins: "O deep and brimming wellspring of ignorance"; Cavalcanti wrote that Florence had been led astray by "their false and lying geometry," and there were many more such attacks.

7 See note 6.

8 A. Bosse, *Maniere universelle de M. Desargues de pratiquer la perspective* (Paris, 1648).

9 *Inferno,* VII, 82: "for there is ever in my mind, and still moves me deeply, the dear and good fatherly image of you, as you taught me in the world, time and again, the way in which man makes himself eternal."

10 See Dionysius the Areopagite, as quoted in Galileo's letter to Dini.

11 *Discoveries and Opinions of Galileo,* tr. Stillman Drake (New York, 1957).

12 Galileo also calls them "affections," using the old scholastic term, to mark his position with respect to the past.

13 MacCurdy translates "conceives" which is quite misleading. The original has *disegnate.* See E. MacCurdy, *The Notebooks of Leonardo da Vinci,* I (New York, 1938), 76.

14 G. Bergmann, "Some Remarks on the Philosophy of Malebranche," *Review of Metaphysics,* X, No. 2 (Dec., 1956), 207–26.

15 A. Grabar, "Plotin et les origines de l'esthétique médievale," in *Cahiers Archeol.* (Paris, 1945), 15–34.

16 L. H. Heydenreich, "Spätwerke B.s," *Jahrb. d. Preuss. Kunstsamml.* (1930).

17 P. Sanpaolesi, *La cupola di S. Maria del Fiore* (Roma, 1941).

A. C. CROMBIE

on the Papers of Rupert Hall and Giorgio de Santillana

I should like to make three general comments, directed first to Dr. Hall's paper. My difficulty as a critic of this paper is that I agree wholeheartedly with its main thesis. Within this perhaps inconvenient framework for an occasion where disagreement would be a better stimulant to discussion, I want to look first at some aspects of the characterization of the Scientific Revolution and the assumptions about the nature of modern science that are made in historical analyses of the origins of modern science. Most historians would, I think, agree in seeing the Scientific Revolution as the triumph of the new "experimental-mathematical philosophy" as it was called in the seventeenth century, and in seeing the most outstanding result and justification of this new philosophy in the Newtonian system. This, after all, is how contemporaries, especially in the second half of the seventeenth century, saw events, and there seems every reason to accept this part of their conception of what had been effected in their time. It means that we take mechanics as the model of the new scientific thought and the mechanization of the world as its driving force, and this view is supported by observing that other sciences (for example physiology with Harvey and Descartes, optics with Kepler and Descartes, and chemistry with Boyle) were all consciously recast on the mechanistic model. Taking a longer view, we can see seventeenth-century physics, especially as developed by Newton, as the completion of the first great new model of scientific thinking since the Greek model of geometry. Indeed one of the central points of the discussions of scientific method that are so characteristic of the whole period covered by "the origins of modern science," going back to the thirteenth century, concerned the inadequacy of the geometrical model for physics and the necessity, for scientific thinking in physics, of striking a new balance between sense experience and experiment on the one hand, and deductive mathematical-physical theory on the other. A series of quotations could be given showing an increasingly clear grasp of this problem, culminating in the classical descriptions of the new model of scientific thinking in physics given by Newton and by Huyghens. It may be of value to remind ourselves of the content of the famous passage from the preface of Huyghens' *Treatise on Light* (1690), which covers the essential points:

There is to be found here a kind of demonstration which does not produce a certainty as great as that of geometry and is, indeed, very different from that used by geometers, since they prove their propositions by certain and incontestable principles, whereas here principles are tested by the consequences derived from them. The nature of the subject permits no other treatment. It is possible nevertheless to attain in this way a degree of probability which is little short of complete certainty. This happens when the consequences of our assumed principles agree perfectly with the observed phenomena, and especially when such verifications are numerous, but above all when we conceive in advance new phenomena which should follow from the hypotheses we employ and then find our expectations fulfilled. If in the following treatise all these evidences of probability are to be found together, as I think they are, the success of my inquiry is strongly confirmed and it is scarcely possible that things should not be almost exactly as I have represented them. I venture to hope, therefore, that those who enjoy finding out causes and can appreciate the wonders of light will be interested in these varied speculations about it.

Now if we accept a highly sophisticated characterization of this kind of the science of the Scientific Revolution, a logically and experimentally intricate inquiry for theories that both predict quantitatively and explain causally, we determine in advance the direction in which we are going to look when we ask *who* made the Scientific Revolution and what influences, intellectual and social, were involved in the event. Dr. Hall, I think rightly, has strongly emphasized the highly intellectual and theoretical character of the Scientific Revolution, and it is not surprising that he has identified the architects of the revolution as the "scholars" who alone were capable of thinking at that level, and who drew into their highly sophisticated system of scientific thought both the problems and *ad hoc* techniques of the craftsmen and also conceptions like atomism from nonscientific philosophers. If we accept this view then we agree with the seventeenth-century scientists themselves in seeing their work as a new philosophy and not simply as a new technology, although a new technology did indeed eventually follow from the new philosophy.

I have specially drawn attention to this point because in many discussions of the origins of modern science, differences in historical interpretation seem to have been determined wholly or partly by different implicit evaluations of the scientific content of the Scientific Revolution. It is one of the merits of Hall's paper that he makes his position in this respect perfectly explicit. The interpretations that have made changes in social circumstances enhancing the social importance of the craftsman and of technology the primary factor in the Scientific Revolution are particularly revealing from this point of view. It has been argued, for example, that the mechanization of industry and the division of work among several workers collected under one roof led to the idea of the mechanization of nature; that the new monetary economy of early capitalism, in which goods of many different kinds were submitted indifferently to the same quantitative measure, manipulated by the new techniques of commercial arithmetic and double-entry bookkeeping, suggested an analogous mathematical idealization of nature and its reduction to exact, quantitative measure. It has been argued that the rise of science is linked with the rise of protestantism, with its alleged quantitative interest in commercial output and profit, in place of catholicism with its ideal of qualitative perfection; that the development of the concept of laws of nature and the rise of rational mechanics are linked with the rise of autocratic government and especially of autocratic monarchy.

An immediate difficulty in all these arguments is that, for example, quantitative empiricism and attempts at the mathematization of nature existed long before the social events here adduced to explain them. There is also the difficulty of the extreme shortage of evidence, in the documents of the history of science as such, that these interesting social events did in fact have any influence on any scientific thinker or that any scientific concepts were in fact derived from their alleged social models. But quite apart from these difficulties, there is also the question whether the authors of these interpretations themselves have in mind a conception of science and of scientific law and explanation that, for example, Galileo or Descartes or Newton, that is the scientists the origins of whose thought is being explained, would recognize and welcome as their own. A recent example may be found in Joseph Needham's *Science and Civilisation in China* (Cambridge, 1956, vol. 2, p. 542) in which he summarizes part of the argument of a well-known paper by the late Edgar Zilsel, surely one of the most interesting advocates of the sociological interpretation, as follows:

Zilsel sees one essential component in the development of 17th-century laws of Nature in the empirical technologies of the 16th century. He points out that the higher craftsmen of that time, the artists and military engineers (of whom Leonardo da Vinci was the supreme example) were accustomed not only to experimentation, but also to expressing their results in empirical rules and quantitative terms. He instances the small book *Quesiti et Inventioni* of Tartaglia (1546) in which quite exact quantitative rules were given for the elevation of guns in relation to ballistics. "These quantitative rules of the artisans of early capitalism are, though they are never called so, the forerunners of modern physical laws." They rose to science in Galileo.

So far as one can see the conception of "physical law" envisaged here is of propositions that enable us to predict but not to explain, that is, they enable us to calculate what will happen in given situations but make no attempt to offer an account of the causes of what happens in terms, for example, of matter-in-motion or any other conception of the nature of the physical world. Nor as laws are they very general. This is a kind of positivist or pragmatic conception of science, and presumably its advocates would trace the beginnings of scientific thought back to the pragmatic techniques of Babylonian arithmetic and astronomy and not in the Greek conception of deductive causal explanation and geometrical proof. Yet it was just such causal explanations that seventeenth-century scientists like Galileo, Kepler, Descartes, and Newton insisted were the goal of scientific inquiry. In Galileo's case, Zilsel's and Needham's account has left out the "mathematical Platonism" which Koyré has shown to be so essential a part of his scientific thought; it begs the essential question in the last sentence, for how did the empirical rules become general laws? and in general it leaves out the search for explanations in terms of the "mechanical philosophy" which is so characteristic of seventeenth-century physics. It is of course an open question to decide what in the end is the most adequate conception to have of science, but if we are looking for the origins of the conception of it held by the great scientists of the seventeenth-century, then it is quite clearly misleading to find these only in the techniques of craftsmen and to overlook the influence of mathematical Platonism, of philosophical atomism, of philosophical discussions of the logic of science, of mathematical questions concerning infinitesimals, limits, variable quantities and dependence which had no immediate practical application, in fact to overlook the questions that preoccupied the "scholars" to whom the Scientific Revolution is generally accredited. The techniques of craftsmen and the social and economic factors influencing their development have an interesting history of their own, but, as Hall has rightly pointed out, these are only one of the elements contributing to the development of modern science. They are quite inadequate to explain its whole content.

Accepting the highly sophisticated and theoretical character of the Scientific Revolution, I want now to look at some aspects of the historical appraisals of the various activities of the "scholars" who brought it about. Hall contrasts the emphasis allegedly laid by the "medievalist" appraisal on scientific method with the emphasis laid by the "humanist" appraisal on the content of science, and he quotes me in connection with the "medievalist" view. Certainly I have emphasized the fundamental importance in the genesis of modern science, and in the transformation of the Greek geometrical model into that of seventeenth-century physics, of ideas on the form and structure of scientific theory and of methodological insights into the roles of experiment, mathematics and a mathematical conception of nature. But I do not think that ideas on scientific method can be effectively formulated, let alone effectively used, apart from the content of science, apart from actual scientific problems, theories, conceptions and techniques, especially mathematical techniques, which scientific method is designed to investigate and use. All these elements have reacted on each other in a most complicated way in the development of modern science, since the twelfth century.

It is certainly true that the humanists made a very important contribution to the content of scientific knowledge in the fifteenth and sixteenth centuries by collecting, translating, and printing Greek scientific writings, but it must not be forgotten that the first introduction of the West to Greek science was brought about by the medieval scholars who made the translations from Greek and Arabic in the twelfth and thirteenth centuries. Moreover, Lucretius was known at least at second hand to Hrabanus Maurus in the ninth century and fairly radical atomistic ideas were put forward by William of Conches in the twelfth century and by Nicholas of Autrecourt in the fourteenth century (see G. D. Hadzsits, *Lucretius and his Influence* [London, 1935]; Sir J. E. Sandys, *A History of Classical Scholarship* [3rd ed., Cambridge, 1904], Vol. I, pp. 631–33); and Clagett has shown that the old belief that Archimedes was unknown in the middle ages is mistaken. The work of Gerard of Brussels on kinematics is an outstanding example of his influence in the thirteenth century (see Clagett's article in *Osiris,* XII [1956]). There are of course many problems still to be solved concerning the continuity of scientific knowledge from

Greek Antiquity through the middle ages into the seventeenth century.

Looking at the whole period from the first translations of important Greek and Arabic scientific writings into Latin in the twelfth century, down to the completion of the Scientific Revolution in the second half of the seventeenth century, one can see a pattern of potentiality and promise at the beginning giving way, by a series of changes of interest and competence culminating in a sudden burst, to realization and fulfilment at the end. Medieval learning was dominated by metaphysical interests and humanist learning by philological interests, to both of which science was really ancillary. Certain significant changes took place in the course of time, for example the twelfth century was dominated philosophically by Platonism (especially the *Timaeus*), the thirteenth by Aristotelianism, the fourteenth by empiricism or "nominalism." The fifteenth century saw revivals involving all three philosophies and the sixteenth saw a renewed interest in Greek mathematics. If it was only in the seventeenth century that modern science emerged fully confident of its proper methods as a distinct form of learning different from other branches of philosophy and scholarship and from Greek science, very definite moves towards such emergence were made in the medieval universities.

This historical pattern enables us to put the medieval and humanist contributions to the genesis of modern science in what seem to me to be their proper places. I think that the evidence shows that the insights of medieval philosophers into the theoretical structure and methodological requirements of modern science were a primary factor in this process, and in my own paper I shall discuss what seem to me to be their main contributions and shortcomings. The latter seem to me to be related to the fact that their interests were primarily philosophical rather than strictly scientific, so that in consequence they showed in practice, for all their philosophical talk, a certain irresolution in equipping themselves with an adequate knowledge of mathematics, in applying empirical criteria, and in grasping the need for measurement, with the further result that they failed to develop a conception of nature that would provide adequate explanations of physical phenomena. Some of these defects were overcome with material provided by the humanist scholars and others by material provided by the craftsmen, but the final grasp of the fully modern conception of science was an act of intellectual creation performed upon all these materials. A most revealing measure of the difference this act made can be taken by comparing the methods of almost any medieval physicist with the strong empirical discipline with which Kepler applied the tests of experiment and measurement to speculations as deeply metaphysical and theological in content as any of theirs and with the powerful mathematical techniques and concepts by means of which he built his theories. It is surely an essential part of the historical phenomenon we are trying to analyze that the "break-through" to modern science should, in each set of problems, have been so long prepared but actually made by a genius, a Galileo, a Kepler, a Descartes, a Harvey, belonging to almost the same generation.

One further point may be made here. In applying his five types of contribution of craftsmanship to science in the sixteenth and seventeenth centuries, Hall's treatment of biology seems to me somewhat inadequate. For example, the first type certainly applies in part to Harvey's work on the circulation of the blood and also to Descartes' mechanistic physiology, both the first and the second apply to the attempts, for example by Harvey and Sydenham, to formulate theories of disease, and through art, surgery and pharmacy the third also applies to anatomy and botany.

Dr. Hall raises the question whether the impulse to a reform of science, the preparation and the actual break-through, was spontaneously generated among scholars within their own immediate field of philosophical interests, or was to some extent, large or small, stimulated by doubts about their methods and conceptions raised by the comparative success of contemporary technology. He concludes that the genesis of modern science does indeed owe a considerable debt to technology, a debt created by the change in attitude of the philosophers themselves in contrast with that taken in antiquity, and that the craftsmen provided a technological setting in which a scientific revolution became possible. Professor de Santillana introduces further significant aspects of this theme. While I agree with the conclusion just stated, I think that it is necessary to date some of these factors much earlier than either author has done.

Just as many of the characteristic methods and contents of seventeenth-century science are found to have had a long preparation going back to the twelfth or thirteenth centuries, so also does the association of technology and learning. After the basic settlement of the barbarian invaders upon the soil of Western Europe, a rapid increase in population in the eleventh and twelfth centuries, accompanied by the clearing of forests, the development of towns and monastic orders, the enlargement of churches, the crusades, saw also a modest but important technological revolution with the harnessing of animals, water, and wind as sources of power and the introduction of mechanisms for mutually converting reciprocal and continuous motion. At the same time there was an enlargement of the content of learning, and we find at the root of scholastic educational theory, in the influential writings of Hugh of St. Victor and Domingo Gundisalvo dating from the first half of the twelfth century, that an important place has been made for the study of technical subjects that give control over nature. A distinction is made between theoretical and practical subjects, most significantly between theoretical and practical mathematical subjects, and the principles learnt in the former are to be applied in the latter. Among the latter are practical geometry, which includes the work of land-surveyors, carpenters, masons and black-smiths, and the science of engines or engineering (*de ingeniis*), which includes building, the construction of elevating machines, military weapons, musical instruments and burning mirrors. Of this Gundisalvo wrote: "The science of engines teaches us how to imagine and to find a way of arranging natural bodies by an *ad hoc* artifice conforming to a numerical calculation, so that we can get from it the use we want" (Dominicus Gundissalinus, *De Divisione philosophiae*, ed. L. Baur, *Beiträge zur Geschichte der Philosophie des Mittelalters*, IV, 2–3 [Münster, 1903]).

Here is very early evidence, contemporary with the earliest important translations from Greek and Arabic, of scholars turning their interest to technology. This is not the occasion for a parade of facts to show that learned interest in technical and useful subjects persisted thereafter, but we may recall, for example, the high proportion of technical treatises among the works translated into Latin in the twelfth century, the importance of the calendar as a practical problem for astronomy and its influence on the university *quadrivium*, the development of astronomical and other instruments of practical mathematics by scholars in the thirteenth and fourteenth centuries, the establishment of medicine with its ancient empirical tradition as a subject ranking with theology and law in higher university education. We may recall also the interest of scholars in the practical problems of alchemy and of preparing medicines and drugs, the interest of thirteenth-century scholars like Albertus Magnus, Vincent of Beauvais and Raymond Lulle in the activities of artisans such as peasants, fishermen and miners, and Roger Bacon's strong plea for experiment and mathematics, as he understood them, as means of obtaining useful power over nature and his praise for his practical interests of

Peter of Maricourt, who had said that science was impossible if discursive reasoning were not combined with manual dexterity.

Mr. Guy Beaujouan in a recent lecture, *L'Interdépendence entre la science scolastique et les techniques utilitaires (xiiᵉ, xiiiᵉ et xivᵉ siècles)* (*Les Conférences du Palais de la Découverte*, Série D, No. 46 [Paris, 1957]), has given further examples of the practical application of theoretical notions learnt at the university, and of problems, information, techniques and apparatus contributed by technology to scholarly science in the middle ages. We would like to know more about the relations between theoretical statics and building, between theoretical dynamics and ballistic machines and cannons (Buridan mentioned cannons in his *Meteorology*), between academic and commercial arithmetic, between alchemy and the serious study of metals and the development of chemical apparatus and techniques, between academic trigonometry and navigational methods, between theoretical optics and the development of lenses and of linear perspective, between the study of anatomy and botany and the practical needs of art, surgery, and medicine.

Two phenomena encouraged contacts between scholars and craftsmen, already bearing fruit by the middle of the thirteenth century: the rise in the social scale of the specialized artisan, and the enlargement of scientific culture by the universities. A good example is the rise of the architect-engineer. When Gundisalvo was describing the *scientia de ingeniis,* the artisan known as an *architector, ingeniator,* or practical *geometricus et carpentarius* was primarily a military engineer. By the thirteenth century the *architectus* was using his knowledge of practical geometry to combine engineering with building, declaring himself superior to the mason, and we find him expressing himself with confident articulation in the notebook of Villard de Honnecourt. Nicolaus Pevsner (*Speculum,* XVII [1942], pp. 558–61) has pointed out that in the twelfth century Hugh of St. Victor and Otto of Freising excluded the architect, along with other artisans, "ab honestioribus et liberioribus studiis" and considered the profession as fit only for "plebei et ignobilium filii," but that in the thirteenth century the *architectus* was distinguished, for example by Villard's contemporaries Albertus Magnus and Thomas Aquinas, from the artisans who now worked under him, and had become the *theoricus* as distinct from the *practicus* of his art. It was by studying the liberal arts that the architect was thus raised up both intellectually and socially. No doubt this is one reason why so many of the architects of the Gothic churches were clerics

74

(even when we omit those who were only the patrons and not the actual builders), but the geometrical knowledge shown by Villard himself was substantially that of the *quadrivium* of the cathedral schools and the universities. It required considerable precision in using the various rules of proportion and seems to have been at least partly derived from the study of Euclid. (See Lynn White, Jr., "Natural Science and Naturalistic Art in the Middle Ages," *American Historical Journal,* LII [1947].)

In the light of these facts and of the theoretical and metaphysical conceptions behind Gothic architecture, of which Otto von Simson (*The Gothic Cathedral,* [London, 1956]) has recently given an account, I do not think it possible to accept without considerable qualification de Santillana's blunt characterization of Brunelleschi, whom he makes central to his thesis, as the first professional engineer who mastered not only his craft but also the philosophical and social position of his subject. De Santillana's comparison of the bold theoretical ideas that preceded the construction of the dome of the Cathedral at Florence with Galileo's method of procedure is profoundly interesting. His attempt to elucidate the cultural position of science in this period by a comparative study of the influence of Platonism and of mathematics in science and in art is much to be applauded, and he is fully aware of the dangers. My criticism is that if he had taken his analysis further back in time the results would have been even more illuminating.

Von Simson describes how the Gothic architects designed their churches in accordance with strict rules of proportion (although some allowance has to be made for artistic deviation) which in turn had a definite metaphysical meaning. Behind this theory of design lay the conception of art as a mirror of ultimate reality and the numerological "Pythagorean" cosmology and philosophy of beauty of St. Augustine, based on the text from the *Wisdom of Solomon:* "thou hast ordered all things in measure and number and weight." This was developed especially in the twelfth-century cathedral school at Chartres. In his account of the building of the choir of St. Denis, consecrated in 1144, Abbot Suger, who was closely in touch with that school, begins with a metaphysical vision of the harmony of the Divine Reason and concludes with an account of the moral significance of his new church. There are few records of the technical methods used in design by the Gothic architects, but they were made public towards the end of the fifteenth century by Matthew Roriczer, the builder of Regensburg Cathedral, and the documentary evidence has been confirmed by measurements made on Chartres Cathedral and other important buildings. Beginning with a single basic dimension the Gothic architect developed all the other magnitudes of his ground plan and elevation, and of other details such as the thickness of walls and the proportions of statues, by strictly geometrical means. There were two kinds of geometrical canons, one relating the parts as multiples or submultiples of a common linear measure, and the other relating them by proportions derived from certain regular polygons, especially the square and the equilateral triangle. Examples are the *sectio aurea* and the proportions obtained " according to true measure," in which lengths are related to each other as are the sides of a sequence of squares whose areas increase or decrease in geometrical proportion.

Building according to geometry not only made the result a symbol of the harmony of the creation, to the glory of whose Author the churches were dedicated, but distinguished architecture as *scientia* from mere *ars*. God himself was described by Alan of Lille as the *elegans architectus* and was depicted (for example in the *Bible moralisée,* No. 15, in Vienna, Cod. 2554, f. 1) as an architect, a *theoricus* creating the universe according to geometrical laws. The distinction between *ars,* mere practical knowledge such as that of a bird in singing, and *scientia,* theoretical knowledge of music that could be applied in composition and appreciation, came from St. Augustine's *De Musica;* other treatises by St. Augustine kept alive the same distinction in architecture, a sister child of number, the one echoing and the other mirroring the eternal harmony. The Gothic architects were as emphatic in this distinction between theoretical and practical knowledge as in their submission to geometry: "ars sine scientia nihil est" as one delegate proclaimed at one of the architectural conferences that took place from 1391 during the construction of Milan Cathedral (von Simson, *Gothic Cathedral,* p. 19).

Yet with all this talk of *scientia* in Gothic architecture where do we find any exact scientific knowledge in the sense of engineering, any measurements of thrust and weight and the strength of materials? The geometrical *scientia* of the Gothic architects does not seem to have gone beyond the realm of the aesthetic and the symbolic. They seem to have relied on the practical skill, pure *ars,* of their masons to carry out their instructions in stone and wood. This seems to me a mental attitude of a piece with the relative indifference of contemporary natural philosophers to actually making measurements. Did the artists and architects of the *quattrocento* change it? Where, before Galileo's *Two New Sciences,* do we find a quantitative study of the strength of materials? Who first *measured* thrust and the engineering requirements of buildings?

In spite of their apparent indifference to engineering there is no doubt that the Gothic architects made their craft into a learned, theoretical, and, in a sense, a scientific activity. This is no isolated phenomenon in the "mechanical arts," although our knowledge of the whole subject is far from complete. The interest of scholars in technology and the rising articulateness of the higher artisans is shown by a long series of technical treatises written from both sides, extending backwards to the sources behind the twelfth-century chemical treatises of Adelard of Bath and of Theophilus the Priest, and forwards through the writings for example of Walter of Henley, Peter of Crescenzi, Guido da Vigevano, Giovanni da San Gimignano, Cennini, Konrad Kyeser, Fontana, Alberti and the other fifteenth-century German and Italian treatises on military engineering and architecture. The series continues through the sixteenth-century descriptions of crafts and inventions by such writers as Agricola, Biringuccio, Erker, and Ramelli, down indeed to the Royal Society's projected *History of Trades,* the volumes on *"ars et métiers"* published by the *Académie des sciences,* and the plates of the *Encyclopédie.* (See Lynn Thorndike, *History of Magic and Experimental Science* [New York, 1923], Vol. I, Ch. 33; Charles Singer *et al., History of Technology* [Oxford, 1956], Vol. II, Ch. 10.)

To summarize my comments, I believe that the evidence shows that the elements, both intellectual and practical, that came to produce the Scientific Revolution had been incubating in Western civilization for several centuries, and that in our analysis of the eventual break-through we must take into account the philosophical principles and the attitudes that were adopted before these were realized in any outstanding results. From the intellectual element came insights into the requirements of scientific thought in experimental science and from the practical element came a strong urge for useful knowledge. We are accustomed to hear that medieval speculations far exceeded proper empirical control and certainly this is often true. On the other hand the excessive medieval desire for immediate utility and the excessive respect for immediate observation, characteristics of the outlook of both technology and of philosophical empiricism, of the artist-engineer like Leonardo in contrast with the scientist like Galileo, throws into even stronger relief the highly intellectual and

conceptual nature of the moves, in the experimental as well as the mathematical and theoretical aspects of the study of the world, that made up
the revolution in scientific thought that culminated in the seventeenth
century.

THE SIGNIFICANCE OF MEDIEVAL

DISCUSSIONS OF SCIENTIFIC METHOD

FOR THE SCIENTIFIC REVOLUTION

A. C. Crombie

We all know Kant's famous comment on the Scientific Revolution, his declaration in the preface to the second edition of the *Critique of Pure Reason* that with the conception and methods of science put into practice by Galileo and his contemporaries "a new light flashed upon all students of nature" and "the study of nature entered on the secure methods of a science, after having for many centuries done nothing but grope in the dark." Kant's view of the Scientific Revolution, which he shared in broad outline with eighteenth-century historians of civilization like Voltaire and Hume, was essentially that of the seventeenth-century scientists and philosophers of science themselves. Certainly Francis Bacon asserted vigorously that the methods used in the study of nature *should* make a clean break with the past and a fresh start, and the scientific innovators of the period, conscious of doing something new, were no less emphatic in declaring that they had in fact done so. Moreover, the leaders of the scientific movement stated very clearly and explicitly their views of the nature of the changes in scientific thinking and methods they were bringing about. It scarcely needs to be said that for the historian, attempting himself to chart the course of those changes, these statements are indispensable evidence. Until about half a century ago the conception they presented of the immediate past was by and large accepted at its face value.

It was the heroic adventures of Pierre Duhem in medieval science, following the campaigns in which in the last three decades of the last century Denifle and Ehrle and their followers brought back into the realm of scholarship such wide new territories of medieval thought and removed so many previously accepted features from the landscape, that made the first general challenge to the conception of the Scientific Revolution as a clean break with the immediate past and made what may be called the "medieval question" a major historical issue. Duhem and the scholars who have followed him have in the first place discovered and put in order an imposing amount of factual data, so that the broad outlines of the history of scientific thought can now be traced with some confidence from the end of the classical period into the seventeenth century. But even when these have been accepted by both sides in the debate, the problem of the relation of seventeenth-century science to medieval science still remains a *questio disputata*.

If we formulate the problem as the discovery of the origins of modern science, it is clear that we immediately introduce some conception, explicit or implied, of the essential characteristics of the modern science whose origins we have set ourselves to discover. In other words a philosophical interpretation of the nature of modern scientific methods and thinking enters into our historical interpretation of the course of events. We place ourselves in a position similar to those taken with such confidence by Francis Bacon and by Kant. There are obvious dangers in this position, for it is a temptation to read history backwards, to regard as significant in the scientific thought of the past only those features that seem to resemble the features favored by our own conception of what science should be and to discount everything else as simply mistaken; at the crudest, it is a temptation to see the past as so many anticipations of, contributions to, or recessions from the present. We do not have to be told that these temptations are essentially anti-historical. Yet the nature of science is such that the historian must at every stage of his historical inquiries ask questions that are non-historical, though not anti-historical. He must for example ask, as an historian, whether something is good or bad science, thereby introducing a non-historical evaluation, based on his own superior scientific knowledge, that is nevertheless essential for his historical judgment of the course of events. In other words, if it is the first duty of an historian of science to see the problems as they were seen in a given period by those who were then facing them, it may be argued that it is equally essential to his understanding for him to appreciate the aspects of the problem that have been revealed only later and that perhaps could not have been seen in the period in question. In this way the development of science itself can continue to shed new light upon history.

It seems evident that a large part of the debate over the significance of medieval thought for the origins of modern science turns on questions of philosophical interpretation rather than of historical fact. Granting, for example, that the two and a half centuries from the *floruit* of Adelard of Bath to the death of Oresme saw the re-education of Western Europe in the sophistication of Greek philosophical thinking, it may be argued that the relatively poor showing in immediate scientific results indicates that scientifically the West had made a false start, and that this is proved by a closer examination of the irresolution with which experimentation was practiced after lengthy discussions of scientific method or of the ultimately ineffective manner in which conceptions like *impetus* or "the intension and remission of forms" were formulated. Their purpose, it may be argued, was clearly not scientific in the seventeenth-century sense. We must ask then what we are to accept as scientific thinking and methods, or at least as relevant to the development of scientific thinking and methods? What were the medieval conceptions of the aims, objects and methods of science, how do they compare with those of the seventeenth century, and what is the historical connection between them? These are our problems.

I should say at the outset that I regard natural science as a highly sophisticated form of thinking and investigating and one that has been learnt only through a tradition. I do not think that the opinion that science is organized common sense or generalized craftsmanship and technology survives comparison with the actual scientific tradition, a tradition which seems to me to be essentially Western and to begin with the Greeks. Impressive as are the technological achievements of ancient Babylonia, Assyria, and Egypt, of ancient China and India, as scholars have presented them to us they lack the essential elements of science, the generalized conceptions of scientific explanation and of mathematical proof. It seems to me that it was the Greeks who invented natural science as we know it, by their assumption of a permanent, uniform, abstract order and laws by means of which the regular changes observed in the world could be explained by deduction, and by their brilliant idea of the generalized use of scientific theory tailored according to the principles of non-contradiction and the empirical test. It is this essential Greek idea of scientific explanation, "Euclidian" in logical form, that has introduced the main problems of scientific method and philosophy of science with which the Western scientific tradition has been concerned. These seem to me to be of three general kinds.

First there are those concerning the character of the natural order by means of which events are explained, for example, whether it consists of substances determined teleologically and qualitatively or whether all changes are reducible to quantitative changes of matter in motion, whether action at a distance is conceivable or whether space must be filled with a continuous medium. It is these "concepts of nature," or "regulative beliefs" as we may call them after Kant, that determine the kinds of abstract entities that are used in a theory; they are the investigator's presuppositions about what he expects to find by his analytical methods; they control the form his explanations will take, the questions he will ask and the problems he will discover; and by suggesting certain types of hypotheses but not others they may assist inquiry in some directions but limit or prevent it in others. They are essentially unfalsifiable by observation; changes in them are brought about by re-thinking.

Secondly, there are the problems concerning the relation of theories to the data explained, the classical problems of "scientific method" as such. They concern, for example, the inductive rules for collecting and ordering data and for establishing causal connections, the conception of causation, the processes of discovery, the use of hypotheses, the criteria according to which a theory is to be accepted or rejected or one theory preferred to another, the questions of the existence of theoretical entities and of the place of concepts of nature in scientific explanation.

Thirdly, there are the problems of scientific procedure, the experimental and mathematical techniques.

It is natural that problems of scientific method should have a special importance in times of great intellectual change when the direction, objectives and forms of scientific inquiry are still being determined or redetermined. The re-introduction of the scientific tradition into the Latin West through the translations from Greek and Arabic made the twelfth, thirteenth, and fourteenth centuries such a time. This is certainly one reason, though not the only reason, why the natural philosophers of the period felt it necessary to become philosophers of science first and scientists only secondly. It is as philosophers that I think we should most justly consider them, and it is as such that the development of their ideas finds an intelligible center and that their differences from seventeenth-century scientists become reasonable. We may then ask what place they have in the scientific tradition. We can provide an answer, I think, by looking at scientific thought in the middle ages and in the seventeenth century from the point of view of the three aspects of scientific method which I have just distinguished.

The influence of the translation into Latin of Euclid's *Elements* and of Aristotle's logic in the twelfth century is an excellent illustration of the transmission of science by a learned tradition. Before these works were mastered, a number of twelfth-century scholars wrote of the investigation of nature as being in principle empirical and even experimental—*nihil est in intellectu quod non prius fuerit in sensu*—and showed that they were aware of the possibility of mathematical proofs. But neither in natural science nor in mathematics were scholars able to transcend the rule-of-thumb methods of the practical crafts; a pathetic example from the eleventh century may be read in the correspondence published by Tannery and Clerval between Ragimbold of Cologne and Radolf of Liège, who vainly tried to outdo each other in attempting to prove that the sum of the internal angles of a triangle equals two right angles, and in the end suggested verifying the proposition experimentally by cutting out pieces of parchment.[1] It is important for the scientific tradition that scholars should have held so strongly in mind a theory of empiricism, but it was not until they had learnt from Greek sources the idea of a theoretical scientific explanation and of a geometrical proof, and had been shown many examples, that they were in a position to begin to possess themselves of the whole tradition. If we turn from the writings of, for example, Adelard of Bath or Hugh of St. Victor to those of Fibonacci or Robert Grosseteste or Jordanus Nemorarius, we have a measure of the difference these sources made.

We may take Robert Grosseteste and Roger Bacon as examples of what thirteenth-century philosophers made of the scientific tradition. In their writings we find a systematic attempt to determine what natural science should be, as distinct for example from mathematics or metaphysics or other kinds of inquiry of which the new translations brought instances. I do not question that it was possible at that time to produce excellent science simply by carrying on with the problems and techniques the Greeks and Arabs had developed from the point where they left off, without bothering with methodology at all. Indeed some of the best work of the period, for example that of Jordanus or Gerard of Brussels or Peter of Maricourt, was produced in just this way. But it is essential to the scientific tradition that its representatives should have sought to possess themselves not only of "know how" but also of understanding of what they are doing. At times and in problems in which there is uncertainty about what a scientific explanation should be and how a scientific investigation should be made, methodological inquiries become indispensable and may change the course of what is done.

I have argued at length elsewhere[2] that the contribution of Grosseteste and Roger Bacon and their successors to the scientific tradition of their time was to formulate, from the theoretical empiricism of the twelfth century and the deductive form of scientific explanation learnt from Euclid and from Aristotle's logic, a conception of science that was experimental, mathematical, and deductive. From one point of view we can see their work as an attempt to combine the form of scientific thought imposed by Greek geometry and expounded by Plato with the empirical requirements insisted upon by the other great tradition in Greek methodology, that of medicine and of Aristotle. The medieval philosophers were searching for a model and a method of scientific thought applicable to natural science, to physics, distinct from the geometrical model. It is relevant to look at some details. First, Grosseteste saw that the problem of discovering and identifying the causes of events was a sophisticated problem, and with the aid of Aristotle he began to develop the logical procedures of "resolution and composition" for analyzing a complex phenomenon into its elements and for discovering the "common nature" of a group of events. He conceived of the cause of an event as something, the "nature," in which the event was prefigured, but the cause was hidden from our direct inspection, and the results of our analysis might leave us with an ostensible plurality of causes. Here he contrasted natural science with mathematics, in which premises and conclusion (cause and effect) were reciprocal. In natural science we may never be able to reduce the ostensible plurality to one actual cause in which the effect is univocally prefigured, but in order to eliminate false causes and so far as possible to identify true causes, he proposed a method of "verification and falsification" whereby proposed causes were tested by comparing consequences deduced from them with observation.

It seems to me that the discussion of the form of natural science and of these inductive procedures by Grosseteste and Roger Bacon, by various optical writers and astronomers, and by Duns Scotus and Ockham, are an important contribution to the history of scientific method. They insisted on the *a posteriori* and probable character of natural science, the dependence of its conclusions on sense experience, its difference from the model of geometry.[3] They came to understand that experimental tests applied to the consequent enable us to deny the postulated conditions, but not to affirm them as more than probable. For choosing between different hypotheses of equal empirical standing they gave a fundamental importance to the principle which in some cases led to a radically conventional conception of theories. In effect, they reached the idea of a cause as the conditions necessary and sufficient to produce an event. A clear statement was given by Ockham in his commentary on the *Sentences* (book I, distinction xlv, question 1, D):

Although I do not intend to say universally what an immediate cause is, nevertheless I say that this is sufficient for something being an immediate cause, namely that when it is present the effect follows, and when it is not present, all the other conditions and dispositions being the same, the effect does not follow. Whence everything that has such a relation to something else is an immediate cause of it, although perhaps not *vice versa*. That this is sufficient for anything being an immediate cause of anything else is clear, because if not there is no other way of knowing that something is an immediate cause of something else. . . . It follows that if, when either the universal or the particular cause is removed, the effect does not occur, then neither of them is the total cause but rather each a partial cause, because neither of those things from which by itself alone the effect cannot be produced is the efficient cause, and consequently neither is the total cause. It follows also that every cause properly so called is an immediate cause, because a so-called cause that can be absent or present without having any influence on the effect, and which when present in other circumstances does not produce the effect, cannot be considered a cause; but this is how it is with every other cause except the immediate cause, as is clear inductively.[4]

A large amount of excellent scientific research has always been carried out at the somewhat "Francis Baconian" or "John Stuart Millian" level indicated by the analysis of the observable conditions necessary and sufficient to produce an event. In the thirteenth and fourteenth centuries the production of rainbows and allied phenomena were explicitly investigated in these terms, and the results were stated in the form that if, for example, you postulate a certain refracting medium, rain drops, at a position at which the incident sunlight makes an angle of 42° with the line connecting the raindrops and an observer, then you will expect a rainbow to be seen. But science at this level remains merely descriptive. Any attempt to offer a general explanation connecting, for example, different optical phenomena and the conditions of their production will depend on postulating a general theory of light, and this will depend in turn on the conceptions we have of the basic entities and principles of which light consists and on our ability to formulate the mathematical laws according to which these behave. In other words a method that aims at finding explanations is of little use in science except in association both with a conception of the kinds of causes, principles and laws the method is expected to discover and with a knowledge of necessary mathematical and experimental techniques, and in fact all discussions of scientific method have presupposed such a "conception of nature" and such techniques.

A fundamental difficulty, as we can now see it, was made for the scientific philosophers of the middle ages by the conception of nature, and consequently of explanation, which they inherited from Aristotle. Indeed one can see the discussions from Grosseteste to Galileo as a long and finally successful attempt to escape from this aspect of Aristotle's conception of science. Aiming principally at establishing the definition of the "nature" or essence or "form" of a thing, Aristotle's conception of explanation meant that when this had been discovered the explanation had been given and so no further questions in this direction need be asked. The difficulty of escape into a new range of questions was enormously magnified by the logically connected series of answers Aristotelian philosophy could provide in terms of the well-elaborated concepts of matter and form, potentiality and actuality, efficient and final causality.

We all remember Galileo's comment on explanations of this kind, and his brisk assertion that they may be answers indeed, but not to the questions he wanted to ask. The definition of the "nature" or essence, as he pointed out in a famous passage on falling bodies in the *Dialogue concerning the Two Principal Systems of the World,* was merely a summary of observed behavior; as such it was useful for classification, but it could tell us no more than the observations it summarized. So "gravity" was not an explanation but merely "the name that has been attached to [the falling of bodies] by the many experiences we have of it a thousand times a day. We don't really understand what principle or what power it is that moves a stone downwards, any more than we understand what moves it upwards after it has left the projector, or what moves the moon round."[5] We all remember how Galileo himself attacked these problems, how, in another famous declaration, in the *Two New Sciences*, he rejected any consideration of the physical cause of the acceleration of freely falling bodies and turned to a purely kinematic approach, "to investigate and demonstrate some of the properties of accelerated motion, whatever the cause of this acceleration may be."[6] These passages indicate a major change in the objectives of scientific inquiry. Instead of searching for the essence, Galileo proposed to correlate certain accidents. He proposed moreover to measure them and to correlate them by means of a mathematical function. The last was fundamental. The kinematic law of falling bodies was his personal triumph in this approach, and as he pointed out in connection with another of his mathematical laws, that describing the trajectory of a projectile, such a law extends our knowledge by enabling us to predict events we have not yet observed. But that was not the end of Galileo's conception of scientific explanation. We all recall yet another famous passage, from *Il Saggiatore* (question 48): "I hold that there exists nothing in external bodies for exciting in us tastes, odours and sounds except sizes, shapes, numbers and slow or swift motions." It was in terms of these "primary qualities" that he conceived of the ultimate composition of the physical world, and in terms of these and the *particelle minime,* the primary particles characterized by them, of which he held matter to consist, that he conceived of explanations of the phenomena described by the mathematical predictive laws. In other words, for Aristotle's conception of nature he substituted another in which all phenomena were to be reduced to matter in motion. Like Newton's his approach was to establish the descriptive mathematical laws before considering their causes, but the mechanical causes remained his ultimate objective in science.

I have given this brief sketch of Galileo's conception of the objectives of scientific inquiry because with him we find a confidently successful solution of a methodological problem that occupied Western scientific philosophers from the thirteenth century. The success of seventeenth-century physics depended on the adaptation to each other of a "mechanical" philosophy of nature, a methodology of experimental science with functional correlation as its aim, and effective algebraic and geometrical techniques. We can, I think, see the beginnings of this combination in the conception of science presented by Grosseteste and Roger Bacon: a conception of physical nature in which the essence or "form" itself is mathematically determined, and a conception of the immediate objective of inquiry as mathematical and predictive laws instead of the Aristotelian essential definition. In the fourteenth century we find the development of mathematical techniques designed to take advantage of the new mathematical methodology and conception of explanation.

It was through his Platonically-inspired "metaphysics of light" that Grosseteste made his fundamental move towards the mathematicization of the concept of nature and the shift from the "form" to the "law" as the object of scientific inquiry. Since *lux* was the fundamental "corporeal form," the primary basis of the extension of matter in space and of all motion, it followed that optics was the fundamental physical science, and the study of optics was "subordinate" to geometry. Hence he could conclude:

The usefulness of considering lines, angles and figures is the greatest, because otherwise it is impossible to understand natural philosophy without these. They are completely efficacious in the whole universe and in parts, and in related properties, as in rectilinear and circular motion. They are efficacious also in action and in being acted upon, and this whether in matter or the senses. . . . For all causes of natural effects have to be expressed by means of lines, angles and figures, for otherwise it would be impossible to have knowledge of the reason (*propter quid*) concerning them. (*De lineis, angulis et figuris.*)[7]

Hence these rules and principles and fundamentals having been given by the power of geometry, the careful observer of natural things can give the causes of all natural effects by this method. And it will be impossible otherwise, as is already clear in respect of the universal, since every natural action is varied in strength and weakness through variation of lines, angles and figures. But in respect of the particular this is even clearer, first in natural action upon matter and later upon the senses . . . (*De natura locorum.*)[8]

There is no need to insist that Grosseteste's execution of this program, and the theory of the "multiplication of species" which it had in view, made little contribution to the permanent body of science. It is the program itself that is significant. And it is relevant to draw attention to a further aspect of it. In developing his theory of the multiplication of species we find him distinguishing between the physical activity by which the *species* or *virtus* were propagated through the medium and the sensations of sight or heat which they produced when they acted on the appropriate sense organs of a sentient being. His geometrization or mechanization of physical nature in effect involved him in distinguishing between primary and secondary qualities. This distinction became methodologically significant in physics when the primary qualities were conceived, as in his suggestions that light was a series of pulses and heat a motion of particles, as a physical activity that need not be directly observable but whose nature physics had to discover. A number of optical writers in the thirteenth and fourteenth centuries, for example Witelo and Theodoric of Freiberg, proposed explanations of the qualitative differences observed in light in terms of quantitative differences in the light itself, for example relating color to angle of refraction. One, John of Dumbleton, was even to attempt to formulate a photometric law relating intensity of illumination to distance from the source.

With Roger Bacon the program for mathematicizing physics and the conception of laws of nature in the seventeenth-century sense become perfectly explicit. He wrote for example:

All categories (*praedicamenta*) depend on a knowledge of quantity, concerning which mathematics treats, and therefore the whole power (*virtus*) of logic depends on mathematics.[9]

In the things of this world, as regards their efficient and generating causes, nothing can be known without the power of geometry.[10]

The language he uses in a discussion of the "multiplication of species" seems to relate this general program unequivocally to a shift in inquiry from forms to laws. He wrote for example: "That the laws (*leges*) of reflection and refraction are common to all natural actions I have shown in the treatise on geometry."[11] He claimed to have demonstrated the formation of the image in the eye "by the law of refraction," remarking that the "species of the thing seen" must so propagate itself in the eye "that it does not transgress the laws which nature keeps in the bodies of the world." Here the "species" of light were propagated in straight lines, but in the twisting nerves "the power of the soul made the species relinquish the common laws of nature (*leges communes nature*) and behave in a way that suits its operations."[12]

It is characteristic of these medieval philosophers that Grosseteste and Bacon should not have combined the methodological insight which they showed in their fundamental notion of the multiplication of species with any attempt at precise mathematical definition, at expressing the amounts of change quantitatively in numbers, or at measurement. One reason for this was no doubt that they had at their disposal no other mathematical concepts beyond those occurring in elementary geometrical optics, but clearly they were also uncertain about how to make another move beyond the position they had reached.

The final stage in the new medieval conception of science is marked by the methodological criticism of Ockham and the methodological and technical attempts by mathematicians in the fourteenth century to express different kinds and rates of change in mathematical form. One of the principal objects of Ockham's logical inquiries was to define the criteria by which something could be said to exist. He concluded that nothing really existed except what he called *res absolutae* or *res permanentes,* individual things determined by observable qualities. This led him, in the debate over the physical cause of motion, to cut the ground from under the feet of all the other parties, for he declared in effect that the problem was illusory. He recognized nothing real in motion beyond the observable bodies that were moving; to say that a body was moving was to say no more than that from moment to moment it was observed to change its spatial relations with other observable bodies. He held that the abstract terms of the debate were misleading. "If we sought precision," he declared, "by using words like 'mover,' 'moved,' 'movable,' 'to be moved,' and the like, instead of words like 'motion,' 'mobility,' and others of the same kind, which according to the way of speaking and the opinion of many are not seen to stand for permanent things, many difficulties and doubts would be excluded. But now, because of these, it seems as if motion were some independent thing quite distinct from the permanent things."[13]

This is not the occasion to discuss the relevance of Ockham's conception of motion to the formation of the concept of inertia; I will say only that I think that the theory of *impetus* was at least as relevant. Nor is it the occasion to discuss the advantages or disadvantages of "nominalism" for science. But I think that it is extremely relevant that Ockham, and others like him, should have so explicitly shifted the ground of the discussion of motion, and of physics in general, from "why" to "how." The scientific treatment of motion was reduced to giving an accurate description of how bodies changed their observable spatial relations with neighboring bodies. The same move that he made as a logician was made contemporaneously by Bradwardine as a physicist, in his attempt to produce an algebraic law relating velocity, power and resistance. The two great technical developments in the fourteenth-century treatment of motion, Bradwardine's "word-algebra" and the algebraic, geometrical, and graphical treatment of "intension and remission of forms" in Oxford and Paris, had origins separate from "nominalist" or Ockhamist methodology, but there is evidence that it was under the influence of the latter, as well as of Bradwardine, that Oxford mathematicians like Richard Swineshead and Dumbleton specifically rejected dynamical explanations of motion and confined themselves to kinematic mathematical description. This was a limited objective which their technical equipment gave them some hope of reaching, and in pursuing it they showed considerable sophistication and mathematical skill. They showed a complete mastery of the technique, perhaps learned from Archimedes, of posing problems and theorems *secundum imaginationem* as possibilities for purely theoretical analysis, without immediate empirical application. In pursuing their object of giving quantitative and mathematical expression to change and to the factors producing it, they grasped the fundamental idea of functional dependence and, especially in Oresme's graphical method, made notable progress in the treatment of variable quantities, which the Greeks had never exploited. In finding expressions for rates of change, they formulated sophisticated concepts like those of acceleration and instantaneous velocity, used infinitesimals, and reached important results like the Mertonian Mean Speed Law.

I have attempted to show how it is possible to trace in the thirteenth and fourteenth centuries the development of a conception of natural science similar in several fundamental respects to that found in the seventeenth century. Following the recovery of the "Euclidean" form of science, the medieval theory of science embraced three broad aspects. First, there was the analysis of the logical relationship between theories and the data explained and of the criteria and methods for testing and accepting or rejecting a theory. These included the empirical principle of verification and falsification or exclusion and the rational or conventional principle of economy, as exemplified in "Ockham's razor." Secondly, there was the conception, neoplatonic in inspiration, that nature was ultimately mathematical and could be explained only by mathematical laws. This introduced, in place of the Aristotelian "form" with its irreducible qualitative differences between different substances and their movements and behavior, a new concept of universal "laws of nature" as the proper object of scientific inquiry. And third, there were the new techniques, especially mathematical techniques, introduced in exploiting this program and ultimately to transform it.

It may seem at first sight that what I have just said would leave no room for any important changes in scientific method between the fourteenth and the seventeenth centuries. A glance at the results obtained in the two periods would be sufficient to warn us against any such conclusion. The manner in which many medieval scientific investigations were actually carried out and the results obtained often seem absurdly inadequate in comparison with the intelligence displayed in discussing principles. One often has the impression that one might have missed the point. If one were expecting simply a seventeenth-century point I think that one's impression would be correct. The difference seems to me to mark a major change in the center of philosophical interest between the fourteenth and the seventeenth centuries, a change of which the immensely superior technical equipment and efficiency of seventeenth-century science is at least partially a consequence.

A striking example of this change and of its consequences for scientific method, an example which has been rather thoroughly studied by historians, can be seen in the treatment of motion. The fourteenth-century scholars who worked out the mathematical functions and kinematic theorems for motion combined a truly impressive mathematical skill and resource with an apparent lack of interest in finding means to apply their results to actual motions. They went through the whole discussion without making a single measurement, although certainly leading scholars like Bradwardine and Albert of Saxony were looking for true laws of motion, and although they formulated their problems mathematically in such a way that they could find the answer only by measuring the concomitant variations in the parameters involved.

Buridan himself commented on the work of Bradwardine and Swineshead: "these rules are rarely or never made to be deduced to the effect" (*quod istae regulae raro vel nunquam inventae sunt deduci ad effectum*). But he added: "Nevertheless it is not necessary to say that such rules are useless or artificial because, even if the conditions which they postulate are not fulfilled in nature, it could happen that they will be realized by the omnipotence of God."[14] These philosophers had before them examples of the measurement of concomitant variations in astronomy and in the accounts by Ptolemy and by Witelo of the systematic increase in the angles of refraction of light with increasing angles of incidence. They even discussed measurement in theory and even units of measurement. But something seems to have kept their gaze turned in a direction different from that which seventeenth-century science has taught us to seem obvious. We can see what this was by looking at the context of the discussion, which was never simply mechanics in the seventeenth-century sense, but problems of motion as they occurred within the more general philosophical and methodological problem of how to find quantitative expressions for changes of all kinds, quantitative or qualitative. The interesting mathematical and physical results that came out of the discussion of the "intension and remission of forms" were really by-products of an essentially philosophical and methodological inquiry.

The direction of interest is quite different when we turn to Galileo. Instead of methodological principles and mathematical possibilities for their own sake, he firmly took the motions actually found in nature as the object of his inquiries. "For anyone may invent an arbitrary type of motion and discuss its properties," he wrote in a famous passage in the *Two New Sciences.*[15] "But we have decided to consider the phenomena of bodies falling with an acceleration such as actually occurs in nature and to make this definition of accelerated motion exhibit the essential features of observed accelerated motions." Galileo's experiments with a ball rolling down an inclined plane to measure the relation between "space traversed in time passed" distinguish his discussion of free fall from all its predecessors. They illustrate a general change of interest between the fourteenth and the seventeenth centuries from scientific method as such to scientific method applied to the facts of nature. The faults of the medieval conception of scientific method are largely the faults of principles insufficiently tested by application. Systematic, technical application in the seventeenth century brought about a far-reaching change in the grasp and precision of the principles themselves. Surely it is significant that the idea of functional dependence and the measurement of concomitant variations, which were really as essential to the thirteenth- and fourteenth-century conception of a scientific law as to that of the seventeenth-century, were *explicitly* discussed as part of scientific method for the first time only by Galileo and Francis Bacon. At the same time several other fundamentally important aspects of scientific method came into the field of discussion, for example the role of hypotheses in predicting new knowledge and guiding research, the conception of controls and, on the technical side, the grasp of the need for precision in measuring instruments.

Although it is my purpose simply to chart the course of an intellectual change rather than to attempt to explain it, it is relevant to point to the institutional context of the medieval discussions. The central doctrines of medieval science were developed almost entirely within the context of academic discussions based on the books used in university teaching, especially in the faculty of arts.[16] It is true that the applications of academic sciences, such as of astronomy in determining the calendar, of arithmetic in the work of the exchequer and of commercial houses, or of anatomy and physiology and chemistry in surgery and medicine, were put into practice outside the universities. It is true also that in other fields of technology, and in art and architecture with their increasing tendency to naturalism, developments took place that were to be of profound importance for science. But it was the clerics of the universities who were responsible for the main theoretical conceptions of science, and their interest in them was in general epistemological or metaphysical or logical rather than strictly scientific in the modern sense. Although it would be false to suggest that scientific problems were always discussed in a philosophical context in the Middle Ages, the ambience of the discussions and the habit of presenting them in commentaries on Aristotle or other philosophical or theological writers certainly put pressure in that direction. This would help to explain the striking contrast between the sustained interest in the logic and theory of experimental science from thirteenth-century Oxford to sixteenth-century Padua[17] and the comparative scarcity of actual experimental investigations. It also helps to explain the development of science, as distinct from merely philosophy of science, from the end of the sixteenth century in scientific societies outside the universities, even though the members were predominantly university men. And it gives point to the rejection by these new scientists of the purely philosophical academic methodology that had lost touch with the actual problems of science, and of the medieval discussions that lay behind it.

It may be asked whether the medieval conception of scientific method can really be claimed to have any more significance for the scientific tradition than the inadequate science with which it was associated, and also whether it did in fact have any influence in the sixteenth and seventeenth centuries. (See the comments by Drabkin and Nagel, below.) In answer to the first question there seems to me no doubt that a process of philosophical orientation was as necessary a preliminary to the scientific performance of the seventeenth century as it was to that of the Greeks. If this is true then the significance of this philosophical stage is clearly much greater than the immediate scientific results that it produced, although certainly this is not to say that the medieval philosophers appreciated the full significance of their methodological insights or saw all the problems which their seventeenth-century successors saw and solved. Indeed their very deficiency in performance shows that they did not. It is we, looking back on the whole story, who are in a position to judge the significance of its various episodes. Among these the renewed appreciation in the sixteenth and seventeenth centuries of the scientific performance of Archimedes was certainly something that allowed Galileo and his contemporaries to get scientific results far beyond the range of philosophical talk. The point of my argument is that the significance of Archimedes' methods, and of the mathematical-Platonic conception of nature, for scientific inquiry could be grasped because the preceding discussions of scientific method had shown, at least partly, what to look for. After the successes of the seventeenth century, philosophers were in a position to assess the actual achievements of science, and since then discussions of scientific method have been largely an analysis of scientific work already achieved. But before the middle of the seventeenth century the general lines of scientific thinking had not yet been firmly established by results, and it seems to me to be one of the most important characteristics of the whole period of the origins of modern science that it contained so many discussions and programs of scientific thinking, from those of Grosseteste and Roger Bacon to those of Galileo, Francis Bacon, Descartes, and Newton.

I have not in this paper dealt specifically with the important question of the actual influence, direct or indirect, of medieval writers which it is possible to trace in the scientific writings of the sixteenth and seventeenth centuries.[18] Certainly most of the principal medieval doctrines were available in printed editions and were taught in various universities, especially in Padua and Paris and in Germany, in the sixteenth century. But when one asks more specifically what, for example, Galileo or Descartes actually knew and what use they made of the dynamics of *impetus* or of fourteenth-century Oxford kinematics or of Oresme's graphical methods, the evidence becomes difficult and unsatisfactory. That they had some knowledge of the earlier work in these fields is beyond doubt, and we can trace it specifically in Galileo. Sometimes he uses old language in a new context, as when he uses *impeto* within his new "inertial" dynamics or when he describes an "Archimedean" procedure for establishing functional relationships in the language of the *metodo resolutivo e compositivo*. Linguistic inertia is evidence of continuity with earlier forms of thought, whatever changes the requirements of successful scientific practice may have brought about.

One of the profoundly influential and fruitful events that made up the scientific revolution of the seventeenth century seems to me to have been the restoration of full contact between science and scientific methodology.

In spite of the promising conception of science that began in the thirteenth century, medieval scientists in the end succeeded only in revising some isolated sections of the physics and cosmology they inherited from the Greeks and Arabs. The successful revision by seventeenth-century scientists of the whole of the theoretical framework and assumptions of physical science was effected by a close combination of philosophical and technical maneuvers. The conception they formed of scientific method, of the counters available in terms of conceptions of nature, hypotheses, logical procedures and mathematical and experimental techniques, was closely measured by the technical success of the results and by their universality. It was proved by results that the mathematicization of nature, the "mechanical philosophy,"[19] the use of abstract theories, the experimental method, the conception of functional dependence, the new algebra and geometry, were successful elements in a scientific method. It was proved by results that motion was the right problem in which to find the key to physics. It was for the very important reason that it succeeded, and this could only be discovered by trying, that seventeenth-century mechanics became the principal model for the methods of science as a whole. It seems to me that their ability to be good judges of scientific success, and to be impressed by it, is a measure of the distance between the natural philosophers of the seventeenth century and their medieval predecessors with whom they shared so many ideas and aspirations in common.

The forms and methods of modern science are the product of a long and complicated intellectual struggle but as in so many aspects of the Scientific Revolution the final stages, so long prepared, were taken rapidly by men of genius in the seventeenth century. But to become a good judge of science was something that required resolution and sophistication even after the main way was clear. Francis Bacon, heir in a sense to the empirical tradition of Greek medicine and Aristotle, and Descartes, heir in a sense to Euclid and Plato, both proposed versions of a second great model of scientific thought that was to replace the first great model, that of Greek geometry, or as Bacon mistakenly thought, of Aristotle's *Organon*. Galileo's version is nearer the mark in physics. But the polemics between the Cartesians and Newton show that even towards the end of the seventeenth century the proper relationship to be understood between observational data, laws, hypotheses, and mechanical explanations in terms of the reigning concept of nature was still a matter of dispute. The continuation of the debate down to the present moment, and the transformation of the seventeenth-century model itself as a result of more recent developments in physics, in non-Euclidian geometries, in the mathematical theory of probability, and in logical analysis, show with the most unequivocal directness that thought about the forms and methods of science is an inseparable part of the progress of scientific thought itself. I have spoken only about the history of physics, because this has provided the leading model. If we complicated our inquiry by including in it, as indeed we should, other and especially the biological sciences, we would find other species of model (although recognizably of the same scientific genus) but an exactly parallel story.

References

1 P. Tannery and M. l'Abbé Clerval, "Une correspondance d'écolâtres du XI° siècle," *Notices et extraits des manuscrits de la Bibliothèque Nationale*, XXXVI (1899), 498.

2 *Robert Grosseteste and the Origins of Experimental Science, 1100–1700* (Oxford, 1953).

3 E.g. concerning physics: "videndum est de modo procedendi, considerandi et demonstrandi in hac scientia. Circa quod dico: esse sciendum quod, licet in hac scientia sint demonstrationes propter quid, sicut in ceteris scientiis, tamen ordine doctrine qui incipere *debet* a notioribus et facilioribus communitati, cui tradenda est scientia, procedendum est ab effectu ad causam . . ." (William of Ockham, *Summulæ in libros Physicorum,* [Rome, 1637], I, 5). "Cum ergo scientia naturalis habeat considerare de compositis, sequitur quod ad considerationem eius pertinent partes compositi et causae eiusdem. Iste autem partes componentes per se sunt materia et forma. . . . Primo dicendum est quod sint a posteriori, quia a priori hoc probare non possumus." (*Ibid.,* 1, 7.) Cf. Grosseteste: "Similiter in naturalibus est minor certitudi propter mutabilitatem rerum naturalium." He places natural science with other disciplines in which we can reason rather "probabiliter quam scientifice, licet in his sint scientia et demostratio sed non maxime dicta. In solis enim mathematicis est scientia et demonstratio maxime et particulariter dicta. . . . In doctrinis [i.e. mathematics] autem est facilior et brevior resolutio usque ad principia quam in dyalecticis vel aliis . . . quia in doctrinis magis convertuntur termini" (sc. premisses and conclusions). (*Commentaria in libros Posteriorum Aristotelis,* I, 11; see my *Robert Grosseteste,* p. 59, n. 2, p. 82, n. 4.)

4 For the original Latin of this passage see my *Robert Grosseteste,* p. 173, n. 5.

5 Galileo Galilei, *Le Opere di Galileo Galilei* (Edizione nazionale; Firenze, 1890–1909), VII, 260–61. Newton made the same point in his attack on specific qualities which "tell us nothing" (*Opticks,* Query 31. [4th ed.; reprinted London, 1931], p. 401), and so did Molière in his mockery of the "virtus dormitiva" in *Le Malade Imaginaire.*

6 Galileo Galilei, *Opere,* VIII, 202. Kepler's photometric law, stating that intensity of illumination varies inversely as the square of the distance from the source, is a good contemporary example of a functional relationship. (*Ad Vitellionem Paralipomena,* I, 9 [Frankfurt, 1604], p. 10)

7 "Utilitas considerationis linearum, angulorum et figurarum est maxima, quoniam impossibile est sciri naturalem philosophiam sine illis. Valent autem in toto universo et partibus eius absolute. Valent etiam in proprietatibus relatis, sicut in motu recto et circulari. Valent quidem in actione et passione, et hoc sive sit in materiam sive in sensum. . . . Omnes enim causae effectuum naturalium habent dari per lineas, angulos et figuras. Aliter enim impossibile est sciri 'propter quid' in illis. Quod manifestum sic: Agens naturale multiplicat virtutem suam a se usque in patiens, sive agat in sensum, sive in materiam. Quae virtus aliquando vocatur species, aliquando similitudo, et idem est, quocunque modo vocetur; et idem immittet in sensum et idem in materiam, sive contrarium, ut calidum idem immittit in tactum et in frigidum. Non enim agit per deliberationem et electionem; et ideo uno modo agit, quicquid occurrat, sive sit sensus, sive sit aliud, sive animatum, sive inanimatum. Sed propter diversitatem patientis diversificantur effectus. In sensu enim ista virtus recepta facit operationem spiritualem quodanmodo et nobiliorem; in contrario, sive in materia, facit operationem materialem, sicut sol per eandem virtutem in diversis passis diversos producit effectus. Constringit enim lutum et dissolvit glaciem." *Die Philosophischen Werke des Robert Grosseteste,* ed. L. Baur. (*Beiträge zur Geschichte der Philosophie des Mittelalters,* IX [Münster, 1912], pp. 59–60.)

8 "His igitur regulis et radicibus et fundamentis datis ex potestate geometriae, diligens inspector in rebus naturalibus potest dare causes omnium effectuum naturalium per hanc viam. Et impossibile erit aliter, sicut iam manifestum est in universali, quando variatur omnis actio naturalis penes fortitudinem et debilitatem per varietatem linearum, angulorum et figurarum. Sed in particulari magis est manifestum istud idem, et primo in actione naturali facta in materiam et postea in sensum, ut complete pateat veritas geometriae." *Ibid.,* pp. 65–66.

9 Roger Bacon, *Opus Maius,* ed. J. H. Bridges (3 vols.; Oxford, 1897), Vol. I, p. 103. "In quo probatur per rationem quod omnis scientia requirit mathematicam."

10 *Ibid.,* Vol. II, p. 143–44.

11 "Que vero sint leges reflexionum et fractionum communes omnibus actionibus naturalibus, ostendi in tractatu geometrie, tam in Opere Tertio quam Primo; sed principaliter in Opere separato ab his, ubi totam generationem specierum, et multiplicationem, et actionem, et corruptionem explicavi in omnibus corporibus mundi." (Roger Bacon, *Un fragment inédit de l'Opus Tertium,* ed. P. Duhem [Quaracchi, 1909], p. 90.) Grosseteste used the word "regulae" for the "laws" of refaction which he proposed: see *De natura locorum,* ed. Baur, p. 71 and *De lineis,* p. 63.

12 "Deinde quia tolleretur visio, nisi fieret fractio speciei inter pupillam et nervum communem, in quo est communis sectio nervorum, de qua superius dixi; et dextra videretur sinistra, et e converso; ideo demonstro hoc per legem frac-

tionum, in geometricis expositam, ut sic salvetur visio. Et nichilominus tamen oportet quod species rei vise multiplicet se novo genere multiplicationis, ut non excedat leges quas natura servat in corporibus mundi. Nam species a loco istius fractionis incedit secundum tortuositatem nervi visualis, et non tenet incessum rectum, quod est mirabile, sed tamen necesse, propter operationem a se complendam. Unde virtus anime facit speciem relinquere leges communes nature, et incedere secundum quod expedit operationibus ejus." Roger Bacon, *Un fragment inédit de l'Opus Tertium,* ed. Duhem, p. 78. I am indebted to Dr. M. Schramm of the Johannes Wolfgang Goethe University, Frankfurt, for drawing my special attention to this passage. Cf. *Opus Maius,* Vol. II, p. 49.) The following passage occurs in a Constitution of the Emperors Theodosius, Arcadius, and Honorius: "Sufficit ad criminis molem naturae ipsius leges velle rescindere, inlicita perscrutari, occulta recludere, interdicta temptare." (James Bryce, *Studies in History and Jurisprudence* [Oxford, 1901], II, p. 119, n. 2.) For further late classical references, especially in Stoic, Jewish and Christian writers, see Marshall Clagett, *Greek Science in Antiquity* (New York, 1956), pp. 122–23, 137–44, 148. Francis Bacon sometimes used "form" and "law" as synonyms, writing in the *Novum Organum,* II, 2: "Licet enim in natura nihil vere existat praeter corpora individua, edentia actus puros individuos ex lege; in doctrinis tamen, illa ipsa lex, ejusque inquisitio, et inventio, atque explicatio, pro fundamento est tam ad sciendum, quam ad operandum. Eam autem *legem,* ejusque *paragraphos,* formarum nomine intelligimus; praesertim cum hoc vocabulum invaluerit, et familiariter occurrat." There is a similar linguistic inertia in Galileo's use of the word "definizione" for his mathematical law of free fall: below, n. 15. Francis Bacon's conception of "form" presupposed the explanation of all physical phenomena in terms of matter in motion; similarly Robert Boyle in *The Excellency and Grounds of the Mechanical Hypothesis:* "The philosophy I plead for . . . teaches that God . . . establish'd those rules of motion, and that order amongst things corporeal, which we call the laws of nature. Thus, the universe being once fram'd by God, and the laws of motion settled . . . the phenomena of the world, are physically produced by the mechanical properties of the parts of matter." (*Works,* abridged by Peter Shaw [London, 1725], I, 187.) An interesting collection of further quotations is given under "Law" in J. A. H. Murray, *New English Dictionary* (Oxford, 1908). Newton used "axiomata seu leges motus" in the *Principia Mathematica,* and he made a sharp distinction between experimentally established laws and hypotheses advanced to explain them by mechanical causes: see A. Koyré, "Pour une édition critique des oeuvres de Newton," *Rev. d'hist. des sciences,* VIII (1955), 19–37; I. Bernard Cohen, *Franklin and Newton* (Philadelphia, 1956), Appendix 1. For a sociological interpretation see E. Zilsel, "The Genesis of the Concept of Physical Law," *The Philosophical Review,* LI (1942), 245 ff., and the comment by M. Taube, *ibid.,* LII (1943), 304 ff.

13 "Propter autem modum loquendi multe sunt difficultates que mihi videntur secundum principia Aristotelis de motu magis vocales quam reales. Si enim uteremur precise istis vocabulis: movens, motivum, mobile, moveri et huiusmodi, et non talibus: motus, motio et consimilibus, quæ secundum modum loquendi et

opinionem multorum pro rebus permanentibus non videntur supponere, multæ difficultates et dubitationes essent exclusæ. Nunc autem propter talia videtur quod motus sit alia res secundum se tota distincta a rebus permanentibus." Ockham, *Summulae in lib. Physicorum* (Rome, 1637), III, 7. Cf. *The Tractatus de Successivis attributed to William Ockham,* ed. P. Boehner, Franciscan Institute Publications, i [St. Bonaventura, New York, 1944], p. 47.) In the *Summa Logicae,* I, 49, ed. Boehner (Franciscan Inst. Publ., Text Series ii), St. Bonaventura, 1951, I, 141, Ockham wrote: "praeter res absolutae, scilicet substantias et qualitates, nulla res est imaginibilis, nec in actu nec in potentia." Cf. *Tractatus de Successivis,* p. 33: "Omnis res una vel est substantia vel qualitas, sicut alibi est ostensum et nunc suppositum."

14 See A. Maier, *Die Vorlaeufer Galileis im 14. Jahrhundert* (Rome, 1949), p. 101, n. 41.

15 Galileo Galilei, *Opere,* VIII, 197.

16 During the thirteenth and fourteenth centuries the scientific content of the arts course was expanded far beyond the four traditional mathematical subjects, geometry, astronomy, arithmetic and music, forming the *quadrivium.* Not only Euclid's *Elements* and Ptolemy's *Almagest,* but treatises on practical astronomy, Aristotle's writings on natural science, Alhazen's and Witelo's optical treatises, and other scientific works found their way into the curricula of various universities.

17 See J. H. Randall, Jr., "The development of Scientific Method in the School of Padua," *Journal of the History of Ideas,* I (1940), 177 ff.; P. O. Kristeller and J. H. Randall, Jr., "The Study of Renaissance Philosophy," *ibid.,* II (1941), 490 ff. The medical faculty was important at Padua and arts was a preparation primarily for this instead of for theology as at Paris and Oxford.

18 For bibliographical studies of the medieval writings available in print see especially A. C. Krebs, "Incunabula scientifica et medica," *Osiris,* IV (1938), and G. Sarton, *The Appreciation of Ancient and Medieval Science during the Renaissance (1450-1600)* (Philadelphia, 1955). For some studies of special cases the following may be consulted: E. Gilson, *Etudes sur le rôle de la pensée médiévale dans la formation du système cartésien (Etudes de Philosophie Médiévale,* XIII) Paris, 1930; C. Boyer, *The Concepts of the Calculus* (New York, 1939); A. Koyré, *Etudes galiléennes (Actualités scientifiques et industrielles,* Nos. 852-54 [Paris, 1939]); A. Maier, *Zwei Grundprobleme der scholastischen Naturphilosophie* (2nd ed.; Rome, 1951), *An der Grenze von Scholastik und Naturwissenschaften* (2nd ed.; Rome, 1952); E. A. Moody, "Galileo and Avempace," *Journal of the History of Ideas,* XII (1951); and my *Robert Grosseteste.*

19 The exploitation of the hypothesis or "regulative belief" that all forms of change could be reduced to matter in motion and therefore explained mechanically gave the seventeenth century a powerful guide to inquiry which combined exactly with the experimental and mathematical method. It could provide a foundation for explanations in a way that medieval concepts of nature, even Grosseteste's and Roger Bacon's "physics of light," could not. Cf. M. Boas, "The

establishment of the mechanical philosophy," *Osiris,* X (1952), 412 ff. For a test-case of a leading scientist whose inquiries were hindered by the lack of a "mechanical philosophy" Harvey's lack of progress in understanding the functions of the blood and of respiration may perhaps be cited. In this respect Descartes' criticism in the *Discours* of Harvey's failure to account for the cause of the action of the heart is also to the point. When he allowed himself to advance beyond the experimental results Harvey's conception of explanation was essentially Aristotelian. Nearly all the leading seventeenth-century scientists, for example Galileo, Kepler, Descartes, Boyle, Newton, were consciously preoccupied with scientific method, especially with the relationships between facts, laws, theories, and the mechanical philosophy, and to use Boyle's phrase, with the "requisites of a good hypothesis" (cf. M. Boas, "La Méthode scientifique de Robert Boyle," *Rev. d'hist. des sciences,* IX [1956], 120–21; R. S. Westfall, "Unpublished Boyle Papers Relating to Scientific Method," *Annals of Science,* XII [1956], 116–17; also E. A. Burtt, *The Metaphysical Foundations of Modern Physical Science* [2nd. ed.; London, 1932].) In the purely descriptive sciences like botany and zoology there was an analogous methodological preoccupation with the relation of "natural" to "artificial" systems of classification.

THE PHILOSOPHY OF SCIENCE
AND THE HISTORY OF SCIENCE

Joseph T. Clark, S.J.

There are, it seems to me, at least two significantly different but basically complementary ways in which to work in the field of the history of science. The first way I call *die von unten bis oben geistesgeschichtliche Methode*. I mean thereby a research policy which prescribes as its own point of departure the earliest accessible inception date of the scientific enterprise and sets as its goal the attempt to reconstruct in as comprehensive and complete detail as possible exactly how contemporary science of any given date has come to be what it is. In the vertical shaft of history this methodology thus works from the bottom to the top. It is therefore more or less compelled to organize its researches according to the structure of a logically and systematically irrelevant pattern of standardized chronological divisions or to adopt with some minor adaptations conventional but alien periodizations already established in the field of general history.[1] One further hazard of this *von unten bis oben* methodology is that it leaves its devotees open to invasion by the *precursitis* virus, an affliction which differs principally from bursitis in the fact that whereas the latter causes pain to the victim and excites sympathy in the spectator, *precursitis* exalts and thrills its victim but pains the observer.[2] Despite its many analytical and systematic limitations, however, this *von unten bis oben* methodology is nevertheless capable of achieving spectacular results. One outstanding example of the success of this procedure is the monumental *opus* of George Sarton's *Introduction to the History of Science*.[3]

The second way in which to work in the field of the history of science I shall call *die von oben bis unten geistesgeschichtliche Methode*. I mean thereby a research policy which prescribes as its own point of departure the logically sound conviction generated by the creditable results of a rigorously analytical philosophy of science that the scientific endeavor of mankind has finally come of age in the twentieth century. This conviction implies that science has already achieved such a sufficiently clear self-conscious awareness of the intellectual structure of its own enterprise that while future operations will surely disclose new and unexpected results through novel and unpredictable techniques, the *logical* pattern of their production must and will remain basically the same as those of today. This conviction does not, of course, imply that science is finished.[4] It merely claims that in an adequate philosophy of science it is now possible to state and to prove a theorem of completeness with respect to an inventory of the *kinds* of conceptual and logical elements both necessary and sufficient for the successful operation of scientific procedures.[5] In the vertical shaft of history this methodology works from the top to the bottom. It therefore sets as its goal—not merely to reconstruct in as comprehensive and complete detail as possible exactly how contemporary science of any given date has come to be what it is—but rather to disclose more by analytical than cumulative procedures exactly how it happened that the history of the internal development of science is as long as it is. This *von oben bis unten* methodology is therefore free to organize its researches according to the structure of a logically and systematically relevant pattern of central ideas and to invent its own periodizations independently of the conventional framework of general history. It may, for example, decide to rewrite the history of science in the West in two broad divisions, that of the closed world and of the unbounded universe.[6] It may again after trial find it more instructive to reconstrue the history of physics by epochs that are uniquely determined by the relative supply of mathematical resources at its disposal for any given era. A further advantage of this method is its value as a means of immunization against the *precursitis* disease. Because this *von oben bis unten geistesgeschichtliche Methode* in the field of the history of science has not yet seriously been tried on any significant scale, it is impossible to exhibit samples of its success. But that is perhaps one good reason to contribute to this Institute a tentative and exploratory paper on the philosophy of science and the history of science.

The purposes of this essay thus determine its contours. There are two broad and fairly equal divisions: one abstract, the other concrete. The first will attempt to depict briefly but adequately three valid insights of contemporary philosophy of science. The first of these will be concerned with the precise logical structure of scientific theory. The second will diagnose the exact logical function that mathematics fulfills in the scientific process. The third will isolate and define the precise logical points of significant correlation between the realm of mathematicized theory and the world of experienced or experimental fact. In the light of these three main disclosures of the first section the second major division will undertake to analyze three selected episodes in the history of science. In order to illuminate the area of theory construction, the work of Copernicus in astronomy will be reviewed. In order secondly to appreciate the full significance of the role of mathematics in the procedures of physical science, an analysis of the work of Oresme on the *latitudines formarum* will be undertaken. In order finally to realize the logical anatomy of significant correlation between mathematicized theory and experimental fact, it will be helpful to re-examine a special instance of misunderstanding between Beeckman and Descartes on the fall of bodies in space.

THE PHILOSOPHY OF SCIENCE

The Logical Structure of Scientific Theory

The logical structure of scientific theory is determined by one central process of inference and two coördinated operational procedures. The single central process of inference is hypothetico-deductive methodology. The first coördinate operational procedure is a meticulously critical attention to the concepts and the sentences which constitute its relevant hypotheses. The second allied procedure is a conscious awareness and a deliberate use of the logical fertility latent in the mathematical apparatus embedded in the theory.

At every random cross-section of successful contemporary science modern analysis by philosophers of science discloses that the logical structure of scientific thinking conforms to the following paradigm:

$$p \supset q_1 \cdot q_2 \cdot q_3 \cdots q_n : q_1 \cdot q_2 \cdot q_3 \cdots q_n : \supset p.$$

Therein 'p' stands as surrogate for all the conjoint suppositions of a specimen scientific theory and 'q_1,' 'q_2,' 'q_3,' 'q_n' represent respectively the statements of its at first deductively implied and tentatively predicted and then later tested and sufficiently confirmed consequences. The schema may thus be read in this context as follows: "If it is the case that if p, then q_1 and q_2 and q_3 on to q_n, and it is the case that q_1 and q_2 and q_3 on to q_n, then it is the case that p." This is indeed precisely how successful scientists think. Those who know whereof they speak are explicit on the point. Conant for example states:

Science we have defined . . . as "an interconnected series of concepts and conceptual schemes that have developed as a result of experimentation and observation and are fruitful of further experimentations and observations." A conceptual scheme when first formulated may be considered *a working hypothesis* on a grand scale. From it one can deduce, however, many consequences, each of which can be the basis of reasoning, yielding deductions that can be tested by experiment. If these tests confirm the *deductions in a number of instances, evidence accumulates tending to confirm the working hypothesis on a grand scale, which soon becomes accepted as a new conceptual scheme.* Its subsequent life may be long or short, for from it new deductions are constantly being made, which can be verified or not by careful experimentation.[7]

Conant here deftly depicts the central process of hypothetico-deductive inference characteristic of scientific theory.

But if some one or other scientific hypothesis '*p*' is deductively to imply some one or other testable consequence '*q*,' there must be present some logical connection between the antecedent and the consequent. That the principal source of this logical fertility in the scientific use of hypothetico-deductive methodology lies in the mathematical apparatus embedded into the theory is clearly realized by Rapoport and Landau for instance. They state that

The aim of mathematical biology is to introduce into the biological sciences not only quantitative, but also deductive, methods of research. The underlying idea has been to apply to biology the method by which mathematics has been successfully utilized in the physical sciences. This method can be briefly described. First, the actual situation, the biological problem presented by nature, is replaced by an idealized model. This is done because one cannot hope to deal with all aspects of reality at once, and also because some of these aspects may be irrelevant to the question in hand. Next, the idealized model is stated in mathematical terms, and the consequences of the mathematical statement are derived. Finally, these deduced results must be interpreted in terms of the original biological problem.[8]

Hence, given a scientific problem P with factors a, b, c, and a developed mathematical system M with elements x, y, z such that n theorems are true of them, then if theorem j is true of the a, b, c interpretation of x, y, z, then so too should be the other interpreted theorems of M.

But if the mathematically generated and logically deduced consequences of an hypothesis '*p*' in scientific theory are to be empirically relevant, it is indispensable that an adequate and accurate inventory be made beforehand of both the concepts and assumptions which together constitute the hypothesis in question. On this point Northrop counsels wisely:

> Long ago . . . the natural scientists learned that the way to be objective is not to try to get along without concepts and theory, but to get the concepts and theory which one is using out into the open, so that nothing is smuggled in surreptitiously and everyone can see precisely what the ideas and concepts are. The scientific method for doing this most effectively is the method of deductively formulated theory. This method forces one to specify the primitive concepts designating the elementary entities and relations in terms of which all the observed and inferred data in one's subject matter are being described. In short, the way to keep facts from being corrupted by faulty ideas and bad theory is not to be hard-boiled and attempt to get along without ideas by restricting oneself to pure fact alone, but to pay as much attention to the ideas or concepts one uses to describe the facts as one does to the facts being described.[9]

For an unanalyzed concept that conceals within its deceptive simplicity reference to multiple factors in the scientific situation is sure to destroy the logical connection that must prevail between antecedent 'p' and consequent 'q.' And an unrecognized assumption must ultimately operate to break the logical linkage between hypothesis and deducible consequence in a given scientific theory.

One may therefore be led to suppose that the logical analysis of scientific theory conducted by contemporary philosophers of science establishes the following rule: In hypothetico-deductive methodology empirical relevance of the consequent is assured if but only if all concepts ingredient in the theory are reducible to observation statements as empirical terms. But such a supposition is unwarranted. As Hempel concisely puts the matter:

> The history of scientific endeavor shows with increasing clarity that if we wish to arrive at precise, comprehensive, and well-confirmed general laws, we have to rise above the level of direct observation. The phenomena directly accessible to our experience are not connected by general laws of great scope and rigor. Theoretical constructs are needed for the formulation of such higher-level laws. One of the most important functions of a well-chosen construct is its potential ability to serve as a constituent in ever new general connections that may be discovered; and to such connections we would blind ourselves if we insisted on banning from scientific theories all those terms and sentences which could be "dispensed with." . . . In following such a narrowly phenome-

nalistic or positivistic course, we would deprive ourselves of the tremendous fertility of theoretical constructs, and we would also often render the formal structure of the expurgated theory clumsy and inefficient.[10]

Abstract concepts therefore and sentences which contain them are admissible in the hypotheses of scientific theory construction.[11] The only rigorously logical requirement is that they severally identify themselves for what they are and do not refuse to stand up and be counted.

But if one now rejoins the central process of hypothetico-deductive methodology and the two coördinated operational procedures of conscientious concept formation and consciously engrafted mathematical apparatus, how in total effect does the logical structure of scientific theory appear to scrutinizing intelligence? Hempel elsewhere describes it well:

> A scientific theory might . . . be likened to a complex spatial network. Its terms are represented by the knots while the threads connecting the latter correspond, in part, to the fundamental and derivative hypotheses included in the theory. The whole system floats, as it were, above the plane of observation and is anchored to it by rules of interpretation. These might be viewed as strings which are not part of the network but link certain points of the latter with specific places in the plane of observation. By virtue of those interpretive connections the network can function as a scientific theory. From certain observational data, we may ascend, via an interpretive string, to some point in the theoretical network, and thence proceed, via definitions and hypotheses, to other points, from which another interpretive string permits a descent to the plane of observation.[12]

Such, indeed, is the authentic logical structure of successful scientific theories, disclosed by the analytical researches of contemporary philosophers of science.

On this view a successful scientific theory is a pure invention of active intelligence. It is guided of course by past successes and consciously relevant to known experience. But it remains a creatively designed reconstruction of experienced events and not merely an ideographic reproduction of the experientially real. It is in short quite literally a guess at the way in which the world is made, but an educated and severely disciplined one. To test impartially its relevance to matters of fact, one selects as many strategic consequences of the deductively fertile conceptual scheme of 'p' as are considered theoretically pertinent and revealing. Hence the n of the consequential 'q_n' of 'p' may be very large or very small, contingently upon contextual circumstances. Such logically significant predictions, if countermanded by recalcitrant experiences through a sufficient series of unequivocal *not-q* results, relentlessly require the constructive revision of '*p*' into a new form '*p''* that may or may not recognizably resemble the original parent '*p*.' If however a sufficient set of successful predictions confirms the empirical relevance of '*p*' more and more strongly, then '*p*' gradually obtains the respectable status of assured physical fact and the honorable position of serving as confident point of departure for the construction of a new and wider p_2.

There are therefore at least four logically relevant criteria for appraising the worth of various scientific theories. The first has reference to the degree of clarity and precision with which the theories are formulated and with which the logical relationships of their elements to each other and to expressions phrased in observational terms have been rendered explicit. The second concerns the relative degree of explanatory or predictive power with respect to observable phenomena. The third centers on the criterion of formal simplicity and assays how much predictive power is obtained per unit element of theory. The fourth and last estimates the extent to which the respective theories have been confirmed by experimental evidence.

The Logical Function of Mathematics in Science

In order to understand the logical function of mathematics in science, it is first necessary to comprehend with authentic insight the logical structure of the mathematical enterprise itself. But such authentic insight into the rational anatomy of mathematics is neither easily nor often achieved by persons outside the fraternity of professional mathematicians. The main cause for such general failure to comprehend the genuine character of mathematics is not far to seek. Dubisch correctly reports that

. . . mathematics remains in a comatose condition all through grammar school. All too often, in fact, it barely comes to life in high school or in freshman and sophomore college work. Thus, even in college, the student frequently does not come into contact with living mathematics—the mathematics of the contemporary mathematician. Instead he largely studies formal manipulation, useful in solving problems of physics and engineering but not useful in fostering an understanding of modern mathematics which is so highly unmanipulational. For real mathematics is a constantly expanding subject in which *ideas* and not manipulations play the dominant role.[13]

A relevant case in point may help to illustrate and emphasize this deplorable situation and highlight the gigantic difference which exists between conventional understanding of mathematics and mathematics itself.

A contemporary professional educator for example teaches teachers to teach children that "multiplication is a short process of addition."[14] It is true, of course, that the sum of four plus four plus four and the product of three by four are both twelve. But the precise mathematical significance of this computational coincidence is to the effect that relations which link antecedents to an identical consequent need not therefore themselves be identical. For if one may speak loosely of adding one half to itself four times and getting two, one cannot speak intelligently at all of adding four to itself *one-half-of-a-time* and also getting two. The first mathematically important consequence of this general interpretation of multiplication as merely syncopated addition would be the disastrous conclusion that the multiplicative relation does not possess the property of commutativity. In actual fact the additive and multiplicative relations, whether constructed in professional mathematics by Frege or Russell or Peano or Grandjot or Kálmar, are always constructed independently of one another. Hence if both relations appear trivially similar in effect in the case of the natural numbers but not so in the case of the fractional and negative and complex and irrational numbers, this fact discloses to the genuine mathematician rather an incidental character of the natural numbers as such and not a property of the multiplicative relation as such. A more serious reason and one which is technically decisive is this: A principal ideal is not always the set of all ring multiples of a fixed ring element.[15] Contrary therefore to popular assumptions, the authentic mathematical fact is that multiplication is *not* repeated addition. Hence equivalence in computational manipulation may not be construed as an identity of concepts or ideas.

School mathematics of traditional type is thus not a safe basis for a sound and accurate analysis of the logical structure of mathematics. It is therefore a mistake to construe the science of professional mathematicians as merely conventional school mathematics, writ large. Projective geometry for example does not start where elementary high school geometry stops. In fact it does not presuppose among its logical resources any of the results of that elementary geometry. Projective geometry stands on its own constructive foundations and is developed logically from its own initial and relatively primitive propositions. Not only is there complete absence of developmental continuity from elementary geometry on to a more complicated version of the same in projective geometry, there is also a surprising and instructive reversal of logical role. For the projective geometry which usually comes after elementary geometry in curricular sequence is logically prior to it. The fact is that only at the tail end of a comprehensive projective geometry does there emerge the logically less significant, mathematically less interesting, and strangely singular case of Euclidean elementary metrical geometry.[16]

Similarly modern abstract algebra is in no way a rectilinear extension of conventional secondary school algebra. For this latter subject is in effect nothing more than a heavy-handed version of an inadequately formulated universal arithmetic. Modern algebra on the contrary is a fascinating study of set-theoretical structures, founded immediately upon the logic of relations, and immeasureably removed from a mere computation with letters from the alphabet instead of with digits from the multiplication table. Here is how Jacobson introduces his readers to the subject:

> We begin our discussion with a brief survey of the fundamental concepts of the theory of sets. Let S be an arbitrary set (or collection) of elements a, b, c. . . . The nature of the elements is immaterial to us. We indicate the fact that an element a is in S by writing $a \varepsilon S$ or $S \ni a$. If A and B are two subsets of S, then we may say that A is *contained* in B or B contains A (notation: $A \subset B$ or $B \supset A$) if every a in A is also a B. The statement $A = B$ thus means that $A \supset B$ and $B \supset A$. Also we write $A \supset B$ but $B \neq A$. In this case A is said to contain B properly, or B is a *proper subset* of A. If A and B are any two subsets of S, the collection of elements c such that $c \varepsilon A$ and $c \varepsilon B$ is called the intersection $A \cap B$ of A and B . . .[17]

And so on. But the sequel in the text is spectacularly significant. For modern abstract algebra in a manner more impressive still than projective geometry succeeds in embracing under one comprehensive logical point of view not only the entire relational content of elementary algebra but also the theory of invariances in geometrical transformations as well. It is undeniable therefore that the total result of the conventional curriculum is to distort correct perspective on the logical structure of mathematics as a whole.

For it is now possible for twentieth-century man to know precisely what mathematics is and exactly what it means to be a mathematician. It is for example now clear that mathematics may initiate its constructions from any alternative choice of relational situation whatever. There is no absolute and uniquely privileged point of logical departure. The elements of any set have all but only those properties with which the relatively primitive statements of the axiomatic system endow them. There is thereafter only one technical objective: to elaborate the logical consequences of such initial statements. To succeed in accomplishing this objective without committing incidental fallacies of concealed contradiction in the process is by that very fact to create a branch of mathematics. To begin at the given point of departure and to follow the subsequent deductions without misconstruing elements and links in the sequence, this is precisely what it is to learn a department of mathematics. It is clear to the contemporary mathematical conscience that the primary definitions and postulates relative to any given axiomatic system are neither rigidly uniform deliverances from a homogeneous human experience nor inescapable necessities of all rational thought. The mathematician is entirely free within the widely variable limits of his own creative intelligence to devise whatever logically consistent pattern universes he desires and has the technical competence to construct. In the conscientious performance of this professional task the mathematician is in no sense under the illusion that he is discovering by the wizardry of abstract and metempirical methods the fundamental structures of our extended world. Nor does the mathematician at work presume that he is dealing with a set of absolute and eternal truths. For all mathematical truths are innocently timeless, not mystically eternal.

In short, the mathematician is both a fertile creator and an incisive critic of the consistency and rigor and elegance of patterns of ideas. The elements with which he deals are of a purely constructive nature. As such, they have no significant relevance to objects of sense experience. If the mathematician speaks a language that is embroidered on occasion with words that carry empirical overtones, such as "transformations in space," it is only because he has to name his creations for purposes of intrascientific communication with his colleagues. And the only names available to man must inevitably bear historically inveterate and tell-tale traces of the only world that is directly accessible to him. It would however be a tragic error to misconstrue these faint echoes of experience. As Keyser phrased it:

... as employed in mathematics, the great terms—Relation, Transformation, and Function—mean essentially the same thing: that, in other words relation theory, transformation theory, and function theory are but three linguistically different expressions or embodiments of one and the same content or substance—a single doctrine set forth now in one and now in another of three different manners of speech.[18]

For the truth of the matter is that the mathematician irrevocably remains uniquely dedicated to one clearly defined objective: the exploration of the relational morphology and comparative anatomy of abstract mathematical patterns, structures, orders, designs, rhythms, harmonies.

It is not therefore difficult to discern precisely the logical function of mathematics in science. Wherever and whenever a branch of mathematics is engrafted upon a scientific theory, it puts at the disposal of the scientist all the dialectical resources, all the implicational architecture, all the deductive fertility, all the relational structure, and all the established theorems of its developed axiomatic system.

But the logical structure of the grafting process whereby an independently prefabricated mathematical system becomes vitally fused with the conceptual schemata of a scientific theory to issue into a spectacularly effective correlation between theoretical predictions and experimental fact has sometimes been a puzzling problem to contemporary philosophers of science. Laso for example puts it as follows: "All modern physicists employ mathematics as a working tool with spectacular effect. But philosophers who do not admit the abstraction of the concepts of mathematics from the real world confront a problem that is insoluble: *how and why a science that is pure and abstract applies to the concrete world of sense perception?*"[19]

The question is a sound one and merits a decisive reply.

The Logic of Correlation between Mathematicized Science and Experiential Fact

The core of the answer supplied by a responsible philosophy of science lies in the fruitful concept of *isomorphism*. By "isomorphism" is here meant a one-to-one correspondence C between the elements and relations of a prefabricated mathematical structure M and the objects and relations of a physical structure M', where relations of order n correspond to relations of order n, and such that whenever a relation R holds between the elements of M, the corresponding relation R' holds under C between the corresponding objects of M' and conversely.

In arithmetic, for example, which is the science of number and not the art of computation, the study of relational characters and not the rote identification of related terms, there exists by constructive definition and mathematically inductive proof the additive relation whereby, let us say, $c = a + b$. Two observations at this point are of central logical importance.

The additive relation is uniquely and unalterably a *mathematical* relation which can therefore admit only mathematical entities as its related terms, and in particular only numbers of a consciously constructed and rigorously defined set. The further relation above between '*c*' and '*a+b*' is not just any sort of a vague relation of similarity or equivalence or even of equality but precisely the stark and severely sharp relation of mathematical *identity* of its related terms.

Some further logical characters or properties of each relation are also here crucially important. The additive relation can be rigorously proven to be both associative and commutative and such that $a+1>a$ for every *a*. Hence everywhere in arithmetic if $c=a+b$, then $c=b+a$, and if $c=(a+b)+d$, then $c=a+(b+d)$ where the parentheses serve as indices of an initially and tentatively presumed but in effect logically irrelevant priority. Finally it can be rigorously proved that the relation of mathematical identity is transitive and such that if $a=b$ and $b=c$, then $a=c$.

In the metricized physics of length, for example, to choose the simplest case as paradigm, there exist in terms of operationally defined experimental fit both the *juxtapositive* relation between lengths of some material and the *satisfactorily-approximate-coincidence-of-end-faces* relation whereby, let us say, $C\ [=]\ A\ (+)\ B$ where *C*, *A* and *B* indicate lengths of some physical material to which have been arbitrarily but systematically assigned appropriate numbers-of-given-units-of-metrical-scale, and where the identity sign is enclosed in brackets and the addition sign in parentheses to avoid in each case the logically and notationally inexcusable solecism of employing a single and undifferentiated symbol to represent two different relations at the same time. For it is undeniable that the *physical* and juxtapositive relation between given lengths of some arbitrary material and the *mathematical* and additive relation between certain appropriate numbers are not the same relation. It is also instantly clear that the *mathematical* relation of identity between a given pair of numbers and their unique sum is not the same as the *physical* relation of satisfactorily-approximate-coincidence-of-end-faces between lengths of some given material. For no two lengths of physical material can possibly be identical with a given third.

But it does turn out *under certain clearly specifiable conditions and to a certain recognizable degree* that (1) the physical and juxtapositive relation between measured lengths is in a formal sense *isomorph* to the mathematical and additive relation between numbers in so far as both relations are associative, commutative, and such that C is longer than A (+) B exactly when $c > a+b$, and (2) the physical relation of satis-factorily-approximate-coincidence-of-end-faces of lengths of material is in a formal sense *isomorph* to the mathematical relation of identity between a given pair of numbers and their unique sum in so far as both relations are transitive, and such that if A is just as long as B and B is just as long as C, then A is just as long as C, in precisely the way that if $a=b$ and $b=c$, then $a=c$. Two central conclusions emerge from this conscientious analysis. The first is this: The physicist has reasons of his own which he sometimes revises in the face of recalcitrant experience for believing that numbers a, b, c are not only unambiguously assignable to lengths A, B, C, but even so assignable that $A(+)B[=]C$ exactly when $a+b=c$. The second conclusion is this: It is technically economical but semantically elliptical to say that one "adds" measured lengths, or to write in strictly mathematical notation without the protection against ambiguity that brackets and parentheses afford: $A+B=C$.

The central and decisive conclusion for the philosophy of science now appears: *More elaborate instances of successful interpretation in more complex mathematicized theories of science are simply more elaborate isomorphisms established under certain discoverable boundary conditions between* (a) *certain ideal and prefabricated mathematical relations and* (b) *certain real and physical relations.* It is however not the case that all kinds of physical systems enter by successful interpretation into all kinds of mathematical relationships. Densities for example do not enter into relations of the additive type. And it is often enough the case in contemporary physics that more refined experiences require revision in previously established and hitherto satisfactorily-reliable-up-to-a-certain-point isomorphisms. The superposition of velocities for example was long considered isomorph to the simple additive relation between the numbers that measured them. But under Einstein's special relativity theory it came to be considered isomorph to the addition theorem of a certain hyperbolic function. But the fact remains that throughout the entire range of science the logic of the correlation between mathematicized theory and experiential fact is exactly the same. The future progress of all science therefore depends critically upon the creative productivity of professional mathematicians and the inventory at any given moment of developed relational systems available for concrete isomorphic interpretation.

Such then in brief are the three significant insights provided by a responsible philosophy of science which the *von oben bis unten* methodologist brings to his analytical researches in the field of the history of science: (1) hypothetico-deductive theory construction and in particular concept formation, (2) the expressly constructive character of mathematical systems, and (3) isomorphism. It now remains to apply the first insight on theory construction to Copernicus, then to apply the mathematical insight to the *latitudines formarum* of Oresme, and finally to trace out isomorphisms as apprehended in diverse ways by Beeckman and Descartes.

⁜ THE HISTORY OF SCIENCE

Theory Construction and Copernicus

In order to comprehend the relevance of philosophic theory-construction insights to an historical appraisal of the scientific work of Copernicus, one must first understand both (a) the authentic mathematical astronomy of Ptolemy and (b) the largely Aristotelian context of physics and cosmology in which it was conventionally embedded. In expediting this task of preliminary explanation pictures are superior to prose and will be employed wherever pertinent.

The explicit assumptions upon which Ptolemy in the *Almagest* based his system are five in number and read as follows:

... (1) that the heaven is spherical in form and rotates as a sphere; (2) that the earth too, viewed as a complete whole, is spherical in form; (3) that it is situated in the middle of the whole heaven, like a center; (4) that by reason of its size and its distance from the sphere of the fixed stars the earth bears to this sphere the relation of a point; (5) that the earth does not participate in any locomotion.[20]

The first assumption is further elucidated to imply that "the aether which encloses the heavenly bodies, being of the same nature, is of spherical form, and because of its composition out of homogeneous parts, moves with uniform circular motion."[21] Ptolemy further notes that

First of all we must premise, as regards the name ("fixed stars"), that since the stars themselves always appear to keep the same figures and the same distances from each other, we may fairly call them "fixed," but on the other hand, seeing that their whole sphere on which they are carried round as if they had grown upon it, appears itself to have an ordered movement of its own in the direct order of the signs, that is, toward the East, it would not be right to describe the sphere itself also as "fixed."[22]

In substance, therefore, the two fundamental assumptions of Ptolemy's celestial kinematics are the geocentric hypothesis and circular motion.

The principal geometrical devices, furthermore, by which Ptolemy attempted to "save the phenomena" of observed planetary motions in an astronomical system thus based upon the joint hypotheses of geostatic center and circular motion are (a) *the eccentric circle,* (b) *the deferent circle with epicycle,* (c) *the equant,* and their more complex combinations.[23]

The geometrical structure and functions of the *equant* may be exhibited as follows. Describe an eccentric circle with fixed center C and another point E, representing the earth (Figure 1). On the circumference of this circle select a point P to indicate a planet. On the apse-line ECP select another point Q, such that $EC=CQ$. Then as P revolves on the circumference of the eccentric circle, let $<PQR$ increase uniformly, that is, increase equally in equal intervals of time. P will always remain at an equal distance from C. But *the point with reference to which its motion is uniform will not be the center C, but the equant Q.* Hence the equal

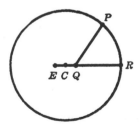

Figure 1 (Rosen)

increases in equal times of $<PQR$ will not be subtended by equal measures of arc. The resultant motion of P about C therefore will not be of uniform velocity, but sometimes slower and at other times faster.

With these technically ingenious devices of circular motion and their several more complex combinations Ptolemy constructed his mathematical theory of planetary paths. The completed structure may be diagrammati-

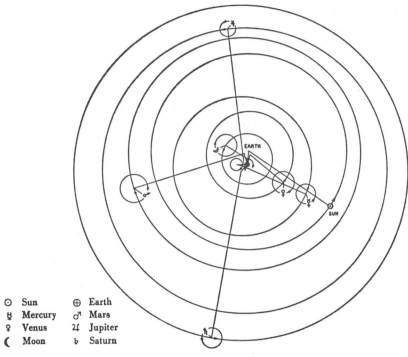

⊙	Sun	⊕	Earth
☿	Mercury	♂	Mars
♀	Venus	♃	Jupiter
☾	Moon	♄	Saturn

Figure 2 (Stahlman)

cally represented as in Figure 2. The figure exhibits samples of each geo-metrical device: eccentric, epicycle, equant. It also displays the significant difference between the merely geostatic and geocentric functions of the earth.

But at the time of Copernicus this mathematical framework of Ptolemy was conventionally embedded in a largely Aristotelian context of physics and cosmology. The total picture may be suggested as in Figure 3. There are eleven concentric spheres, partitioned for theoretical considerations of physics and cosmology into two groups of four inner and seven outer spheres respectively.

The first group of four inner spheres comprises the sublunary world of Aristotelian physics. At the center of the cosmos is set the static earth, one of the four basic elements in its undiluted state and *natural habitat*. Contiguous with its outer surface is the three-dimensional spherical shell of water, second of the four basic elements in its pure state and *connatural home*. Surrounding the sphere of water is the spherical blanket of air,

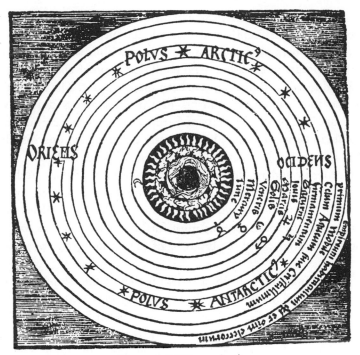

Figure 3 (1508 woodcut)

third of the primitive elements and likewise in its pure state and natural position. Beyond the sphere of air but contiguous with it and still concentric with the earth is situated the realm of fire, fourth and last of the terrestrial elements and also in its naturally pure state and *natural location*.

Such is the skeletal blueprint of the conventional *sublunary world* of ancient physics in its *ideal* and static condition. For all the globs of earth in the universe, as naturally heavy and heavier than water, *belong* by natural destiny in a dense homogeneous cluster about the fixed center of the universe. And all the drops of water in the cosmos, as also naturally heavy but less heavy than earth, *belong* in a continuous shell of uniform thickness about earth. And all the winds of the world, as naturally light but less light than fire, *belong* within the confines of a sphere around the region of water. And all the flames of the universe, as also naturally light but lighter than air, *belong* in a realm superior to that of air.

But the *real* state of the world of experience is vastly different. Large land masses of earth protrude above great ocean depths of water. Disturbances on the ocean floor release explosive bubbles of air which reveal the presence of winds imprisoned beneath its turbid surface. Combustion discloses that both air and fire are contained within wood, as well as earth which reappears in the ashes. And ebullition is proof enough that air was contained in what seemed to have been merely water. The fact is therefore that the four sublunary elements of the Aristotelian physics were all mixed up. The physical objects of familiar experience were not masses of elements in their pure state, but variously compounded and distributed in forced exile from their natural locations of existence and rest.

In order to understand how the ideal condition of static and mutually exclusive regions for earth, water, air and fire, prescribed by physical theory for the sublunary world, degenerates into the actual heterogeneity of the realm of experience, one must further understand the *cosmological* principles that governed the behavior of the second group of seven upper spheres.

The eleventh and outermost sphere is that of the Empyraean Heaven, "the abode of God and all the Elect." Contiguous to its under surface is set the sphere of the Prime Mover which revolved from East to West and imparted motion to the entire system of concatenated spheres. Below in spheres of uniform thickness lie in turn the Crystalline Heaven, the sphere of the fixed stars, and those of Saturn, Jupiter, Mars, the Sun, Venus, Mercury, and the Moon. The total effect of this interlocking system of the seven upper spheres is to disturb the ideal equilibrium of the regions of fire, air, water and earth and thus distribute the elements into the mixed state of combination that characterizes the heterogeneous world of physical experience. The dynamism of the science of physics, therefore, once intermeshed with the strands of this cosmological context, is nothing more nor less than the instinctive and spontaneous effort of the displaced elements to *return* to the natural levels of rest from which the smoothly geared rotations of the celestial spheres have forcibly exiled them. This is the meaning of natural motion in Aristotelian mechanics. This is the paramount problem of the ancient science of physics.

In brief, then, there were two irreducible classes of moving objects in the Aristotelian scientific context into which Ptolemaic mathematical astronomy was embedded at the time of Copernicus: (1) the imperishable celestial bodies above the region of fire; (2) the terrestrial elements and their soluble compounds below the lunary sphere. Both moved and *under the constant application of a constant force*. But the motions *natural* to each were (a) irreducible to the other, and (b) unresolvable into geometrically more simple components. The simple motion characteristic of the celestial bodies was circular and under the propulsive power of their nature or *form*. The simple motion characteristic of the terrestrial elements was perpendicular to the plane of the horizon and either vertically upward or vertically downward, dependently upon the natural lightness or relative heaviness of the exiled elements or compounds concerned. Here too the propellent force was nature or *form*.

To this syncretic *Weltanschauung*, thus compounded of Ptolemaic mathematical astronomy and Aristotelian physico-cosmology, what in effect was the reaction of Copernicus? As I read and analyze him, his reaction fixed principally on two points and developed in two phases. The first point and the first phase concerned the mathematical astronomy of Ptolemy. It was an intellectually irritating dissatisfaction with the *theoretically* inconsistent device of the equant. The second point and the second phase referred to the physico-cosmology of Aristotle. It was the insight that substantial form was unnecessary as a constant moving force for the spheres and consequently dispensable in a *theory* of celestial mechanics.

As early as the *Commentariolus* Copernicus stated:

> . . . the planetary theories of Ptolemy and most other astronomers, although consistent with the numerical data, seemed likewise to present no small difficulty. For these theories were not adequate unless certain equants were also conceived; it then appeared that a planet moved with uniform velocity neither on its deferent nor about the center of its epicycle. Hence a system of this sort seemed neither sufficiently absolute nor sufficiently pleasing to the mind.
>
> Having become aware of these defects, I often considered whether there could perhaps be found a more reasonable arrangement of circles, from which every apparent inequality would be derived and in which everything would move uniformly about its proper center, as the rule of absolute motion requires. After I had addressed myself to this very difficult and almost insoluble problem, the suggestion at length came to me how it could be solved with fewer and much simpler constructions than were formerly used, if some assumptions (which are called axioms) were granted me.[24]

These assumptions included, of course, the motion of the earth about the sun.

Some years later this same abiding concern is reflected in the enthusiastic *Narratio prima* report of his student, Rheticus, who writes:

> Furthermore, most learned Schöner, you see that here in the case of the

moon we are liberated from an equant by the assumption of this theory which moreover corresponds to experience and all the observations. My teacher dispenses with equants for the other planets as well by assigning to each of the three superior planets only one epicycle and eccentric; each of these moves uniformly about its own center, while the planet revolves on the epicycle in equal periods with the eccentric.[25]

And in his summary of the principal reasons why astronomers should accept the systematic revisions of Copernicus, Rheticus declares: "Fourthly, my teacher saw that only on this theory could all the circles in the universe be satisfactorily made to revolve uniformly and regularly about their own centers—an essential property of circular motion."[26] In the *De revolutionibus* Copernicus himself objected thus to the use of the equant in the lunar theory of his predecessors:

> For when they assert that the motion of the center of the epicycle is uniform with respect to the center of the earth, they must also admit that the motion is not uniform on the eccentric circle which it describes. . . . But if you say that the motion uniform with respect to the center of the earth satisfies the rule of uniformity, what sort of uniformity will this be which holds true for a circle on which the motion does not occur, since it occurs on the eccentric?[27]

Again in criticizing the planetary theory of the astronomers Copernicus comments: "Therefore they admit in the present case also that the motion of a circle can be uniform with respect to some center other than its own."[28] For Copernicus thus internal consistency in principle and external confirmation by fact are equally indispensable criteria for the acceptability of a scientific theory.

The technical triumph of Copernicus was therefore the construction of a *theoretically consistent* system of planetary motions, without the device of the equant (Figure 4). In the diagram the point C represents the momentary center of the earth's orbit. Around C cluster the actual momentary centers of rotation of the planets. Thus in order consistently to preserve the principle of pure circular motion, Copernicus was forced to refer that motion to a point C other than the center of the universe. There, indeed, at the geometrical center of the sphere of the fixed stars, the sun is prominently located. But its systematic function is merely optical, as a source of light, not dynamical.

The second point and the second phase of the reaction by Copernicus

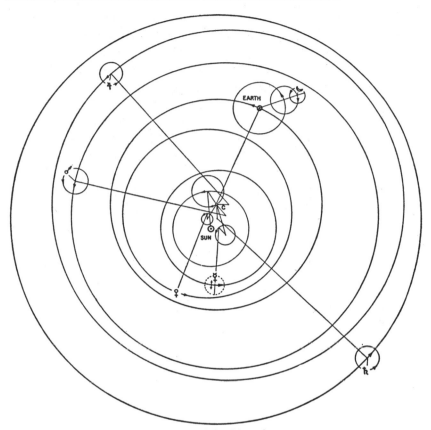

Figure 4 (Stahlman)

to contemporary astronomy was the systematic insight that an economical theory could dispense with the postulate of substantial forms as movents of the spheres. This insight reveals itself in the very structure of the first book of the *De revolutionibus*. The concept of "sphericity" dominates the introductory treatment. Chapter one contends "that the universe is spherical." Chapter two belabors the point "that the earth also is spherical." Chapter three argues that the presence of water on the earth does not detract from its total sphericity. Chapter four maintains "that the motion of the heavenly bodies is uniform, circular, and perpetual, or composed of circular motions." This chapter opens with a full disclosure of the insight: "We now note that the motion of the heavenly bodies is circular. Rotation is natural to a sphere and by that very act is its shape expressed. For here we deal with the simplest kind of body, wherein neither beginning nor end may be discerned nor, if it rotates ever in the same place, may the one be distinguished from the other."[29] The disclosure is quietly and subtly made but its significance is momentous. With a stroke of the pen Copernicus transposes substantial form into geometrical form. In the Aristotelian physico-cosmology it was the specific character of the substantial form, joined to an appropriate matter, which determined the motion of the celestial bodies. In the Copernican revision of Ptolemaic theory geometric shape now performs that dynamic function. *Sphericity, as such, naturally and necessarily entails uniform circular motion.* Since sphericity, moreover, as a geometrical shape, is common both to the earth and to the conventional celestial bodies, the same inevitable motions apply to all alike. The standard distinction between terrestrial and celestial physics thus disappears and a single universe takes its place.

At this point the two insights of Copernicus come together to provide the dynamics of his system. The elimination of the equant cleared the field everywhere for uniform circular motion. And everywhere the spherical shape of the bodies supplied the source of such motion and guaranteed to maintain its continuance. No external mover was necessary. Nor was a body required to serve as momentary center and physical support for such uniform rotation. A mathematical point C was sufficient. From this viewpoint the system of Copernicus was heliostatic, indeed, but not heliocentric. For Copernican spheres, precisely because they were spherical, were well able to take care of their own natural, necessary, inevitable, uniform, circular motions.

And yet in Chapter ten which treats of the order of the heavenly bodies, the earth included, Copernicus concludes a long series of arguments as follows:

Given the above view—and there is none more reasonable—that the periodic times are proportional to the sizes of the orbits, then the order of the spheres, beginning from the most distant, is as follows. Most distant of all is the sphere of the Fixed Stars, containing all things, and being therefore itself immovable. It represents that to which the motion and position of all the other bodies must be referred. Some hold that it too changes in some way, but we shall assign another reason for this apparent change, as will appear in the account of the earth's motion. Next is the planet Saturn, revolving in 30 years. Next comes Jupiter, moving in a 12 year circuit. Then Mars, who goes round in 2 years. The fourth place is held by the annual revolution [or orbit] in which the earth is contained, together with the orbit of the Moon as on an epicycle. Venus, whose period is 9 months, is in the fifth place, and sixth is Mercury who goes round in the space of 80 days.[30]

In the manuscript of Copernicus the diagram printed here as Figure 5

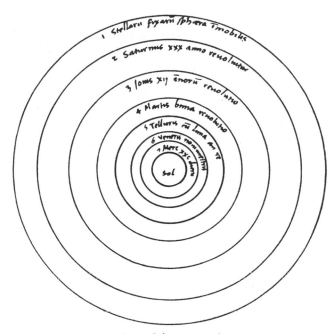

Figure 5 (Copernicus)

accompanies the text. In Copernicus' own hand is clearly written: *Stellarum fixarum sphaera immobilis*, "The immovable sphere of the fixed stars."

But this is the same Copernicus who only a few manuscript pages before had confidently written: *Mobilitas sphaerae est in circulum volvi, ipso actu formam suam exprimentis* . . . "Rotation is natural to a sphere and by that very act is its shape expressed."[31] At this precise point the very same Copernicus who began his original scientific career by a flight from the theoretical inconsistency of the Ptolemaic equant, introduces into the structure of his own system the theoretical inconsistency of the *immobilis sphaera naturaliter mobilis*, "the sphere which naturally must move (or surrender its geometrical identity), but which does not move and cannot move." The main reason for the introduction of the inconsistency is a further *undisclosed* assumption of Copernicus that everything was somewhere, and that "place," as Aristotle had defined it, "is the innermost *motionless* boundary of what contains."[32] It is this hidden assumption that is operative when Copernicus writes: *stellarum fixarum sphaera, seipsam et omnia continens, ideoque immobilis, nempe universi locus,* "the Sphere of the Fixed Stars, containing itself and everything else, and therefore immovable, since it is the place of the universe."

It is not to be construed as a fatuous complaint against history but merely as a comment by a philosopher of science, interested in applying theory-construction insights to an historical appraisal of Copernicus in the spirit of the *von oben bis unten* methodology, to remark that had Copernicus gotten *all* his assumptions out into the open, as the rigorously axiomatic procedures of the hypothetico-deductive method prescribe, he might rather have been the first to break open the closed world than the last to preserve it. For one has only to eliminate the *sphaera immobilis,* as theoretically inconsistent with the prior sphericity-motion postulate, and instanter the fixed stars twinkle above the planetary paths, afloat in unbounded space.

Mathematics and Oresme

The first major insight, therefore, which contemporary philosophy of science puts at the disposal of the *von oben bis unten* research worker in the field of the history of science is a clear vision of the logical structure of hypothetico-deductive theory and the relevance thereto of critically controlled processes of concept formation. The second such major insight is a clear perception of the freely constructive and completely relational character of mathematics. In order to exhibit the value of this insight to scholars in the history of science, an examination of the mathematical import of Oresme's doctrine *de latitudinibus formarum* may be useful.

This choice is in a certain sense a delicate and critical one. For the work of Oresme is one of the two central logical pillars upon which rests a current thesis in the history of science. For in his *History of Science* Crombie writes:

. . . The most striking advances were made by Oresme, who was an original mathematician: he had conceived the notion of fractional powers, afterwards developed by Stevin, and given rules for operating with them. Oresme represented *extensio* by a horizontal straight line (*longitudo*) and made the height (*altitudo vel latitudo*) of perpendiculars proportional to *intensio*. His object was to represent the "quantity of a quality" by means of a geometrical figure of an equivalent shape and area. He held that properties of the representing figure could represent properties intrinsic to the quality itself, though only when these remained invariable characteristics of the figure during all geometrical transformations. He even suggested the extension of these methods to figures in three dimensions. Thus Oresme's *longitudo* was not strictly equivalent to the abscissa of Cartesian analytical geometry; he was not interested in plotting the positions of points in relation to the rectilinear co-ordinates; but in the figure itself. There is in his work no systematic association of an algebraic relationship with a graphical representation, in which an equation in two variables is shown to determine a specific curve formed by simultaneous variable values of *longitudo* and *latitudo*, and *vice versa*. Nevertheless, his work was a step towards the invention of analytical geometry and towards the introduction into geometry of the idea of motion which Greek geometry had lacked. He used his method to represent linear change in velocity correctly.[33]

Crombie then quotes from part 3, chapter 7, of Oresme's treatise *De configurationibus intensionum* as follows:

Any uniformly difform quality has the same quantity as if it had uniformly informed the same subject according to the degree of the midpoint. By "according to the degree of the midpoint" I understand: if the quality be linear. If it were superficial, it would be necessary to say: "according to the degree of the middle line.". . . We will demonstrate this proposition for a linear quality.

Let there be a quality which can be represented by a triangle, *A B C* (Figure 6). It is a uniformly difform quality which at point *B* terminated at zero.

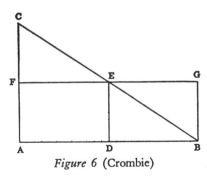

Figure 6 (Crombie)

Let *D* be the midpoint of the line representing the subject; the degree of intensity that affects this point is represented by the line *D E*. The quality that will have everywhere the degree thus designated can then be represented by the quadrangle *A F G B*. . . . But by the 26th proposition of Euclid, Book I, the two triangles *E F C* and *E G B* are equal. The triangle which represents the uniformly difform quality and the quadrangle *A F G B* which represents the uniform quality, according to the degree of the midpoint, are then equal; the two qualities which can be represented, the one by the triangle and the other by the quadrangle, are then also equal to one another, and it is that which was proposed for demonstration.

The reasoning is exactly the same for a uniformly difform quality which terminates in a certain degree. . . . On the subject of velocity one can say exactly the same thing as for a linear quality, only instead of saying: "midpoint," it would be necessary to say: "middle instant of the time duration of the velocity." It is then evident that any uniformly difform quality or velocity whatever is equalled by a uniform quality or velocity.[34]

In order to guarantee complete perspective on the issue, it may help to present a more comprehensive and perhaps more representative sampling of the full contents of that same treatise.[35]

The treatise has not only a pedagogical but also a doctrinal purpose. The pedagogical purpose is to put into systematic form his ideas about the graphical representation of uniform and difform rates of change in qualitative alteration. The doctrinal purpose is to make clear once and for all the import and the bearing of the technique. There are three parts: (1) the first discusses the geometrical shape and the physical efficacy of qualities that abide in the subject after acquisition and are acquired at either a uniform or difform rate; (2) the second deals with the geometrical shape and the physical efficacy of qualities which are such that the acquisition of a new degree entails the loss of prior degrees; (3) the third is concerned with the acquisition and the measurement of qualities and velocities.

Chapter one of part one has for its theme the scalar continuity of degrees in qualitative alteration. With the unique exception of numbers, everything else subject to processes of measurement by proportional comparison is construed as a continuous quantity. Hence to achieve effective measurement of them, one must have recourse to the device of points, lines, and planes and make use of the geometrical theorems that are true of them. For to these elements measurements or proportional ratios properly belong. But the mind of man may use them as analogical referents for the measurement of other items. Even if, moreover, one believes that continua are not composed of indivisibles, yet one must make believe that there are such points and lines if one is to carry out the mathematical operations of measuring things and discerning their proportional relations.

It is a rule of this method to represent graphically every successive degree of intensity by a line erected perpendicularly on either a point in space or on the thing which is affected by that changeable factor, such as a quality. The reason is that so long as we are dealing with the same type of changeable phenomenon, whatever be the proportion between any two degrees, the same proportion will obtain between the corresponding lines, and conversely. The device of lines is chosen because proportions between lines of different lengths are more easily perceived. Such lines, moreover, are drawn perpendicular to the subject to which they are applied for reasons of visual convenience. The correspondence between certain degrees of intensity and specified lengths of lines is exact but proceeds always on the basis of proportionality: equal lines for identical degrees, double length for double intensity, etc. Such lines are only mathematical fictions and could really be drawn in any direction and be incident on any point. But it is more convenient to draw them perpendicularly to the subject that is informed by the quality.

Chapter two discusses the "latitude" of a quality. It opens with the observation that every such line representative of a degree of intensity of a quality should properly be designated as the *longitudo* of that quality. The reasons are three. First, wherever there is succession, there too is *longitudo*. Second, the concept of qualitative change implies a continuous scale of degrees but does not necessarily entail the existence of an extended entity. Third, in mathematics the concept of *latitudo* implies *longitudo* but that of *longitudo* does not entail *latitudo*. But the conventional term is *latitudo* and is here so used.

Chapter three discusses the "longitude" of a quality. The extent of a quality, if any, ought to be called its "latitude," and represented graphically by a horizontal line, constructed on the subject of the quality. And on it should correspondingly be drawn the vertical line representative of the degree of intensity of the quality. In any case since both extension and intension have to be considered and since latitude and longitude are here correlatives, if intension be termed "latitude," then extension becomes "longitude" and conversely. But precision yields to usage and since the results are equivalent, let extension be called the "longitude" of the quality, and intension its "latitude" or "altitude."

Chapter four introduces the concept of the "quantity of qualities." A "point quality" is defined as "a quality which exists in an immaterial and unextended entity." Since such a quality lacks extension, its appropriate geometrical representation is a single line indicating the sole dimension of its intension. One cannot therefore speak of the "quantity" of a "point quality" but only of its degree. A "linear quality" is defined as "a quality whose extent in space or time is representable by a horizontal line of longitude." There are likewise a "surface quality" and a "corporeal quality."

The "quantity of a linear quality" is therefore representable in graphic form by the area of a plane figure the longitude or base line of which is the horizontal line which defines it as a linear quality, and the latitude or altitude of which is the line perpendicularly erected upon it and representative of the degree of intensity of the quality. If the altitude of the vertical lines of intension is different at different points along the base line but not by the same amount, then the line which joins their summits will be a broken line and the resultant plane figure not easily classifiable. If however the quality in question decreases in intensity at a constant rate from some given degree to zero or conversely, the appropriate plane figure representation will be a right angle triangle with constantly decreasing or increasing altitude. The value of this geometrical representation as a visual aid is clear. For the result is that one more easily understands the irregularity, the structure, the shape, the size and other properties of such a quality. Another consequence is that one can thus prove that the quantity of a uniformly difform quality is equal to its midpoint, or in other words the quantity of the quality is the same as it would be if the quality were uniform and at the degree of intensity of the midpoint.

There are therefore such qualities in the world as *qualitates triangulares, semicirculares, pyramidales,* etc. This point is important to Oresme. For the physical efficacy in action of various bodies in the world is a function of their shape. The same should therefore be true of the above geometrical shape of their qualities. Take two qualities the quantity of which is equal. Let one be pyramidal in shape. Let the other be completely uniform and in the shape of a rectangular solid. Then the former will be capable of more penetrative action than the latter. Or take two qualities, both pyramidal in shape, but differing in the relative length of the base and/or altitude. Then under comparable conditions the more sharply pyramidal quality will be more effective in activity.

This is also the more probable explanation of a curious phenomenon in experience. For of two qualities of the same kind and of equal intensity one is found to be more active or pungent than the other. Their respectively different *configurationes* may just be the answer to this puzzle. It is also the more likely solution to another puzzle, if what people say is really true. For some persons believe that certain figures and shapes, made at stated times and under certain astrological conditions, possess marvelous powers and virtues. But such powers or virtues more probably derive from the natural configuration of an active quality than from their induced geometrical form. This opinion, moreover, is philosophically more respectable. For philosophers agree that quantity is inert. What holds for the natural shape of active qualities is also true for passive qualities or susceptibilities of a body. The rate of speed with which a body in one state acquires the contrary quality is a function of the natural figure of the passive quality.

Oresme's thesis also transcends physics and enters into the world of biology. The nature of animals and plants determines for them a certain physical shape and form. The same factors also determine a certain shape and form for their connatural qualities. The first shape or figure of such qualities comes from the physical contours of the subject. The second configuration derives from the geometrical shape which their intensity chart takes on one of our graphs. In each case the form of the entity sees to it that the proper configuration is present. The parallelism therefore holds: Just as nature requires a different physical structure for lion and eagle, so too the intensivity graph in our system of the natural heat of lion and eagle or of eagle and falcon will show different geometrical features. This difference, furthermore, is not merely pictorial. It entails universal physical consequences. For the actual physical differences in the natural heat of a lion or an ass or a bull derive in large part from the geometrical differences in the intensivity charts of such heat qualities. The same analysis applies to all of their other qualities, as well as to those of all other natural bodies.

Such is a fair sample of both the text and the context of Oresme's work. In the perspective of the *von oben bis unten* research worker in the field of the history of science how much genuine mathematics does it even germinally contain? Analysis of the logical structure of Oresme's contributions reveals the three following decisive insights: (1) a basic misconception of the true nature of mathematical entities vitiates the entire enterprise; (2) it is in no sense an embryonic analytical geometry, and (3) it is at least questionable, despite the assured availability to Galileo of suggestive works in the authentic *latitudines formarum* tradition and his own acquaintance with the teachings of that tradition, whether the work of Oresme may fairly be interpreted as a logical anticipation—and not merely an historical antecedent—of the later work of Galileo with which it is often correlated.

The basic misconception which vitiates Oresme's mathematical enterprise is the assumption, no doubt derived from Aristotle's theory of the abstraction of mathematical entities from objects of concrete experience in the physical world, that there exists a relation of *material* identity between the relations and relata of a mathematical system and the relations and relata that obtain in the physical universe. Such abstraction theory accounts for the fact that the triangular shape of the intensivity graph of the *calor naturalis* of a lion, for example, is not hot to the touch of the geometer-physicist. But it requires and expects that in point of physical fact such *calor naturalis* is really *in rerum natura* literally a *qualitas triangularis*. The fundamental outlook of Oresme is that there is in fact only one world open to scientific investigation: the world of physical experience. But he also believes that *that* world may be explored in two epistemologically different ways: one concrete wherein both the qualitative and the quantitative properties of things are considered; the other abstract wherein one systematically ignores all but the quantitative determinations of things. The decisive cue to this outlook is given by Oresme in the preamble to his essay: "the first part deals with the geometrical shape and *the physical efficacy* of the uniformity and difformity of permanent qualities." But if, as the logical analysis of mathematics by philosophers of science seems to show, the world of mathematics is a consciously constructed universe of *ideal* entities, psychologically indeed but not logically grounded on physical experience, and only isomorphically relatable to the *real* world, then the work of Oresme is not mathematics at all, but rather an abstract and unsuccessful physics.

This is also perhaps the basic reason why Oresme's work with the *latitudines formarum* and the *configurationes intensionum* is in no acceptable sense an embryonic analytical geometry. For analytical geometry is in essential logical structure the conscious exploitation of a perceived isomorphism between the mathematical realms of algebra and geometry. But such isomorphism Oresme excludes in fact and denies on principle. Integral, furthermore, to the methodology of analytical geometry is a reference system of coördinate axes, related to a specific point of origin. A second inspection of Figure 6 will reveal that despite superficial resemblances to one quadrant of a coördinate system, the zero point for Oresme is at *B* and not at *A*. And at *B* there is no latitude or pseudo-ordinate at all. A third fundamental difference is that Oresme is exclusively concerned with *areas* while analytical geometry deals with lines or curves. Since these assumptions in Oresme are such that analytical geometry can be invented, not by clarifying their obscurities nor developing their insights nor improving their symbolism and operational techniques, but only by denying some and deserting them all, it is at least questionable whether Oresme is truly to be counted among those who took the first steps toward the creation, out of the statically conceived Greek mathematics, of the algebra and geometry that were to transform science in the seventeenth century.

Isomorphism: Beeckman and Descartes

The third major insight which the philosophy of science puts at the service of the historian of science is the central concept of isomorphism which illuminates the precise logical structure of the correlation which may happen to exist between a set of mathematical relations and relata and a group of physical relations and relata. Such an isomorphic correlation of purely formal identity requires a one-to-one correspondence between the respective relations in each area. The selection of this correspondence: Which mathematical entities and relations are to be set into correspondence with what physical entities and relations, is the vital decision which determines the degree of empirical validity with which a scientist may reason "mathematically" about matters of experiential fact. For mathematics, as it should be, is physically neutral. In order to grasp the relevance of this precise point to research in the history of science, it may prove useful to review briefly a notorious misunderstanding between Beeckman and Descartes.[36]

Beeckman requested Descartes to reply to the following question: "If one concedes that I am correct in assuming the principle that a body once set in motion will move forever in a vacuum, and if one further supposes the existence of such a vacuum between the earth and a falling stone, is it possible to determine the distance which a falling body will traverse in one hour, if one already knows the distance that it covers in two hours?"[37] It is important to grasp precisely the point of this inquiry. Beeckman is not asking Descartes why it is that bodies fall at all. He knows the answer to that question. Bodies fall because the earth "attracts" them in a Gilbertian or Keplerian sense. Nor does Beeckman ask Descartes why such descent is an accelerated motion. He knows the answer to that question also. Falling bodies are accelerated because at each new instant of fall there is a new measure of attraction so that, so long as the motion persists, each new measure of attraction confers a new degree of movement. The questions, therefore, to which Beeckman already knows the answers, are all questions of *physics*. The point therefore of Beeckman's actual inquiry is a *mathematical* one. From the purely physical principles which he already possesses Beeckman is unable to infer the entailed consequences. He is unable to hit upon the generalized and therefore logically fertile mathematical formula which will allow him to calculate the speed and the distance traversed by a descending body.

Descartes records the incident as follows:

A few days ago I had the good fortune to enjoy the company of a very talented gentleman. He asked me the following question: "A stone falls from point A to point B in one hour. The earth attracts it always with the same force. But it never loses at any $n+1$ instant the speed which it had acquired at the nth instant." It was his opinion that whatever is set in motion in a vacuum will continue to move forever. His precise question is as follows: "in what time interval will a given distance be traversed?"[38]

The two versions of the presumably identical question do not completely coincide but they are recognizably and sufficiently familiar and similar. Descartes, furthermore, records his reply:

I worked out the solution to the question proposed. In a right angle isosceles triangle the area $A\ B\ C$ represents the motion. The difference in area at chosen points from the point A to the base $B\ C$ represents the difference in the motion. The distance $A\ D$, therefore, is traversed in an interval of time which is represented by the area $A\ D\ E$. The distance $D\ B$, however, is covered in an interval of time represented by the area $D\ E\ B\ C$. Here it must be noted that the smaller area represents a slower motion. Now the area $A\ E\ D$ is a third part of the area $D\ E\ B\ C$. Therefore the falling body will cover the distance $A\ D$ three times more slowly than the distance $D\ B$. . . .[39] (See Figure 7.)

Figure 7 (Koyré)

The geometrical reasoning, as such, is flawless. But at this juncture it is the *physical* interpretation of the geometrical relata and relations that is of prime concern. In particular it is important to note that for Descartes the line *A B* is physically interpreted to represent the path or trajectory of the falling body.

Beeckman's version of Descartes' solution to the proposed question is as follows:

> When the intervening space is a vacuum, the way in which things move downwards toward the center of the earth is as follows: in the first instant the body covers as much space as possible in proportion to the force of the earth's attraction. In the second instant not only does the prior motion remain but a new motion by attraction is superimposed upon it. Hence the distance traversed in the second instant is double the distance covered in the first instant. In the third instant not only does the speed by which double the space was covered remain, but there is superimposed upon it by the attraction of the earth a third. Hence in a single instant the space traversed is triple the distance covered in the first instant.[40]

It is important for our purposes to keep both versions constantly in view. Descartes thought that he had said that "the falling body will cover the distance *A D* three times more slowly than the distance *D B*." Beeckman thought that what Descartes had said was that "in the third instant the space traversed is triple the distance covered in the first instant."

Beeckman now continues to reason mathematically on the issue as follows:

> Since, however, these instants are individual units of time, the space which the body will traverse in one hour is represented by the area *A D E*. The distance through which it will fall in two hours doubles the proportion of the time factor, i. e., *A D E* to *A C B* which is double the proportion of *A D* to

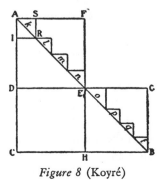

Figure 8 (Koyré)

A C. For let the value of the space through which the body falls in one hour be of some definite magnitude, say *A D E F.* Then in the interval of two hours it will cover three times that amount, namely *A F E G B H C D.* But the area *A F E D* is composed of the sum of the areas of *A D E* and *A F E,* whereas the area *A F E G B H C D* is comprised of the sum of the areas *A C B* and *A F E* and *E G B,* i. e., with the double of *A F E* [Figure 8].

Hence if the value of the space for any given instant be *A I R S,* there will obtain a proportion of space to space, such as *A D E* together with *k l m n* to *A C B* with *k l m n o p q t.* But *k l m n o p q t* is double *k l m n.* And yet *k l m n* is much smaller than *A F E.* Since therefore the proportion of the space traversed in one time interval to the distance covered in another time interval consists here of the proportion of one triangle to another, to the sides of which equal increments of area have been added, and since these equal added increments of area decrease exactly as the time intervals of fall are smaller, it follows that these added areas will have no assignable size when the interval of time of fall has no specifiable length. But that is precisely what is meant by "an instant of fall" in the descent of the body through a space. The conclusion therefore stands: the space through which a body falls in one hour is to the space through which it falls in two hours as the area of the triangle *A D E is* to the area of the triangle *A C B.*[41]

The solution is mathematically deft and empirically relevant. For the proof discloses that distances traversed are proportional to the square of lapsed time intervals. But what is important for present purposes is the observation that for Beeckman the line *A C* is physically interpreted to mean the lapsed time interval of total fall.

For Beeckman, therefore, since the line *A C* represents the lapsed time intervals, the triangles *A D E* and *A B C* represent the distances traversed in the time intervals *A D* and *A C* respectively. For Descartes, however, since the corresponding line *A B* represents the trajectory of the falling body, the area of the triangle *A D E* and the area of the trapezium *D E B C* represent the amount of "motion" for the corresponding parts of the trajectory, *A D* and *D B.* But the same areas are therefore also describable as representing the "sum of the velocities" for *A D* and *D B.* In the logic of this interpretation Descartes correctly concludes that since the "sum of the velocities" in *D E B C* is triple the "sum of the velocities" in *A D E,* then the distance *D B* on the trajectory *A D B* will be traversed three times more rapidly than the distance *A D.*

Because of this basic difference in the physical interpretation of what Descartes labels *A B* and Beeckman names *A C*—a mathematical entity which Beeckman isomorphically sets in correspondence with time elapsed whereas Descartes isomorphically correlates it to the path of trajectory—the same type of mathematical reasoning leads to two empirically different conclusions. Beeckman's conclusion is to the effect that the uniformly accelerated motion is a motion the speed of which is proportional to lapsed time intervals. Descartes' conclusion is to the effect that the uniformly accelerated motion is a motion the speed of which is proportional to distances traversed. *And there is no mathematical way to decide the issue between them.* For the mathematical relations which exist between the lines $A D$ and $B D$ and the respective areas $A D E$ and $D E C B$ are no less valid than the mathematical relations which exist between the lines $A D$ and $A C$ and the respective areas $A D E$ and $A B C$. What is decisive here, as always in mathematicized science, is the selection of the isomorphic correspondence in hand: which mathematical entities and relations are to be set into correspondence with what physical entities and relations. Mathematics does not determine the choice and is forever indifferent to the empirical validity of one which is made. Nature does not prescribe the choice but can and does report on the empirical validity of one which is selected for experimental test. The choice of mathematical and physical isomorphic correspondence depends uniquely upon the initiative, insight, or inventive genius of the mind of man. For man stands midway between two worlds: the ordered world of the universe which the Creator made for him and the ordered world of mathematics which man has constructed for himself. The more patterns of order he can construct in the world of mathematics, the more patterns of order he will be equipped to perceive, to understand, and to control in the world of physical experience. The future history of science thus depends centrally upon the intelligence of man. As did also the past history of science. For the history of science, when explored with the analytical tools of the philosopher of science—concerning hypothetico-deductive theory construction, mathematics, and isomorphism—is nothing else but a partial, indeed, but significant phenomenology of the mind of man, questioningly confronting the world of his experience. And that is perhaps why the humanizing value of the discipline of the history of science is so surpassingly high.

References

1 Some recent samples are: Carl B. Boyer, "Analytical Geometry in the Alexandrian Age," *Scripta Mathematica,* XX (1954), 30–36, 143–54; Paul H. Kocher,

Science and Religion in Elizabethan England (Marino, Calif., 1953); Norman Davy, *British Scientific Literature in the Seventeenth Century* (London, 1953); Shelby T. McCloy, *French Inventions of the Eighteenth Century* (Lexington, Kentucky, 1952); J. D. Bernal, *Science and Industry in the Nineteenth Century* (London, 1953); George A. Foote, "Science and its Function in Early Nineteenth-Century England," *Osiris,* XI (1954), 438–54. An especially interesting example in which the subtitle of the book is far more illustrative of its *analytical* contents than its title is René Dugas, *La Mécanique au XVII° siècle: des antécédents scolastiques à la pensée classique* (Paris, 1954).

2 For instances of this serious defect in historical research see Pierre Duhem, *Etudes sur Léonard de Vinci* (3 vols.; Paris, 1906, 1909, 1913). And compare the sober judgment of Anneliese Maier, *Die Vorläufer Galileis im 14. Jahrhundert* (Roma, 1949), p. 1: "Grundsätzlich hat Duhem sicher recht, wenn er in der Naturauffassung des 14. Jahrhunderts eine Vorstufe und Vorbereitung der klassischen Physik sehen will; nur hat er im einzelnen die scholastischen Lehren oft in zu modernem Sinn interpretiert und zu viel aus ihnen herausgelesen." The fever has not yet abated. See Jean Rostand, "Un précurseur inconnu de Mendel: le pharmacien Coladon," *Revue d' Histoire des Sciences,* VIII (1955), 171–73.

3 Three volumes; published in Baltimore (1942–1947).

4 All the thought processes of science in the West, for example, have heretofore been conducted within the framework of a two and only two-valued logic. So fundamental a change in scientific thinking as a shift to a three-valued or *n*-valued logic would nevertheless fit naturally into the logical structure of scientific thought, as analyzed by the philosophy of science. On this point see Rudolf Carnap, *Foundations of Logic and Mathematics* (Chicago, 1946), p. 38.

5 As was first done for the logic of the propositional calculus by E. L. Post, "Introduction to a General Theory of Elementary Propositions," *American Journal of Mathematics,* XLIII (1921), 163–85, and the method revised and perfected later by W. V. Quine, "Completeness of the Propositional Calculus," *Journal of Symbolic Logic,* III (1938), 37–40.

6 In the spirit of the titles of the two following books on the history of science: Marjorie H. Nicolson, *The Breaking of the Circle* (Evanston, Illinois, 1950), and Alexander Koyré, *From the Closed World to the Infinite Universe* (Baltimore, 1957).

7 James B. Conant, *Science and Common Sense* (New Haven, 1951), p. 47.

8 Anatol Rapoport and H. G. Landau, "Mathematical Biology," *Science,* CXIV (1951), 3.

9 F. S. C. Northrop, "The Importance of Deductively Formulated Theory in Ethics and Social and Legal Science," in *Structure, Method and Meaning,* ed. E. J. Henle *et al.* (New York, 1951), pp. 99–114, p. 106.

10 Carl G. Hempel, "The Concept of Cognitive Significance: A Reconsideration," *Contributions to the Analysis and Synthesis of Knowledge,* Proc. of the Amer. Acad. of Arts and Sciences, No. 80–81 (Boston, 1951), pp. 61–77, p. 73.

11 Compare *ibid.,* p. 74: "Significant systems range from those whose entire extra-

logical vocabulary consists of observation terms, through theories whose formulation relies heavily on theoretical constructs, on to systems with hardly any bearing on potential empirical findings."

12 Carl G. Hempel, *Fundamentals of Concept Formation in Empirical Science* (Chicago, 1952), p. 36.

13 Roy Dubisch, *The Nature of Number* (New York, 1952), pp. v–vi.

14 G. Newton Stokes, *Teaching the Meanings of Arithmetic* (New York, 1951), p. 138.

15 For an explanation of these technicalities see Neal H. McCoy, *Rings and Ideals* (Philadelphia: The Mathematical Association of America, 1948).

16 For a history of the transition from elementary to projective geometry with just the right analytical emphases at just the right places see Ernest Nagel, "The Formation of Modern Conceptions of Formal Logic in the Development of Geometry," *Osiris,* VII (1939) 142–224.

17 Nathan Jacobson, *Lectures in Abstract Algebra* (New York, 1951), I, 2.

18 Cassius Jackson Keyser, "Three Great Synonyms: Relation, Transformation, Function," *Mathematics as a Culture Clue and Other Essays* (New York, 1947), p. 213–34, p. 226.

19 José Alvarez Laso, *La Filosofía de las Mathemáticas en Santo Tomás* (Mexico, 1952), p. 117.

20 I follow here the translation provided by Morris R. Cohen and I. E. Drabkin, *A Source Book in Greek Science* (New York, 1948), p. 122.

21 *Ibid.,* p. 125.

22 *Ibid.,* p. 115.

23 For a more complete and a more detailed discussion of these and other technical points of astronomical theories see Edward Rosen, *Three Copernican Treatises* (New York, 1939), especially pp. 34–53.

24 *Ibid.,* pp. 57–58.

25 *Ibid.,* p. 135.

26 *Ibid.,* p. 137.

27 *Ibid.,* p. 39.

28 *Ibid.*

29 I employ here the translation of the Preface and of Book One by John F. Dobson and Selig Brodetsky, published as "Occasional Notes" by the Royal Astronomical Society, No. 10 for May, 1947, and reproduced by I. Bernard Cohen in a preliminary edition of his *The Nature and Growth of the Physical Sciences* (New York, 1954), pp. 85–106, p. 90.

30 Dobson and Brodetsky in Cohen, *Nature and Growth,* p. 104.

31 *Ibid.,* p. 90.

32 Aristotle, *Physics,* Bk. IV, Ch. 4 (212ª 20–21).

33 A. C. Crombie, *Augustine to Galileo: The History of Science: A. D. 400–1650* (Cambridge, Mass., 1953), pp. 260–61. See also the same author's *Robert Grosseteste and the Origins of Experimental Science* (Oxford, 1953), pp. 1, 9, 178.

34 Crombie, *Augustine to Galileo,* pp. 261–62.

35 The relevant parts of the Oresme manuscripts in the Vatican Library have been

edited and published by Anneliese Maier, "Nicolaus von Oresme's Lehre von den *Configurationes Intensionum*," in *Zwei Grundprobleme der scholastischen Naturphilosophie* (Roma, 1951), pp. 89–109. I have found it too bulky to append to my translations or paraphrases the key portions of the original text upon which each is based.

36 The latest discussion of this incident and one of the best is that of Alexander Koyré, *Etudes Galiléennes* (Paris, 1939), II, 25–40.

37 *Oeuvres de Descartes,* ed. Chas. Adam and P. Tannery (Paris, 1897–1913), X, 60.

38 *Ibid.,* p. 219.

39 *Ibid.*

40 *Ibid.,* p. 58.

41 *Ibid.,* pp. 58–61.

I. E. DRABKIN

on the Papers of A. C. Crombie and Joseph T. Clark

I am no expert in medieval science or in the philosophy of science and I cannot deal as an authority with the papers of such experts as Father Clark and Professor Crombie. But I should like to state my impressions as an amateur and to ask some questions in the hope of provoking discussion that may clarify the issues. Let me say at the outset that both papers were most welcome and informative since they deal with such broad and basic matters. The papers actually intersect at many points, not merely in such details as whether Oresme influenced Galileo or not, but in fundamentals. The problem of historical continuity and discontinuity and the real meaning of precursorship and influence are central to both of them. It is all very well for some to say that such problems are verbal: The fact remains that most of our disagreements are on just such questions. Galileo as a student spends years absorbing certain ideas; later he comes to see their inadequacy for his purpose and he finds new ideas, new methods, new ways of looking at things. Some will say: "The first set of ideas had decisive influence, for he would not have thought about the problems at all unless he had studied these ideas, even if he did ultimately reject them." Others will say: "The first set of ideas was without significant influence; for they do not appear in his own formulations." Of course, a penetrating analysis of the case of each scientist would define the issue and elicit the facts more realistically, but I venture to say that, even on an agreed statement of facts, two competent historians of science might paint wholly different pictures because they weigh the facts differently. Is it a matter of pro-medieval or anti-medieval bias, or some other such conditioning? Now Father Clark proposes a program for what he holds to be an objective analysis of a scientist's achievements and shortcomings, so that the relation of one scientist to another may be better defined. But let us leave that discussion for later, and take up Professor Crombie's paper first.

The title of Professor Crombie's paper is "The Significance of Medieval Discussions of Scientific Method for the Scientific Revolution"; that is, we are primarily concerned here not with substantive science but with discussions of scientific method. Now, of course, Professor Crombie points out that method does not operate *in vacuo* and certainly in the actual literary formulation the two things are intimately linked. But it may help to clarify our discussion if we see that they are not inseparable, that actual scientific activity is a different activity from formulating scientific method, and that the influences of these two activities on the course of history may be wholly different. This point will be discussed later.

Professor Crombie has summarized the material he so ably treated with such rich detail in his recent books—the formulations by men like Grosseteste, Roger Bacon, Ockham, Oresme, and the others, on methods of studying nature, on the concept of explanation and cause, on framing hypotheses and testing them empirically, on the application of mathematics to problems of nature, and so on. What was the significance of these discussions for science in the sixteenth and seventeenth centuries? Here Professor Crombie is very cautious—no one is more fully aware of the complexity of the influences at work than he is; he stresses, for example, the changed approach to scientific problems—e.g. (in the case of Galileo), the emphasis on the *measurement* of *actual* motions, rather than speculation about methodological possibilities; he recognizes that there is a difference in general outlook between practical scientists or technologists, on the one hand, and men who are interested in scientific method as logical and philosophical studies. He stresses the paucity of actual scientific achievement in the universities from the thirteenth to the sixteenth centuries, and, indeed, he sees in the sixteenth century something of a divorce between science and the discussion of scientific method, to be followed in the seventeenth century by a restoration of such contact, with methodology now enriched by the fruits of a century of brilliant scientific activity—so that seventeenth-century mechanics becomes the model of scientific methodology. What then, in Professor Crombie's view, is the influence of the medieval predecessors, from thirteenth-century Oxford and Paris to sixteenth-century Padua? Surely, as Professor Crombie says, many of the earlier ideas and aspirations were still considered valid— the distinction between description, particularly mathematical description, and explanation by physical or metaphysical reasoning, the emphasis on empirical verification or falsification of hypotheses, principles of simplicity and economy, and the distinction between primary and secondary qualities. To the extent that such ideas persist, one can readily agree that there was influence, even though the center of interest shifts from logic and philosophy to what we now call science.

We are not concerned with the *originality* of the medieval formulations of scientific method (that is, how much they went beyond a mere reworking of Greek thought; that in itself is a large question). Nor are we concerned with the *validity* of these formulations in the light, let us say, of modern logical theory. Neither of these considerations bears on the question of later influence.

One question we must face (i.e., in assessing significance for the scientific revolution) is the relation of the working scientist to explicit formulations of scientific method. Professor Crombie at several points remarks on the necessity that the scientist be aware of what he is doing. But does that mean awareness of explicit formulations of the kind that Professor Crombie is concerned with? How far is the working scientist actually conscious of such formulations? Is an expert knowledge of such formulations (and a diligent attempt to apply them) any guarantee of scientific discovery? And for one unacquainted with a field of inquiry, not knowing where to start an investigation, are the canons of induction of any avail? Is even the possession and use of the practical scientific techniques of mathematics, experimental design and procedure, and all the rest, any guaranty of significant discovery? My impression is that science often progresses independently of discussions of methodology. I have found in my limited contact with young creative scientists, especially in mathematical physics, no particular interest in philosophic discussions of methodology (or in history, for that matter). This sort of discussion is more congenial to mature scientists, or those whose peak of creativity is past. (There are obviously exceptions; I state what I think is the normal situation.) That is, in assessing the significance of the medieval formulations, must we not remember that an interest in such formulations might more often be a result of scientific activity than a stimulus to it?

What produces scientific discovery seems to be familiarity with the phenomena in a field, a talent for imagining possible correlations, and for deciding on certain possibilities for investigation, a talent for analyzing the possible components of a complex phenomenon, for inventing fruitful analogies in the form of mechanical models, or relations of a mathematical sort—such talents, rather than familiarity with explicit formulations of inductive or deductive procedures, seem to me decisive.

Are we then to say that the discussions of men like Grosseteste, Roger Bacon, and the others are of little important significance for the future development of science, and are merely to be admired for their own sake as achievements of the human intellect, but operating, so to speak, *in vacuo?* I would not take such a position. Inextricably woven into the fabric of these medieval discussions is a concern with actual scientific problems. In fact, I am inclined to think that this interest in the substantive problems of science was more significant for the future development than were purely methodological concerns. There were problems in optics, astronomy, statics, and, especially the unsolved problems of motion and dynamics. Admittedly, no very fruitful contributions to their solution were forthcoming, but the mere fact that these problems were under such constant discussion in the universities (and later in nonacademic circles, too), that they were in the air, so to speak, and formed the environment in which the genius of men like Copernicus and Galileo was nourished —surely this cannot be discounted. And when we remember that the discussions were generally carried on in the spirit of reason and science, that authority was constantly being challenged, that the scientific questions were often dealt with as separate from theology and metaphysics, that kinetic or mechanistic, rather than teleologic explanations were sought, we can see that this is not a movement of negligible significance.

How much significance is another question. Here I suppose the career of each scientist must be studied individually to separate the strands of influence—Harvey's case would not be the same as Galileo's, and later on, Lavoisier's would be different from both of these. If we consider Galileo's position as central in the scientific revolution—and in a sense any other view misses the point—we have to consider various factors that went into the preparation of an environment ideally suited to exploit so fully the genius of the man—to begin with, the dissemination of printed Latin editions of the works of the Greek scientists, Euclid, Archimedes, Apollonius, Ptolemy, Pappus, Hero, and Aristarchus. (The influence of this tradition on Galileo could be documented by many far more adequately than by me—I am inclined to believe that Galileo felt a closer spiritual kinship with Archimedes than with any of his own medieval or Renaissance predecessors.)

Previous papers have developed at some length the influence of the technological and industrial environment, which was the source of many of the problems that engaged Galileo's mind, an environment that gave enlarged scope to an idea that had existed long before his time (and was perhaps stronger in other minds than his own)—namely that nature was not only to be contemplated but controlled. Since there was rather full discussion of sociological influences in this connection, I shall not pursue this line of thought, except to ask to what extent Galileo or his contemporaries were influenced directly or indirectly by the new printed editions of the Greek writers on mechanics, pneumatics, and other branches of applied science, e.g., by Hero's *Pneumatics,* by Vitruvius, Frontinus, and Pappus. Here is a tradition which originated in antiquity from the collaboration of skilled craftsmen and theoretically trained mathematicians. Possibly the author of the Aristotelian *Mechanica* belongs to it. Certainly Archimedes is linked to it, despite his unwillingness to write about mechanical matters. In fact, in its craft aspects, it must have remained largely unwritten, though what we do have about the theory of simple machines and descriptions of more elaborate machinery, pumps, and pneumatic devices is extremely interesting and not wholly for purposes of amusement, as the books often say. The stories of Archimedes at the siege of Syracuse, and the experimental work on ballistics and the design of the projecting devices (in Hero and Philo) come to mind. And, if I may digress a bit, Mr. Derek Price has called our attention, on the archaeological side, to a remarkable piece of geared machinery, dating from the beginning of the Christian Era, which he thinks may have actually operated a planetarium of the kind that Archimedes built. Theory plays a great part in this tradition, and there are links, as you recall, with atomism.

I mention all this because it is a part of the Greek scientific tradition that must, at least in its literary form, be considered among the lines of influence. And we should also have to assess the influences of the ancient philosophic traditions, especially Platonism, and the Neoplatonic movement. This is something of a controversial subject—and I'm hardly the man to deal with it, with so many Galileo specialists on hand. But the notion of finding true reality in ideal entities which are not given in experience, the idea that the mathematical relations of physics reveal the essence of nature better than do the most minute observations, or the most acute sensory perceptions—this seems to me to be very characteristic of

Galileo's thought, whether you call it Platonism or Pythagoreanism or by any other name. The Aristotelian logic of existential classes is uncongenial to this sort of approach, and Aristotle's own attempts at equations of dynamics were notably unfruitful (however successful his logic might have been for biological and other classifications). The medieval discussions that Professor Crombie deals with do themselves often reflect a feeling that the mathematical approach goes to the essence. But the actual beauty and power of this approach that could be seen in the works of Archimedes may have had more immediate response in Galileo than all the talk about it by the philosophers.

In any case, I should certainly not agree with Whitehead who says that "It is a great mistake to conceive this historical revolt [i.e., the scientific revolution as reflected in the work of Galileo] as an appeal to reason. On the contrary it was through and through an anti-intellectualistic movement. It was the return to the contemplation of brute fact; and it was based on a recoil from the inflexible rationality of medieval thought." [*Science and the Modern World,* p. 9] It was, in my opinion, not mere observation, but observation disciplined by reason and enlightened by imagination, and by this sense of the primacy of the mathematical relations, that enabled Galileo to find the key to his problem by fixing his mind on an ideal case never observed in nature. Professor Crombie says that "the mechanical causes remained his [i.e. Galileo's] ultimate objective in science," but I don't think that that really alters the point I am making.

And we must not neglect, in this connection, the other features of the anti-Aristotelian movement—the undermining of the dichotomy between celestial and terrestrial substance and the rejection of the hierarchic distinctions among the forms of terrestrial substance.

Besides all these factors, we certainly cannot forget (and, indeed, Professor Crombie would emphasize) the centuries of scientific and philosophic discussion of nature carried on in the universities, discussions that converged on the problem of motion, and Galileo's study of that tradition, his awareness of the approaches that his predecessors had made to these problems and, after his initial acceptance of one of these approaches, his gradual realization of the inadequacy of them all. All this we must keep in mind, unless we are willing to look upon Galileo's work as the sudden inspiration of a genius analyzing a complex phenomenon and arriving at results from which others would soon explicitly frame the law of inertia. Not that I think Galileo's work is the end product of the scholastic development—it is rather, I believe, the beginning of a new development. In the words of Koyré, we are dealing with "a well-prepared revolution, but nevertheless a revolution."

Each of us will assess differently the elements that went into that preparation. And, of course, one who is concerned with nonmechanical and nonmathematical branches of science, e.g., biology and medicine, will stress other factors. My own emphasis on the dissemination of Greek science in the sixteenth century may merely reflect the parochialism of a classicist. In my opinion, it was not because of any inherent methodological defect, not because of a lack of understanding of experimental method, not because of an exclusive concern with the static and geometric, not because of any inability to distinguish science from philosophy, that Greek science reached a dead end. Greek science in antiquity was by and large a poorly organized and precarious enterprise in which relatively few participated, in a society which, except for a few brief periods, was generally indifferent to their work, and which provided few lines of communication and coöperation among them. But after a thousand years or more of arrested development, the achievements of Greek science, thanks to the printed editions, became widely disseminated over an area that enjoyed a flourishing university tradition in which thousands participated, at a time when social and political conditions were favorable to intellectual activity, and the work of the scientist could be highly valued. Under such circumstances the importance of the contribution of Greek science to the new scientific movement can be readily understood. In this development the university tradition that both Professor Hall and Professor Crombie emphasize was a very necessary link; perhaps we are pursuing a will o' the wisp in trying to tell which was the strongest link —the important thing is that the chain held together.

Let us now turn to Father Clark's paper, which certainly goes to the essence of our subject—how should the historian of science do his job? As I understand it, Father Clark would have us pay more attention to the type of study which takes current conceptions of the theoretical structure of science as a frame of reference and views the history of science against that frame. This method he contrasts with the type of study which considers a certain period or certain scientific personalities without benefit of that frame of reference.

I state the contrast in this explicit way because his terminology "von unten bis oben" and "von oben bis unten" may be misleading. One is at first tempted to translate these terms in some such way as "prospective" and "restrospective" but it is obvious that this is not the contrast Father Clark has in mind. Anyone who writes about a period or about an individual scientist has not only the preceding but also the subsequent historical development in mind. He chooses to study that period or person because the subject has a certain significance for the future development; and he treats his subject in such a way that this special significance is stressed. Surely he must view his subject both prospectively and retrospectively. One who writes about the forerunners of Galileo starts from a conception of Galileo's achievement and tries to show how earlier views differed from, foreshadowed, or actually influenced Galileo. So far as the historian's thought processes are concerned, the arrow points both forward and backward. But this is not what Father Clark has in mind; he would not call such a study truly "von oben bis unten" unless the ideas under examination were criticized from the viewpoint of current conceptions of sound physical theory—clarity, precision, and explicitness of the concepts and assumptions, predictive and explanatory power of the hypotheses, their susceptibility of mathematical formulation, the economy and simplicity of the means they employ, the degree of verification they receive by experiment, and the design and techniques of such experiments.

Father Clark has given us some illuminating examples of what he has in mind. There are, for example, the large subjects that cut across chronological periods; such subjects as "a history of controlled experimentation" or "a history of the application of statistical methods in the social sciences," might be treated in this way. But the distinction is not a matter of the type of subject—it is in the method of treatment. For example, a history of the mathematical formulation of physical laws might contain a chapter on Archimedes, with a critical analysis of his scientific achievement on the basis of Father Clark's criteria. But the value of a book on the life of Archimedes would be very much enhanced if it too contained that same material. My point is that the important distinction is not between writing on subjects within a limited chronological framework and writing on subjects that cut across conventional chronological divisions. Any study, if it is good history of science must deal analytically and critically with its subject, but it must do so without losing sight of the historical situation. The two methods that Father Clark discusses are, as he says, complementary, but, while he seems to indicate that they are essentially independent of each other, I believe that good historiography in our field requires that they be tied together. This is another way of saying, what has often been said, that the historian of science must have some competence in both history and science.

To go back to our example of a study of the mathematical formulation of physical laws, the sort of subject Father Clark has in mind, one chapter might deal with Aristotle's laws of motion—but to deal with these laws apart from the context of Aristotelian cosmology would be very incomplete and unsatisfactory. My point is that in the analytical approach there is the risk of giving insufficient consideration to the historical situation in which the particular idea under consideration was born, and the meaning of the idea in that context. Essentially it is the same kind of risk as attends what Father Clark would call the conventional type of treatment, with its tendency to speak glibly of precursorship. The error in each case is in historical judgment.

This is very much like the error into which Aristotle himself fell when he wrote the history of philosophy as if all previous discussions could be viewed as faltering and imperfect steps toward his own doctrine. How completely he misrepresented the various presocratic doctrines is demonstrated by modern studies in that field.

I may say, in passing, that when a scientist writes history without the requisite historical knowledge, his work nevertheless may contain important insights on what, as a scientist, he considers important. We can learn much from writers like Mach, Jeans, and Osler, despite their inadequacy on the historical side.

On the other hand, to the extent that Sarton's *Introduction* does not sufficiently deal with the logical and systematic relation of ideas from age to age, it cannot be considered (and was surely not intended to be) a fullscale History of Science, but rather an Introduction, a repertory of materials necessary to the writing of such a history.

Father Clark seems to be exclusively concerned with current conceptions as criteria with which to judge the work of the past. To be sure, the validity of an idea can be considered quite apart from its history. But there is the same risk of missing the meaning of the idea by reading it in terms of modern notions and categories rather than in terms of its historical context and its meaning in that context. With due care, proper attention can be given both to the meaning of an idea in its historical setting and to its relation to ideas that came after it. But is Father Clark's mid-twentieth-century frame of reference demonstrably perfect? Will it never be superseded? What was sufficient proof for Thales was not sufficient for Euclid; what was sufficient for Euclid is no longer sufficient. Have we reached the limit? Surely Father Clark will agree that some fine analytic work of the kind he has in mind was done before the rules he relies on were formulated—work by men like Meyerson, Cassirer, and Duhem, though perhaps one might wish to substitute other names.

Father Clark is particularly dissatisfied (and rightly so) with uncritical discussions of questions of precursorship. He would subject the views of a scientist to critical analysis on the basis of our current understanding of scientific method, and would assess his work accordingly. No one could criticize such a program, if, as I have indicated, the historical realities are kept in mind, so that the meaning of the work is not misread. This kind of analysis has been practiced perhaps more widely than Father Clark leads us to believe; certainly many of us seek to practice it. He is undoubtedly right in emphasizing its importance, for it enables us to see more clearly the precise relation of scientist A's ideas to those of scientist B. Father Clark holds that many assertions of precursorship will be proved false by such analysis. Here I think he may be mistaken. Precursorship—the influence of one individual on another—is not a well-defined term. It is not proved or disproved by the logical analysis of two ideas. The question is not only logical, but psychological and historical. Similarity or identity of ideas does not prove influence; nor does the fact that someone read a book prove that the book influenced the reader. On the other hand, A may be confused and bungling, but may have the germ of an idea that B clarifies and renders fruitful and sets in its proper frame. Father Clark's analysis showing wherein A fell short and B succeeded is very valuable; it is the essence of our subject, which seeks not only to chronicle but to relate ideas. But this analysis does not and cannot disprove A's influence on B, and his contribution to the resulting success. It can, however, make the precise meaning of that contribution clearer and more meaningful.

Let me take as an example Father Clark's discussion of the relation between Galileo's graphical representation of the so-called Mertonian rule and Oresme. I believe that Oresme's idea, crude though it was, may have given a fruitful suggestion to abler intellects who succeeded him; certainly in the hands of Galileo the idea was precise and could be used as a powerful tool. With all its faults, the original notion of Oresme was undeniably an attempt to deal with nature quantitatively, to exhibit information graphically, to consider the intension-extension relationship as a two-dimensional system, and to treat the area bounded by the graph as a sum. The attempt may have been marred by errors and misconceptions, which Father Clark does well to point out; but I think it would be wrong to deny, on the basis of this analysis, any relation, direct or indirect, between Oresme's work and that of Galileo or Descartes. On this question I would agree with Professor Crombie.

Father Clark takes up Copernicus' failure to dispense with the sphere of fixed stars, in view of their non-participation in circular motion; he sees here a case in which the scientist failed to analyze all his assumptions (since the assumption of a sphere presumably implied circular motion). Was Copernicus' decision connected with his unwillingness to assert that the universe was actually infinite rather than immeasurable? Was Copernicus just not psychologically prepared to think in terms of empty space, and did he therefore adhere to the Aristotelian notion that everything must be in its proper place, and that even if they did not move circularly the fixed stars, to keep their position relative to each other, must be fixed on a sphere, even an immeasurably thick sphere, whether physical or conceptual? After all, the idea of an imaginary sphere of fixed stars is still current in elementary discussions. And how, in Copernicus' thought, would stars "afloat in unbounded space," to use Father Clark's phrase, keep their position relative to each other?

The importance of psychological considerations leads me also to refer to Father Clark's strictures on the teaching of mathematics. The subject is not really relevant to his main argument, but I wonder whether he does not throughout tend to neglect psychological as contrasted with logical considerations. Even if projective geometry is logically anterior to traditional Euclidean metric geometry, that would not be an argument for teaching it first. The theory of sets may logically precede the solution of linear and quadratic equations, but the question of order of teaching is a pedagogical and psychological question, not exclusively a question of logic. Pedagogical order should not in general correspond to the logical order of a finished deductive system. (So also the order of the discovery of propositions in mathematics is not, in general, the same as the order of those propositions in a finished deductive system.) The point I make is that even if Father Clark is right on the question of the proper order of teaching, the supporting argument would have to be psychological and pedagogical, not merely a showing of logical structure. When I say this, I do not wish to be understood as saying that I think all is well with the teaching of mathematics.

In one brief statement Father Clark said that such work as Theodoric's (and that would apply to Oresme's too) might, for all its methodological error, contain some useful suggestion.* But is not that precisely the point? That suggestion kindles an idea in the mind of a subsequent reader, and are we not back to the matter of precursorship? At any rate, as I see it, there are two diseases, not one—*precursitis* (to use Father Clark's term), the tendency to see continuity where none exists, and what we may call *vacuitis,* the failure to see continuity where it *does* exist. I'm sure that Professor Crombie and Father Clark will agree that both diseases should be avoided, but I'm afraid that in any specific case they might not agree as to who is sick.

* This statement was deleted from the final draft of Father Clark's paper.

ERNEST NAGEL

on the Papers of A. C. Crombie and Joseph T. Clark

It is only by extreme courtesy that I can be counted as a historian of science, and I am acutely conscious of my limited qualifications to evaluate the historical content of the papers by Dr. Crombie and Father Clark. Both papers deal with aspects of medieval intellectual development of which I have little first-hand knowledge. However, Dr. Crombie himself observes that a good part of the current debate over the significance of medieval thought for the origins of modern science turns less on issues of historical fact than on questions of philosophical interpretation. I am, therefore, perhaps not completely miscast as a discussant of the papers.

My comments on Dr. Crombie's interesting contribution are quite brief, and consist of but one question. I raise it in the spirit of Father Clark's brilliant attempt to introduce a critical note into the discussion of the contributions of medieval thought to the rise of modern science. Dr. Crombie's paper is a valuable summary of the considerations that have persuaded him that the thirteenth and fourteenth centuries witnessed the formulations of a conception of science which is both experimental and deductive and which is similar in several important respects to conceptions of science found in the seventeenth century. My hesitations in accepting this thesis arise from the circumstance that while Dr. Crombie cites some of the apparently confirming evidence for it, he also notes that the men he is considering made little contributions to the permanent body of scientific knowledge, and that their actual researches are absurdly inadequate to the notions of scientific method they ostensibly entertained.

Now I find this discrepancy between alleged conception and actual performance puzzling. For one would expect that if fourteenth-century thinkers really did hold ideas on scientific method similar to those of the seventeenth century, their actual scientific achievements would testify rather consistently to their grasp of such ideas. The contrary appears to be the case, and Dr. Crombie's only explanation for this is that "something seems to have kept the gaze [of the medieval thinkers] turned in a direction different from that which seventeenth-century science has taught us to seem obvious."* Now, as has been often pointed out, there is frequently a considerable gap between what a man does and what he says he does. However, if there is a persistent incongruity between a man's actual performance and his account of what he does, it is not unreasonable to ask whether the individual does have a clear grasp of what it is he is saying, or alternately, whether the man's words mean what we suppose them to mean.

The question I wish to put to Dr. Crombie is, therefore, this: What would be the effect upon his thesis if the texts he cites in support of it were construed in terms of the way his thirteenth- and fourteenth-century thinkers actually tackled substantive scientific problems? In particular, would those texts reveal a conception of science similar to that of the seventeenth century, if the sense of their language were understood in the light of the practice of medieval scientific inquiry, rather than in the light of the practice of seventeenth-century science? More generally, the issue I am raising is whether it is sound procedure to interpret methodological commentaries out of the context of particular substantive problems to which those commentaries are addressed. My query therefore is whether the interpretation Dr. Crombie places on the writings of the medieval schoolmen really is the sense which those men attached to their discussions, and in any case whether the medievals clearly understood what it is they were presumably saying. I am of course not prepared to argue that Dr. Crombie's own reading of the texts is mistaken. But it seems to me that until the question I am raising is settled, the material Dr. Crombie cites in support of his major thesis does not constitute admissible evidence.

I have already expressed my whole-hearted sympathy with the critical tenor of Father Clark's paper. I must also state my admiration for the effective way in which he has employed logical analysis upon historical materials. In my opinion, his essay is in many ways a model for the practice of the analytico-historical method. However, I cannot escape the task, expected of a formal discussant, of indicating at what points in his discussion I find myself in difficulties.

Father Clark's account of two methods for doing the history of science seems to me both intriguing and important, though I am not sure that I fully grasp the import of the distinction he introduces. In one sense, his distinction between the methods merely baptizes the difference between

* As it was framed in the first draft of Dr. Crombie's paper.

approaching the history of science with ostensibly no presuppositions (and in consequence with no clear ideas as to what are the characterizing features of modern science), and approaching that history with relatively adequate notions concerning the method and the substantive content of scientific inquiry. Since in my judgment it is quite impossible to engage in significant historical study with no standards of evaluation and no doctrinal commitments, Father Clark's distinction between his two methods perhaps boils down simply to the difference between historical inquiry which is controlled by explicit principles of interpretation, and historical inquiry which is at the mercy of unavowed assumptions. Everyone will agree with him that the former approach is the preferable one.

However, Father Clark undoubtedly intends something more by his distinction than this obvious one, though it is not certain just what more is intended. He describes the second method as one which is pursued by employing "the creditable results of a rigorously analytical philosophy of science" of the twentieth century; and he enumerates three such "valid insights of contemporary philosophy of science." I have three comments on this part of his discussion.

First, it is not evident why anyone practicing Father Clark's *von oben bis unten* method as here described is any more immune to the disease of "precursitis" than one who practices the other method he mentions.

Moreover, the *von oben bis unten* method, as Father Clark describes it, has its own special dangers, not least the danger of using the materials of history merely for the purpose of illustrating antecedently assumed first principles. Indeed, it is debatable whether there is in fact a body of firmly established "creditable results" in the philosophy of science, upon which all competent contemporary students are in generally good agreement. Accordingly, there is a serious risk that in claiming certain results to be the outcome of a "rigorously analytical philosophy of science" one may be expressing only the views of a particular philosophic coterie, and of thereby importing into the study of the history of science the warfare of contemporary philosophical schools. I do not wish to deny that in competent hands the approach Father Clark is recommending can be most illuminating. But I think it well to note that this approach does not necessarily yield information that would normally be recognized as a bit of history. Many of us are interested in the history of science partly because it provides knowledge about ideas not already familiar to us, and partly because it supplies information about the genesis of ideas and the chains of their influence and use. These interests are not readily satisfied if we pursue historical study with the sole intent of using the materials of the past as illustrations for currently recognized methodological principles.

And third, I am frankly puzzled by Father Clark's claim that the methodological "insights" he cites are the products of contemporary philosophy of science. The view that the procedure of theoretical science exhibits the pattern of the hypothetico-deductive method was stated by William Whewell more than a century ago, and by W. Stanley Jevons more than seventy-five years ago. The conception of the nature of pure mathematics Father Clark embraces was advanced by George Peacock, Augustus De Morgan, and other members of the mid-nineteenth-century school of Cambridge mathematicians. The notion that a system of applied mathematics must, if sound, be isomorphic with the intended domain of fact, was already argued by Spinoza. Surely, therefore, Father Clark is using the word "contemporary" in some unspecified Pickwickian sense.

I turn now to Father Clark's defense of the hypothetico-deductive method against the charge of involving the formal fallacy of affirming the consequent.* He suggests that those who make this accusation argue from two tacit premises: that there is just one theory which is uniquely "true-to" a given range of phenomena, and that the aim of science is to find that theory for each range. According to Father Clark, however, all that science does assume is that there is some theory, not necessarily a unique one, which is "true-enough-of" a domain of phenomena. He maintains, moreover, that the number of logically possible and physically relevant alternatives to any given theory entertained at a given time is quite small; that the number of such alternatives tends to decrease toward zero with advances in knowledge, and that in consequence, science approaches asymptotically "the valid rigor of strict equivalence" between a theory and its experimentally confirmed consequences. He therefore concludes that in the limit the use of the hypothetico-deductive method "safely escapes the pitfalls of fallacy."

What are we to make of all this? Though Father Clark is here grappling with an important problem manfully, it does not seem to me that he has quite vanquished it. In the first place, it is doubtful whether those who have charged the hypothetico-deductive method with a formal fallacy,

* The next two paragraphs refer to material deleted from the final form of Father Clark's paper. But since they bear fundamentally on Father Clark's theses, they have been here retained.

have in the main argued from the premises he imputes to them. This imputation is certainly not accurate in the case of Jevons, as well as of many others who could be cited. On the contrary, those who have called attention to the alleged fallacy have usually done so in order to make plain that science does not seek to find theories which are uniquely "true-to" the phenomena. In the second place, though Father Clark makes an excellent point in noting that the available alternatives to a theory at a given time are usually quite small in number, it does not follow that new relevant alternatives may not emerge with the advance of knowledge. As I read the history of science, at any rate, the development of fresh alternatives to accepted theories is frequently the rule rather than the exception. Moreover, I do not find evidence in that history for the comforting notion that though theories may be revised and replaced by others, the sequence of theories tends to converge toward some limiting theory, whose content is equivalent to the set of its experimentally confirmable consequences. It seems to me, in consequence, that the formal fallacy in the use of the hypothetico-deductive method is undeniable, although I do not think that the use of this method is therefore unsound. But neither do I think that Father Clark's gambit in defense of the method is a successful one.

I must take a little space to comment on the stress Father Clark places on the notion of isomorphism as the clue for understanding the relation between mathematized science and experiential fact. Much that he says about the nature of pure mathematics, and about the importance of distinguishing mathematics from physics, seems to me excellent. But I think he overplays the significance of isomorphism as the key to the problem he seeks to unravel. Consider his own example of the use of arithmetic in connection with the measurement of physical lengths. He argues quite correctly that the juxtapositive relations between lengths are not identical with the arithmetically additive relations between numbers. But he goes on to say that the former set of relations are nevertheless isomorphic with the latter set. I believe he is demonstrably mistaken in this latter claim. For the numbers normally assumed in the mathematical formulations of physics are the so-called "real" numbers, while the numbers used to represent the magnitudes of lengths identifiable by overt laboratory measurement are the rational numbers. However, while the rational numbers are denumerably infinite, the set of real numbers are non-denumerable, and it is impossible to establish a one-to-one correspondence between the two sets. It therefore follows that there is in fact no strict isomorphism between the mathematics of theoretical physics and the domain of ascertainable physical lengths.

Now to be sure, Father Clark explicitly weakens his thesis about the need for an isomorphism between a mathematized theory and experimental fact; for he introduces the qualification that the isomorphism obtains only "under certain clearly specifiable conditions and to a certain recognizable degree." But this qualification is really not helpful, and tends to conceal the important point that the connection between theory and fact is much looser and vaguer than the language of isomorphism suggests. Indeed, though I cannot develop the point here, I believe that by describing this connection in terms of isomorphism, grave misconceptions easily arise which become standing invitations to intellectual wild-goose hunts. I must content myself here with the flat assertion that the relation of such a physical theory as the kinetic theory of gases or quantum mechanics to experiential fact can be construed in terms of the notion of isomorphism only at the price of radical misapprehensions concerning the function of theory in science.

Let me turn, in concluding, to Father Clark's three historical examples upon which he deploys the *von oben bis unten* method. The Copernican theory is an excellent example to illustrate the use of the hypothetico-deductive method, and Father Clark illuminates the structure of the Copernican system by exhibiting the role of this method in the articulation of the system. He suggests, moreover, that if Copernicus had employed that method more rigorously (by making explicit all his assumptions), Copernicus would have been able to recognize that his view of the fixed stars as having their place in an immovable sphere is inconsistent with other assumptions of his system. Waiving the textual question whether Copernicus did involve himself in an inconsistency, I wish to query Father Clark's suggestion as resting on an unhistorical counsel of perfection. It is in general not feasible to make explicit all of one's assumptions, and the difficulty of doing so is increased in those cases where the assumptions form a part of a complex set of beliefs which constitute the substance of a dominant and unquestioned intellectual heritage. Moreover, even if Copernicus had succeeded in fully axiomatizing his assumptions, it is not credible that he would have surrendered his views on the finitude of space and of the universe, as Father Clark intimates Copernicus might have. For no one knows better than Father Clark that these views were part and parcel of a fairly coherent and interlocking set of comprehensive beliefs concerning nature. Unless, therefore, Copernicus had been able to develop a system of physics radically different from the physics of Aristotle —unless, that is, Copernicus had been a Galileo or Newton—it is beyond reasonable likelihood that he would have done what Father Clark thinks he might have done. I suspect that at this point the *von oben bis unten* method has betrayed Father Clark into identifying a logical with a historical possibility. But such an identification, though extraordinarily tempting, is nonetheless erroneous, since what is logically possible is often a historical impossibility.

Father Clark's analysis of certain aspects of the thought of Oresme is perhaps the most challenging of his historical discussions. I have found this part of his paper particularly illuminating, for it makes evident how cautiously one must proceed in seeking evidence for such a thesis as the one advanced by Dr. Crombie. Nevertheless, I am not entirely convinced that Father Clark has fully established a case against the scientific merits of Oresme's thought.

One of Father Clark's conclusions is that Oresme's mathematics is "not mathematics at all, but rather an abstract and unsuccessful physics." Now I am not qualified to take sides in any debate on the proper interpretation of Oresme's texts. Nevertheless, the construction Father Clark places on Oresme's words does not seem to me to be the uniquely plausible one. For example, he interprets Oresme to say that the triangular shape of the intensitivity *graph* of a quality signifies that the quality itself is triangular, so that Oresme is found guilty of confusing mathematics with physics. However, his account of Oresme's views also shows that the latter regarded construction lines in a mathematical representation of nature as "only mathematical fictions," and that he held certain graphical representations as more appropriate than others simply because of their greater effectiveness as visual aides. It is therefore not obviously evident whether, when Oresme ostensibly declares that there are triangular qualities in nature, he is not using language analogically—comparable to a modern physicist's talk of harmonic motions. Father Clark shows convincingly enough that Oresme did not have pellucidly clear notions concerning the nature of mathematics. But in this respect he was not unlike many more distinguished figures in the history of science who lived centuries after him.

Father Clark also questions the cogency of Oresme's famous proof that any uniformly difform quality (e.g., a uniformly increasing velocity) can be equated to a uniform quality (e.g., to the constant mean value of the uniformly increasing velocity). His ground for his doubts is that Oresme proceeds by constructing a line which bisects the line *AB* in the diagram (representing the time required for acquiring the maximum velocity), rather than proceeding (as did Galileo) by constructing a line which bisects the line *AC* (representing the maximum speed acquired). It is claimed that Oresme's construction indicates an incoherence in his thought, but I fail to see the force of this contention. Surely there are many alternate ways of constructing a cogent proof, as Father Clark himself admits. Moreover, a published demonstration does not necessarily reproduce the sequence of the ideas which have gone into the invention of the proof. Nor do I find any substantive difference between saying, as Oresme appears to have done, that a uniformly increasing velocity acquired during a period of time can be equated to a certain uniformly constant velocity (equal to the mean value of the uniformly increasing velocity), and saying, as Galileo did, that a certain uniformly constant velocity can be equated to a uniformly increasing velocity acquired during a period. My reservations about this part of Father Clark's discussion do not, of course, prove him to be wrong in claiming that one is suffering from a touch of precursitis if one holds Oresme to have been among those who "succeeded in taking the first steps toward the creation of the algebra and geometry of change that were to transform science in the seventeenth century." My only point is that Father Clark has not quite succeeded in showing that those who do hold that view are really ailing from this disease.

In Father Clark's final example, he aims to exhibit the methodological significance of the notion of isomorphism, by explaining the famous misunderstanding between Beeckman and Descartes in terms of that notion. I do not contest Father Clark's claim here, for I admit the misunderstanding can be explained in this way. But is the notion of isomorphism really indispensable in this context, and does its use illuminate the situation in some special way? I fail to see any merit in an affirmative answer to such questions, for the historical misunderstanding can be made intelligible much more simply—namely, by noting that Descartes obtained a false conclusion because he started from a false assumption.

Despite my doubts about some points in Father Clark's paper, I cannot conclude these comments without once more expressing my admiration for the substance of his essay. The paper illustrates an illuminating approach to the history of science; all of us who are concerned with the future of the subject are grateful to him for having written it.

THE ORIGINS OF CLASSICAL MECHANICS

FROM ARISTOTLE TO NEWTON

E. J. Dijksterhuis

Of the double title of this essay the first part calls for an explanation and the second for a motivation.

The first part refers to the origins of *classical* mechanics. Here *classical* is not to be taken in the sense of "relating to Greek and Roman Antiquity" (for this we shall use the word ancient), but in the sense in which present-day physicists speak of *classical* as contrasted with *modern* science. And mechanics does not signify, at least not primarily, what it expresses etymologically and originally indeed only meant, namely, the theory of the μηχαναί, machines, but a general theory of motion, with rest as a particular case of it, to be divided into kinematics (which is purely mathematical) and dynamics, in which motions are considered in relation to their supposed physical causes and which includes statics, or the theory of equilibrium, as a particular case.

Is it not really superfluous to say these things? Probably it is. But then one can never safeguard oneself too much against the numerous risks of misunderstanding which the use of words involves; one has to be doubly careful if, as in the present case, a difference of idiom may also play a part; and finally one should be trebly on one's guard now that the term mechanics is concerned, a term which, with its derivations—mechanism, mechanical, mechanization—is found to constitute such an inexhaustible source of linguistic confusion.

This much for the first part of the double title. The second, "From Aristotle to Newton," is clear without any further explanation, but raises questions of a different character. Aristotle, indeed! it may be objected. Is it really necessary for one who wishes to understand the origins of classical, that is, Newtonian mechanics, to go back in the past to twenty centuries before Newton's time, and if so, is it not likely that an attempt to give a survey of this whole period in a single essay of short compass will result either in superficiality or in overtaxing the reader's attention?

The first question can at once be answered in the affirmative. Every inquiry about the sources of present-day knowledge is bound to lead ultimately to Hellas; it does so immediately where the foundation of mathematics and natural science are concerned, and in the latter case it will undoubtedly come up against Aristotle, who like no other Hellene, perhaps like no other scholar of any age, dominated the evolution of scientific thought.

And as for the second question: I can only promise to do my utmost to avoid the two dangers mentioned and have to leave it to the reader to judge to what extent I succeed in this.

Of the departments into which we now divide mechanics only statics led a somewhat independent existence in Antiquity, mainly in connection with the very simplest and most fundamental of all machines, the lever. As in all other sectors of mathematics and science, in this case, too, the Greeks naturally were able to benefit by an age-long experience of older civilized nations. But just as elsewhere, they were the first to make that which had long appeared trivial to others the subject of theoretical reflection. It is this capacity of wondering at everyday, familiar, apparently quite unproblematic things, coupled with a remarkable faculty of conceiving ideas, tracing connections, seeing relations, which has given them their unique position in the history of science. And so we find them seeking a theoretical foundation for the long-known fact that on a long lever arm only a small force has to be applied in order to support a heavy weight on the short arm or to move a big load. In the treatment of this problem two views are to be distinguished, each of which has left a permanent trace in physics. The one is associated with the name of Aristotle, the other with that of Archimedes.

Aristotle directs his attention at the simultaneous displacements suffered by the points of application of the two forces in the case of a straight lever in equilibrium under the influence of two weights when the bar is conceived to pivot about the fulcrum. These displacements are proportional to the lengths of the lever arms, and since they take place in the same time, the same applies to the velocities of the points of application. The inverse proportionality of effort and arm, known from experience, can therefore also be formulated by saying that the product of weight and velocity has the same value for both the weights suspended on the lever. According to peripatetic dynamics, however, this product in general is a measure of the force required to move a body, and the condition imposed on the weights thus expresses that they exert equal influences on the lever, so that no pivoting occurs.

The striking feature of this argument is naturally that whereas in the law of dynamics the product of weight and velocity was a measure of the force which causes the motion of a body, it here serves to express the motive power of a body. We have to bear in mind, however, that each of the two bodies strives to set the other in motion and that it therefore functions as effort in relation to the other body considered as load. The exertion required to move the load, however, is conditioned by the product of weight and velocity, and it is obvious to see in this, also, a measure of the ability of the effort to cause that motion. In any case, this deduction, which is merely outlined in the *Quaestiones mechanicae,* was afterwards generally taken in this sense. That the element of time, which is quite irrelevant, was brought in, was evidently because it thus became possible to speak of velocity and to establish the connection with dynamics.

The argument of course is very much open to question and not very convincing. This does not alter the fact that it contains the germ of a general principle which was afterwards to perform an important function in mechanics under the name of the principle of virtual displacements (or, by another name reminiscent of its origin in the above line of thought —the principle of virtual velocities).

Fundamentally different is the train of thought followed by Archimedes in his study of the equilibrium of the lever. He is every inch a mathematician, and for him Euclid's *Elements* constitute the prototype of a scientific treatment. In his work *On the Equilibrium of Planes* he therefore starts to found the science of statics on axioms to be accepted as evident (for example, that a symmetrically loaded lever is in equilibrium) and succeeds in giving faultless logical proof of the inverse proportionality of effort and arm.

In two respects this idea was to have no less great a future than the Aristotelian principle of virtual velocities. First, it underlies a conception which dominated science for many centuries and, in spite of frequent criticism, was never definitely defeated, namely, that the fundamental chapter of physics which treats of motion, has to be dealt with as a branch of mathematics, and not of experimental physics. And second, it led to the application of concepts and methods borrowed from mechanics to purely mathematical, notably geometrical, problems. Archimedes himself had already applied it as such in the quadrature of the parabola, and since the publication in 1906 of his work *Ephodos,* which had been believed lost, we know what an essential function the barycentric method fulfilled in the discovery of his principal geometrical results (such as the area and the volume of the sphere).

Of the two methods I have briefly outlined here only the first, that of Aristotle, seems to have found further applications in Antiquity, namely, in the theory of the five *dunameis,* the five fundamental machines discussed by Hero of Alexandria: wheel and axle, lever, system of pulleys, wedge, and endless screw. The Aristotelian principle here appears in the form, Force is to force as time is to time inversely, which under the name of Golden Rule of Mechanics will continue to form a permanent part of elementary statics.

This is not the place to go into Hero's ingenious technical applications of these simple machines and the way in which he manages to turn meshing gear-wheels to account for this. Only two remarks of a theoretical character have to be made here: (1) The theory of the machines remains confined to the case that effort and load balance, and it does not get to the question how much the effort has to be increased in order actually to move the load. It thus remains what is nowadays called a quasi-static theory; (2) It is curious that the Greeks did not by either of their methods arrive at the proportion of effort and load for the inclined plane.

In contrast with the position in statics, no independent sciences of dynamics and kinematics can properly be said to have existed in Antiquity. Any discussion of these subjects forms part of physical speculations, which appear in philosophical writings, but do not constitute a separate *corpus doctrinae.* For the sake of clearness we represent the matter in this way in discussing it; we therefore speak, for example, of the fundamental law of peripatetic dynamics and of Aristotle's law of falling bodies as if they were explicitly formulated somewhere. If we further formulate these laws, with the aid of modern algebraic symbols, in a general form as proportions, properly speaking we commit an anachronism which is none the less dangerous because we practically cannot dispense with it. With all the necessary reservations, we represent the said fundamental law, which correlates the force which moves a body over a certain distance with the motion produced, by formula I:

$$v = f \cdot \frac{F}{R} \text{ or } F = c \cdot v \cdot R$$

Here v stands for the velocity which the body receives under the influence of the force F, while R comprises everything that resists the motion, consequently inertia (mass) as well as friction and resistance of the air. And the law of falling bodies is represented by formula II:

$$t = f \cdot \frac{D}{W}$$

Here t is the time taken to fall a given distance, W the weight of the body, and D the density of the medium through which the motion takes place.

However, we have to guard against considering every conclusion we can draw from these formulas as a statement actually occurring in the theory we represent by them. Thus formula I does not apply for $R > F$, for in that case there is no motion at all, and it is very doubtful whether Aristotle, if he had been asked this question, would have accepted the inference to be drawn from formula II, namely, that a body which is a hundred times as heavy as another must take one hundredth of the time to fall a given distance. Now that the history of science is gradually growing to be a subject of higher education, there is a danger that an in itself praiseworthy desire to summarize ancient theories briefly in as simple a form as possible may give the student a far too sharply defined idea of such theories and may induce him to put much more into them than they ever really contained.

The above remarks naturally do not exclude that in the great majority of cases the conclusions to be drawn from the formulas *do* express the views held by Aristotle. If this were not the case, they would be altogether pointless. Thus the conclusion: From $F = 0$ it follows that $v = 0$, that is to say, a body which is not acted upon by any force is at rest, expresses a fundamental Aristotelian conception, which we may call the ancient principle of inertia—a body which has been removed from all external influences is at rest. As was already observed above, we can also understand the Aristotelian deduction of the lever principle in this way: When F is given (according to I), the product $v \cdot R$ is constant, while with the lever, R can be interpreted as the weight to be supported or moved.

The two above-mentioned laws were to influence mechanical science to a high degree, if not to dominate it, down to the seventeenth century. This applies to no less an extent to two other elements of the Aristotelian doctrine, namely, (1) the theory of projectiles; (2) the antithesis between heaven and earth.

Concerning the first: The fact that it is possible to throw a heavy body vertically, upwards, or sideways, confronts Aristotle with a problem. One of his fundamental dynamic principles is, *omne quod movetur ab alio movetur* (all that moves is moved by something else). In particular a non-natural motion of an inanimate body calls for the continual action of a motive cause in contact with that body, a *motor conjunctus* (connected with the *mobile*). Now what performs this function for projectiles? In order to answer this question, Aristotle frames the hypothesis that a projectile is kept in motion by the surrounding air, namely, in the following way: While a body is being thrown, the *projector* is initially still in contact with it, and so he himself first functions as *motor conjuctus*. During this period he now sets in motion, together with the projectile, the adjacent layer of the medium, but—and this is the essential point—he moreover imparts to it a *virtus movens,* a power to set something else in motion; he transfers his *projector's* function to that layer of the medium. And this layer then does during the next period the same three things that the original *projector* did during the first period—it sets the projectile in motion and transfers motion as well as *virtus movens* to the next layer. Thus the projectile finds in every point of its path the *motor conjunctus* that is required for maintaining the motion. The motive power, however, is somewhat weakened with every transfer; a moment will arrive when the next layer of the medium is still set in motion, but no longer acquires a *virtus movens.* The projectile now proceeds to perform its natural motion, and the layer of the medium last set in motion comes to rest, together with the last but one.

Concerning the second: The Aristotelian theory of the elements (which I am not going to discuss here), in connection with the distinction between natural and enforced motion, led to the conviction that the sublunary world, in which heavy and light bodies, composed of earth, water, air, and fire, by their nature perform limited, non-uniform rectilinear motions, was bound to differ *toto genere* from the superlunary world, in which celestial bodies, consisting of ether, by their nature carry out unlimited uniform circular motions. On this account, the idea of applying theories framed in connection with terrestrial phenomena to celestial phenomena was excluded *a limine.*

We have now become acquainted with a number of ideas in the realm of what was later to be called mechanics, as clearly expressed by Aristotle himself. There are numerous other questions to which he refers inci-

168

dentally and about which he does not speak in clear terms. I wish to mention two examples. First: If the principle *omne quod movetur ab alio movetur* is to be taken seriously, the question should also be asked: What is the *motor* of a body falling or rising respectively in a natural motion? Second: Natural falling and rising motions appear to take place with increasing velocity. How is this possible, since they are caused by a constant heaviness and lightness respectively?

On these two points, as on numerous others, it is often a problem in itself what Aristotle really means by the remarks he devotes to them. No other ancient author stands so much in need of a commentary as he does. Plenty of such commentaries have been written, with the consequence that a good many of the rather vague statements in his works have only been defined more precisely and made susceptible of discussion by his commentators. However, they often disagree seriously about the question of how this more precise definition is to be effected, and accordingly each of his obscure or vague remarks gives rise to a controversy. It is thus usually almost impossible to compare that which the commentators read in Aristotle with his own statements or verify it in this way. In fact, one gets to know him through the intermediary of his commentators, and anyone who should decide to read Aristotle's work itself with an unbiased mind, would soon come to the conclusion that he was merely adding another to the existing commentaries.

Aristotle-commentaries are of three kinds: ancient Greek, Arab, and Latin-Christian commentaries, of which the second group continues the work of the first, and the third that of the two preceding groups. The study of ancient mechanics therefore shifts to the Christian Middle Ages, in particular to the French and English universities.

Before we proceed to speak of this, it is desirable to make separate mention of one of the ancient commentators, on account of the independent standpoint he takes towards Aristotle, in the very domain with which we are here concerned. On the strength of experience the commentator Johannes Philoponus denies the assertion that a body falls more rapidly according to its weight, for in fact bodies not greatly differing in weight *do* appear to have approximately the same rate of fall. He further rejects the theory which seeks the *motor conjunctus* of a *projectum separatum* (released by the thrower) in the surrounding medium; he holds that a certain immaterial motive power which the *projector* has imparted to the *projectum* while throwing it is to be regarded as *motor conjunctus*.

When we now direct our attention at the European Middle Ages, we find ourselves transported to the sphere of problems of Greek mechanics again. The Aristotelian conceptions about gravity, and about falling bodies and projectiles, are now in a still higher degree becoming the subject of extensive discussion, and the results of ancient statics appear again to constitute an independent subject, referred to by the name of *Scientia de ponderibus,* Science of Weights. I will first discuss the latter.

Our knowledge of medieval statics is based in the first instance on the pioneering work which the French physicist, Pierre Duhem, performed in the first few years of the present century by deciphering numerous manuscripts; his results were laid down in his work *Les Origines de la Statique.* For many years nothing of any importance was added to this until Ernest Moody and Marshall Clagett gave a new impulse to the research by the publication in 1952 of their work *The Medieval Science of Weights,* which owes its value in the first place to the fact that here all the available texts, which Duhem had always disclosed only in fragments and in a French translation, have been made accessible in the original language, and secondly, to the introductions and notes by which the editors explain the published works.

The study of this edition confirms what we could already suspect with more or less certainty on the strength of Duhem's publications, namely, that in medieval statics the two ancient methods of treatment mentioned above, lived on. Writings from ancient or Arab sources were studied, in which statics was treated in the Euclidean manner, or new treatises in this strain were written. In this way the concept of the static moment of a force in relation to a point was gradually reached, by which means the lever problem in its most general form was ultimately solved. Side by side with this, in the school that is usually called after Jordanus Nemorarius, the treatment of problems of equilibrium by consideration of virtual displacements was continued. The Aristotelian principle of virtual velocities, however, was now replaced by what is sometimes called the principle of Jordanus: That which is capable of moving a load over a given vertical distance is also capable of raising a load n times as great as the first over $\frac{1}{n}$th of the vertical distance. With the aid of this principle, Jordanus, or whoever else it may have been, succeeded not only in essentially improving the Aristotelian deduction of the lever-law (namely, by leaving the time-element out of account and considering only the simultaneous vertical displacements of the points of application of the weights), but also in providing the oldest known deduction of the law of the inclined plane. For the first time West European science achieved a definite advance over Greek science.

In the department of medieval dynamics, too, the pioneering work was performed by Duhem, namely, in his books *Etudes sur Léonard de Vinci, ceux qu'il a lus et ceux qui l'ont lu* and *Le Système du monde.* Since 1939, however, a considerable correction and extension of this was given by Miss Anneliese Maier, to whom everyone who is concerned with medieval science owes a debt of great gratitude. In five books and numerous treatises, among which I would specially mention *An der Grenze von Scholastik und Naturwissenschaft* and *Die Vorläufer Galileis im 14. Jahrhundert,* she brought to light a large body of material previously buried in manuscripts, and on a great many points gave occasion to a revision of particular notions made current by Duhem.

Owing to the work of Duhem and Miss Maier, and, though to a lesser extent, of the Polish scholar Michalsky (whose writings were chiefly published in Polish and consequently became less widely known), a keen interest is now universally evinced in medieval science in general, and that of the fourteenth century in particular. At the moment this interest mainly centers in the figure of Nicole Oresme, and I shall therefore say a few words about him and his time.

As I have already observed, it was only in Scholasticism that full justice was done to the problems of gravity, falling bodies, and projectiles. The consideration that the fall of a heavy body is a natural motion, with the secondary thought that it does not therefore call for any further explanation, appeared incapable of completely satisfying the desire for a causal explanation.

Indeed, it remained a curious thing: A *corpus inanimatum* (inanimate body), for which, consequently, motion *a se* (by its own effort), such as animate beings can perform, was ruled out, seemed to move without the action of an external *motor* linked up with the body. The cause could naturally be sought in the *generans,* that is to say in the cause, whatever this might be, which at one time gave rise to the *grave,* which impressed its substantial form on matter, and thus called into existence all the associated accidents. This *generans,* however, can only be an *agens remotum* (remote agent); the substantial form is *agens proximum* (immediate agent); but a substantial form can act only through the associated qualities and capacities; the function of *agens instrumentale* (instrumental agent) is now performed by the accident *gravitas* (gravity), through which the body is potentially in its natural place about the center. In order, however, that the falling motion may actually occur, the *impedimentum* (the impediment in the form of a support or a cord on which the *grave* is suspended), which first prevented the falling of the body, has to be taken away; the *removens impedimentum* (that which removes the impediment) thus acts as *motor accidentalis*.

But scholastic thought, which was very exacting on the point of causality, was by no means satisfied with this as yet. Although in this way it may be understood how the motion of a falling body originates, there remains the more difficult question: How is it maintained, that is to say in particular, what acts as *motor conjunctus* during the fall? There were authors who continued to seek this *motor* in the *generans;* others considered *gravitas* itself as such and consequently assumed an internal tendency towards the natural place of the *grave*. There was also a theory which did not speak of a tendency towards a place but of an *inclinatio ad suum simile,* an inclination of the body to be united with the whole of that to which it is similar. Some spoke of an attractive action assumed to proceed from the world center, others of a repulsive action supposed to be exercised by the lunar sphere, but these two attempts at an explanation were not very successful in view of the inconceivability of an *actio in distans*. Further, there was also the notion of a field of force filling the whole sublunary sphere, in which the *grave* found in every place the *motor* which impelled it to the world-center.

This enumeration is not complete. Anyone who by experience is to some extent acquainted with scholastic thought will understand that all sorts of variants of the explanations outlined above were possible, and that others could also be placed beside them. But, in any case in this enumeration, we have seen all the principal conceivable possibilities appearing: the cause of the fall is either in or beyond the body. In the first case it is a tendency either to a geometric place or to a physical body; in the second case it is either an attraction or a repulsion or a force that resides everywhere. If in this way a real advance of scientific thought had been attainable, it would undoubtedly have been attained. That this did not happen constitutes a fact of great epistemological interest. Natural science is also an empirical science in *this* respect in that it is a matter of experimentation to find out how it is to be studied. The right method of dealing with phenomena has to be learned by experience just as much as the phenomena themselves.

With regard to projectiles there existed a similar dichotomy of possibilities. Aristotle had assumed an external cause which resided in the medium. Philoponus had spoken of an internal motive power, and this notion, which had never been forgotten (the scholastics were slow to forget a conceivability; the mentioning and combating of other people's opinions was to them at least as important as the expression and vindication of their own views), was revived in the fourteenth century among the Paris Terminists, especially through the work of Jean Buridan, in the theory of the *impetus*. A projectile remains in motion after the contact with the thrower has been broken because the latter, in throwing it, had impressed an *impetus* in it, a *vis impressa,* as Galileo was later significantly to call it.

This was a fruitful idea: It appeared capable of doing the best thing that may be expected of a theoretical conception—being applicable to other phenomena besides those for which explanation it has been conceived, and consequently being able to include apparently unrelated facts under one point of view. The other facts were: (1) the still unexplained phenomenon that the motion of a falling body is an accelerated motion and, (2) the motion of the celestial spheres.

As to the first point, the motion of the falling body becomes faster and faster because the body, while falling, receives an *impetus,* which impels it in conjunction with gravity and, since the *impetus* itself constantly grows, does so more and more rapidly. For this reason the *impetus* is sometimes called *gravitas accidentalis.* This explanation of the acceleration of the fall is essentially identical with the one that was to be given in the seventeenth century by the founders of classical dynamics, especially Isaac Beeckman and Christian Huyghens, and which was to become superfluous apparently only when, upon the logical development of the subject, instead of making it intelligible how a constant force can produce a constant acceleration (and not a constant velocity), scholars would be simply content to postulate it.

As to the second point, the explanation of the perpetual uniform motion of the celestial spheres is given with the aid of the *impetus* by supposing that with the creation, God gave to each sphere a certain *impetus,* which is preserved unchanged because no resistance operates and the spheres do not have a tendency to another motion. Consequently it is no longer necessary to assume that they are kept in motion by special intelligences. The explanation, by means of the *impetus,* had the great historical significance that a concept borrowed from terrestrial mechanics was applied to celestial phenomena, so that the fundamental antithesis which Aristotle had set up between heaven and earth was broken in principle. For both the reasons here mentioned the framing of the *impetus*-theory forms an important fact in the history of the evolution of science.

There has been a good deal of dispute about the question of what *impetus* really is. Some say that it is *quantitas motus,* Quantity of Motion, momentum, but others recognize in it inertia or kinetic energy, and when seeing it applied to the celestial spheres they are tempted to think of moment of momentum.

The correct answer seems to us to be that *impetus* is a little of all these things, but none of them altogether. It was a still vague conception, which was to be specified in the course of the history of mechanics; in the process it was to be differentiated into the different mathematically well-marked concepts I enumerated just now, but for that very reason it cannot be identified with any of them.

On the other hand, it *is* permissible to ask what is its most obvious, approximate specification, and then it would hardly seem doubtful to me that this is the *quantitas motus,* in the first place because one receives very clearly the impression that it is conceived to be proportional to the quantity of matter of the body (later called mass) and to the first power of the velocity; and secondly because Quantity of Motion was the first sharply defined mechanical notion to occur in the future development.

However, the view is also met with that the attempt to establish a relation like that between germ and fruit between the *impetus* of the Terminists and any concept of classical mechanics, or even to speak of their resemblance, is fore-doomed to failure because of a fundamental difference in interpretation of the motion-concept between ancient-medieval and classical science. This difference is often expressed in the formula that in the former, motion is a process and in the latter, a state. So long as we confine ourselves to uniform rectilinear motion, this formula indeed expresses a correct view. In fact, according to the ancient conception this motion as well as any other, requires the constant action of an external *motor;* according to the classical conception it is the state of motion of a body that has been removed from all external influences. Related to this is the fact that according to the former conception rest is the opposite of motion, and according to the latter it is a particular case of motion, while it does not even depend on the body itself whether this case does or does not present itself, but only on our choice of the frame of reference. It is merely a difference *in modo concipiendo*.

One may admit all this fully without renouncing the right to compare the term *impetus* with the classical term Quantity of Motion and to establish a close relationship between the two. For in the first place the antithesis in question between the two conceptions of motion exists only so long as uniform rectilinear motions are considered; all other motions are processes in classical mechanics as well, that is to say, they equally require the constant action of an external force, even though its effect is evaluated quite differently from the way it was done in Antiquity. And secondly, it depends entirely on a man's philosophical attitude whether he will be able to assent to the thesis that inertial motion is no process in classical mechanics. When he believes in the existence of an Absolute Space and an Absolute Motion in the Newtonian sense, regarding the latter as change of absolute place, when then, on the ground of the causality principle, he inquires after the cause of this change, and assigns Quantity of Motion as such, he will not have the least objection to setting up a very close relationship between Quantity of Motion and *impetus,* unless he prefers to consider the cause of persistence in uniform rectilinear motion to be inertia. In the latter case he may even appeal to Newton as his authority, for the latter ascribes inertial motion explicitly to a *Vis Inertiae,* Force of Inertia, residing in the body. If the view that the fundamental difference between ancient-medieval and classical mechanics consists in the distinction between motion-as-a-process and motion-as-a-state is to be maintained consistently, it thus appears necessary to consider Newton as not yet an exponent of classical mechanics, so that it would also become a necessity to make a distinction between Newtonian and classical mechanics.

All that I have said so far applied to all the Terminists. Specially associated with the name of Oresme is the introduction and application of the method of the graphic representation of changes in intensity of qualities. It is known how great was the preoccupation of the scholastics with the difference between the fact of a quantity growing larger and smaller on the one hand and that of a quality growing more or less intense on the other hand, and what pains they took to understand what really happened in the latter case, for example, when the degree of heat of a body or the intensity of illumination of a wall varied. For mechanics it was of great consequence that the concept of *intensio* and *remissio,* of change of intensity, also included the change of the instantaneous velocity of a motion, owing to which such questions as that about the distance traveled in a uniformly variable motion in a given time already began to be asked. At present it is known for certain that problems of this kind were already discussed before Oresme in the school of Merton College at Oxford, and that in those quarters the thesis was already enunciated which Duhem proposed to call the rule of Oresme, namely, the thesis that the distance traveled in a uniformly accelerated motion in a given time is equal to that covered in a uniform motion of the same duration, the velocity of which is equal to the instantaneous velocity of the uniformly accelerated motion at the middle instant of the time in question. This name therefore could hardly be maintained and has indeed been generally replaced by that of Mertonian Rule. However, it has not yet been found that before Oresme the deduction of this rule by means of a graphic representation was also already known, so that subject to further information he may certainly be credited with having made a considerable contribution to the development. In particular, the recognition that the area of the velocity-graph represents the distance traveled, may be placed to his credit.

Nowadays there is a tendency to make light of these things. But let us not forget that it is only a few decades since these same results were considered original and brilliant contributions of Galileo and that it was not in the last place on this that the great fame he enjoyed was based. That a truly fundamental discovery has had to be dated about 300 years earlier is, after all, a fact of great historical importance, and it is by no means clear why the appreciation of this discovery should now become smaller. Or can this be an after-effect of the belief held so persistently in the nineteenth century, namely, that the Middle Ages were devoid of all significance for the evolution of science?

One of the criticisms which tend to be provoked in our time by an appreciation of Oresme's merits that is considered excessive, consists in the opinion that Oresme indeed knew that the area of the velocity-graph represents the distance traveled, but that this knowledge may not be regarded as legitimate, because he was unable to define the concept of instantaneous velocity exactly, that is to say, as derivative, nor to regard the distance traveled as the integral of the variable instantaneous velocity. The idea at the root of this is that a man cannot be considered to have actually known something so long as he could not prove it in a manner satisfying present-day requirements of mathematic rigor. This can, of course, be maintained, but then the far-reaching consequences resulting from it for the history of mathematics and the natural sciences will also have to be accepted. It will no longer be possible to say that Newton and Leibniz created the calculus, and a sharp distinction will have to be made again between Newtonian and classical mechanics. Indeed, Newton still regards instantaneous velocity as an intuitively clear concept; he does not define it as the *fluxion* of the distance considered as *fluens,* but on the contrary describes the fluxion of a *fluens* as the rate of its variation. The consequences to which the view in question (which is endorsed by no less a person than Miss Anneliese Maier) will lead (just imagine what changes it would cause in our appreciation of the Bernoullis and of Euler) are sufficient to reject it as unacceptable.

So long as the ideas of Duhem governed the investigations into medieval mechanics to an almost unlimited extent, the conceptions of the Paris Terminists attracted much greater notice than the so-called *calculationes* of the school of Merton College, though the influence exercised by the Calculator *par excellence,* Suisset, on the later evolution was known. This situation also has changed, again largely owing to the work of Miss Maier. In particular we are nowadays greatly interested in the remarkable attempt of Thomas Bradwardine to interpret the fundamental principle of peripatetic mechanics in a different way from that which had so far been usual. If, again with all the reservations previously made, we formulate his conception algebraically, it means that he replaces the relation

$$v = c \cdot \frac{F}{R} \text{ by } v = c \cdot \log \frac{F}{R},$$

which is expressed in words by saying that v follows the ratio $F : R$. The condition $F > R$ has now, as it were, been incorporated in the fundamental principle, because for $F < R$ the logarithm would take a negative value.

This so-called relation of Bradwardine naturally deserves our full attention as a symptom of the growing tendency of the fourteenth century to treat mechanics with the aid of mathematical concepts. It should, however, be emphasized—a fact which is nowadays apt to be overlooked—that the conception concerned had a perfectly speculative character and was not based on any experience, while it was not confirmed by the subsequent development. This constitutes a cardinal difference from the theories of the Paris Terminists.

When we survey the picture of the history of the evolution of mechanics as it has now broadly presented itself to our view, it greatly satisfies our desire for the combination of continuity and renewal, a desire which as a rule spontaneously directs our historical inquiries. This applies to a much lesser extent to the next period, which separates fourteenth-century scholasticism from the flourishing period of science which began in the sixteenth century and produced such brilliant results in the seventeenth century. It is indeed quite likely that an uninterrupted line of evolution linked the schools of Oxford and Paris *via* the University of Padua with Galileo, but for the time being this notion is a program of research rather than an already guaranteed result of investigation. A great many more investigations about details of science in the fifteenth and sixteenth centuries in the manner of Clagett's study on Marliani and of Koyré's treatment of Buonamici's work *De Motu* will have to be made, and in particular much more attention will have to be devoted to the writings of the Paduan school before we shall be able to express ourselves on this matter with greater certainty.

Even more problematic still, however, is the possible relation between the school of Jordanus Nemorarius and the new impulse which statics received at the end of the sixteenth century from the work of Simon Stevin. Duhem's contention that Leonardo da Vinci formed an important link in this respect has not been confirmed by later investigations. On the contrary, it is highly probable indeed that here as well as in the field of pure mathematics a considerable influence was exercised by the publication of certain works of Archimedes, especially by the appearance of the complete edition which Thomas Gechauff Venatorius accomplished in 1544. Stevin's works, *Weeghconst* (Art of Weighing) and *Waterwicht* (Hydrostatics), continue, as it were, without a break, the observations in the Archimedean writings *On the Equilibrium of Planes* and *On Floating Bodies*. But it also seems likely that he belongs in a line of evolution of statics which originates from the medieval Science of Weights. This appears at once from his terminology. Thus the Dutch neologism, *staltwicht,* introduced by him, which means, among other things, the component of the weight parallel to an inclined plane, is quite clearly a translation of the medieval term *gravitas secundum situm*. And it also follows from the fundamental criticism he offers on the Aristotelian principle of virtual velocities. He thinks it absurd to deduce a condition of equilibrium from considerations of motion, the possibility of which is excluded by the very fact of the equilibrium. This criticism is unfair; he overlooks the circumstance that the motion concerned is a virtual motion, which only exists in thought and is not caused by the acting forces. It is also inconsistent that, in spite of this criticism, he accepts the principle of Jordanus as "gemeene weeghconstighe reghel" (general static principle). The main thing, however, is that these are subjects belonging in the Science of Weights.

Stevin did not deal with the problems of falling bodies and projectiles, with one important exception. In the *Weeghconst* of 1586 he describes an experiment which he performed in conjunction with Jan Cornelis de Groot, burgomaster of Delft and father of the famous jurist Grotius, in order to find out if it was true, as the Aristotelians asserted, that the time taken by heavy bodies to fall a given distance varied inversely as their weight. It appeared that lumps of lead of 1 and 10 pounds took exactly the same time to fall a distance of 30 feet. This happened five years before the moment at which a long-exploded legend, which still lingers on in many historical works, makes Galileo perform tests with falling bodies from the top of the leaning tower at Pisa, which, in the florid expression of Favaro, dealt peripatetic philosophy a blow from which it was never to recover.

In the meantime at Pisa and Padua, Galileo passed through that remarkable evolution of his thought which was to have such a powerful and fruitful effect on science, and hence on the whole of European culture. When reading his juvenile writings, one may still imagine that one is listening to Jean Buridan lecturing to his students. His *vis impressa* is a twin-brother of the *impetus,* with the sole difference that the prosaic matter-of-fact character of the scholastic argumentation has been replaced by a vivid mode of expression, fascinating in its antithetic eloquence, which makes it clear why the Italians also mention Galileo in the history of their literature.

Apparently Galileo soon felt that ingenious speculations were not sufficient to solve the enigmas presented by falling bodies and projectiles, and that a radical change of method was necessary. He realized that one has to know a phenomenon before one can try to account for it, and that a dynamic treatment of falling bodies and projectiles consequently had to be preceded by a kinematic description of their motion. He therefore made it his object to exclude for the time being any question about the cause and first to describe the course of the motion of falling bodies and projectiles in an exact mathematical manner. The results of this investigation, which are communicated on the third and the fourth day of the *Discorsi,* constitute the true foundation of the evolution of seventeenth-century mechanics, which was to end in Newton's definitive founding of classical mechanics.

Galileo's treatment of the uniformly variable motion of falling bodies forms a continuation on a higher plane of that of Oresme. The intermediate period is characterized, along with the raising of the level of West European mathematics caused by the publication of the works of Archimedes, by the growth of the recognition that uniformly variable motion, treated *in abstracto* by Oresme, forms an idealization of the natural phenomenon of bodies falling freely from rest.

It is interesting to compare the reasonings of Oresme and Galileo in detail. One is then struck by the fact that while Oresme, without giving any further explanation for it, considers the area of the velocity-graph as a measure of the distance traveled, Galileo, avoiding this statement, arrives at the same result by observing the aggregate of degrees of velocity in a given interval of time, for which he follows the same train of thought that was applied in mathematics by Cavalieri in those days. The way in which this subject was treated between Oresme and Galileo, in general the history of the theory of indivisibles between Archimedes and Cavalieri, and in particular the relation between Galileo and the latter, still requires elucidation.

The limitation on principle to the kinematic treatment of the motion of falling bodies which Galileo imposed upon himself naturally makes it impossible that he could also have advanced the dynamic conception of this phenomenon and thus could have given an answer to the old question why it is that a constant gravity can produce a uniformly growing velocity. The still encountered interpretation of Galileo as the creator of classical dynamics is therefore already untenable on logical grounds. One cannot possibly, as is still done, commend Galileo on the one hand for the wisdom which induced him to leave the dynamic aspect entirely alone and on the other hand respect him for the manner in which he treated this aspect.

This does not, of course, alter the fact that he has greatly contributed in an indirect way to the birth of classical dynamics. No other author has helped so much to get the new inertia-conception and the relativity of the motion-concept accepted, though on the other hand it has to be borne in mind that his own principle of inertia does not relate to the undisturbed motion of a point in an infinite void but to motion on a spherical surface with the center of the earth for its center, and that in connection with this he does not yet arrive at a recognition of the true relativity-principle of classical mechanics as Huyghens was to formulate and apply it.

On all these accounts the picture we now form of Galileo differs very widely indeed from that generally current in the nineteenth century and still widely prevalent even at present. It also does so in another respect. We can no longer regard Galileo, as used to be done exclusively, as the founder of the experimental method in physics. We know that it was not by measuring tests that he found the law of falling bodies, and also that the experiments he describes in the *Dialogo* and the *Discorsi* are largely mental experiments, of the kind also appearing in Scholasticism and in Nicholas of Cusa.

Nor is it possible for us any longer to employ the nineteenth-century expression of Galilean-Newtonian mechanics, by which the impression is created that after Galileo not very much remained to be achieved to make possible the systematization of mechanics in Newton's *Principia*. In reality a good deal more had still to be done, among which especially the work of Christian Huyghens was of great consequence. He evolved the theory of the cyclodial motion of falling bodies, with its application in the construction of the pendulum clock, was the first to formulate the relativity principle of classical mechanics, and made an ingenious use of it in the treatment of impact; in his theory of the physical pendulum he laid the foundations of the mechanics of solid bodies; he taught the dynamic treatment of uniform circular motion and of harmonic motion, and through generalization of an axiom of Torricelli he introduced a general mechanical principle which is equivalent to the principle of preservation of mechanical energy in the gravitational field of the earth, conceived as homogeneous.

But after all this, however important and admirable, the work of Newton still had the effect of a great and original achievement. Not so much because of his foundation of classical mechanics on axioms, which still left much to be desired from a logical point of view and from which only in the eighteenth and nineteenth centuries a more satisfactory treatment was to evolve, nor on account of the further mathematization of the subject at his hands, which indeed bears evidence of great genius, but which ceases quite to satisfy the ancient demands of rigor which Galileo and Huyghens still had complied with and does not yet by any means meet with those which were to be made in the nineteenth century, but because Newton's undying merit is based above all on the synthesis of terrestrial and celestial mechanics in the theory of gravitation, by which the old Aristotelian antithesis between earth and heaven was definitely abolished and the astronomical work meanwhile performed by Copernicus and Kepler at last received its definitive consummation and culmination.

The historical survey of the evolution of classical mechanics is now coming to an end, and such a thing always acts to some extent as an anti-climax. In fact, the story finally leads to a result which is of an elementary character to the present-day scientist; he has grown up in this sphere and consequently is apt to consider it as a matter of course. He can hardly feel the same pleasure and admiration that filled him as long as he was still looking at things from the historical point of view and was therefore still mentally evolving towards the final result pursued. This is the essential psychological difficulty which attaches to every historical discussion of a mathematical or scientific subject, but which has to be overcome if the History of Science is to serve its purpose. For this purpose is to re-experience things that have apparently become elementary and trivial as the new discovery they constituted at one time.

This is important not only for the sake of re-experiencing the joy of discovery, but to no less an extent because it is associated with an intensi-fication of our understanding of the problems from which the discovery in question resulted and of the methods that led to this discovery. The History of Science forms not only the memory of science, but also its epistemo-logical laboratory. It not only recalls the work of the predecessors without whose exertion and ingenuity our present-day science would not exist, but also makes it clear what course had to be followed in order to make it possible.

There are few fundamental theories of science for which as soon as one has succeeded in escaping from the paralyzing effect of the notion infused by education that one has to do with something simple, with a matter long since definitely settled, the profitableness of historical study is so clearly evident as for classical mechanics in general and Newton's theory of gravitation in particular.

But a close examination of both derives its great importance also from another cause: There are few occasions on which it is realized so vividly how every solution of a scientific problem raises new questions, which are partly of a scientific character again, but partly also of an epistemological nature.

The school child repeats it thoughtlessly after his teacher: the stone falls because the earth attracts it. But what *is* this attraction, and how does it happen? May it also be said that the stone tends towards the earth, or is impelled towards it? When the pupil has advanced a little further, he learns to say that every body perseveres in its state of rest, or of uniform motion in a straight line, unless it is compelled to change that state by forces impressed upon it. But in relation to what frame of reference does this statement apply? In relation to absolute space? If so, what *is* absolute space and how can we establish absolute motion? Finally the student becomes acquainted with the general principle of gravitation, and thus learns to explain the motion of the planets about the sun, that of the moon about the earth, the tides, and the motion of bodies falling on earth. But what does "explain" mean here? To what extent does that which is announced as an explanation satisfy man's desire for causality? What do we understand of the phenomenon now? Is this understanding something else than a description in mathematical terms? If so, what is this something else? If not, do we have to conclude that to understand a thing is no more than to subsume it under a general notion with which one has become familiar?

These are fundamental questions such as science raises on all hands, but which nowhere force themselves on our attention in so acute and clear a form as during our reading of the work of Newton. For this reason it is still extremely instructive for a scientist who, though not historically minded, is interested in epistemology, to read the *Principia*. He will repeatedly find things in it which are absent from the current Newton-picture of our time, but which are of essential importance for the just appreciation of his scientific personality. Under the influence of more than two centuries of classical mechanics modern man is sometimes inclined to consider the Newtonian concept of force as a kind of ultimate explanatory principle. In particular he tends to speak of attractive forces as if their existence and action were so perfectly intelligible that one need ask no further now. This manifests itself, among other things, in the fact that it has become more or less customary, instead of saying with Newton that two bodies gravitate towards each other, to use the expression that they attract each other, and consequently to speak of gravitational attraction instead of gravitation.

It is true that Newton also occasionally uses this attraction-terminology, but he repeatedly points out that he does not feel committed by this expression and that one can say just as well that a body is impelled towards something or that it tends towards something. Compare, for example, Definition V: "A centripetal force is that by which bodies are drawn or impelled, or any way tend, towards a point as to a centre." And read the explanation of Definition VIII:

I . . . use the words attraction, impulse, or propensity of any sort towards a centre, promiscuously, and indifferently, one for another; considering those forces not physically, but mathematically: wherefore the reader is not to imagine that by those words I anywhere take upon me to define the kind, or the manner of any action, the causes or the physical reason thereof, or that I attribute forces, in a true and physical sense, to certain centres (which are only mathematical points), when at any time I happen to speak of centres as attracting, or as endued with attractive powers. . . . For I here design only to give a mathematical notion of these forces, without considering their physical causes and seats.

However, I do not intend to go into this any further; it might provide subject matter for a separate discussion, partly historical and partly epistemological. But this does not form part of the theme I have been asked to deal with, so that I will here conclude my exposition.

CARL B. BOYER

on the Paper of E. J. Dijksterhuis

⚎ Given the title of Professor Dijksterhuis' attractive paper, a tyro might have attempted a bird's-eye view of this vast field; but Professor Dijksterhuis wisely has resisted this temptation. Where Duhem covered late medieval and early modern statics in two substantial volumes (*Origines de la statique*, 1905–1906) and late medieval and early modern dynamics in three still more substantial tomes (*Études sur Léonard de Vinci*, 1906–13); where Dugas devoted a weighty book to the mechanics of a single century (*La mécanique au XVIIᵉ siècle*, 1954), an *Überblick* covering the period in question here could only have resulted in superficiality of the material or in superannuation of the audience. Professor Dijksterhuis has cleverly avoided both of these extremes, as he has avoided also the two afflictions of "precursoritis" and "vacuitis."

Often those who have approached the history of science through a single subject-matter field in science, with its omnipresent emphasis upon the superiority of current theory over earlier schemes, look over the past with the idea of singling out figures who were prescient enough to anticipate some aspect of modern science, conceding to others only a disdainful smile. Not so Professor Dijksterhuis. His analysis has been remarkably sympathetic from beginning to end, and the early steps in dynamics found in the Peripatetic *Problems of Mechanics* are not held up to ridicule in comparison with the Newtonian synthesis two millenia later. On the other hand, he has not succumbed to the temptation to regard the history of science as an old curiosity shop in which items are equally cherished with no attempt at evaluation. So authoritatively has the author treated his subject, and with such unerring judgment, that I have no quarrel with his conclusions. My comments will therefore take the form of suggesting some questions for further study which come to mind in connection with his essay.

One question is suggested by the title of the essay—"The Origins of Classical Mechanics *from* Aristotle to Newton." Does not this very title whet one's curiosity as to what came before Aristotle? While it is indeed true that all roads lead back to Aristotle and Archimedes, there undoubtedly were paths leading to them from the other direction. Dugas, in his *History of Mechanics* (now available in an English translation), opens with the words: "For lack of more ancient records, history of mechanics starts with Aristotle." So hazardous is it in history to use the words "start" or "invent" or "discover" that it has been said facetiously that it is the triumph of the historian of science to prove that no one ever started or invented or discovered anything. Without succumbing to such an extreme case of *precursoritis,* I would nevertheless point to a problem which might be exploited. In Greek mathematics the *Elements* of Euclid is, with some trivial exception, the oldest extant work; yet no one would dream of writing, "For lack of more ancient records, history of mathematics starts with Euclid." Quite the contrary! It probably is safe to say that, despite the lack of documents, more has been written on pre-Euclidean mathematic than on the whole of the Alexandrian Age. If, in mathematics, the lack of documents has been a challenge to reconstruct, why should this not be true of mechanics? It may be argued that mathematicians have Proclus' *Eudemian Summary* as a springboard, with nothing comparable in the field of mechanics. This may be granted without prejudicing the thesis that a reconstruction of mechanics before Aristotle might well be attempted. In mathematics the period before Euclid is known as the Heroic Age. Was there no Heroic Age in mechanics?

The Golden Age of mechanics centered about Aristotle and Archimedes, as the Golden Age in mathematics found its focus in Euclid, Archimedes, and Apollonius (*ca.* 325 to 225 B.C.). In mathematics there was a Silver Age centering about Ptolemy, Diophantus, and Pappus (*ca.* 125 to 325 A.D.), and one wonders whether there was anything comparable in mechanics. Professor Dijksterhuis went from Archimedes to Philoponus, with brief mention on the way of Hero's practical work on machines. Perhaps the intervening years warrant further study. One should expect that a concern for astronomy would be accompanied by some interest in dynamics. In astronomy we know of the work of Hipparchus through that of Ptolemy; but there seems to have been no equivalent post-Hipparchan synthesis in dynamics. This is regrettable, for it is reported that Hipparchus differed with Aristotle in the theory of the acceleration of falling bodies.

A Silver Age may not be discernible in dynamics, but one may make out a case for such an age in statics if one considers the law of the inclined plane. Professor Dijksterhuis has called attention to the strange failure of the Greeks to discover this law, and it may not be amiss to hazard here a conjecture concerned with this failure. The laws of the lever and the inclined plane are so similar in statement that one should expect that the ancients would have hit upon the second as soon as they were aware of the first, yet this was not the case. These laws may be formulated as follows:

(1) Weights balancing each other on a lever are to each other inversely as the lengths of the arms.

(2) Weights balancing each other on the sides of inclined planes (of equal height) are to each other as the lengths of the planes.

In spite of the apparent similarity, the first was known at least from the days of Aristotle, whereas the second eluded the greatest thinkers of antiquity.

It often is said that science is nothing but trained and organized common sense, but one is tempted to suggest that advances in science are at least as likely to stem from uncommon as from common sense, and this seems to have been the case with the inclined plane. As one approaches a hill common sense tells one that the hill impedes progress, uncommon sense tells one that the hill dilutes gravity. And common sense suggests *rolling* a log up the hill, while very uncommon sense tells one to think of an unbelievably smooth soapbox *being held* on an uncommonly smooth plane. And Greek thinkers, choosing the path indicated by common sense, were badly misled about the inclined plane.

Dugas in his *Histoire de la mécanique* (1950) wrote that Pappus "appears to be the only geometer of Antiquity who took up the problem of the motion and equilibrium of a heavy body on an inclined plane." In this he, like Duhem, overlooked the contribution of Hero of Alexandria. Combining geometry and common sense, Hero reasoned that a cylindrical log balanced on an inclined plane can be thought of as divided into two segments by a vertical plane through the line of contact of the cylinder with the plane, the larger segment tending to make the cylinder roll down the plane and the smaller segment tending to roll the cylinder upward along the plane. The resultant of these two tendencies he thought was the force needed to hold the cylinder in equilibrium on the plane. This hypothesis leads to the law $F = W(2\theta + \sin 2\theta)/\pi$, instead of to the correct law $F = W \sin \theta$, where W is the weight of the cylinder, θ the angle of inclination of the plane (in radians), and F the force needed to hold the cylinder in equilibrium. Hero's law is correct, of course, only for $\theta = 0$ and $\theta = \pi/2$. For small angles Hero's formula comes close to the correct result if 4 is substituted for π.

I. E. Drabkin some years ago pointed out that the traditional impression that Greek science was hampered by an approach which was excessively rational and abstract is misdirected as far as dynamics is concerned; and one can say much the same thing about the law of the inclined plane. What was needed was *less* of the common sense empirical approach and *more* of the uncommon or idealized attack.

Professor Dijksterhuis has called attention to two approaches to the law of the lever—the Aristotelian dynamic formulation and the Archimedean static expression. Somewhat the same distinction can be seen in the case of the inclined plane. The reasoning of Hero, with attention focussed on merely holding the cylinder on the plane, corresponds to that of Archimedes. A couple of centuries later Pappus tried to solve the law of the inclined plane by a comparison of the force needed to roll a sphere up a hill with that needed to move it along a horizontal surface. Common sense seems to have had a stronger hold on Pappus than on Hero, and an unfortunate application of Aristotelian dynamics further vitiated Pappus' work. Without clearly justifying his steps, Pappus concluded that the force needed to roll the sphere up a hill of inclination θ is to the force needed to move it along the horizontal as the exsecant of θ. Using this result he held that if forty men can move a given weight along the horizontal, then it will take 300 men to move the spherical weight up an incline of 60°. Greek scientific rationalism indeed! Both statics and dynamics suffered from a lack of just this, and it was the thirteenth and fourteenth centuries which did much to put them back on the right track.

Professor Dijksterhuis judiciously called attention to the significant innovations in dynamics in the fourteenth century, advances which have become well known through the work of scholars contributing to this symposium. These innovations seem to have overshadowed the discovery in the thirteenth century of the correct law of the inclined plane, given by Jordanus Nemorarius. Fortunately for our problem, the works of Hero and Pappus were virtually unknown to the Latin medieval world; and the scholars of the thirteenth century turned to another source for inspiration—the science of weights. They did not ask the practical question, "How difficult is it to roll a weight up a hill?" They asked a very impractical one, "How heavy is a body on an inclined plane?" Jordanus derived from this query the general principle, mentioned by Professor Dijksterhuis, of *gravitas secundum situm*—the idea that, in the study of the force exerted by a heavy body, the measure of its "weight" is the change in the vertical distance as it moves along the inclined plane. The "gravity according to position" of a weight on a horizontal plane would be zero, common sense notwithstanding. In terms of this static principle the law of the inclined plane is immediately obvious. Given two unequally inclined planes of the same height, and given weights on each which are assumed to have the same gravity according to position, it is clear that the weights must be to each other as the lengths of the planes.

The reasoning of Jordanus failed to convince his successors. His work seems not to have been entirely forgotten, but the views of Pappus and Hero seem to have been more popular, perhaps because more vivid and closer to common sense. In Leonardo da Vinci an argument similar to that of Pappus rubs elbows with one making use of the idea of gravity according to position. At one point the rolling spheres of Pappus are replaced by two non-spherical weights balanced on two inclined planes of equal height of which one is twice as long as the other, and the ratio of the weights is correctly given as two to one; but elsewhere results are incorrectly given.

It has been argued by Duhem that the tradition of Jordanus can be traced through Leonardo and Cardan to modern times, but Professor Dijksterhuis has pointed out that later investigation has not confirmed Duhem's contention. Cardan's crude idea on the inclined plane—that the force needed to hold a weight on an inclined plane is proportional to the angle of inclination—certainly is far from the views of Jordanus and Leonardo. On the other hand, Professor Dijksterhuis shows that Stevin, besides being influenced by Archimedes, belongs in the tradition deriving from the medieval science of weights. If Stevin was not the first one to state correctly the law of the inclined plane, he nevertheless was the one who first gave it a vivid representation. Who among us is not familiar with the wreath of spheres with which the law became immediately obvious? Perhaps one should keep in mind that science progresses not only through great discoveries but also through devices which impress these vividly upon the minds of others. This is seen not only in the work of Stevin on the inclined plane, but also in the law of Oresme. Professor Dijksterhuis has pointed out that the latter law had been given earlier by the Mertonians at Oxford, but it generally is ascribed to Oresme who enabled us to "see" the law through his graphical representation. It may be that this same factor is at work in our evaluations of ancient astronomy, where the arithmetical schemes of Mesopotamian astronomers appeal to us far less than the easily visualized geometrical representations of the Greeks. And that the law of motion of Bradwardine achieved little popularity may have been due to the difficulty men had in picturing it. Professor Dijksterhuis has depreciated the law, but graphs of the laws of Aristotle and Bradwardine would show that the latter conforms somewhat more closely to experience.

In closing I should like to call attention to the exceptional modesty of Professor Dijksterhuis in presenting the work of those two great representatives of mechanics in the Low Countries, Stevin and Huyghens. He has not insisted unduly on the originality of Stevin, but he has clearly pointed out that, while so much ink has been spilled over whether or not Galileo dropped weights from the tower of Pisa, such an experiment actually was performed by Stevin and de Groot at least as early as 1586. Perhaps we need a study in public relations of scientists to determine why the undoubted experiment of Stevin went virtually unnoticed while the putative experience of Galileo has had such a vigorous tradition. The explanation does not seem to lie entirely in the fact that Stevin wrote mainly in Dutch.

RUPERT HALL

on the Paper of E. J. Dijksterhuis

It is one of the most elementary, and yet one of the most exacting, of literary exercises to summarize a large subject in few words. So many volumes have been written on the history of mechanics since Mach laid down his pen, that one would not envy Professor Dijksterhuis the task of which he has so ably acquitted himself. It is a most interesting essay, one which has not merely surveyed a large and vital field of study but has opened many enticing avenues of interpretation.

At the end of the seventeenth century, mechanics was the most dramatic of the sciences, for it had in the *Principia,* leapt, as it were, to take the heavens as well as the earth within its grasp. It was the most advanced of the sciences in its logical structure, in its mathematical development, and in the certitude of its experimental confirmation. It was the most fundamental of the sciences, not only as the base of the mechanical or corpuscular philosophy that seemed to offer a clue to the more involved phenomena of nature, but as deriving directly from the most primary of nature's laws. Mechanics thus affords an outstanding example of the success of the scientific revolution; and not merely of its success, but of its capacity to change men's ideas.

It follows that this science must be a prime object of historical scrutiny. If we would understand anything of the growth of scientific explanation, if particularly we would understand the origins of modern scientific ideas and methods, then we need to know—Professor Dijksterhuis has briefly delineated the answer—how this one branch of science grew from its ancient to its Newtonian shape. We are faced, then, with four problems of the first order, and these have provided the skeleton of Professor Dijksterhuis's discussion. They are, first, the extent, nature, and limitations of the Greek accomplishment; secondly, the assessment of medieval contributions to mechanics, and of their origins—here above all historical opinion has changed in the last fifty years; thirdly, the transition from medieval to Galilean mechanics—the one phase of this history which as it seemed to me the paper passed over somewhat lightly; and, fourthly, the study and criticism of Newtonian mechanics itself.

To begin with the Greeks, with Aristotle, the *fons et origo* of mechanics as of so much in science. It was really unnecessary for Professor Dijksterhuis to apologize for going back to Aristotle, for how could the historian do otherwise? Aristotle, as he properly reminded us, was no less important for the Middle Ages in statics than for his more notorious principles in dynamics, though his statics was to be overlaid by the more compelling reasoning of Archimedes. And it was pointed out that there was a profound difference in the treatment of these two departments of mechanical science, for whereas statics early became an independent study, highly geometrized, dynamics was embedded in natural philosophy and did not receive the same clear, logical treatment. The difference was not fully resolved before the seventeenth century. It seems to correspond to a difference in the degree of involvement of the problems. The problem of the lever was, after all, a cross-word puzzle one; it could be isolated from everything else, and nothing of major philosophical weight hung upon it, though it was the source of the theory of the five simple machines, so-called. Problems of dynamics, on the other hand, led straight to the largest questions about the universe. They involved—or necessitated— the distinction between earthly and celestial phenomena; they brought in the great theory of the elements; cosmology and the science of motion were entwined. This was not a fact suddenly discovered during the scientific revolution when Aristotle was violently called in question; it was understood and implicit from the first. Motion was no less the basic phenomenon of nature for Aristotle than it was for the atomists, or the seventeenth-century corpuscularians. The *Physics* is a treatise on the science of motion in the most general terms. Consider for example a sentence that comes very near the opening: "The principles [of the science] in question must be either (a) one or (b) more than one. If (a) one, it must be either (i) motionless as Parmenides and Melissus assert, or (ii) in motion, as the physicists hold, some declaring air to be the first principle, others water" (184b). And again, "We physicists . . . must take for granted that the things that exist by nature are, either all or some of them, in motion—which is indeed made plain by induction" (185a). The old verdict that the ancients had no science of dynamics should rather be expressed by saying that dynamics lost its identity by its very universality, through its unification with physics and cosmology.

Perhaps we might not too unjustly contrast the endeavor of Newtonian science to build up physics from the basis of dynamical relationships, with that of the ancients who tried to extract dynamical laws from a web of physical theory. And, naturally, the more one saw motion as ubiquitous, the more difficult it was to single out one element for detailed study, as was successfully done in statics with the lever. Moreover, since the object of physics and cosmology was the explanation of natural phenomena in accord with a rational world-order, it resulted that the principal entrance to the science of motion was through problems of causes. Thus we discover that the dynamical propositions in Aristotle that Professor Dijksterhuis has described were almost incidental to the application of Aristotle's ideas about motion to physical phenomena; they are products of a system of ideas, not the foundations of such a system. The idea, for instance, that the speed of a body's fall (other things being equal) is proportional to its weight, is so patently untrue that it could only serve as a conclusion, and not as a premise; nothing was built upon it. Similarly with Aristotle's explanation of the motion of projectiles (in which, from his rather tentative language, he himself appeared to have no great confidence): could anything be more *ad hoc?* These matters were not fundamental to the world-picture, and therefore they were not of the highest importance. The weaknesses of the dynamical (or kinematical) relationships are, however, more comprehensible when set against their background; they belong to a science of motion that was not strictly dynamical in the Newtonian sense at all, but was physical and cosmological. Whereas the "laws of motion" in Newtonian mechanics are concerned with mass, force and acceleration and apply strictly to a mass-point particle, the "laws" of Aristotle are not the quasi-mathematical relations stated by Professor Dijksterhuis but rather his distinction between the heavenly and sublunary spheres, between violent and natural motions, or his conception *omne quod movetur ab alio movetur,* in other words the wide scientific and philosophical principles to which he has drawn our attention. It is in this sense that, as he remarked, it is misleading to draw direct comparisons between formulae deducible from Aristotle's treatise and those of Galileo and Newton.

Perhaps I may here interpolate a remark on a difficulty I felt at one point in Professor Dijksterhuis's essay, when he referred (as we should all find it necessary to do!) to the more than occasional unintelligibility of Aristotle's ideas. He instanced (i) the problem of the nature of the

motor in natural motion, and (ii) the question of the cause of the acceleration observed in such motion, as in the fall of a heavy body. I confess
to feeling that the discussion of these topics was left somewhat indetermi-
nate. Aristotle does, for example, declare boldly enough that all bodies
in natural motion are moved either by themselves or by something else;
in the second group he specifically mentions light things and heavy
things, "which are moved either by that which brought the thing into
existence as such and made it light or heavy, or by that which released
what was hindering and preventing it" (*Physics,* 255b–56a).

I shall not attempt to follow the author among the medieval writers
on statics, but I note that here Professor Dijksterhuis spoke of "Arabic
sources" as being among the authorities studied by Latin writers, while
in relation to dynamics he appeared to record nothing of an Islamic participation in the tradition between Philoponos and the earlier Latin theorists. I should certainly be interested to know whether he thinks that
there was any such Islamic contribution of importance. [Professor Dijksterhuis later replied that he recognized the importance of the Islamic
writers, but had had insufficient time to consider them.] I remember
that this topic was discussed in an article in *Isis* some years ago (S. Pines,
1953), and also I think by Miss Maier. There is no doubt a great deal of
work still to be done on the development of mechanical ideas between the
time of Philoponos in the sixth century and that of Peter Olivi in
the thirteenth.

If I might add a gloss to what Professor Dijksterhuis has said of the
impetus concept, it would be to emphasize its duality: it afforded on the
one hand an explanation of certain effects, while on the other it offered
the possibility of making calculations, whether arithmetical or geometrical. The latter aspect was, of course, an extension of the former; if the
causal role of impetus in the acceleration of falling bodies were rejected,
then the possibility of calculating vanished also. Thus the development
of the impetus theory covers a very important transition, from the problem "Why does the arrow fly from the bowstring?" to the question "In
what manner does the stone fall?" or even "What are the characteristics
of a projectile's trajectory?" It was this transition that made it necessary,
not merely to challenge Aristotle's explanation of a particular and somewhat incidental phenomenon, but to strike at the roots of his ideas on
motion. A mathematical theory of impetus was a far more destructive
weapon than a merely causal one.

I must add, readily admitting that the failure may be my own, that I do not altogether see the necessity for Professor Dijksterhuis's concern with the likeness or analogy between the concept of impetus and that of quantity of motion, or some other in seventeenth-century mechanical theory, though as he remarked the question was not first raised by himself, and his attitude to it is one with which I should not wish to quarrel. Nevertheless, it seems to me that there is a very considerable difference between the ideas of the seventeenth century, which were at least groping towards a conservation principle, however imperfectly, and those of the Middle Ages. Certainly, Descartes' view of this matter was faulty, and his assertion of the constancy of the amount of motion in the universe may be regarded as metaphysical rather than physical. Others, like Huyghens, brought rigor into the discussion of this problem, which involved the formulation of the laws of impact. Impetus mechanics, on the other hand, seems to have lacked the notion of something's being conserved; motion was simply dissipated, and (as with Aristotle) no attempt was made to strike a balance-sheet to show where the lost motion had gone to. Again, the concept of impetus did not traverse the ancient categories of violent and natural motion—which seems to place it in a very different world of theory from anything post-Galilean.

The thorny problem of precursorship, suggested once more by Professor Dijksterhuis's comparison of Nicole Oresme and Galileo, is one that has already been debated at some length, so that I can abbreviate my remarks here. It was aptly said, as it seemed to me, that "Galileo's treatment of the uniformly variable motion of falling bodies forms a continuation on a higher plane of that of Oresme"—provided we give due weight to the qualification "on a higher plane." I am sure that the author would be the first to agree that Oresme left much for Galileo to discover, not merely in the realm of conceptualization, but in the application of mathematical reasoning to kinematical problems. I should have thought (taking a rather contrary view to his) that, thanks to the efforts of medievalist scholars, appreciation for the magnitude of Oresme's achievement had never been greater than at present: at least, the medieval "precursorship" of Galileo has never been more widely recognized than now. Galileo, on the other hand, has suffered more than one hard knock in recent years. . . And I do not believe we need admire Galileo less, because we admire Oresme more than people did in the days when everything in the *Discorsi* was assumed to have sprung unheralded from his brain. No one could object to Professor Dijksterhuis's revision of over-superlative nineteenth-century estimations; here also, it would seem, a more just evaluation of the richness of the preceding tradition has become fairly general.

I hasten to add that I am very much in sympathy with Professor Dijksterhuis's anxiety for further study of the history of mechanics—both statics and dynamics—in the still rather dark period between Oresme and Galileo. The effect of Duhem's concentration upon Leonardo da Vinci in this period has perhaps not yet been wholly overcome.

With Galileo kinematics at least takes on a coherent form. As in statics the theorem of the lever provided a firm nucleus, so in the new mathematical science of motion the law of falling bodies, in its relation of time, velocity and distance, was both comprehensive and dramatic. It was the calculatory and predictive value of Galileo's work that at once fixed the attention of contemporaries, partly through the emphasis given to it by Mersenne, while Galileo's dynamical ideas were sharply criticized by Descartes and others. Galileo's science of motion was imperfect, in its phenomenalism, in the lack of precision of its concepts, in its mathematical presentation. One must certainly be cautious in order not to read into it more modernity than it holds. Once more one must applaud Professor Dijksterhuis's observation that the transformation of Galilean into Newtonian mechanics was no straightforward or routine operation. For example, it is obvious enough that Galileo's level of treatment is macroscopic, he is concerned with gross bodies; Newtonian mechanics, however, is only necessarily or primarily related to gross bodies in so far as these must be the subject of experiments. And it is as a molecular science that dynamics enters on its modern stage of development.

Reflecting on the closing paragraphs of Professor Dijksterhuis's paper it is difficult to do other than re-echo, as he has done, our lack of depth of insight into the nature and manner of Newton's work in science. A century ago the Victorians had a confident, if complacent, image of what that was. We have rejected that image because we are more sophisticated, and perhaps more philosophically alert, although in fundamental scholarship the works of Edleston, Rigaud, Brewster, and Rouse Ball still stand unrivalled. One may well doubt whether further spade-work, though very necessary, is not less essential than some great capacity for penetration and synthesis—the kind of ability, devoted to a more particular problem, that Professor Dijksterhuis has so conspicuously displayed in his swift, illuminating survey of the whole history of mechanics to Newton's time.

CONTRA-COPERNICUS:

A CRITICAL RE-ESTIMATION OF THE

MATHEMATICAL PLANETARY THEORY

OF PTOLEMY, COPERNICUS, AND KEPLER

Derek J. de S. Price

The chief purpose of this paper is to throw the spotlight on the mathematical astronomy of Copernicus. As one of the most popular subjects for writings on the history of science, there has been an enormous bulk of appreciations of the importance of *De revolutionibus,* and in all these there has been stated an appropriate expression of admiration for the undoubted worth of Copernicus as a mathematical astronomer.[1] I feel that in many ways this is a dangerous myth, more serious perhaps than those of Newton and the Apple, or of Galileo and the Tower of Pisa.

There is always some difficulty in writing historically about a subject with highly technical mathematical ramifications. It is inevitable that the works of Ptolemy, Copernicus, Kepler, and Newton have suffered in this way,[2] just as the work of Einstein will undoubtedly suffer also if the only due estimations must be based on more or less popular non-mathematical accounts. Is the theory of relativity to be reduced to a little statement that all motion is relative? It is much more difficult to explain the way in which Einstein developed new techniques of tensor calculus to meet new physical concepts and experimental results.[3]

It is just such a trouble we must find in estimating the work of Copernicus. Clearly he was a great figure, certainly his work is one of the great landmarks in the history of scientific thought. It is conceded that he was no great observational astronomer like Tycho Brahe; it follows therefore, in the minds of many people, that he must have been a great theoretician, and in all these treatments of the Copernican Revolution one finds an undercurrent of respect for the mathematical ingeniousness that surely must lay behind the spectacular advance he made in cosmology.

It is my thesis that the real cosmological changes proposed by Copernicus were quite unattended by any mathematical complexity and quite independent of any new mathematical techniques of more than the most simple and unsophisticated variety. Furthermore, even these simple changes had no effect whatsoever on the actual processes of calculation; they therefore did not increase or decrease the accuracy of the theory and, in fact, since the theories were formally equivalent there was no possible observation that could decide whether one was true and the other false.

Perhaps it is rather unfortunate historically that the same Copernicus who made this striking cosmological step also made several changes in the mathematical theory of planetary motion, quite separate from the cosmology and independent of it. I shall try to show that these mathematical changes are on the whole such as to increase the complexity of the system without increasing the accuracy; indeed, one may show that the accuracy was diminished by the process. Moreover, these changes were highly conservative, lacking all mathematical brilliance and insight. It would have been much easier for historians if these changes would have been made by, let us say, Regiomontanus, just before Copernicus, or by Tycho Brahe, just afterwards. There is no reason, other than historical coincidence, to associate them with the important cosmological step. Without them there would have been little danger of the erroneous suggestion that Copernicus was a greater or original mathematician, and his new theory would have been readily seen as a question of non-mathematical cosmology and incapable of support or refutation from all observations then accessible to naked-eye astronomers.

Lest it be thought that this is fruitless denigration of a great scientist, I must explain that only by clearing away these stumbling blocks can we arrive at a clear appreciation of the stature of Copernicus and the radically new features of his theory over and above the purely philosophical device of allowing the Earth to move. It is also of the greatest interest to show that Copernicus himself did not really understand what he was doing. He believed that the change from geostatic to heliostatic would make the theory more accurate, whereas it leaves it precisely unchanged. In one glaring case he points out Ptolemy's error in failing to account for the variation in brightness of Venus without realizing that his own theory contains exactly the same error. Even when Copernicus correctly claims the advantage of extra philosophical simplicity for his ideas, he confuses the issue by counting circles[4] rather than going to the heart of the matter; he did not realize that he was the first inventor of a mathematical planetary system as distinct from a mathematical theory of the individual planets.

♈ THE PRE-COPERNICAN RIFT
BETWEEN ASTRONOMY AND COSMOLOGY

Perhaps the first and most essential difficulty in re-estimating the work of Copernicus lies in a confusion about the relation between the mathematical and non-mathematical parts of astronomy. Since the time of Galileo and Kepler we have become very acclimatized to the idea that cosmology and non-mathematical, popular astronomy is but a reflection of the more technical and mathematical parts of the subject. It is therefore difficult to put on a different thinking-cap and realize that before this time the two parts of astronomy were so detached, one from the other, that they formed almost separate, mutually exclusive subjects having little bearing on each other. They had a common interest but little mutual influence.

On the one hand was the subject of descriptive and physical cosmology, predominently Aristotelian, with its objective of understanding the heavenly universe and relating it to the rest of man's knowledge; its methods included at the most quite simple and unsophisticated mathematical techniques.

On the other hand was the highly complex mathematical theory concerned with the phenomenological determination and prediction of the irregularities of planetary motions—the grossest regular motions being already accounted for by the simple uniform rotations of the descriptive cosmology. We shall call this mathematical theory "Ptolemaic Theory," though it must be remembered that it had its beginnings long before Ptolemy[5] and carried through Islam and the European middle ages up to and including the time of Copernicus. There are many widespread misconceptions about this Ptolemaic Theory, and we must now turn our attention to these.

First, there never was such a thing, in this sense, as a Ptolemaic *System*. The complex mathematical theory is arranged so as to deal with each planet separately and individually. There is no single mathematical connection between these several models, only a general similarity in the methods used for each. Not before the work of Copernicus was any such mathematical link devised that welded the whole into a mathematical system rather than a cosmological one. This is, in fact the most important aspect of Copernican Theory—the invention of a mathematical planetary system, rather than the change from geocentric to heliocentric. It must be kept in mind, however, that welding the parts into a whole did not necessarily make any changes in the format of the mathematical techniques used. The absence of any essential change means that there could be no change in accuracy or facility of calculation, and no possible observation of planetary positions which could be used as an *experimentum crucis* to decide between the rival theories. The position has, of course, been complicated by the fact that Ptolemaic Theory was set in a matrix of its contemporary cosmology—the nests of concentric spheres so often illustrated—but this setting had the status of philosophical belief rather than of a theory related to observations in an exact scientific way. It is my belief that Ptolemy himself and the other mathematical astronomers were particularly sensitive and guarded about this lack of relation between their technical studies and general cosmology. In the *Almagest* (IX, 1) there is an unequivocal statement that the parallaxes of the planets cannot be detected, so that no information can be obtained to decide their absolute or relative distances, and that one therefore must rely on the opinions of the "early mathematicians" even for the order of the planets— such hypotheses as one might make having no effect whatsoever on the subsequent mathematical treatment of the individual planets.

The fact that planetary parallaxes cannot be detected by the naked eye should make it clear that all observations and hence all planetary theory was concerned only with the angular motion of the planets. Indeed it was concerned only with their apparent motion on and about the arbitrary unit circle constituted by the ecliptic. Thus we must point out that in devising various models of deferent and epicycle circles, the mathematical astronomers were concerned only with finding the spherical coördinate angles of the planet on the unit sphere and not the length of its radius vector from the Earth. Contrary to popular belief, the technical astronomer was careful (in any doubtful case) to treat the proper point on the epicycle, not as the body of the planet, but only as a marker, which when projected on the ecliptic yield the position of the planet in that circle. We must not therefore make the mistake of thinking that the mathematical astronomers regarded the epicyclic loops traced out by the combination of deferent and epicycle as being in any way the real path of the planet in space. The orbit in space was not a question which could be resolved from observation alone, only by the importation of cosmological ideas not capable of experimental proof or disproof. One must allow that in many ancient, medieval, and Renaissance popular expositions, the planetary "marker" on the epicycle was taken as the body of the planet, and it was often supposed that the real planet was actually carried in such a path in space. It is a readily understandable error, perhaps a little hair-splitting to correct. Nevertheless it is an important point in the history of astronomy, for clearing away this misconception it becomes evident that the Copernican invention of a planetary *system* was the first step in enabling the technical astronomer to conceive of the mathematical device of circles as representing the orbit of a physical planet in space. However, there is no indication that this was clearly realized at the time, and it was not before the work of Kepler that it became quite certain that the mathematical treatment yielded an actual orbit in space, rather than a marker which was significant in angle but not in length of radius vector.

Having now made this point we must partially retract it by showing that although the Ptolemaic Theory was not designed to yield correct values for the variation in radius vector of the planets it did in fact do so. Indeed, it had to do so, for only by effectively simulating the true state of affairs was it possible to give an effective representation of the angular variation itself. In proving this we shall also be opposing the common contention that the Ptolemaic theory was merely a sort of Fourier synthesis, representing and successfully predicting periodic planetary motions only by the empirical superposition of enough circular components. It might also be remarked that neither the Ptolemaic nor the Copernican theory was unnecessarily complex because of any fetish about uniform motion in a circle being the only permissible and pure motion for celestial purposes.[6] The planetary motions, as seen against the background of the ecliptic circle, were evidently periodic and circular. They could not confirm in any way to motion in a straight line—the planets quite obviously "went round." Furthermore, their rotation was uniform, increasing by constant angles in constant times rather than undergoing any general difform acceleration. To the mathematical astronomer, uniform motion in a circle had the implication of what we would now call uniform angular velocity; only to the cosmologist and the popularizer did it carry also the connotation of a real circular orbit in space. One does not need to invoke any mystic considerations of purity of circular form; it was quite evident to the mathematical astronomer that the only conceivable theories of planetary motion must be based ultimately on some arrangement involving uniform angular velocities—there was no other

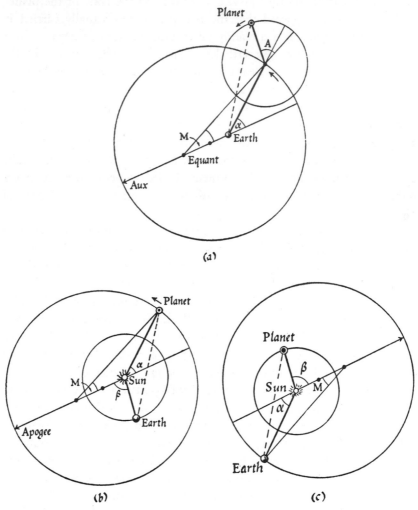

(a)

(b)   (c)

Figure 9.—Equivalence of the Ptolemaic and Copernican hypotheses; the diagram illustrates the two possible cases. (a) Ptolemaic construction. (b) Heliocentric construction for Mars, Jupiter and Saturn. (c) Heliocentric construction for Mercury and Venus.

M = Mean center, Mean anomaly. A = Mean Argument. β = A + M.

reasonably convenient form of any mathematical use and tractibility.

Returning now to the manner in which Ptolemaic theory gave an accurate representation of the actual motions of the planets we must first establish the fundamental principal of geometrical similarity which shows that angles, distances, and indeed all mathematical techniques are quite unaffected by any change from geostatic to heliostatic systems (see Fig. 9). The case illustrated is indeed only a special case of a more powerful theorem which might be used to consider three points in space, S(un), E(arth) and P(lanet), all moving in any paths whatsoever. If one considers S to be at rest, the ends E and P, of the vectors $\overline{SE}$ and $\overline{SP}$ will trace our paths which we may call the orbits of E and P respectively. Now, if one takes E as being at rest instead, S will appear to move in an orbit marked out by the vector $\overline{ES}$. Since this is exactly the negative of $\overline{SE}$, the orbit will be geometrically identical but turned through 180°. In this "geostatic" system, the orbit of P will now be given by the motion of the sum of the vectors $\overline{ES}$ and $\overline{SP}$, the former having just been discussed, the latter now becoming an "epicycle" added to this orbit. It should be noted that this argument is perfectly general and not restricted to a plane; it shows that planetary latitudes as well as longitudes are included in the principle of geometrical relativity between geostatic and heliostatic systems. Provided that suitable lines and planes of reference are maintained there is not a single mathematical technique or calculation which is peculiar to the geostatic or to the heliostatic system alone.[7]

It follows from this principle that the use in Ptolemaic theory of a geostatic deferent with epicycle is strictly equivalent to a heliostatic system in which the epicycle, transferred to a central position, becomes a second "orbit." For an outer planet the deferent acts as the planetary orbit and the epicycle as that of the earth; for an inner planet the roles are reversed. As a direct consequence we find that the variation in length of the radius vector from the earth to the planetary marker on the epicycle must actually agree with the real variation in distance. If it did not, the angles could not be brought into agreement. Without a correct simulation of the eccentricity of the planetary deferent, for example, the epicycle could not be made to subtend an appropriately varying angle at the earth. It is for this reason we have maintained that although Ptolemaic theory set out only to "save the appearances" in accounting for the angular variation of the planets, it had necessarily to represent correctly also the distance variation and hence the complete orbit of the planet with respect to the earth. With this established, we can turn our attention away from problems of geo- and helio-centricity and examine the way in which eccentric circles and epicycles may be used to correspond to real planetary orbits.

It is well-known today that the actual path of a planet around the Sun is accurately represented, perturbations excepted, by a Kepler ellipse. That is to say, the planet moves in an ellipse with the Sun at one focus, and it moves at such a rate that the radius vector sweeps out equal areas in equal times. Unfortunately, many people seem to have a most inaccurate picture in their minds when they think of this law: They imagine the ellipse to be rather elongated—perhaps taking their conception from the often used illustrations of the path of Halley's comet. In fact the planetary orbits are all very nearly exact circles in shape; the orbit of Mars for example is flattened by less than 0.5 per cent of the mean radius though the Sun is displaced from its center by more than 9 per cent of the radius. It follows from this that the actual shape of planetary orbits can be very well represented by eccentrically placed circles—it is certainly much more correct qualitatively than any sort of elongated ellipses.

The central difficulty of planetary theory is not however with the shape, as expressed in Kepler's first law, but with the angle variation, as expressed in the second. It is readily seen that the motion of the planet on the Kepler ellipse is very far from being uniform, whether one takes the Sun at the focus as center or whether one considers this motion as being about the center of the ellipse. In fact, even with modern mathematical methods there is no closed expression for stating the position of the planet on the ellipse as a function of time. For this purpose it is necessary to invoke the complication of elliptic integrals and their tabulation for each special case. In view of the extreme intractibility of the mathematics it is extraordinarily fortunate that an elegant and very accurate approximation exists. Perhaps the most extraordinary feature of it is that it is not exact but only a curious mathematical coincidence. One might well maintain that the entire success of Ptolemaic theory was due to the existence of such a coincidence. It gives rise to what one might call the fundamental theorem of pre-Keplerian astronomy.

The approximation may be stated in two slightly different ways. In the way in which it was known during the Renaissance it relates to the Kepler ellipse and states that the planet appears to move with uniform angular velocity about the *empty* focus (the *kenofocus*). It thus provides an answer to the usual question of the student who learns that a planet moves on an ellipse with the Sun at one focus and immediately demands to know what is at the other focus; it is the center of uniform rotation! In its classical and medieval form, relating to the approximation that the planet moves in an eccentric circle, the kenofocus becomes the Ptolemaic *equant*—a center of uniform rotation which mirror-images the displacement of the Earth within the eccentric circle (see Fig. 10). In either case

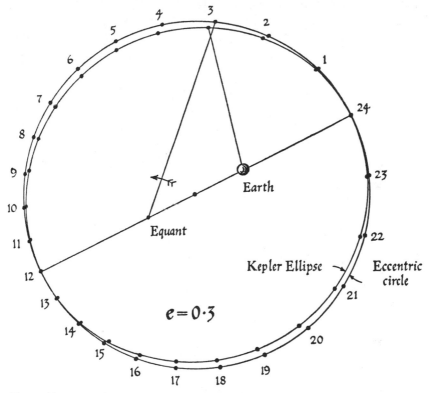

Figure 10.—Equivalence of a Kepler Ellipse and an eccentric circle. The error has been exaggerated by taking $e = 0\cdot3$. Since the errors in angle and radius vector are proportional to e^2, the orbits and positions would be scarcely distinguishable on a diagram this size constructed for Mars ($e = 0\cdot1$).

the approximation is surprisingly good, being well within the best attainments of naked eye astronomy (see Table I).

Using this "eccentric and equant" model for a Kepler ellipse we may now set up ideal schemes for both geostatic and heliostatic planetary theories. By comparing these with Ptolemaic and Copernican practice respectively, we may then see how near these systems attained the highest accuracy possible in consistence with the available techniques of using circles and uniform rotations. An ideal heliostatic system would consist of an appropriate eccentric circle for each of the planets in turn, the Sun being at one of its foci and the equant being at the empty focus. The circles must, of course, correspond in size with the real planetary orbits, and to yield correct variations of planetary latitudes, the planes of these orbits must severally be tilted so as to deviate slightly from the plane of the ecliptic. The angles and directions of tilt and axis for these circles may be considered as fixed or subject, at the most, only to a small secular motion.

For an ideal analogue of Ptolemaic theory we must select from the heliostatic system the eccentric circles corresponding to the orbits of the Earth and of *one* other planet. Using the principle of geometrical relativity we must now convert this from heliostatic to geostatic, and with just the same mathematical devices we arrive at a scheme in which the larger of the two orbits becomes the Ptolemaic deferent, while the smaller becomes an ideal version of the Ptolemaic epicycle. It should be noted that this ideal epicycle is to be mounted eccentrically on the already eccentric deferent and that either the epicycle or the deferent is to be conceived as being moved in a plane which preserves a slight angle of deviation from that of the ecliptic; for outer planets the deferent remains so inclined, for inner planets the epicycle.

The chief respects in which Ptolemaic theory falls short of this ideal lie in its use of central instead of eccentric epicycles and in not keeping the orientation of the epicyclic plane constant in space, but fixing it instead to the radial arm on which the center of the epicycle moves. Neither of these difficult questions was ever successfully solved by Copernicus or by any other astronomer totally immersed in the Ptolemaic tradition; it awaited the quite new ideas on real orbits, as introduced by Kepler, before the previously complex issues could be seen as essential simplicities. For Copernicus the matter was especially complicated by his most uninspired insistance on the use of central rather than eccentric circles. This caused him to take the center of the Earth's orbit, rather than the Sun, as center of the whole system and as the point through which the planetary planes passed. It caused him also to avoid the simplicity of the eccentric deferent circles and substitute instead the geometrical equivalent of a central circle together with a central epicyclet[8] having the same periodic rotation—an equivalence long familiar in Hellenistic astronomy, but usually avoided as inelegant. The use of epicyclets as well as epicycles cost Copernicus a great deal in pictorial simplicity for his system, though, of course, it made no difference whatsoever to accuracy of representation or ease of calculation.

The Ptolemaic use of central epicycles is a much more difficult matter to assess in importance. It is, however, of peculiar interest, not only in estimating the accuracy of that system, but also in showing how certain chance features of our solar system conspired to make this powerful system workable. It so happens that the orbit of the Earth has a very small eccentricity, so also has Venus, one of the two inferior planets. The other inferior planet, Mercury, has a very high eccentricity, but again it so happens that this planet is so near to the Sun that it cannot be observed by the naked eye except near its maximum elongation. We shall now show that if the Earth or Venus had a larger eccentricity, or if Mercury had been observable throughout a larger part of its orbit, the Ptolemaic system would have been unallowable in its elegant form.

First, let us consider the superior planets, Mars, Jupiter, and Saturn. For each of these the Ptolemaic theory invokes the use of an epicycle which moves with an annual period[9] to represent the relative motion of the Sun and Earth. Ideally, this epicycle should have a small eccentricity, fixed in space (see Fig. 11) and amounting to about 2 per cent of its radius. The figure (which has been exaggerated for clarity) shows that it is possible, to a large extent, to take this eccentricity away from the epicycle and add it instead to the larger eccentricity already present in the deferent circle. The vector addition of these eccentricities is precise, and carries the corollary that the *aux* or apsidal line of classical and medieval astronomy does not correspond numerically with the *apogee* line of Copernican and modern astronomy—there is a small angular difference caused by the vector addition of the eccentricity of the Earth. It should also be noted that although the vector addition is exact, it would result in the loss in colinearity of the Earth, equant and center deferent. If a linear approximation to this state of affairs be taken (as with actual Ptolemaic theory) the positions of Earth and equant can no longer be accurate and the theory will therefore become approximate; the error however is considerably smaller (about 1/10) than that caused by complete neglect of the eccentricity of the epicycle.

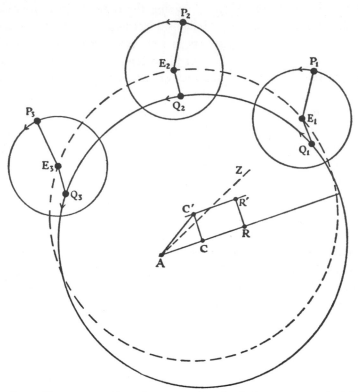

Figure 11.—Incorporation of an eccentric epicycle into an eccentric deferent. $AC=CR=C'R'$; $CC'=RR'=E_1Q_1=E_2Q_2=E_3Q_3$. The eccentricity in both deferent and epicycle has been exaggerated to show the effect clearly.

Secondly, we must consider the remaining possible case of the inferior planets, Mercury and Venus. The ideal geostatic model would require the use of a deferent corresponding to the motion of the Earth, and an epicycle corresponding to that of the planet. Now, Venus has an eccentricity even smaller than that of the Earth (about 0.5 per cent of the mean radius) so that once again the eccentricity is so minute that no damage is done by the use of an epicycle which is not eccentric. In this case, the deferent can, of course, be eccentric, so that the motion of the Earth is accurately represented without further approximation. For Mercury, the position is vastly different; the eccentricity amounting to nearly 20 per cent— an impossibly large amount to take into the deferent. Luckily however, Mercury can only be observed in the region of its maximum elongation, and the theory had therefore to account only for such positions and none in between. This leads to a very neat and novel solution. The deferent of Mercury (representing the motion of the Earth) was taken and distorted so that the angular variation was left unchanged, but the radial variation was changed to give the orbit an oval rather than a circular shape. This permitted the epicycle to vary its distance from the Earth in exactly the way required so that its angular diameter (on which depends the angle of maximum elongation) accorded exactly with observation.[10] Thus, because of all these special features of our world, a Ptolemaic theory was possible. It is perhaps interesting to reflect that Martian and Mercurian classical astronomers, should they exist, would never be able to start their civilization with a planetocentric theory; they would have had to evolve a heliocentric, Copernican-type system, or perish in the attempt.

ᛦᛦ THE PRACTICAL ACCURACY
OF PTOLEMAIC AND COPERNICAN THEORIES

Up to this point we have been considering the general agreement which exists between the actual laws of planetary motion and the kinematic models established by Ptolemaic and Copernican theory. It is already clear that neither system contained errors so gross as to lead to any readily observable divergences between theory and observation—not, at least, in the principal objective of accounting for the motion of the planets in longitude. We know however that the use of eccentric circles instead of Kepler ellipses involves some degree of approximation, and it is therefore necessary for us to examine the magnitude of this and other errors. This is of importance because both Copernicus and Kepler claimed the practical improvement of accuracy for their theories as well as the additional blandishments of increasing mathematical elegance and philosophical or physical truth. We are not here concerned with the philosophical battles over the systems but with the purely scientific ones (in the modern sense) about which theories corresponded better with the observations. We shall show that in this respect the claim of Kepler to have detected and removed inaccuracies was just, but that of Copernicus was false—a reasonable explanation, it may be allowed, of the gradualness with which the Coperican doctrine was accepted, even by his peers in mathematical astronomy.

It has already been demonstrated that the essential approximation of pre-Keplerian astronomy was connected with the use of eccentric circles instead of ellipses and this caused an error firstly by virtue of the angle in the deferent and secondly by virtue of the error in radius vector which affected the angular subtense of the epicycle. And beyond this there is the additional error caused by forcing the deferent to carry an additional eccentricity which ought to belong to the epicycle. These three quantities and their sum are calculated for the several planets in Table I, below. The special method necessary for Mercury makes the third type of error inapplicable here—if calculated in the same way it would amount to nearly 1°—hence the necessity of this special oval deferent which probably does its job to an accuracy better than 6′ of arc.

TABLE I

Maximum error (in minutes of arc) involved in use of eccentric circles and epicycles.

	Error caused by use of circle instead of ellipse in deferent.		Error caused by neglect of eccentric (epicycle)	Total error
	Direct angular error	Radial error (epicycle)		
Mercury	0.5	0.5	*	*
Venus	0.5	0.5	5.0	6.0
Sun	0.5	. .	. .	0.5
Mars	9.0	12.0	9.0	30.0
Jupiter	2.0	1.0	3.0	6.0
Saturn	2.0	0.5	1.5	4.0

* Principle not applicable; special model used to account only for positions near maximum elongation.

It should be remembered that the components of error discussed above have different periods and do not generally come to a maximum together; usually the error will be less than the values recorded in the last column of this table. Thus one can safely say that under the best conditions, a geostatic or heliostatic system using eccentric circles (or their equivalents) with central epicycles can account for all angular motions of the planets to an accuracy better than 6′ of an arc, excepting only the special theory needed to account for the seldom visible Mercury and excepting also the planet Mars which shows deviations of up to 30′ from such a theory.

Now, an accuracy of 6′ of arc coincides reasonably well with the best precision attained by naked-eye astronomers before Tycho Brahe.[11] It certainly is better than the accuracy of 10′ which Copernicus himself stated as a satisfactory goal for his own theory. Considering the influence of errors of refraction (already 1′ at an elevation of 45°) and the various perturbations which actually exist and modify the ideal Kepler orbits, such an accuracy is commendable if it could in fact be achieved. The systematic errors were, however, much greater than might appear from this argument, which has proceeded on the false assumption that the parameters used in the theory would be the best possible. Unfortunately these parameters were often highly inaccurate, partly because of the retention of old values rendered false by slow secular change, more often by the way in which these parameters were obtained by reasoning based on a single observation of some critical planetary configuration which could yield that parameter especially well. Tycho Brahe was the first astronomer to derive parameters by considering the entire cycle of the planet rather than a single special position of it.

It is difficult to estimate the accuracy with which the essential planetary parameters were known at any time; of course there was much variation from author to author and from time to time. Experience with medieval and early Renaissance texts however leads me to suggest as a working hypothesis that this inaccuracy was such as to lead to planetary positions which deviated by at least 30′ from true places, thus exerting more influence than the error inherent in the use of circles instead of ellipses. The Ptolemaic errors which Copernicus set out to rectify were thus almost certainly due to bad values of parameters rather than faults inherent in the theory. Certainly Copernicus did his share in trying different—often widely different—values of the planetary constants.[12]

Once the main source of error had been removed by Tycho Brahe's accurate recording of the entire cycles of planetary motion, the most outstanding divergence between theory and observation was in the case of Mars. It is especially important to note that Brahe improved the accuracy of parameters and observations at the same time, so as to obtain, for the first time in history, a clear discrepancy between the place of a planet and that predicted by "circular" theories. In the past such happenings had been swamped by all the other errors and uncertainties. Using Brahe's results, Kepler could prove in this one case (but in no other) that circular theory must break down. This is exactly the reason why Kepler ellipses could not be suggested before the work of Brahe, and indeed why Kepler had to develop his theories on the basis of the study of Mars.

It thus appears that after two thousand years of observation with the naked eye, the diligence of Brahe was a step that led Kepler to the culminating achievement of that art, a proof that circular movements could not accurately represent the heavens. We know how that enabled him to build his theory of real elliptic orbits in a heliostatic system; a theory later raised by Newton from the realm of kinematics to that of mechanics and physical explanation. It is therefore one of the greatest and most enigmatic coincidences in the history of science that this final culmination struck almost in unison with the new dawn following Galileo's first telescopic discoveries. It is well known how these new observations not only gave immediate qualitative proof of the soundness of the heliostatic system, but also the telescope itself was soon to provide a path, much easier than that followed by Brahe, to increasing the accuracy of observations to limits not conceived possible. It is almost like a man who escapes from prison on the very last day of his long sentence.

ꗣ THE BRIGHTNESS OF VENUS, A COPERNICAN CONFUSION

We return now to discussing the relation between Ptolemaic and Copernican theory, and consider a hitherto unnoted peculiarity of the case. In the Ptolemaic theory, the largest planetary epicycle was that of Venus, it being about three-fourths the size of the main deferent circle of that body. This size was obviously essential to account for the fact that Venus was restricted to a maximum elongation of about 45° from the Sun, a fact that could readily be appreciated by the non-mathematical cosmologists. Not only did such a large epicycle consume a tremendous amount of space if it was considered as a physical entity, but (again supposing it to be more than a mathematical fiction) it implied that this planet should undergo an enormous variation between its maximum and minimum distances from the Earth. The first difficulty is specifically pointed out by Copernicus in Book I of *De revolutionibus,* the second is given the greatest publicity at the very beginning of Osiander's foreword to the book. It is a much quoted passage, showing that if Ptolemaic theory were correct there should be this huge variation in distance and an even more drastic variation in the brightness of Venus because of the inverse square law of light. It so happens that although the other planets vary in brightness reasonably well in accordance with the expected or supposed fluctuation in distance, Venus, the planet with this large epicycle, varies surprisingly little. Certainly it does not fluctuate by anything so great as a factor of sixteen which

Osiander conservatively estimates to be required by Ptolemaic theory.

If this were the whole story, Copernicus would have a most valuable point in his armory, a proof that Ptolemaic theory was blatantly incorrect and needed his rehabilitation. Alas, there are many hidden barriers here. In the first place, and perhaps most curiously, neither Copernicus nor Osiander seems aware that the new theory predicts just the same fluctuation in distance between Venus and the Earth. There is therefore little point in upsetting the old theory to replace it by a new one containing the same fault. As we have shown, there is no possible observation of angle or distance which can lead to any blatant difference between the rival systems.

The explanation of this curious phenomenon of the too steady brightness of Venus was one of the things revealed by Galileo's telescope. Because its orbit is so near to that of the Earth (i.e. it has a large epicycle), Venus shows phases like the Moon, and more evidently than any of the other planets. Mercury also shows such phases, but, as we have noted, this was observed only near maximum elongation and therefore always in about the same phase. The phases of Venus are necessarily arranged so that it is "full" when furthest from the Earth and "new" when nearest. It so happens that the variation in distance is very largely compensated by the variation in amount of disc illuminated. Thus, although Mars, having a slightly smaller epicycle, varies through four stellar magnitudes of brightness in a cycle, Venus varies through only one (see Fig. 12).

There is no evidence that this explanation was ever conceived before the observations of Galileo. The planets, it was thought, were either self-luminous bodies, or, if they had their light from the Sun, it was reflected in a diffuse way from the body of the planet. The example seen so readily on the face of the Moon does not seem to have been heeded by anyone. Perhaps, after all, it is just as well. If astronomers had been impressed by the fact that the Moon always presented the same face to the Earth, it might have been taken as a powerful argument for the physically central position being occupied by the Earth instead of the Sun; it is a good anti-Copernican argument. But then, the Moon was geocentric in the Ptolemaic as well as Copernican theory—a reason why we have not discussed it above. It must however be conceded here that the Ptolemaic theory of the Moon was vitiated by essential errors and that these were efficiently corrected by Copernicus. It has, however, recently been shown that the method of adding a secondary epicyclet, as practiced by Coperni-

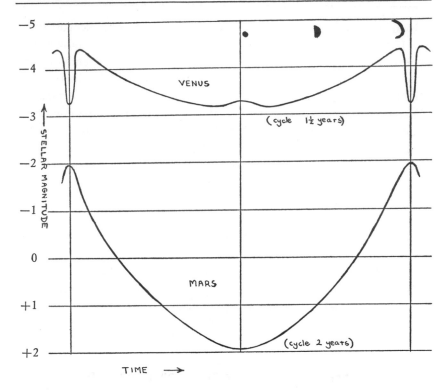

Figure 12.—Cycle of brightness for Venus and Mars (based on Nautical Almanac tables 1957/58). Note: Planet not visible to naked eye near ends and middle of cycle.

cus, had already been anticipated some 150 years before by the Arabic astronomer, Ibn ash-Shātir.[13] It may well have been this success with additional epicycles that led Copernicus to his cumbersome multiplicity of epicyclets in the other planetary theories.

⊞ THE REAL ACHIEVEMENT OF COPERNICUS

We are now in position to summarize the negative part of this argument and state what Copernican theory did not do. It did not, and could not use essential mathematical techniques or make any sort of calculation not already possible within the old Ptolemaic theory. Except for slight re-determination of parameters it could not provide different predictions or make calculation less difficult or arduous. It could explain just the same range of phenomena, and failed in just the same way to explain the anomalous variation in the brightness of Venus. At first sight it did not provide a more simple or a more elegant model of the planetary geometry—for although the change from geocentricity to heliocentricity transformed away the five planetary epicycles, Copernicus had added a large number of extra epicyclets to get rid of eccentrics. The new system may have shown an internal cohesion, but the old had the advantage of being split into separate units of less complexity and some uniformity.

To all this we might venture some criticism of *De revolutionibus* as a book. Again we need say nothing about the part of the book—its first few pages—which is concerned mainly with the thrilling philosophical arguments for mobility of the Earth. After this section, Copernicus begins with the real business of technical mathematics, and at this point the book becomes little more than a re-shuffled version of the *Almagest*. One might, in fact, call it a plagiarization of the *Almagest,* if it were not for the undisputed fact that Copernicus had no intention whatsoever to deceive. He had however the best of reasons for wishing his new ideas to appear, clothed in the most respectable and conservative form for traditional astronomy. It is for this reason that the essentially cosmological contribution of Copernicus duplicates not only the mathematical machinery of Ptolemy but also the method and structure of his book. Chapter by chapter it has the same format and language, the same arrangement of subject-matter with only the slightest changes as dictated by the change to heliocentricity. Thus the star catalogue is made to precede the sections on Solar and Lunar theory instead of following them. The *Almagest* is not an easy book to read, but it was at least original in many of its parts. The *magnum opus* of Copernicus does not have that distinction beyond its first few pages. It contains a few new observations and computations based on them, but beyond that it paraphrases Ptolemy, putting his work into a somewhat transparent new cloak.

What then did Copernicus achieve? Certainly he re-opened the question of mobility of the Earth and showed that no mathematical damage was done by the hypothesis. Beyond that his sole significant mathematical contribution lay in recognizing a common feature in the separate planetary theories of Ptolemy and in using that common feature to form the whole into a single system. It was already well known that, in addition to its own peculiar periodicity, each planetary theory contained an annual element; for the superior planets it was in the epicycle, for the inferior, in the deferent. Copernicus saw that this common element could be separated and made to correspond to the annual motion between Sun and Earth. In the course of this separation the remaining circle of each planet (deferent to superior planets, epicycle for inferior ones) would be in proportion to the size of that planet's orbit relative to that of the Earth. In Ptolemaic theory all distances had been relative. In the new theory they were each related to the common element of the Sun-Earth system, and thereby related to each other. Conveniently, on this system, the planets were ranked in order in the traditional fashion (or rather, the most popular of the traditional fashions), according to their angular velocities. As an additional point, the new scheme left plenty of room for the geocentric Moon and kept the size of the universe within reasonable bounds. As a mathematical system for the planets as a whole it was decidedly more attractive than a theory which needed special application to each of the planets in turn. In this sense it was a better theory because it was more economical. In actual practice though, as we have shown, extraneous reasons caused Copernicus to increase the apparent complexity of his system so that it must have seemed much less economical than the old and more familiar ideas.

In some ways the proposals of Copernicus must have seemed at the time rather similar to those set out by Einstein before observations of relativistic effects in Mercury and eclipses had confirmed his conjectures. The new theory unified several parts of the old; it was more complicated but more economical; it explained various phenomena, though these could otherwise be explained by more traditional hypothesis; finally it could not be proved or disproved by any observation available at that time. No wonder good scientists remained skeptical until the new and decisive evidence was forthcoming.

For this reason, I contend that although Copernicus made a fortunate philosophical guess without any observation to prove or disprove his ideas, and although the elementary mathematics necessitated by his change in the cosmological picture shows competence, his work as a mathematical astronomer was uninspired. From this point of view his book is conservative and a mere re-shuffled version of the *Almagest*. Above all, it introduced many false trails that must have hindered the acceptance of the one good point. In the domain of mathematical astronomy the first major advance after Ptolemy was made, not by Copernicus, but by Kepler.

References

1 It would be invidious to cite a selection of otherwise good books which retreat into blind admiration instead of a critical examination of the work of Copernicus in mathematical astronomy. Until recently there has been no good corrective to this. While this paper was in preparation however, there appeared a new book by Thomas S. Kuhn, *The Copernican Revolution* (Cambridge, Mass., 1957), which provides for the first time in English a brilliantly clear account of the chief technical features of Ptolemaic and Copernican astronomy. The appearance of this book has enabled me to prune from my paper all the background material which otherwise would have been necessary to set the scene for my own special criticisms of Copernicus. A good mathematical analysis of the technicalities of Ptolemaic theory is given by Norbert Herz, *Geschichte der Bahnbestimmung von Planeten und Kometen* (Leipzig, 1887–94); in two parts.

2 As an indication of the way in which Ptolemy, Copernicus, and Kepler have suffered it may be noted that for none of these authors is there a good English translation available.

3 Although difficult, a good explanation of this sort is given by René Dugas, *A History of Mechanics* (Neuchatel and New York, 1955), pp. 463–534.

4 In point of fact, Copernicus actually finished with more circles than Ptolemy because of his machinery for avoiding the use of eccentrics.

5 For a comprehensive account of the work before Ptolemy, especially that of the Babylonian astronomers, see Otto Neugebauer, *The Exact Sciences in Antiquity* (2nd. Ed., Providence, 1957).

6 Professor Neugebauer has pointed out to me that, in fact, however, Copernicus over and over again polemises against Ptolemy for his use of motions which are not of constant angular velocity. Copernicus sees a major advantage of his system in the *philosophical* restoration of uniform circular motion to the whole planetary system. Neugebauer feels it is one of the main complaints against Copernicus that he *mixed up* the cosmological section of astronomy with its mathematical methodology.

7 Perhaps one might dare to add at this point that the whole problem implied by "sun at rest" is undefined, even within our galaxy, not to speak of larger problems and relativity (O. N.).

8 I use the term "epicyclet," corresponding to what is otherwise called a "minor epicycle," in contradistinction to the major epicycles of planetary theory. The major epicycles are usually much larger and are governed by a period quite different from that in the main deferent circle. The epicyclets are usually small and have periodicity exactly equal to that in the deferent or (rarely) just twice this period. The use of an epicyclet to replace eccentricity of the deferent is noted in *Almagest* iii 3, and in *De revolutionibus* iii 15.

9 This annual periodicity is slightly concealed by the use of a line of reference which is not fixed in space but turns as a radius of the deferent. The angle in the epicycle is therefore equal to that in deferent minus that of the annual motion. An explicit statement recognizing this rule occurs in the tentatively Chau-

cerian text, *The Equatorie of the Planetis,* edited by the present writer (Cambridge, Eng., 1955), line G.d., p. 30.

10 The Ptolemaic theory of Mercury thus provides an excellent example of a scientific theory that was true only in cases where it could be tested, and was in fact false for all other cases. Thus the position of Mercury as calculated for horoscopes was only correct if that planet happened to be near maximum elongation, and the whole astrological theory of the influence of Mercury proceeded on this false theory.

11 It appears that this ideal limit of accuracy is valid even as far back as Babylonian astronomy (O. N.). See also my discussion of instrumental accuracy in *A History of Technology,* ed. Singer *et al.* Vol. III (Oxford, 1957), p. 583.

12 As an example of this one might cite the all-important value for the eccentricity of the Earth's orbit as used by Copernicus. In the *Commentariolus* it is taken as 1/25, but in *Derevolutionibus* it is taken first as 1/31 and subsequently as varying between 1/24 and 1/31.

13 I am indebted to Professor E. S. Kennedy of the American University at Beirut for communicating this discovery to me. An account of it has been published by V. Roberts of that same university in *Isis,* XLVIII (1957), pp. 428–32.

Acknowledgments

I must thank the Cambridge University Press for permission to reproduce figures 9–11 which first appeared as figures 16–18 in my book, *The Equatorie of the Planetis.* I should like also to express my deepest thanks to Professor Otto Neugebauer for his comment and criticism of this paper.

FRANCIS R. JOHNSON

on the Paper of Derek J. de S. Price

⚕ Bearing in mind Dr. Price's interpolation in his paper in which he described Copernicus' work as a "plagiarism of Ptolemy," I cannot resist the temptation to point out that his provocative title, "Contra-Copernicus," is itself in the same sense a "plagiarism" of the title of a late sixteenth-century pamphlet. Then, as Dr. Price did, I will retract the word "plagiarism" and substitute a word much more suitable and accurate in its Renaissance connotations: "imitation."

Dr. Price's lucid paper should contribute greatly to the therapy of those writers on the history of science suffering from a special form of the disease described previously by Father Clark as *precursitis*—writers who express surprise at the slow acceptance of Copernicus' heliostatic cosmology—I agree with Dr. Price and with Father Clark that "geostatic" and "heliostatic" are terms preferable to the traditional designations of "geocentric" and "heliocentric." Dr. Price's clear exposition of the defects of the mathematical planetary theories of Copernicus should relieve his contemporaries of the charges of blind conservatism and obscurantism so frequently made against them because they did not immediately enroll themselves as supporters of the new heliostatic cosmology. His exposition makes abundantly clear that they had sound scientific reasons to reserve judgment and to withhold unqualified support until new evidence came to light to demonstrate the superiority of Copernicus' mathematical cosmology, and thus give it the right to supplant that of Ptolemy, from which it did not differ essentially either in method or in basic assumptions.

Dr. Price clarifies a great many points by emphasizing the pre-Copernican rift between mathematical astronomy and descriptive cosmology. The popular textbooks of the sixteenth century were superficial, and few attempted to give a lucid exposition of the mathematics of Copernicus. The advanced texts for which new editions were in demand after 1543 were the standard expositions of the mathematics of Ptolemy; for example

Purbach's commentary on Regiomontanus' *Theorics of the Planets.* These held the field among scientists until the time of Kepler. Purbach and Regiomontanus dealt with each planet separately as a special case; thus, in the sense in which he uses the term, Dr. Price is right in stating that there was no such thing as *the Ptolemaic system;* that Copernicus' distinctive contribution was the invention of a mathematical planetary *system.* In Ptolemy, planetary theory was concerned solely with the angular motions of the planets—with locating the angle of the planet and not in determining the length of the radius vector. The innovations of Copernicus were of the same mathematical order as the constructions employed by Ptolemy. Within the limits set by the accuracy possible to observations by the naked eye, the mathematical constructions of Copernicus could not, by their very nature, be shown to be significantly more accurate in predicting the observed position of a planet than the complex mathematical constructions of Ptolemy. In fact, in the matter of accounting for the variations in the brightness of Venus, Copernicus had exactly the same defect as his predecessor, Ptolemy. Whether Copernicus was or was not aware of this defect is of minor importance. Probably he had overlooked this consequence of his mathematical theory until his critics pointed it out to him. The fact that should be emphasized and re-emphasized is that there were no means whereby the validity of the Copernican planetary system could be verified by observation until instruments were developed, nearly three centuries later, capable of measuring the parallax of the nearest fixed star. For that length of time the truth or falsity of the Copernican hypothesis had to remain an open question in science.

It is true that by introducing a cosmographical assumption incapable of verification, pre-Copernican astronomers did proceed to calculate the distances of the various planets. The assumption was purely mechanical; namely, that each planetary sphere imposed its rotation on the sphere immediately below it by having one point of contact with it, so that the least distance of one planet from the central Earth was equal to the greatest distance of the next planet inferior to it. Hence come the illustrations which accompanied most editions of the *Theorics of the Planets,* portraying annular rings of varying thickness representing the space within which each planet moved in the mechanical model of the planetary spheres which was introduced to portray graphically, for the imagination of the reader untrained in mathematics, a model cosmology which could not possibly be supported by observational evidence. The mathematicians themselves were always clear that their constructions had no provable relation to physical reality. This was not always true of their readers, whether in the sixteenth century or today.

THE PLACE OF THE HISTORY OF SCIENCE

IN A LIBERAL ARTS CURRICULUM

Dorothy Stimson

Readers of these papers, I am sure, do not have to be "sold" on the value of the history of science for men entering professions related to science, nor indeed on its value for those whose later careers will bring them in touch with science. Most of us also are well aware of Dr. Sarton's insistence on the fundamental value of the history of science in education in general. As Professor Max Fisch has summarized Dr. Sarton's key positions, the history of science is "the leading thread in the history of civilization, the clue to the synthesis of knowledge, the mediator between science and philosophy, and the veritable keystone of education."[1] Through it, as Dr. Sarton himself wrote, one traces the "gradual liberation of minds from darkness and prejudice" and sees "the gradual revelation of truth, with man's reactions to that truth."[2]

These are ideals, basic propositions with which we are in accord in greater or lesser degree. What is their relationship to the liberal arts curriculum of today? The conditions of the modern world have compelled people in general and educators in particular to re-evaluate the aims and programs of education in college and university classrooms. How best can the youth of today be trained to live fruitfully in this shrunken but expanded world while under the shadow cast by the atom? Oceans are barriers no longer. What happens in Korea or Taiwan forthwith affects the Mississippi Valley. Insularity and provincialism can no longer be afforded. How can the individual best be helped to maintain his own sanity and develop his best powers while coping with the tremendous issues confronting the nations today? How should the colleges and universities meet such problems?

The Report of the Harvard Committee, on *General Education in a Free Society,* published in 1945, has proved to be one of the most influential as well as profound of the various studies made by educators about these questions. That Committee saw as one aim of education the need

223

"to break the stranglehold of the present upon the mind." For "the civilized man is a citizen of the entire universe; he has overcome provincialism, he is objective and is a spectator of all time and all existence." To them, therefore, "honest thinking, clearness of expression, and the habit of gathering and weighing evidence before forming a conclusion" are much more important than any particular group of introductory courses.[3] Herein lies the value of the scientific attitude of mind.

One of the major fields the Harvard Committee advocated to achieve its goal of the civilized man is that of the history and philosophy of science. Through this is gained "not only specific knowledge and skills, but conceptual interrelations, a world view, a view of man and knowledge, which together constitute the philosophy of science; a history which forms a continuous and important segment of all human history; and writings which include some of the most significant and impressive contributions of all literature." For science, they wrote, is both "the outcome and the source of the habit of forming objective, disinterested judgments based upon exact evidence." As it has always been an important part of the total intellectual and historical process, through its history is gained an insight into both the principles of science and an appreciation of the values of scientific enterprise. They pointed out the "combination of logical analysis, careful observation and experiment, and imaginative insight which has characterized the great scientific advances of the past."[4]

So much for the civilized man; how about the civilized woman and her education today? Are not these values of equal importance for her? She may have no intention of entering the professions involving science. Her major concern may be to be a partner in a happy marriage, and to that end, as a true partner, do her share by earning money while her husband completes his training before she settles down as wife and mother. What values for her has this study as part of her education?

Ten years ago I presented a study of this question as "A Case Report,"[5] based on 273 replies I had received to a questionnaire sent out to the 445 young women who had studied the history of science in my classroom at Goucher College at some time during the previous twenty-five years. Many of them in formulating their replies had perhaps been directed in their thinking by the phrasing of my questions. Most of them had had the perspective of the passage of the years in which to consider relative values. In addition to these 273, three others returned the questionnaire unanswered, one saying it had been sent to her by mistake since she had not taken the course (my records showed she had earned a grade of C in it). But for the 64 per cent who responded, it had made a vivid enough impression for them to link it specifically with their increased awareness of the history of ideas and of historical perspective, their appreciation of scientific method in other fields than just in the laboratory, and their recognition of an integration and correlation of the fields of knowledge beyond the somewhat artificial barriers of college departments. Those were some of the major findings from that study.

Now I wish to add to that study one I have just made on a much smaller and more informal scale. What do women students think are the values of the course while they are still in college? This spring I asked a class of 21 sophomores, juniors, and seniors to give me frank written arguments pro and con about its value to them in a college setting where the history of science had not been taught before. Two-thirds of the group were majoring in science, ranging from pre-medical preparation to geology, mathematics and psychology; the other third as majors were scattered among the humanities and the social studies from Fine Arts to History, Child Study, and English. In numbers they were too few to make a comparable group to the alumnae of the earlier study. On the other hand all twenty-one replied with an independence and freshness of outlook that perhaps compensates at least in part for their smallness in number. Their comments are suggestive as to the values women see in this field.

With one accord this group of students emphasized the correlation this work had made for them with their other studies, bringing them "increased awareness of the interrelationships of areas of cultures." It helped one "to see how the different fields of learning are related to one another, to see them as a whole"; or, as another wrote, "it presents a challenging, dynamic, vivacious and penetrating continuum of the advance of man's control over his universe, his reconciliation of this scientific knowledge with social, political, economic and religious conditions of his time, and of the unbroken chain of the scientific point of view." One remarked that it should be a required course, for it was a survey of thought. Through it the student gets "a feeling of civilization awakening and of opening its eyes to everything around it. It brings everything together," she went on.

Though these students did not know it, not only were they supporting what their predecessors in the classroom had written ten years earlier, but they were bringing out what Dr. Sarton had emphasized, that the history of science is a clue to the synthesis of knowledge. And the one who thought it "vital," and the other, that it should be required, were in a measure considering it as he did, the keystone of education. A third referred to it "as necessary as vitamins usually are in the diet, especially the most narrow or limited diet."

Furthermore, the one who saw the work as "a continuum" and "an unbroken chain" had realized what is the major characteristic and the major importance of the history of scientific thought. In science no man starts afresh; each one builds on the work of others whether in mathematics or in biology. Consequently, science, like an organism, is always changing, ever growing, as Veblen[6] pointed out long ago. Its massive achievements today are the masterpieces not of our contemporaries alone but the cumulative results of thousands of workers through the ages, of many nations, creeds and races, the known and the unknown. They asked questions, they made suggestions, they died. Others, checking those suggestions, asked other questions, made other suggestions, and so the work went on. This, I believe, is a continuity not as true of the history of art, nor of religion even. But rarely in my experience, has an undergraduate consciously realized "the unbroken chain" so early in her experience of the study. Yet this recognition of the continuity in man's scientific thought is a fundamental value in the education and outlook of the fully civilized man, for it fosters humility of mind as well as comprehension and appreciation.

Many of the students went on to make the more obvious comment: If the history of music, of art and of religion, of philosophical and of political ideas, is taught, why not the history of science? Almost all of them asked this question in one form or another. It is "as important as a knowledge of our literary heritage or of our artistic development," wrote a student majoring in history. Another stated it was necessary for every intelligent person to know something about science, and the history of that subject stimulated interest. A chemistry major claimed that it awakened an awareness to the history of scientific ideas and their importance, since "the philosophy of modern society is nurtured on science." "It provides an opportunity to see why other things were studied," wrote a senior, "to respect science, to realize its values and its limitations." "For any science student it correlates the other sciences with her own science and makes her aware of science's place in the literary and philosophical worlds."

A third aspect of the subject, variously stated, might be summed up in the phrase, "It makes one think." "It opens channels" and "aroused curiosity in more fields than I had hoped for," were two of these comments, while a third referred to "limitless areas opened up for research and study." But for one student it revealed as she expressed it, "How thin our ideas are in comparison with thinkers of the past. It is good to see how dedicatedly and how humbly these men of the past worked for the sheer love of work."

It is true that any field of study can make one think, but the history of science with the continuity of its development down through the ages together with the importance of science today gives this study special advantages. For one student, "the development of scientific ideas studied in context, in the framework of culture and civilization, interrelated one culture and one civilized area with another in terms of specific ideas." "It leads," she claimed, "to a better understanding of civilizations outside the West and helps to rediscover the East on its own terms, acknowledging the differences of culture that make for differences of opinion."

We who have worked in this field for many years may have forgotten or have failed to realize how dramatic, even stunning, some of our subject matter may be to the novice making her first acquaintance with some of the ideas it presents. Scattered among these replies are three examples of this. An English major, a senior, wrote of gaining as she phrased it, "a simultaneous ordered consciousness of the world as it changes and develops." Then, she added, "how very old the earth is and how very short a time the race of man has been on it." A physics major wrote: "The thrill of reading Leeuwenhoek's discoveries of the world under the microscope is equally as exciting as reading any of the Elizabethan authors." For a third, "the trials and tribulations over the acceptance or the nonacceptance of the Copernican theory is an amazing story." In more general terms, one or two spoke of developing "great respect for the ancients." For another, the Middle Ages came alive.

Among these various statements there were two that to me were significant in their prescience. A sophomore wrote that the history of science was not "necessarily of immediate value to her now but a subject awareness and understanding of which [I quote her exactly] will grow with the student in years ahead." This student has foreseen a value in the work which alumnae of the course after twenty-five years amply supported in their responses to that questionnaire ten years ago.

The other significant comment was made by a junior, and, may I add,

I know nothing of her circumstances. This study, she wrote, would lead to more financial support of scientific research and fellowships. "I personally," she went on, "would be more willing to donate such at the age of my financial prime after a study of the history of science than before." To me this comment underlines one important aspect of the value of the history of science in a liberal arts curriculum for women. Ultimately many of these women as heirs of their fathers and their husbands will control large financial resources. It is a well-known fact that much of the wealth of this country, both in real estate and in stocks and bonds, is held by women. They, as college graduates, will be members of boards of directors of many kinds of organizations, especially of philanthropic and educational ones. As they comprehend the aims of scientific research and appreciate the ultimate values and the goals of work being undertaken, they can further its progress both by their understanding of the development of science in general and by their financial aid, as this junior foresaw. This may be a low motive for encouraging the study of a subject, but it is not to be disregarded. Furthermore, women's opinions, as well as their financial support, affect the climate of ideas in their communities. It is important that they be trained to think clearly and objectively, to be sensitized to the significance of ideas and to acquire historical perspective as well as some general idea of what science is and how it has come to be. Thus they should be better equipped to judge wisely when they come into positions of responsibility, and science should be the gainer.

If a class of twenty-one young women without previous precedent or exposure can find such values, interest and influence, in a one semester course in the history of scientific ideas, why is such a course not universally offered in liberal arts colleges? One hampering factor has been the lack of translated source materials to which the student might be directed. That handicap, with each passing year, becomes less and less important as scholars are now making available the classic masterpieces of science in translation. Even in the "paper-backs," and so well within the range of a student's pocket-book, one can now get a modern translation of Galileo's *The Starry Messenger*,[7] until this year available only in a very rare out-of-print translation of the eighteen-eighties. And along with this most influential little book is the famous *Letter to the Grand Duchess Christina*, that in 1616 sparked the condemnation of the Copernican theory "until corrected." Important secondary works, like E. A. Burtt's *The Metaphysical Foundations of Modern Science* that were also out of print, are now reappearing in this same inexpensive form. The library situation is notably changed even in the last ten years and is steadily improving as more and more of these translations and reprints are published.

At the risk of trespassing upon the preserves of the next paper, I want to insert a word or two here about the special usefulness for the non-scientific undergraduate of certain recent publications. Three of these books are: Herbert Butterfield's *The Origins of Modern Science* (1949), Marshall Clagett's *Greek Science in Antiquity* (1955), and A. R. Hall's *The Scientific Revolution* (1954), now available as a "paper-back" also. A fourth, of greater difficulty for the undergraduate student, is Alistair Crombie's *Augustine and Galileo* (1952), which has been re-issued as a "paper-back" also. It is to be hoped that Mr. Clagett's proposed second book dealing with science in the Middle Ages may soon be published, for it is needed to be used with Mr. Crombie's, on the period between the end of his first study and the starting points of Mr. Hall's and Mr. Butterfield's books. These books are by no means textbooks in the ordinary sense, but they provide a basis for discussion and a supporting foundation for classroom lectures that of necessity must emphasize major themes at the expense of detailed explanations and they are both readable and intelligible to the undergraduate with a limited training in science.

In addition to these books there is Henry Guerlac's *Science in Western Civilization: a Syllabus* (1952). This is particularly useful as a guide to some of the notable articles appearing nowadays in various learned journals like *Isis* and the *Journal of the History of Ideas*. Other books may provide valuable bibliographies for supplementary reading but rarely do they call attention to studies appearing in periodicals; yet some of these articles are of considerable significance, as, for example, J. H. Randall, Jr.'s "Development of the Scientific Method in the School of Padua" in the *Journal of the History of Ideas* (I, 177–206), or Lynn White, Jr.'s article on "Technology and Invention in the Middle Ages" in *Speculum* (XXV, no. 2, 141–59). Nor should one miss the William Harvey Issue (April, 1957) of the *Journal of the History of Medicine and Allied Sciences*. Such studies not only make available fresh material but also present possibilities for similar research in related fields. Aside from its other values, Mr. Guerlac's guide to some of these articles is particularly useful for undergraduates and their instructors.

But to return to my question: If the values of this historical study of scientific thought are so important, why is it not a part of the liberal arts curriculum for women and men alike? The earlier handicap of limited printed materials is now being rapidly overcome. But a second difficulty faces the colleges, especially those for women that favor women on their faculties. Who is prepared to teach such a course, particularly among the women? Harvard, Cornell, and Wisconsin, in particular among the universities, have been providing graduate work in the history of science for a number of men, but thus far they have attracted relatively few women who might later become teachers of these courses. To my mind, this is the result of a vicious circle: The undergraduate colleges, especially those for women, do not give such a course, therefore the women are not introduced to the possibilities of the field.

The situation is further complicated because of the breadth of preparation and background the really expert teacher should have. Dr. Sarton used to claim that the instructor was adequately trained only if he had had advanced degrees in both some science and also in history, if he knew Latin and Greek as well as the modern European languages, and above all if he had a knowledge of Arabic. Who among us can lay claim to all these qualifications? Fortunately, the true teacher learns as he teaches—though maybe not Arabic—and students are encouraged as they realize their instructor is working along with them as an older guide. With vision, courage, and the scholar's love of learning, teachers of the history of science will grow in their work as they teach. But they have to start teaching it.

Here is another difficulty, one that has changed in character in recent years. College faculties have to be persuaded to approve the introduction of the subject into their programs. Years ago, and possibly in some places still, the science departments were fearful lest it be considered a substitute for a course in "real" science. They believed that no one who was not a scientist could teach a history of science. That doubt seems to have worn thin of late. The interrelations of science with world affairs today may well have helped to modify their thinking and to lead them to encourage their students to consider the history of their subject in order to have a better comprehension of its present meaning and importance. Indeed, members of these science departments are in a number of instances considering how to prepare themselves to introduce and teach the course.

Unfortunately, this receptive attitude is still not usual among the non-science faculties. Never having themselves been exposed to such a history and being absorbed in the virtues of their own subjects, they tend to think that scientific ideas and their history are the concern of the scientists and far less so of students in the humanities, the arts, and the social studies. There are many exceptions to this position, of course, but the instructor introducing a history of science course must be prepared to find himself classed with the scientists on occasion. Until he makes his work known through his students and through his own relationships with his faculty colleagues, he may find that relatively few non-science students are electing his work. Yet the subject has immense value for them as well as for the chemist or the physicist. Patience and considerable faculty education are still much needed before full acceptance is gained. The basic trouble is indifference, not antagonism. As more courses are introduced, this difficulty will subside.

The problem remains, however, of encouraging this study for the women students in the liberal arts colleges. I would be interested to know how many women there are in the courses being given in the coeducational institutions of the West and the Middle West where this subject is now being developed. In the eastern women's colleges there are only a handful. Yet it is an unusually attractive field for the woman college graduate in the variety of opportunities it provides, whether for research and teaching, or for positions in the other professions and in the business world, when accompanied by special skills and training. One line of development is through librarianships in the specialized libraries of today; another, through curatorships in museums of science and technology; a third, through the use of historical information in the advertising of scientific products, while the able writer has almost a virgin field wide open for her talents as a novelist, a biographer or a reporter. Above all, the college woman as a citizen of the world today needs the perspective, the point of view, the understanding of the meaning of science and scientific method, whether she be housewife or professional woman or both. With such a background, what she reads and what she hears discussed have new meanings for her and new attractions, while at the same time she has a basis of fact and training to help her separate fact from opinion and to bring reasoned judgments to bear on some of the major questions of the day. The woman so trained is not likely to be misled by such talk as there was some years ago about the supposed desirability of a moratorium on science because of the terrors arising from splitting the atom. Fuzzy opinions like that are deporable and dangerous.

Before concluding this paper, may I be permitted to give some personal evidence on this general subject? Possibly I am a pioneer teacher in this field, for I first gave a course in the history of scientific ideas in 1922–23 and have been teaching it ever since with the exception of two years when I was on leave for studying and writing. I have taught it in one college for thirty-one years, in two others for a semester each. I have led a week's seminar of daily meetings in one part of the work in a fourth college. I have even given a series of ten lectures in the field for an adult education program for men and women. I was able to undertake this teaching in the first instance through an accident in an academic situation that gave me an opening, even though by Dr. Sarton's standards I was woefully ill-prepared. As a student of history I have never claimed knowledge of science; but I have maintained throughout these years that the history of science is one phase of intellectual history and, as the history of ideas, can be taught with profit to student and teacher alike. The more the instructor studies and learns, the richer his teaching becomes. I have seen my syllabus and reading list grow from four pages of mimeographed sheets to a crowded booklet of twenty pages, three times revised and reprinted, and now alas, again out of date. This paper, therefore, is written largely on the basis of my experience of thirty-five years' teaching in four different women's colleges.

Lest I have fooled myself about the value of the subject that has been my major intellectual interest all this time, ten years ago, as I have already stated, I drew upon the experience of the alumnae of my course and again this year I asked for opinions, this time from undergraduates just completing their work with me. Despite whatever shortcomings their instructor had, the alumnae and the students agree that the history of scientific ideas is for them, in the phrases with which I began this paper, a leading thread in the history of civilization, a clue to the synthesis of knowledge and, as one student called it, the capstone of her education.

The question may well be asked, what are the tangible results of these efforts? Only three or four of these women are known to me to be specifically working professionally in this field. The course is still being given in modified form in one college; in two others there is neither money nor instructor for its continuance at this moment. But faculties have been quickened to the possibilities of the subject just as students have been sensitized and imaginatively stirred by it. The climate of opinion is changing as well; our world of today is very different from that of 1922 when I first undertook this teaching. The very existence of this Institute is proof that as a college subject the history of science is passing out of the pioneer stage into one of general acceptance. The students are ready for it; they say, "If a history of art or religion, why not of science?" Printed aids, whether of sources or of authoritative studies, are available or soon will be. What is needed now are men and women as instructors trained for this work. The openings are at last developing even in the liberal arts colleges for women.

The world today is being shaped by science and its activities as never before. The civilized woman as well as the man must learn to live in that world and not be lost in it. The problem of the colleges and the universities, therefore, is how best to shape their curricula for the training and development of the students who by virtue of this training will be among the leaders of tomorrow. Few among these students, especially among the women, will study science directly in any of its varied aspects. How then can they respond adequately to the demands of their environment? Here is, as I hope I have shown, the especial value as well as the place of the history of scientific thought in the liberal arts curriculum. Through some understanding of the age-old processes of trial, error, and success, of the questions science asks and the questions science cannot answer, its limitations as well as its achievements, the student of the humanities as well as the one in science can gain a perspective, an integration of her knowledge obtainable to my belief in no other way. And in this educational process both students may learn what John Dewey[8] called one of the two articles that remained in his creed of life, "that the future of our civilization depends upon the widening spread and deepening hold of the scientific habit of mind."

Dewey made that statement in 1909 when as chairman of Section L, today the historical division of the American Association for the Advancement of Science, he gave his vice-presidential address on "Science as Subject Matter and as Method." He went on to say—I quote from his concluding paragraphs—"Mankind so far has been ruled by things and by words, not by thought, for till the last moments of history, humanity has not been in the possession of the conditions of secure and effective thinking. . . . The magic that words cast upon things may indeed disguise our subjection or render us less dissatisfied with it, but after all science, not words, casts the only compelling spell upon things. . . . Scientific method is not just a method which it has been found profitable to pursue in this or that abstruse subject—that is what we mean when we call it scientific. It is not a peculiar development of thinking for highly specialized ends; it *is* thinking so far as thought has become conscious of its proper ends and of the equipment indispensable for success in their pursuit." Therein, he stated, lies the importance of education "that brings home to men's habitual inclination and attitude the significance of genuine knowledge and the full import of the conditions requisite for its attainment." This I submit is the place of the history of science in a liberal arts education.

References

1 "Foreword," p. vi, in George Sarton, *The Life of Science: Essays in the History of Civilization* (New York, 1948).
2 *Ibid.*, p. 19.
3 Report of the Harvard Committee, *General Education in a Free Society* (Cambridge, Mass., 1945), pp. 70, 53, 190.
4 *Ibid.*, pp. 222, 50, 227, 224.
5 Dorothy Stimson, "A Case Report on a History of Scientific Ideas," *Scientific Monthly*, LXIV, No. 2 (February, 1947), 148–54.
6 Thorstein Veblen, *The Place of Science in Modern Civilization and Other Essays* (New York, 1919, repr. 1942), p. 38.
7 *Discoveries and Opinions of Galileo*, tr. Stillman Drake (New York, 1957).
8 *Science*, n.s. XXXI (January 28, 1910), 121–27.

HISTORY OF SCIENCE

FOR ENGINEERING STUDENTS

AT CORNELL

Henry Guerlac

This paper follows closely the remarks delivered before the American Society for Engineering Education at Iowa State College, June 28, 1956.

The course which we offer each year at Cornell, under the title of *Science in Western Civilization,* had its eleventh anniversary in the fall of 1957. As I hope the title indicates, its scope is somewhat broader than that of a course in the History of Science properly speaking, and I am not sure that I should have undertaken to discuss it here. The course is, quite frankly, something of a hybrid; but I hope not a completely sterile one. Its peculiarities, at all events, are deliberate, for the course was planned with a clientele of engineering and science students foremost in mind, and particularly with an eye to Cornell's five-year curriculum in the College of Engineering.

Science in Western Civilization is a six-credit course, three credits a semester, running throughout the academic year. It is an upper-class course, for juniors and above. Though it was designed for, and is largely populated by, engineering students, it is open to upperclassmen and graduate students from other parts of the University. My reason for pitching it on this level is simple: I want the students to have completed their elementary science courses and their mathematics through the calculus. But there was a further tactical reason: Not only would the students have a needed maturity, but they would be sufficiently exposed to their technical courses, sufficiently saturated with their own subjects, to be willing to lend an ear to something quite different. We hoped to set this new course apart from the "required humanities" courses which are to be disposed of in the Freshman and Sophomore years, and which often seem to postpone still further the student's contact with his field of major interest. I should like to emphasize a further point, namely that when I say the course is designed for engineers, I do not mean to suggest that it is watered down and simplified—on the theory that the engineers are an unlettered lot—but only that the content has been selected with their areas of interest, needs and requirements foremost in mind. I have tried to make the course as demanding as any other upper-class course in the field of history or political theory. The students, in general, appreciate the fact that they are not talked down to. And they appreciate also that the course is not confined to engineers, and that they share the lecture hall with science students from other fields, with pre-medical students and with a scattering of bold majors in economics, history or philosophy.

Chronologically, the course is ambitious in scope. It begins with the dawn of civilization in the Middle East and comes down to our own century. When things are operating on precise schedule, we reach the time of Kepler and Galileo—the early seventeenth century—at the end of the first term. About half of the first semester is devoted to Antiquity—with an emphasis on the thought and culture of the Greeks—and the other half is divided about equally between the Middle Ages—the Arabic and Byzantine world and the Christian West—and the period of the Renaissance. The principal themes emphasized in this first semester are the following: (1) the rise of Greek rationalism and its culmination in Hellenistic science; (2) the decline of scientific interest in Roman times and in the centuries of early Christianity, and the virtual extinction of the classical heritage in science, mathematics and philosophy; (3) the partial recovery of Ancient learning in the twelfth and thirteenth centuries and the elaboration of medieval Christian philosophy rooted in Aristotelian philosophy, cosmology and physics. An opportunity is afforded to contrast the role of scientific speculation in societies as different from each other as those of Greece, Rome, Islam, and medieval Europe. On the technical side I devote considerable time to a study of Aristotelian natural philosophy (science), its assumptions and methods, in an effort to show its extraordinary ingenuity and range. For without a sympathetic appreciation of its persuasiveness and power over men's minds, and the way in which it was built into the fabric of medieval Christian thought, the Scientific Revolution of the seventeenth century which was directed against it, can scarcely be understood.

The second semester begins with a careful study of the Scientific Revolution of the age of Galileo and Newton and its impact upon the minds of men of the seventeenth century. We discuss in some detail the conditions that brought about this great change in the foundations of thought, and show how the conditions for a coöperative exploration of nature were provided by the creation of the Academies and the first learned journals. Above all, however, we study the underlying assumptions and methods of the new physics, both in the light of modern science and in the light of the Aristotelian view which it replaced. It is, of course, stressed that the accomplishments of Galileo and Newton were not merely scientific in the narrow sense, but constituted a revolution in the theory of knowledge, with implications extending far beyond science itself.

The eighteenth-century part of the course follows the consequences for science itself of the Newtonian synthesis, but focusses primarily on the problem of the impact on the Western mind of this great transformation. The eighteenth is the first century in which science begins to exert its full effects on Western culture, if not yet in the material sphere. Although some time is spent on the early Industrial Revolution, I like to point out that these transformations proceeded with very little help from science until the very end of the century. The most important influence of science was on the minds of men, was its effect on the ideology of the Enlightenment. This is illustrated in some detail from a study of Locke, Voltaire, Franklin, Jefferson and others. Not unimportant is the reaction against science and its supposed materialism which we first find clearly enunciated by Rousseau, by Joseph de Maistre, and even, in his peculiar way, by Goethe.

As you might imagine, the nineteenth century poses a serious problem of selection, so immensely complicated are the developments in the physical and biological sciences. As far as physics is concerned, I concentrate on (1) the experimental developments in chemistry and physics (e.g., spectroscopy) which led to speculations on the structure of matter; (2) the growth of thermochemistry and thermodynamics, and the discoveries in light, electricity and magnetism, and their inter-relationship showing how this led, in the time of Faraday and Maxwell, to field concepts in physics and to the abandonment of exclusive preoccupation with a physics of central forces.

But there are other aspects of the nineteenth century which I try to stress: (1) the tremendous extension of the experimental method into the biological and medical fields (with Claude Bernard, Koch, Pasteur and the rest); (2) the impact of Darwinian evolution as a great unifying force in natural history, and as an exemplification of the tendency to historical or genetic thinking so marked in all fields of thought; (3) the immense progress made in the nineteenth century in extending the scientific spirit into the study of man (archaeology; the study of prehistoric man; the emergence of cultural anthropology and sociology). Three aspects of the concluding years of the nineteenth century and the early years of the twentieth century are singled out: The first is the unprecedented influence of applied science in altering the material conditions of life in the Western World; the second is the emergence of America as a creative force in the world of science; the third is a discussion of what has been called the Second Scientific Revolution: those profound changes in physics which laid the foundations of modern atomic and nuclear physics, which altered our understanding of the classical assumptions of Newtonian physics, and which were accompanied by a revision of the logical and epistemological foundations of science more radical than anything since the time of Galileo. I end the course, or have in the past year, with a serious attempt to present some inkling of the implications of the work of Poincaré, Mach, Duhem, Frege and Russell, and their contributions to a modern philosophy of science.

BASIC PURPOSES OF THE COURSE

To state the problem most generally, we designed our hybrid course with the express purpose of introducing the technically-trained student to the cultural history of Western Civilization by focussing attention upon something in which he is presumed to have an interest, namely *science* and its applications. By discussing the growth of science in a broad historical and social context, it is possible to make excursions into social and intellectual history. A discussion of the influences exerted *upon* the growth of science, and of the influence of modern science upon other aspects of society, provides the opportunity to introduce the student to other cultures, with different patterns of thought and different value systems. And it gives an opportunity to introduce him to the great formative influences upon Western thought: to Plato and Aristotle; to St. Augustine and St. Thomas Aquinas; to Machiavelli, Hobbes, Locke, Rousseau, Freud, and William James.

But more important, in my opinion, is another objective: to give to the American engineering and science student, something that his European and British compeers absorb without effort, namely a sense of history. By this high-sounding term, I do not mean an exaggerated devotion to the past, an admiration of early accomplishments at the expense of the present, or a love of antiquities for their own sake. What I mean is a realization that much of the present is totally incomprehensible without an understanding of the past; that some grasp of the flow of events from a past, even a distant past, into the present illuminates the present; that much that we take to be absolute and beyond question—our manners, beliefs, prejudices, even our dress—are the result not so much of their abstract rightness or fitness, as they are of historical accident or of the historical experience of men like ourselves.

The sense of history cannot readily be taught, but it can be learned. I doubt very much if it can be conveyed in a rapid survey covering all aspects of Western Civilization, however valuable these courses surely are in other respects. But I think it can be conveyed by focussing upon the historical transformation in some one fundamental realm of human accomplishment, such, for example, as the history and development of political or social institutions, the growth of the law, or the changes in economic theory and behavior. Equally well, I'm convinced, it can be conveyed through the historical study of science. And this is true, more especially, because among the basic problems of science itself, is the emergence—in geological thought and biological evolution—of precisely this kind of thinking.

When I began to teach this subject to my engineering and science students, I thought I should be spared the problem of teaching science, though I knew I should have to spend time on historically important matters of which they were ignorant. But experience soon showed me that I was giving them credit for a deeper and more sophisticated understanding of the nature of science, its methods and assumptions and limitations, than they in fact possessed. I soon discovered, what I should have realized, that a detailed discussion of certain of the key concepts and key episodes in science, at first confused them and then gave them something which their science courses had not had time to give. They learned not to think of science as a bundle of ready-made formulas and techniques, which they had learned by rote and often did not understand, but came to realize that science is intellectual inquiry, that it is really highly abstract; and such philosophy of science as they possessed began, I think, to appear naive, rudimentary and hopelessly out-of-date.

Yet acquiring a sense of science as an enterprise of the mind, as a constantly modified body of theory and speculation putting order in our sense-experience, was a necessary condition for everything else this course set out to accomplish. If the students could not respect and value science as an intellectual achievement, how could they be expected to have any sympathy for intellectual activities still more remote from their experience? It soon occurred to me that it was my first responsibility, at all points of the course, to foster a genuine understanding of the complexities and subtleties of scientific thought through well-chosen, if possible crucial, examples. They had to learn not to think of science as a mass of ready-made results or encapsulated findings. They had to learn to think *abstractly,* and not to confuse mere memory and manipulation with abstract thought. This is, I believe, the first step in breaking down the simply vocational orientation of the majority of engineering students.

I. BERNARD COHEN

on the Papers of Dorothy Stimson and Henry Guerlac

The final form of this paper was the joint work
of I. Bernard Cohen and John Murdoch

Perhaps it would be of some help in this area to offer as commentary on the papers of Miss Stimson and Professor Guerlac something of a comparative historical sketch of the teaching of the history of science at Harvard, which spans a continuous history of sixty-eight years, going back to the time when Theodore W. Richards of the Chemistry Department gave a course in the history of chemistry. While this was not the first course in the history of science to be offered in America, it appears to have been the first course in the history of a special branch of science. (See Frederick E. Brasch's article "The Teaching of the History of Science," in *Science*, 1915, *42*: 746–60). This course was also given by Professor Josiah Parsons Cooke. That course was succeeded by one given by Professor Lawrence J. Henderson, which branched out from being a history of the theories of chemistry to a general history of science. When George Sarton came to Harvard on a more or less permanent basis after the First World War, the elementary courses in the history of science were taught by Henderson and Sarton, and for many years there were two half-courses, one given in the fall semester by Henderson on science in antiquity, the Middle Ages, the Renaissance and seventeenth century (up to Harvey and Galileo), and a second half-course given in the spring semester by Sarton on science from the eighteenth century to the present (beginning with Descartes and Newton). The paradoxical situation, in which Sarton who was an expert on the science of the Middle Ages did not give the portion of the history of science devoted to the Middle Ages, probably arose from the fact that Henderson had three major interests in the history of science, Hippocrates, Harvey, and Galileo, and would not surrender the portion of the course dealing with them.

Prior to the Second World War, other introductory courses on the history of science were given at various times by Dana B. Durand, Willy Hartner, and Giorgio de Santillana. By the time of Sarton's retirement, he had instituted a sequence of four half-courses, covering the whole development of science from the earliest times to the present, offered one per semester in a sequence that occupied two years. Since these four half-courses would be the equivalent of one-eighth of a man's college curriculum, very few students took them all.

At the present time, undergraduate instruction in the history of science at Harvard is centered around a single introductory course, covering two semesters, and given by the writers. In the first semester, there are two lectures per week; the students are divided into small sections which meet for discussion every other week. In this course about one-third of the time is devoted to ancient science, a third to the Middle Ages, and a third to the Renaissance and seventeenth century. Reading is assigned in both primary and secondary sources, the major secondary sources being Clagett's *Greek Science in Antiquity,* the volumes of the Oxford *Legacy* series devoted to Greece and Islam, A. C. Crombie's *Augustine to Galileo,* and the first half of A. R. Hall's *The Scientific Revolution, 1600–1800.* Some of the original source material is taken from Guerlac's *Readings in the History of Science,* which is supplemented by mimeographed materials for the ancient, the Islamic, and the European medieval periods. Among the major topics taken up in the discussion sections are: (1) Pythagorean "number theory" and the proof of the irrationality of the diagonal of a square, (2) a comparison of the Euclidean "axiomatic" treatment of geometry with the Aristotelian requirements for definitions, postulates, and axioms in a mathematical science, (3) Aristotle's doctrine of the four causes and an examination of its influence on his "logic" of scientific definitions, (4) Archimedean methods in geometry—analysis of the so-called "method of exhaustion" and a comparison of parts of *The Method* with *The Quadrature of the Parabola,* (5) an investigation of the Ptolemaic and Copernican systems for explaining the heavenly motions, together with a consideration of their relative efficacy in "saving the phenomena," (6) the medieval doctrine of the intension and remission of forms, (7) seventeenth-century science and the microscope (students examine specimens through replicas of Leeuwenhoek instruments), (8) Galileo and the problem of motion. In all of these discussion sections the students are furnished with primary source material.

For the seventeenth century, all students read Harvey's *De Motu*

Cordis, portions of Galileo's *Two New Sciences,* and parts of Newton's *Principia.* Last year students were required to submit three written reports, on the following subjects: (1) a critical analysis of Proclus' judgment that ancient geometry progressed "from sense perception to reason and from reason to understanding" together with an examination of the relevance of this contention to the growth of other areas of Greek science, (2) a discussion of the dependency of some particular topic in Islamic or medieval Latin science upon the prevailing philosophy, and (3) an original comparison of the methods and concepts of the physical and biological sciences in the seventeenth century. During the Reading Period (between the end of the Christmas recess and the beginning of the examination period, during which no classes or lectures are held), all students are asked to read the papers in the *Philosophical Transactions* associated with the publication of Newton's theory of light and color, which are printed in facsimile in *Isaac Newton's Papers and Letters on Natural Philosophy* (Harvard University Press, 1958).

In the second semester, no attempt is made at systematic coverage, but the students are given lectures on selected topics illustrative of the main development of science and its intellectual contacts and social influences. The major secondary source material used includes W. P. D. Wightman's *Growth of Scientific Ideas,* A. E. Heath's *Science in the Twentieth Century,* and Herbert Dingle's *A Century of Science 1850–1950.* The discussion sections for this semester center largely on problems in the development of the philosophy of science, an important aspect of the growth of scientific thought. Among the main topics discussed are: (1) Leibniz's criticism of Newtonian absolute space, (2) the Humean problem of induction in the nineteenth century, and (3) P. W. Bridgman's operational analysis of concept formation.

During the second semester, original source material is assigned in Guerlac's *Readings,* in a special volume of offset reproductions of classical scientific papers, and in selected portions of scientific classics that are readily available—such as Newton's *Opticks,* Claude Bernard's *Experimental Medicine,* and Darwin's *Origin of Species.* The single term paper centers on a comparison and contrast between the aspects of a single topic in its eighteenth and nineteen century manifestations. In the spring Reading Period students are assigned two biographies, usually René Dubos's *Louis Pasteur* and Philipp Frank's *Einstein.*

Harvard undergraduates are also allowed to take the advanced specialized courses, conference courses and seminars given in various aspects of the history of science. One notable feature of the Harvard course offerings is related to the existence of an undergraduate field of concentration called "History and Science." Undergraduates who elect to concentrate in this field are required to take courses in one aspect of history, to do a science program which includes at least three courses in a single area of science (to prevent the possibility of meeting the requirement of science courses by introductory courses alone), and also to take at least two regular half-courses in the history of science—the history of science being the aspect of the field which ties together the work in history and in science. A notable feature of this program is the provision during the three upper-class years of tutorial. Sophomores are tutored in groups of from two to four students, meeting every other week to discuss reading in the growth of the philosophy of science and in various crucial issues in the development of science. In the junior and senior years, the students have individual tutorial, meeting with their tutor once weekly to increase their competence in a special field of the history of science. Junior tutorial, like senior tutorial, is the equivalent of a full course (occupying a quarter of the year's effort), and the student's senior year is marked by the writing of an undergraduate thesis and the taking of a written examination. Only students who are honors candidates, and who maintain an honors level (grades of B or higher) in the field are allowed to concentrate in History and Science. The present number of concentrators is 50.

MARIE BOAS

on the Papers of Dorothy Stimson and Henry Guerlac

🎗 We have had the privilege of having described two different kinds of history of science courses, given by two pioneers in the field. As no pioneer, except in an entirely passive sense, I have great diffidence in attempting to comment on either paper, for one cannot criticize these papers in the ordinary sense of the word. In fact I shall not seriously try to do so. Of Miss Stimson's paper I should like to remark that I missed one subject which I should very much like to have heard her discuss, namely her early training and the interests which led her to concern herself with the history of science, and more especially the circumstances surrounding her first years of teaching a history of science course, including the reactions of her colleagues. (Parenthetically, Miss Stimson, by her remarks on the role of women in the history of science has, I suddenly realize, placed me in the rather peculiar position of wondering about my own identity—for it has always seemed to me that women are fairly well represented in the field, and I was not aware of any reluctance on the part of the universities offering graduate programs to having women students. Paradoxically, perhaps, most women teaching in the field are not in women's colleges.)

Of Mr. Guerlac's course I can only say from my own direct observation and experience that it is a course which "works"—that is, that it is a course that achieves its objectives and that it is a very exciting course to be associated with. In all honesty I must add that I have shamelessly pillaged it for ideas and materials for use in quite different courses.

For the reasons obviously inherent in the above remarks, what I propose to do is to turn from a discussion of the papers already presented to give a brief account of what I have found useful and interesting and practicable in quite a different kind of course from any previously discussed, the kind of course which I have been teaching the past several years, a course presented not to engineers nor indeed to any specific group of students, but a course open to any upper class student in an arts and sciences school.

245

The course I have given and expect to continue to give is roughly similar in outline to the Cornell course, but as its audience is totally different, its aims and methods are different. I do not include purely historical material and I never intentionally teach science, though I often find myself reluctantly compelled to instruct the students in the facts of elementary science which they were supposed to have learned earlier. I must confess that I have not solved the difficulties inherent in having scientists and nonscientists in the same course and that I run the risk always of dealing with material wearisomely familiar to one or the other group. (In general I find the scientists less impatient of elementary exposition of their field than the nonscientists.) But it is possible, if not ideal, to mix the two groups.

Granted that such a course is possible, of what use is it? First I might say that both scientists and nonscientists benefit from discovering the other's conceptions and misconceptions about what science is and what it can do. Miss Stimson has already mentioned many of the advantages of such a course to the nonscientist. She has not, I think, indicated what I, confirmed in this by my students, regard as one of the most important functions of such a course for the nonscientist, namely that a student previously bored and baffled by elementary required science courses can sometimes find an excitement in a history of science course which may lead him—or her—to try to acquire more scientific knowledge; this we would all agree to be a laudable aim. I have had students who, while not really interested in science, have elected the course because they realized that they were living in a scientific civilization and because they felt that some knowledge of how this came about would help them to understand their own era. (As you can see, I have been most fortunate in having at least a few articulate students who told me what I had intended to try to teach them.) There are other possible reasons for interest of the ordinary undergraduate arts student in such a course. I do not know for certain but I am prepared to suggest, not altogether tentatively, that in an age which doubts the certainty or reality of progress, the history of science can be curiously reassuring of the possibility of cumulative experience in at least one aspect of human endeavor. Equally, as history of science deals and must deal primarily with the work of individuals, a course in the history of science makes an interesting and useful foil to a possible surfeit of courses dealing with more orthodox history, stressing, as it usually does, trends and social and economic pressures, rather than individuals.

So much for some of the value which I find for the nonscientist in a course in the history of science which is history, not science. I must confess that I myself am more interested in what the course can do for the scientist and that I have always tried to design my course to attract the scientist rather than the nonscientist. Obviously, as a first approximation, one can see for the scientist all the benefits which Mr. Guerlac has described in connection with the engineer. But there is more than that. Mr. Guerlac has already mentioned the lack of scientific knowledge shown by the engineering student. The same applies to the undergraduate scientist, who is kept so busy learning the facts and tools which will prepare him to be a scientist that he seldom has the opportunity to think about the implications of these facts. I am tempted to say that many science majors never learn science, only scientific facts; and as things now stand only the very brightest students realize this and try to remedy it. A history of science course helps to correct the stultification of ideas inherent in this method of teaching. Some of my best students have told me that the history of science offered them precisely the opportunity they had been looking for, to, as they cocksurely put it, "find out what science is all about." I am, I hasten to say, far from accepting the cheerful youthful optimism which this implies, but I do believe that such students find in a history of science course the opportunity for appraising what they have been taught and of putting it into a shape which permits them to see with some clarity what they are learning and why and where it leads. Many have never thought at all about what they have been taught; one might say that for them science is all law and no meaning. However reluctantly, they will be led—or even forced—to make it mean more if they are made to consider the history of science. I do firmly believe that scientists who have been exposed to the history of science are far less liable to turn out to be illiterate outside their narrow field of specialization than scientists not familiar with the history of their subject except for the simplified, highly positivistic view of science usually found in elementary textbooks. (It occurs to me that if we ever successfully introduce any large numbers of scientists to the history of science, the historical sections of textbooks will inevitably improve and we shall no longer have to deplore their naïvete and error, in itself a quite sufficient aim for teaching the subject.)

Besides this, some young scientists are always gravely concerned with the role of science and of the scientist in the modern world, and these always find it rewarding to discover what has been the role of science in earlier phases of their own civilization and in other civilizations. They find it comforting to discover that the scientist has been faced with similar problems before, and instructive to discover the ways in which he has attempted to adjust his relationship to society. For this reason I always try to have as a conclusion to the course some modern problem in the history of science involving the relations of the scientist and society.

I do not want to sound as if I were merely offering propaganda for a history of science course in an arts and sciences school—though of course I am doing just that—and therefore I shall only mention one problem not touched upon by previous speakers, an obvious point, but one easily overlooked—that is, that we have been talking only about an introductory and general course in the history of science. Rightly, since this is the course most commonly given, and most critically difficult in terms of presentations and audience. I should however like to remark that we should not forget that advanced courses of fascinating variety exist or, better, could exist. We have perhaps convinced our colleagues that a history of science course is desirable—though many institutions have not got that far; I am not sure that outside the few centers of graduate training we have convinced them that many more courses could be taught by competent people. It seems to me most important that we begin to do so, if only for the encouragement of our graduate students.

DUANE H. D. ROLLER

on the Papers of Dorothy Stimson and Henry Guerlac

Two points of view were expressed in the papers of Miss Stimson and Mr. Guerlac, and though not differing particularly, they look from different angles at the subject of history of science courses. Miss Stimson expressed her theme that history of science is indeed as much a study for women as for men, and suggested to us some of the reasons for that theme. And she gave us a number of reasons why the history of science should be in the liberal arts curriculum: the fundamental position of science as a leading thread in the history of civilization; the value of the scientific attitude of mind; the need for history of science to take its place among the histories of other intellectual areas; the correlative influence of history of science courses; and a number of other reasons. Most of these are familiar reasons, indeed the classic reasons given in support of the thesis that history of science should be in the liberal arts curriculum. In respect to what Miss Stimson called a "low motive," namely the influencing of persons to give money to science, I would only say that I have a much lower mind; I would like to see them give to the *history* of science.

Professor Guerlac, like Miss Stimson, has a special sort of student in mind—or at least mostly in mind—his special group, of course, being the engineering students. These technically-trained students can be led, he tells us, *through* a subject of main interest to them—science—to an area that he *wishes* them to become interested in: the cultural, social, and intellectual history of the West; the study of cultures other than his own; the, in his terms, "great formative influences upon western thought"; a sense of history; and last but not least, he tells us, he hopes he can teach them some *science*. Thus it appears that the justification Professor Guerlac offers for teaching the history of science in a liberal arts curriculum is that it is a device for bringing the technically-trained engineering student into contact with history and then perhaps teaching him some science as well and even, in Professor Guerlac's words, "breaking down the simply vocational orientation of the majority of engineering students."

I personally find no serious quarrel with any of these reasons, these justifications for history of science courses in the liberal arts curriculum, either those given by Miss Stimson or by Professor Guerlac. I was, however, struck by the fact that each of the papers was concerned with a special case, Professor Stimson with the education of women, Professor Guerlac with the education of engineers. This in turn causes them to present *special reasons:* For example, Miss Stimson tells us of certain jobs open to women *if* they have an education in the history of science; Professor Guerlac tells us that he can, through the history of science, fill certain regrettable gaps in the education of *engineers.* But of course despite her special interest in the education of the feminine sex, Miss Stimson told us that her justifications for history of science courses applied to both sexes, and I am certain that Professor Guerlac would desire *all* college graduates to have a sense of history and an understanding of science. In brief, history of science courses could, from their arguments, be of value to all students: Thus, even though history students, as a special case, might well obtain a sense of history elsewhere, surely they should have some education in the history of science.

Now one may well ask why there should be *any* discussion whatsoever on this question. The history of science is an area of knowledge; surely it should be taught in colleges and universities. Perhaps the question that we are talking around is whether such courses should be required of students or at least whether students not majoring in history of science should be urged nonetheless to take history of science courses.

Now most of us feel that there is something of importance in the history of science or we would not be here. And those who teach, of course, feel that it is important that students enroll in the history of science courses. A teacher's security may indeed depend upon whether or not students do enroll in his courses. For such reasons, historians of science are hardly in a position to judge whether such courses should be required. Indeed their function is perhaps more to *argue* and *plead* the case than to *decide.* Some teachers teach to earn a living so that they may use the rest of their time for some other purpose, and they may prefer to hold down enrollments, but I think that most persons who teach because they want to teach also want students in their courses and necessarily feel that the students will profit from the experience. How then are they to obtain enrollments?

One method is to offer such arguments as those presented in the papers of Miss Stimson and Mr. Guerlac to administrators in decision-making positions. If history of science courses will teach students to think, to realize their own limited place in time, to understand science, to be broadly educated, then of course no reasoning administrator could fail to insist on such courses for all students. But these reasons for taking certain courses are age-old, although applied to other areas. Of course one must understand science, we are told, but it is the science course where one gets this understanding. There the student studies science, not about science. Of course one must have a sense of history, a feeling of continuity, and to acquire these one takes a course in history. Of course one must learn to think, and practically every teacher says that courses in his area teach thinking.

What I am trying to say is of course apparent; *we* know that history of science courses play these roles in a liberal arts education, but to those who make decisions on the curriculum these ideas may sound like platitudes.

It seems to me that if it be true that history of science courses can play important and indeed vital roles in the education of college students, conviction must come through the mere state of affairs and not through argument. The teacher of the history of science must, *by his teaching,* convince his colleagues and his peers that their students should indeed enroll in these courses.

This is of course not easy to do, but I believe that it can be done. Circumstances are making the task somewhat easier as time goes on. For example, large numbers of faculty members in the sciences and engineering are at the present time extraordinarily sensitive to the idea that the orientation of their students is, in Professor Guerlac's words, "simply vocational"; or as Professor Stimson puts it, "faculties have been quickened to the possibilities" of the history of science. Many of these science teachers are genuinely seeking, for their students and sometimes for themselves, that famous *bridge* from their own specialty to broader areas of human knowledge. Many of them think that the history of science may be that sought-after connection. And the number of scientists with this view should increase as time goes on—particularly if those teachers of the history of science who have science students in their classes properly do their jobs, science faculties may have increasing numbers of properly oriented members. A teacher of the history of science tells me that the faculty of the physics department in his institution is utterly uncoöperative. But to take a somewhat ghoulish view, they can't live forever and if, for example, a replacement to that physics department comes from my institution, he will have had at least three semester hours of the history of science.

However, it will be necessary for the teacher of the history of science to satisfy his students and, most of all, his colleagues, that the history of science courses do indeed serve the desired purpose. In this respect it seems to me that the past has a clear lesson for us: As you all know there have been serious difficulties in the last few decades concerning the teaching of science courses for non-science students. Two major failures have emerged as important contributory causes of those difficulties. In the first place, science teachers have far too often attempted to teach non-science students precisely as they have taught science students. In this way they have often concentrated upon minor techniques and subjects of no consequence to the non-major student, and have often succeeded in sending away such students with an undying hatred of science.

The second major failing in such science courses was the attempt at encyclopedic coverage. We should bear in mind these failings and perhaps should ask ourselves whether history of science courses are perhaps so designed as to drive away students from an excess of technical, scholarly detail. We must remember that many of the virtues we claim for history of science courses have been claimed for science courses as well. In brief, if history of science courses are to become an important part of the educational process of students not majoring in the history of science, they must be carefully designed and taught to fit their place in the curriculum.

Professor Guerlac mentioned that he found that he was also teaching his students science. This of course points toward another and famous type of role for the history of science, namely its use as a tool for the teaching of *science*. I am thinking particularly of the so-called general education courses, usually at the freshman level, in which a vast variety of approaches and techniques is used, including the history of science. Here the problem of obtaining customers is rarely a problem; if I remember correctly, Professor Lawson, at Michigan State, has five thousand students in his course. In such general science courses the role of the history of science is sharply and clearly that of a pedagogic tool; yet I wonder if that is not true in all history of science courses except those designed to train historians of science. Certainly both of the papers told of many aims to be achieved with the student through use of the history of science.

There seems at times, in such meetings as these, nearly a compulsion for commentators to object to the main papers. It seems to me, however, that here one can find no *basis* for objections. By that I mean that there is little in the way of a standard of comparison for history of science courses, at least for courses designed to serve non-major students. We are all currently working out our own salvation, so to speak. In the long run, one might say that the standard of comparison must be the successful courses.

THE *ENCYCLOPÉDIE* AND

THE JACOBIN PHILOSOPHY OF SCIENCE:

A STUDY IN IDEAS AND CONSEQUENCES

Charles Coulston Gillispie

"La distinction," wrote Diderot of *Jacques le fataliste,* "d'un Monde physique et d'un Monde morale lui semblait vide de sens,"[1] a remark which provides this study with a text. For ultimately the subject emerges as an instance of the tension between science and the aspirations of humanity to participate morally and through consciousness in the cosmic process. These aspirations demand a nature different from that embraced by post-Galilean science, the nature not of the atomists (and much less of the Aristotelians) but of the Stoics. Newton's world offered virtue no purchase, and the Enlightenment saw, in consequence, a curious and important effort to transcend Newtonianism in a modernized Stoic physics through which virtue could, as in ancient times, be drawn from Nature.[2]

For such a physics the model of order is the organism, some unitary emanation of intelligence or will, or else identical with intelligence or will. In the long dialogue which science throughout its history has conducted between the unity of nature and the multiplicity of phenomena, this image is the dialectical opposite of that objective order into which analysis ranges whatever particulate entities it discerns as the term of measurement, whether actual or conceivable. And the renewals of this subjective approach to nature make a tragic theme. Its ruins lie strewn like good intentions all along the ground traversed by science, relics of a perpetual attempt to escape the consequences of western man's most characteristic and successful campaign, which must doom to conquer. So, like any thrust in the face of the inevitable, it induces every nuance of mood from desperation to heroism. At the ugliest, it is sentimental or vulgar hostility to intellect. At the noblest, it inspires Herder's vision of history and Goethe's of nature, the poetry of Wordsworth and the philosophy of Whitehead or of any other who would make a place in science for our qualitative and aesthetic appreciation of nature lying all around us.

I shall pursue the theme in historical compass, considering as an example the attempt to alter the image of nature, first by the science and philosophy epitomized by the project of the *Encyclopédie,* and then by political actions expressive of that philosophy, which culminated in the liquidation of the foremost scientific institution of the world in the French Revolution. Since the subject is controversial it would, perhaps, be well to define the terms of discussion at the outset. The aspect of the Revolution with which I am concerned is Jacobinism, its most active element, in Burke's words, political "fixed air . . . broke loose." For Jacobinism will not here be taken as simply a collection of expedients. On the contrary, I think it the most passionate attempt in western history to realize moral ideals of virtue, of justice, equality, and dignity, in political institutions. It was obviously not a system developed *a priori* like its successor, Marxism. But only in this sense was it an improvisation, and system in it can indeed by discerned *a posteriori* by analysis. One must, therefore, distinguish its philosophy of science, like its political philosophy, in its acts as well as in the words which inspired it. I should say further that with certain reservations, I adopt Professor Talmon's recent theory of Jacobinism as messianic democracy.[3] I also accept M. Belin's claims for utility as an *idée-force* in the Revolution.[4] My thesis is that once affairs were engrossed by Jacobinism, science was bound to incur the enmity of the Republic. This was not because scientists were unpatriotic. On the contrary, they pressed into the service of the State on a scale unequalled until the twentieth century. But in its intrinsic combination of assurance and irrelevance, science all unintentionally stood across the cosmic ideals of the Republic in a posture nonetheless insulting for being unassumed. Only by ceasing to be science, could it give the Republic the nature it needed. And the Republic, true to its inspiration, could ask no less.

Elie Halévy remarks somewhere that whoever would be definite about the evolution of religious opinion must fix attention on the sects and organized churches in which alone it takes a form accessible to historical enquiry. Something of the sort may be the chief interest offered by the fortunes of scientific institutions for the history of scientific opinion, and it is in order to body ideas out into reality that the reader's attention will be returned from time to time to the law of August 8, 1793. This abolished the learned academies of France as incompatible with a republic.[5] It may fairly be described as a measure inimical to science and learning. In the historiography of the Third Republic, however, it became canonical to represent the Academy of Sciences as compromised by the undoubted sins —corporatism, favoritism, futility—of the other academies, to the point that the Convention's innate respect for its work and for science only just failed to save it.[6]

The facts are otherwise. The defenders of the Academy had been careful to disassociate it from the stigma of corporatism. They never disputed the justice of the impeachment when directed against the humanistic academies. Ultimately this strategy would save it at their expense by merging it into the republic educational system as the scholarly apex of the nation.[7] To this end, its friends on the Committee of Public Instruction inserted in the decree abolishing the other academies five clauses conserving a provisional existence to the Academy of Sciences.[8] The tactic miscarried. A single speech by David secured adjournment of those clauses. No voice spoke in their favor. This maneuver, therefore, enables us to estimate the attitude of the Jacobin Convention to the scientific community. As a community it was not to be permitted to exist. The attitude was hostile, then, and the question is, what did it spring from, this hostility?

The suppression of the Academies came in the late summer of defeat, alarm, and treason, in the months when the Jacobin regime was establishing its extraordinary efficacy on a rising curve of idealism, patriotism, and terror. It was the successful culmination of a considerable campaign, waged in pamphlets, journals, and popular clubs which since early in the Revolution had been heaping opprobrium on the Academy, certain of its members, and its influence in society. Three distinct themes recur throughout these writings.[9] To take them in the order of increasing political importance, they first of all express resentment for the new chemistry, directed expressly against the person and influence of Lavoisier, who appeared to his detractors not in the humble guise of chemist, nor simply as a financier, but as the arrogant spokesman and evil genius of science. Secondly, a gentle sentiment appears, which might seem inharmonious in this hostile chorus. Enthusiasm for natural history was unanimous. The paradox, however, is only apparent. Finally, there was a political assertion of the sort which rings of injured interests. Science was undemocratic in principle, not a liberating force of enlightenment, but a stubborn bastion of aristocracy, a tyranny of intellectual pretension stifling civic virtue and true productivity.

These were deep emotions, instinctive responses to an image of science as alien to the common man, but their revolutionary expression was patterned rather than informed. To understand their import, and the passion in them, the question must be taken back into the history of the Enlightenment, back to where it was explicit among the makers of opinion who knew why they felt as they did, who saw the point of the world picture of Newtonian physical science, and who rejected it as uninteresting and unsatisfying. The trail leads to the mid-century crisis of philosophy between the surrender of Cartesian science and the rising of the Romantic phoenix. For through the ashes between them runs the bond of dissatisfaction with a poverty in the Newtonian conception of scientific explanation. To the Cartesians, nature was the seat of rationality, and Newton's laws appeared intellectually trivial. To the Romantics, Nature was the seat of virtue, and Newton's laws were morally unedifying. In both views, nature is congruent with consciousness. The work of Diderot's generation, therefore, had to be to preserve the continuity of man and nature by opening its personality to reality rather than its intellect. For if nature *is* congruent with man, if science is the correspondence, it has to be a continuum, a whole, a "tout," as d'Alembert is made to see in his dream. But it is the whole personality which communicates, and not just the heart, because until Rousseau's revolt there is no question of irrationality.

The first thrust back to this more intimate sense of Nature came from chemistry, in an attempt to deepen the concept of matter, to give back body to what physics deprives of every attribute but surface and dimension. In speaking of physics as concerned with bodies in motion, we tend to forget that the word "body" originally implied organization, internal material organization, to which chemistry might properly address itself. The article on that science in the *Encyclopédie* is extremely significant. It is by Venel, a disciple of the elder Rouelle. He invokes a new Paracelsus, who will make of chemistry the science that understands nature and displaces geometry from that pretension. He will be gifted, this Paracelsus, with the sheerly technical insight to penetrate beyond physics, but he will have a spirit and imagination like that of the pre-Newtonian philosophers. For Venel agrees with Buffon that theirs was a less limited genius, a more extensive philosophy, that they "voyaient mieux la nature telle qu'elle est," because, "une sympathie, une correspondance, n'étaient pour eux qu'un phénomène, tandis que c'est pour nous un paradoxe, dès que nous ne pouvons pas le rapporter à nos prétendues lois de mouvement."

Physics, to pursue the comparison, is superficial. Chemistry is profound. Physics measures the gross characteristics of bodies—surface, shape, position, and motion. Chemistry penetrates their essence. Physics confounds abstract notions with verities of existence. One asks for a fact; it gives one a theorem. The physicist uses rigorous calculation to arrive at those exact theories, which experiments then confirm "à peu pres." The chemist, by contrast, never deludes himself by calculations. He apprehends his theories rather by a "pressentiment experimental," and his theories are only approximate. But as a reward for his modest humility, in his case it is the fit with nature that is exact.

The question, then, is nothing less than the structure of nature. Mechanics will never bear the chemist into the heart of things, for the texture of reality is not corpuscular. Mass, the superficial aspect of matter which is the object of physics, doubtless is corpuscular. But—and this is the whole point—the essential merit of chemistry is to take the sting out of atomism. For it allows the masses in which atomism resides no ontological interest. So it is that chemistry carries the empirical answer to Newton's unreal abstractions. It is the qualities in things which impress our senses, our windows on reality, and this reality inheres—not in mass, be it repeated—but in the principles which run through the world as activities, as bearers of qualities and causers of perceived effects. The physicist, therefore, who denies existence to entities like yellowness and fire is simply presumptuous. They are not in his field.

It is not for the physicist to study quality, nor for the chemist to study quantity. Even his laboratory operations will be different from those of physicists like Boyle or—to anticipate—Lavoisier. His laboratory will offer no scene of weighing and measuring. What concerns him is the combinations and separations in matter, its state of interpenetration, and masses do not combine or penetrate. They only aggregate. Principles are what combine in perfect mixture, and the chemist, therefore, will catch glimpses of "la vie de la nature" coursing through his laboratory in phenomena which run all through, around, and under mass: in effervescences and distillations, evaporations and condensations, rarefactions and expansions, elasticity, ductility, malleability, and liquidity. The vision is of stretching and blendings in depth, full of that Faustian sense that nature has an inside. It is not alchemy, but like some alchemist, Venel is always moving in the mind's eye from fermentations in his laboratory through digestions in the animal down into the mineral gestations of the earth, whose cosmic womb is the source of unity in nature. The chemical authority Venel most admires and follows is Becher's *Subterranean Chemistry*. But back in his laboratory, your chemist, true to his calling, wields his materials with art. His hands are gentle. It is the physicist who brutally pulverizes, ignites, and destroys. The chemist does not analyze. He divines.

The chemist's world, then, is a palpable continuum—his science is Cartesianism stripped of geometry with its clear ideas. To replenish the Newtonian destitution of nature, it sees down into a world of matter pulsing with activity. In place of universal attraction between particles, Venel has discovered that the fundamental property of matter is universal miscibility. But chemistry is more than intimacy with nature. It has the common touch. It is everybody's science, the poor man's manual metaphysics, whereby that artisan in whose skills true wisdom lies manipulates reality, not in the humiliating abstractions of mathematics, but with his own hands: "La chimie a dans son propre corps la double langue, la populaire et la scientifique." And all this seems harmless enough until suddenly, out of the *Encyclopédie,* there speaks in one startling sentence, the authentic voice of the *sans-culotte:* "Parlez plus bas," the mathematical physicist is told, "vous feriez rire nos porteurs de charbon, s'ils vous entendoient."[10]

In the *Interprétation de la nature* Diderot calls attention to this science of the chemists and to the work of Benjamin Franklin as examples to be emulated in handling nature with the surety of experimental art.[11] But sentience and organism weave a more grateful veil, and though Diderot drew his conception of material reality from the palpable continuum, he transposed it out of chemistry into the far more plausible terms of natural history, launched it in the flow of time, and incorporated it into a significant *Weltanschauung*. Diderot's prophecy about the decline of mathematics is sometimes taken as a passing slip in a prescient vision of the biological shape of things to come. This seems a misreading. Anticipations in the history of science are curious but by definition almost meaningless. To understand what a man meant, it is always not only more historical, but also more helpful, to look backward rather than forward to Darwin or to Einstein; and to read the connotations of modern biology into the eighteenth century, when the word itself was not yet invented, is to obscure that what distinguished natural history was not just its taxonomic method, but the uncontrollability of its metaphors by its evidence.

Moreover, Diderot's philosophy of science put no confidence in abstract conceptualization of any sort. His rejection of mathematics was fundamental. He objected to its claim to be the true language of science on all grounds, metaphysical, mechanical, and moral. It is not just that geometry idealizes—it falsifies, by depriving bodies of the perceptible qualities in which alone they have existence for an empirical science.[12] The comparison of mathematics to games of chance has sometimes been cited as exemplifying stochastic foresight: "Une partie de jeu peut être considerée comme une suite indeterminée de problèmes à résoudre, d'après des conditions données. Il n'y a point de question de mathématiques à qui la même définition ne puisse convenir." But he goes on: "La *chose* du mathématicien n'a plus d'existence dans la nature que celle du joueur."[13] Chance exists in affairs, not in things.

Even the science of mechanics has been rendered trivial by mistaking mathematics for understanding. To suppose that a body is indifferent to its state of rest or motion is to suppose it purely passive, without activity or force, which is to say without existence. But at the very moment that the physicist annihilates some block of marble with his calipers, it gives him the lie. For it is not still: It is a hive of decay and disintegration, a little world of living forces. Diderot will not even accept on Newton's terms inertia or the inverse square law. To make a universal constant vary with distance is a clear contradiction in terms, and "pesanteur" is no tendency to repose, but to local motion.[14]

The mathematical spirit, in fact, is a blight. Fortunate but rare the mathematician whose own aesthetic sensibilities are not blunted by his subject, which has fallen into aridity and circularity as must any science which ceases to "instruire et plaire." Once idle curiosity is satisfied and novelty wears off, only its power to edify will keep a science living. "Je n'en excepte pas même l'histoire de la nature."[15] But Diderot gives back to the Enlightenment, perhaps from his chemical studies, a more ominous note, which echoes down the whole romantic movement. Mathematics is worse than inhumane. It is arrogant, "orgueilleux." In a sense, no doubt, everyone who has felt himself reach his mathematical frontier, whether at long division or out somewhere beyond the calculus, must know something of the helpless resentment engendered by the hidden beauty of the abstract. But Diderot's own mathematical competence was by no means contemptible, and he fully appreciated its value as an instrument of precision in subordinate matters.[16] His indictment is curious and interesting and not mere petulance. Mathematics is the science by which a finite intelligence purports to plumb the infinite. Now, man aspiring above himself incurs the classic guilt of *hubris,* the Christian guilt of pride, and the prospect of an infinite universe has always disconcerted those who would render science humane. But Diderot was no Pascal to agonize over infinity. We are in the eighteenth century, and he responds with perfect nonchalance. He simply dismisses infinity as uninteresting. Since we shall need some criterion to establish bounds between knowledge and the infinite unknown, why let it be our interests. "Ce sera l'utile qui, dans quelques siècles, donnera des bornes à la physique expérimentale, comme il est sur le point d'en donner à la géometrie."[17] So Diderot restores the mind, in a sense, to a finite cosmos, by wrapping science tight around humanity.

Nor in form are Diderot's writings on nature an artless collection of *aperçus.* To see clear, d'Alembert, the geometer, is put into a dream, almost a delirium, out of which he speaks truths instantly recognized as such, and easily anticipated, by whom?[18] By a doctor, the universal doctor, who sees nature across the perspective of human nature: "Il n'y a aucune différence entre un médecin qui veille et un philosophe qui rêve."[19] And the apparent formlessness of the *Interprétation de la nature* is skillfully adapted to convey the congruence between man and nature. For it is written as a stream of consciousness, a reverie on the Experimental Art, the true route to a Science of Nature, moving out toward three objects: Existence, Qualities, Use[20]—a threefold object, but a single purpose. What is the young man to look for in Diderot's natural philosophy? "Un plus habile t'apprendra a connaître les forces de la nature; il me suffira de t'avoir fait essayer les tiennes."[21]

So he reverses Descartes, who studies himself to know nature. Diderot attends courses on chemistry, he reads Buffon, he studies nature—to know himself.[22] But communication is direct, experiential—it does not lie through mathematics. It lies, instead, through craftsmanship. For Diderot, as for the chemists, truth opens to the common touch, and the importance of right method is that it dispenses the ordinary man from the need for genius. In its pride, genius draws a shroud of obscurity between nature and the people—mathematical in Newton, conceptual in Stahl.[23]

They are in error who say that some truths can never be put "à la portée de tout le monde."[24] Certainly, ordinary people will never see merit in what cannot be proven useful. But in this, they see aright—or rather they are aright to fail to see. For only experiential philosophy is an "innocent study," in that it supposes no prior preparation of the mind.[25] The habit of actually handling materials in dumb, untutored experiments, develops in him who performs the coarsest operations an instinctive *"pressentiment"* which has the character of inspiration. Manual facility gives a power of divination, the ability to "smell out—*subodorer*" how it must be with nature.[26] But how do you know you have this power? How do you know you are right? It is—if the analogy is permissible—like awareness of Grace. It is like Virtue. It is participation in the Truth. You recognize it in yourself, in your own intimacy and more than intimacy, your solidarity with Nature. In such a breast, science and nature are one, the reality of the great organism suffusing for the moment the material consciousness of the little. Not mathematical abstraction from nature then, but moral insight into nature, is the arm of science. And consistently with Diderot's conception of scientific understanding as illumination, he contemns, in an unwonted access of Puritanism, the extravagances, the frivolities, the vanities of those who go down the garden path of abstract reason.[27] Presuming to prescribe as rules for nature his own formulations, the savant in his pride conceals from himself and others that it is not his laws of nature which are simple, but nature herself in her essential unity.

For nature is the combination of her elements and not just an aggregate. Otherwise there is no philosophy: "Sans l'idée de tout, plus de philosophie."[28] And Diderot, therefore, is bound to interest himself in continuity and not in divisibility. "Convenez que la division est incompatible avec l'essence des formes, puisqu'elle les détruit."[29] When he writes of molecules it is of their transience, not their existence. In genetics it is the atomistic as well as the providential implications of *emboîtement* which are unacceptable. For nature knows no limits. The male exists in the female and vice versa. (Hence the curious fascination with hermaphroditism.) Mineral blends into mineral. The qualities of one living species penetrate in some degree the others. Minerals are themselves fused into living matter through the *latus* of the plant which aliments the animal. Individual animals are real eddies of tighter organization, the ultimate but impermanent units, borne along a stream of seminal fluid flowing down through time and out from the matrix womb of nature herself.

Even the physicist will do better to devote attention to what endures and spreads—to resonance, for example, to fire and electricity, to sulphurous exhalation, and to the behavior of standing waves. Diderot, too, has a substitute for the universal attraction of corpuscular physicists: It is universal elasticity.[30]

"Tout change; tout passe; il n'y a que le tout qui reste."[31] And Diderot uses two figures to express this unity. The second is the more familiar, the universe as a cosmic polyp, time its life unfolding, space its habitation, gradience its structure, for this embodies the twin ideas of universal sensibility and of evolution.[32] The latter idea Diderot took from Maupertuis and Buffon, but he treats evolution as a consequence of the indivisibility of time, a time which is that of biological subjectivity and in no way dimensional.[33] But although this is consonant with historicism, there is no serious sense in which it foreshadows Darwinism, the success of which, after all, is vindicated by its reducibility in genetics to material atomism. It is, therefore, not this, but rather the first of Diderot's metaphors which is the more significant. In it he evokes the swarm of bees. For the solidarity of the universe is social. On a cosmic scale, it is that community which the social insects know.[34] "Only the bad man lives alone," Rousseau was told,[35] and in social naturalism there is a more prescient concordance between the whole and the parts, the one and the many, than in reversion to an antique hylozoism.

Neither lingering devotion to an archaic chemistry nor growing enthusiasm for natural history—nor both these influences together, for the point is that they were in fact the same thing—would pose a serious threat to the scientific community. That would come only from political forces of sufficient importance. Nevertheless, these intellectual factors played their part in creating an image of science behind which it could scarcely be defended should such forces arise, as in fact they did. For it happened that the Revolution coincided with the sensitive period in the history of chemistry, when that science was the arena of the scientific revolution. Lavoisier's enemies were quite right to single him out as the epitome of science. The new Paracelsus had appeared, but in response to the summons of Lagrange and in what grievous form—betraying hopes with his material algebra.[36]

The word "enemies" is chosen advisedly, and not critics or opponents, for the important discussion was not about phlogiston nor was it held among scientists. Properly understood, phlogiston belongs to the objective history of chemistry itself, and as M. Daumas has shown us, Lavoisier's conception of oxygen represents much less sharp a departure than is traditionally said.[37] The question was rather what nature is like, agitated not as between scientists, but as between scientists and opponents of modern science, who wanted to see nature humanely through subjective perceptions of quality, beauty, and goodness. The point is explicit, for example, in Venel, who very well understood what he was saying: "C'est que la plûpart des qualités des corps que la Physique regarde comme des modes, sont des substances réelles que le chimiste sait en séparer, et qu'il sait ou y remettre, ou porter dans d'autres; tels sont entre autres, la couleur, le principe de l'inflammabilité, de la saveur, de l'odeur, etc."[38] Once the chemical revolution was under way, the issue was equally apparent in the chemical writings of Lamarck, the most considerable in the qualitative vein, and I have argued elsewhere that his emanationist theory of evolution originated as the transfer to, or refuge in, biology of this old sense of nature for which Lavoisier made chemistry uninhabitable.[39] But this archaic chemistry was not a science recorded in books, or taught by treatises. It partook rather of the character of lore, passed on by word of mouth like the trade secrets of the artisan, or taught in the public lecture courses where chemical artisans, the pharmacists, learnt their art. And for them, it was no abstract theory which offended, but the new nomenclature which came as a deliberate injury, reducing honest craftsmen to dependence on the scientist, making a mystery of their livelihoods.[40]

M. Mornet has sufficiently celebrated the vogue of natural history in the Enlightenment that it is hardly necessary to insist further on the theme.[41] Suffice it to point out that if this sentiment, too, is brought to the institutional test in the Revolution, its reality is abundantly manifest. On June 10, 1793, just two months before the *Académie des sciences* was abolished, the staff of the *Jardin des plantes* secured from the Convention a decree vesting the administration in their own hands, establishing twelve professorships in the different branches of natural history, and changing the name to *Muséum national d'histoire naturelle*—this munificent provision at a moment when virtually no chairs existed for teaching higher mathematics or physics. The distinction of the museum as a center of higher education and biological research dates from this reorganization. Its transformation offers, indeed, an epitome of the French Revolution in botanical microcosm: There was the situation of the *Jardin du roi* in 1788, at the death of Buffon, who had run it with a high and feudal hand and who left it bankrupt and with fine resources; there was the staff—Daubenton, Lamarck, Jussieu, Fourcroy, Thouin, Lacépède, and lesser figures—an active and able group suffering (though not seriously) from unjust arrangements (unequal salaries, a variety of obscure and arbitrary tenures) and who knew what they wanted (control of their own affairs); there was the attack on privilege and sinecures, for Buffon was succeeded not by a scientist but by a courtier, one Billarderie, who had secured the reversion by a private arrangement with Buffon; there was the discovery that favoritism does not automatically disappear with Revolution, for the staff got rid of Billarderie only to find the post given, not to one of themselves, but to the moralist and nature writer, Bernardin de Saint-Pierre; there were the projects and pamphlets, the approaches to the Assembly by people whose interests were threatened or who hoped for some advantage, each resting his case on buoyant principles like the equality of naturalists and the general welfare; there was the problem of safeguarding what was good from the old regime—of preventing, for example, the people from picking the flowers now that the flowers belonged to the people; there was, not to go into narrative detail, the ultimate success—rationalization of resources, equalization of employments, the rendering professional of a public institution and its dedication to public service and education.[42]

There is, therefore, no need to disagree with Mornet—except in thinking all this nature study less a vehicle of scientific culture than an escape from it. For it developed a pastoral picture of science, in the manner of Boucher, as of something charming, inoffensive, dilettantish, comforting, even cozy. If disillusionment came, it could only be sharp. In the physical sciences, too, popularization turned out a mixed blessing. It is not sufficiently appreciated how much the cult of science was a cult of marvels, nor how directly the fad of electricity led to the fad of Mesmer. In imagining the layman's idea of the scientist, we are reminded of Condorcet's description of that class of men who have the secrets in their keeping, and whose corporate interest is to obscure reality from common sense.[43] And the Jacobin mentality was not one to resign itself to what seemed to escape its control. But a similar rejection of mystery, a comparable insistence on imposing standards in circumstances which matter, is characteristic of the scientist. It explains, for example, the extreme reaction of the scientific community when confronted with a Velikovsky, or a Robert Chambers, or a Mesmer, its inability to understand, much less tolerate, the layman's pleasure at what in his willful ignorance he takes to be the discomfiture of science at the hands of someone who has done what he would like to do himself, and has broken the rules. Nothing was so damaging to the popular reputation of the scientific community as its maladresse in handling Mesmer. For at first, in refusing to examine his claims, it took the line that what everyone was tremendously interested in was not worth the attention of serious people, which was true but not tactful; and then it created the impression that the issue was between the self-esteem of a few scientists and the hope held out to suffering humanity by one who had the vision and daring to fly in the face of the pedants.[44]

The appeal of popular science, therefore, and of natural history, had the disadvantage of surrounding scientists with a fringe of enthusiasts not very well qualified by temperament or the nature of their interest to participate in the serious work of science. Consider, for example, the experience of the French Linnaean Society, founded in 1787 by Broussonet, under the banner of Linnaeus and in rebellion against Buffon. After a few months, it became known that anyone who hoped for election to the Academy would do well to disassociate himself from this group. The pressure was sufficient—the society collapsed, and the members separated to nurse their grudges. But not for long, for among the rights of man is that of forming voluntary associations, and at a meeting on May 16, 1790, the naturalists reorganized themselves as the *Société d'histoire naturelle*.[45] It soon undertook regular communication with the authorities on matters of natural history, going over the head of the Academy, and it entered into relations with other popular societies—of inventors, artisans, and artists. New members were recruited, of whom no qualifications were required beyond a love of nature. There were arresting figures among the members of the first Linnaean Society: Lamarck, for example, and Dolomieu. Several became deputies in the Revolution: Broussonet himself, Bosc, who sheltered Roland during the Terror, and Ramond, who had been Cagliostro's laboratory assistant and whose writings did much to bring mountain scenery into favor.[46] Perhaps the most influential was Creuzé-Latouche, who, as a member of the Finance Committee of the Constituent Assembly, was the first to raise in the legislature a serious question about the propriety of academies in the new order.[47] Additional notables appeared in the revived society: the abbé Grégoire, Romme,

Fourcroy, Hassenfratz, and a number of others who have left names, not in the history of science, but on the rolls of the Jacobins in Paris.

It would hardly be worth going to great lengths to demonstrate that men who came to power in the Year II, their ears attuned to Nature by Diderot and Rousseau, did not understand science, for how could it have been otherwise? And we have a hundred indications that they did not: Thibaudeau's remark that the Jacobin leadership regarded intellect as the enemy of liberty;[48] Collot's instinct that excessive erudition is unsuited to a Republic;[49] Robespierre's attack upon Condorcet's educational proposals as tending toward an intellectual aristocracy and his enthusiasm for the Spartan plan of Le Pelletier.[50] In no serious sense can these men be said to have had any experience of science at all, or any interest in it. The real question, therefore, is what images came to their minds in the few minutes which they must have diverted from saving the Republic to approving the liquidation of the Academy of Science? And put this way, it is obvious that the point is not that they failed to understand science, but that they misunderstood it with a peculiarly damaging moral enthusiasm for nature. For any glimpse they got of science itself could come only with the shock of a betrayal of humane values which, so they had been taught by the scientizing moralists, derive through science from Nature herself.

There is an epitome of how chillingly inconsequential such an encounter might be, of what a gulf could open, in the passage between Laplace and Brissot over the claims of Marat. In 1782, Brissot had published in Marat's behalf a dialogue on academic prejudice. It is based on an actual conversation with Laplace, and he makes his straw geometer, all obstinate in his Newtonian idolatry, say that it is by mathematics that he knows Newton to be right. To which the humane skeptic retorts that Newton's critics too have their calculations: "Que faire dans ce chaos de chiffres? Recourir à la nature," says the skeptic, ending the discussion, "Voir le fait. . . . J'aime mieux croire mes sens et la nature, que vos volumes de chiffres."[51] And in his memoirs, written in the full tide of disillusionment with Marat and in hiding from the guillotine, Brissot admits that Laplace may, after all, have been right. "Mais je ne pouvais supporter qu'il traitât avec insolence et despotisme un physicien parce qu'il ne jouissait pas comme lui de fauteuil. Je suivais depuis trois ans les expériences de Marat, et je croyais qu'on devait quelque estime à un homme qui s'ensevelissait dans les ténèbres pour reculer les bornes des sciences."[52] In

268

common justice a man ought to receive some return for making over 6,000 experiments, and for Brissot the issue between Marat and science is one to be settled man to man, according to principles of fair play and equality.

꧁꧂

Such were the patterns of incomprehension in which statesmen responded to an active campaign mounted against the Academy. It was a campaign founded in real interests, and fired by democratic ideology. To appreciate the ideological setting, it will be well to recur once more to the Encyclopedic movement and its relation to science, about which it is time to propose certain second thoughts. The *Encyclopédie* was, of course, a complex work of many hands. Nevertheless, after all exceptions are made, it is possible to group the contents under the two broad headings of liberal ideology and Baconian technology. Diderot's master stroke was to make the latter carry the former—to seem to say that the institutions of the old Europe were absurd and unjust because contrary to nature, not with the voices of Denis Diderot and Jean-Jacques Rousseau, but with the voice of science. Surely, however, it took sleight of hand to invest the utilitarian idea of progress with the high authority of Newton. For when one thinks of it, what had Newton to do with sly or sardonic japing about the Christian religion? How could Newton have inspired the proposition that the route to true knowledge of nature lies through rationalizing the puffing of the bellows, the creaking of the water-wheel, the hammering of the blacksmith? In the preface to the *Principia* he says, indeed, precisely the contrary, that he means to give the mathematical, not the mechanical, principles of natural philosophy.

Writers who describe the *Encyclopédie* as the exemplar of Newton's influence on the Enlightenment might further reflect that the philosophic testament with which Diderot accompanied that work was the *Interprétation de la nature*.[53] Whoever studies the philosophy of the Enlightenment must be struck by its eclecticism and humanitarianism. The interest seldom lies in the originality of the ideas, but in how they are combined in the service of mankind. So it is most strikingly in the work of Diderot. He never drew his image of science from Newton. He drew it from the classics. It is Greek—with one all-important variation. He has abandoned the Greek admiration for vaulting speculation and substituted for it out of Bacon a professed humility (what in Diderot seems even a rather meaching humility) about the human understanding, combined with an equally Christian sentimentalization of honest manual work. He seldom loses an opportunity to strike a contrast between the pathetic and appealing humility of technique and the haughty arrogance of mathematical abstraction.

Certainly—as I have argued in another connection—the *Encyclopédie* represents the education of industry by science.[54] Descriptive science dignifies the arts and trades by bringing them within the ambit of systematic learning, by finding them their rightful place in a great fraternity of human knowledge. Not that the arts and trades were actually languishing in quite the depths of despite from which Diderot and d'Alembert would rescue them, but the *Encyclopédie* did carry out Bacon's injunction when it would have been easier simply to repeat it.

There is, however, another aspect to the problem of science and industry in the *Encyclopédie,* related to the role of that work in the history of populism. Its liberal ideology, couched in innuendo and irony, can never have been other than critical in effect, never have done more than unnerve the aristocracy and titillate the intelligentsia. The people, after all, do not like wit. It was the technology which was truly populist, taking seriously the way people made things and got their livings.[55] Artisans were indeed, in Diderot's words, taught "to have a better opinion of themselves,"[56] an opinion which the politicians learned to respect on the 14th of July, 1789. To the dignification of craftsmanship by science in the *Encyclopédie,* corresponds, therefore, a reciprocal democratization of science which entrained, for a time, a no doubt inevitable cheapening of expectations. For this is the final consequence of that eighteenth-century humanitarianism which would retrieve from the cold abstractions of classical physics a science warming to man. It assimilates the whole of science to its applications, makes it only the rationalization of technology, and seeks in practice to obey Diderot's injunction to keep man at the center, not only man but everyman.[57] It proposed that dream of a citizen's science which the Revolution turned into actual measures.

The Year II gave artisans their opportunity to reverse the aphorism according to which science governs the arts, and a revolt of technology was one among the many rebellions swelling into the great Revolution. To follow its course is to move down into obscure places among people whose history is hard to come by: craftsmen, engineers, inventors, minor manu-facturers—small but solid people in the midst of whom fermented a leaven of cranks and malcontents. Nevertheless, fragmentary traces do remain of the societies which they formed to attain their interests, popular societies of inventors and artisans, of a piece with the famous popular societies of a purely political nature in which transpired the actuality of the French Revolution.

The specific trouble began with the Academy's statutory responsibility for advising the government in administering the encouragements to tech-nological development. These took the form of subsidies, or, more rarely, of monopolies granted to inventive entrepreneurs, and it is not surprising that the Academy's role of referee earned it deep hatred among the arti-sans whose work it judged. A quotation from a writer in the *Journal du point central des arts et métiers* will give the temper. After dismissing the metaphysical sciences as sophistical, he writes:

Les Arts sont plus surs, et leur bienfaisance est plus certaine! Combien n'étoient-elles donc pas coupables ces formes abusives et tyranniques, qui violant la plus sainte, la première des propriétés, celle de la pensée, celle du génie inventif, ou *perfectionnant,* soumettoient les Priviligiés de la nature, les *Artistes,* à ces Loix gênantes, à ces dures épreuves, à ces censeurs inquiets et durs, dont l'ignorance ou la jalousie inquisitoriale n'avoient pour premier but que le soin d'humilier, ou d'écarter le vrai talent!

Combien n'étoient-elles pas cruelles et vexatoires, ces prétentions éxagerées des corps académiques? Combien n'étoit-il pas revoltant cet empire tyrannique et destructeur de l'industrie, que la richesse donnoit à ces vampires usuraires, à ces frélons despotes, qui toujours prêts à dévorer le miel travaillé par les abeilles, profitoient de leur fortune ou de leur puissance, soit pour s'emparer des Ruches, soit pour réduire les Artistes à des compositions avilissantes et ruineuses et les dépouiller même de l'honneur attaché à leurs travaux, en usurpant leurs in-ventions; en fatiguant, en rebutant leur zèle, leur courage, et leur constance par les dégoûts de toute espèce; enfin, en les forçant le plus souvent d'aban-donner des idées, ou des découvertes très heureuses, soit parce qu'elles con-trarioient l'amour propre des premiers privilegiés, soit parce qu'elles nuisoient aux portions d'intérêts données dans les entreprises préexistantes.[58]

Such complaints abound in the archives. But a considerable sampling in the *Archives de l'académie des sciences* of projects submitted to its judgment in the late eighteenth century, yields, in fact, no significant in-stances of meritorious ideas going unrecognized—much less of suppression or sinister interest at work. The only complaint materially justified was of undue delay, and the work of examination was arduous and exasperating. Reports on technological devices consumed more time than scientific memoirs, and the great majority were valueless. The literature surviving from inventors' claims amply testifies to the intensity, amounting almost to fever, of inventive activity in the late eighteenth century. But to turn it over is to enter a curious, indeed a feverish, atmosphere, compounded of illusion and delusion, secrecy and deception, avid hope and bitter recrimination. One has the impression that great drafts of publicity and fresh air were needed, but that the altogether understandable impatience of theoretical scientists was probably unhelpful.

Nevertheless, although the charges of academic obstruction of ingenuity are little more than instances of the tendency to blame disappointments on the authorities, this did not lessen disappointment nor solace injured pride. And it is just here, in what concerns self-respect, that the sources are most eloquent and precise. For there emerges the picture of the worthy mechanic, the salt of France, the hero of the *Encyclopédie*, all unversed in polite ways, standing before a committee of scientists with his new machine, on which he has lavished years of labor and pinned his hopes, twisting his hat and trying to answer incomprehensible questions about the laws of statics and dynamics. The artisans insist with very deep feeling on their rights: not only on their right to private property in the fruits of their ideas, but with even greater feeling on their right to be judged by their peers, who look at mechanical problems from their own point of view and not from some high theoretical plane.[59] And on this score, it seems almost certain that these bitter and frequent complaints result from that sort of real injury to human dignity which, suffered by a whole people in its ablest elements, made the great Revolution all it was. The correspondence of foreign scientists visiting Paris just before the Revolution offers confirming evidence of the impression of a certain arrogance which the French scientific community—particularly its mathematicians—might make upon an outsider.[60] Indeed, the Academy itself became conscious of the effect it was creating. Early in 1789, Laplace, impatient at the quantity of chimerical projects, had proposed that every applicant be subjected to a test in geometry before his designs would be considered.[61] In 1791 the Academy did modify its procedures, not however in that direction, but by abandoning explicit approbations or condemnations to limit itself to a simple recommendation.[62]

Early in the Revolution inventive entrepreneurs began overreaching the Academy to submit requests for grants directly to the municipality of Paris and the National Assembly itself. There they were received with sympathy. Projects which had been rejected by the Academy were successfully appealed, and subsidies were voted by the legislature.[63] In 1791 a technological jury, the *Bureau de consultation des arts et métiers,* replaced the Academy in advising the ministry.[64] Half of the thirty-man panel was composed of representatives of the artisans' societies. Delegates from the Academy—Lavoisier, for example, Lagrange, and Laplace—who composed the other half thus found themselves colleagues of men whom they had rebuffed. This body, too, was in principle more lenient than the Academy, though there is no evidence that French technology benefited—perhaps because the state mitigated its excessive generosity by failing in practice to honor many of the grants.

It was, however, in popular societies that the majority of artisans pressed the interests of their trades. Painters and sculptors, it would appear, were the first and the most fiery in the rebellion against the academic regime. Under the leadership of David, the *Commune des arts* demanded absolute freedom of exhibition, and beyond that a liberation of the artistic spirit soaring out into infinite Shelleyan realms of creativity.[65] The mechanical arts assembled in more restrained fashion in the *Société des inventions et découvertes.*[66] There people of substance—men of the calibre of Fortin, Mercklein, Lucotte—stood at the head of the lesser artisans. In general, this body proposed itself as a free or revolutionary association of technology. In particular, it worked as a lobby to secure first passage and then administration of a patent law in the interests of mechanical innovators. The law of 1791, which remains the basic French patent law, was drawn to its specifications. In the view of the artisans, the influence to be overcome was that of the *Académie des sciences.*

Here is the comment of Boufflers, reporter for the committee of the Constituent Assembly which drafted the law. He dismisses the Academy's view that applications for patents should undergo technical scrutiny and rallies to the inventors' desire for a system of patents to be had for the asking. How, he asks, is it possible for any panel to judge fairly of an invention which, by definition, does not yet exist?

Mais les savans eux-mêmes ne sont-ils pas quelquefois accusés d'être partis au procès? Ont-ils toujours été justes envers les inventeurs? Convenons-en: l'étude a peine à croire à l'inspiration, et des hommes accoutumés à tracer les chemins qui mènent à toutes les connaissances, supposent difficilement qu'on puisse y être arrivé à vol d'oiseau.[67]

Finally in 1791 appeared the strident *Point central des arts et métiers*, "composé de tous artistes vrai sans-culottes,"[68] although in an earlier manifesto the constituents are more largely identified as "une classe encore plus directement utile . . . celle des manufacturiers et chefs d'attelier."[69] The *Point central* set up a constant clamor for replacement of the academies by a democratic administration of support for the arts. This body was taken seriously by the authorities. Two of its representatives sat on the *Bureau de consultation*. It was encouraged to establish a *Lycée des arts,* a perpetual chatauqua of popular science in the court of the *Palais royal*.[70] The fraternity of all the arts was its guiding principle, which could not be realized until they were led out from under the despotism of science and academies. A few passages from another of its communications deserve quotation:

Du haut de leur Montagne nos Législateurs veillent; ils planent; ils guêtent par-tout la malveillance, et si d'une main ils tiennent la foudre toujours prêt á frapper les traîtres, de l'autre ils dispensent les bienfaits. Nous sommes donc assurés qu'il est impossible que l'abandon des Arts échappe à leur vigilance. . . . Laissez aux sociétés libres le soin de reculer, par les perfectionnements, les bornes de nos connaissances.—Que l'industrie pratique réunisse les vrais artistes en assemblées primaires des Arts; et qu'ils choisissent librement des commissions temporaires pour chacune des parties de la nouvelle administration; la liberté fera le reste et les fruits, n'en doutez pas, seront abondants. . . .

Toutes les classifications scientifiques et amphatiques (sic) de nos connaissances, nous les réduisons à six commissions temporaires, et tout ce qu'on nomma *science,* nous le rapportons aux seules connaissances utiles. Enfin, cette nouvelle administration ne coûterait plus d'un million. . . .

Not the sciences, but the arts, will serve as basis of the new education; and the teachers will be not scientists, but artisans:

En conservant les pères des Arts, en les faisant servir à l'instruction publique, vous occuperez, vous endurcirez cette jeunesse bouillante dont les âmes doivent être préparées avant tout au premier devoir du citoyen, celui d'être utile à la patrie.—Voilà les véritables moeurs républicaines.—Les prêtres hypocritiques disaient: *Sachez vaincre vos passions,* et ils appelaient cela de la morale. —Le républicain chaud et actif doit dire: *Laissez les passions aux hommes, mais sachez les diriger.* Ce sont elles qui lui donnent son énergie. Un homme sans passions n'est qu'un fédéraliste modéré, ou un feuillant hypocrite, in-capable de grandes choses. Le véritable Sans-Culotte, c'est celui qui *travaille*.

Is it fanciful to suggest that this is a view of the human aspect of nature congruent with that which Venel's chemistry (or Goethe's or Lamarck's) takes of its material aspect, as a web of activities quickened by sympathy, vivified by the Stoic *tonos?* In the moral philosophy of Robespierre, it is the priest who tears the web by throwing that barrier of obscurity between nature and the people for which Diderot blames the *savant*. In any case, these are passages from the proposal for a new constitution of science which the *Point central* sent to the Convention on September 26, 1793 along with its congratulations on having delivered to the Academies "leur extrait-mortuaire en bonne forme."[71]

For the fall of the academies on August 8th set in train liquidation of the entire structure of French science. Rebellion was followed by purge. On August 10, members of the former Academy resolved to take advantage themselves of the Constitution and to form a *"société libre et frater-nelle pour l'avancement des sciences"* to carry on their work.[72] Lavoisier wrote Lakanal for approval of this course, telling him in overwhelming detail of all the valuables and projects entrusted to the Academy: The Academy is trustee of expensive astronomical instruments, and has begun the construction of new ones; Vicq d'Azyr has undertaken an anatomical treatise, for which 6,000 livres have already been expended; the Academy intends to publish the voyage of Desfontaines along the coast of North Africa, financed by the nation; Desmarets has been given funds for a mineralogical map of France; money has been awarded to Fourcroy for research on alkalis, to Berthollet for work on dyes, to Coulomb for in-vestigations of magnetism, to Sage for mineralogical experiments, to Haüy for studies of crystallography; the entire section of chemistry has a grant for work on the combustion of diamonds; agreements have been made for publication of many manuscripts; the Academy's own Memoirs are three years behind, and the work will be lost if not printed. Most im-portant of all, there is the great metric project. Contracts have been signed with many artisans, whom the Convention surely will not disappoint. Standard weights and measures are under construction. The survey of the meridian is in progress.

Lavoisier still could not believe that all this would be simply aban-doned.[73] But if the government does, indeed, intend to take over the direction, will it please send precise instructions covering every particular? And for a moment the strategy seemed hopeful. The *Société libre* was authorized on the 14th.[74] But on the 17th the members' papers were placed under seal.[75] By September 1st, Lavoisier had at last despaired. His colleagues, he wrote Lakanal, dared not proceed even on a voluntary basis. To do so would flout "l'opinion dominante du Comité d'instruction publique et de la partie prédominante de l'assemblée." He has little hope of a further report promised by the Committee. He fears it impossible to reconcile the interests of science with the politics of the moment: "Nous sommes dans une situation où il est également dangereux de faire quelque chose et de ne rien faire."[76]

For the popular authorities proved very vigilant. In October a committee of agronomists was appointed to advise on the threatened crop failure. It disbanded after three meetings on being warned that it was about to be denounced at the Jacobins as an attempt to revive the *Société d'agriculture*.[77] At the Observatory the fourth Cassini was ousted as Director by four young men whom he had appointed as student assistants. They stood on the equality of scientists, brought off their *coup* by means of their influence in the *Section Saint-Jacques,* and controlled the Observatory for a year and a half. One was later executed for his part in the Baboeuf conspiracy. The others relapsed into obscurity after discovering an allegedly republican comet.[78]

As for the metric project, it seemed for a time that the government would see the work through. Two teams were in the field running the survey, Delambre from Dunkerque and Méchain from Barcelona. Precise determinations of units were under way in Paris. Administrative preparations were being concerted for the shift from *toise* to meter.[79] On September 11 a Temporary Commission was formed of those members of the Academy who had exercised its responsibility until the suppression. Prieur de la Côte-d'Or, emerging as the member of the Committee of Public Safety responsible for technology and war production, sat as representative of the regime. Like many graduates of Mézières, Prieur was an engineer who had indulged vague scientific ambitions with no success.[80] He may well have felt ill at ease in the company of men like Lavoisier and Laplace. On December 23, 1793, they, together with Borda, Coulomb, Brisson, and Delambre, were removed from the Commission by a decree of the Committee of Public Safety. This document was drawn in Prieur's handwriting. The ground was that in the interests of public spirit, the government must assign missions only to men "digne de confiance par leurs vertus républicaines et leur haine pour les rois."[81] The chairmanship was placed in the nerveless hands of Lagrange. Prieur became absorbed in other problems. The work on the metric system was abandoned.

With the scientific institutions of the old regime overthrown or in patriotic hands, the popular societies launched a movement to replace them with a new, republican organization of science. Each measured itself for the mantle of the Academy. The *Société d'histoire naturelle* proposed a definite constitution to the *Société d'inventions et découvertes*.[82] One clause provided that should a discovery occur of which the utility could not be shown, the freedom of science could be preserved by refraining from publication. A century later Berthelot represented the *Société philomatique* as the shelter in which the sciences took refuge.[83] The surviving papers of that group do not bear him out.[84] We may, indeed, catch a glimpse of a week in the life of Jacobin science in the *Lycée des arts*. It opened its courses in the wooden circus which then stood in the *Palais royal* on April 7, 1793, before a great assemblage. The president, Fourcroy, flanked on the dais by four members of the Convention, delivered an address. His subject was utility as the object of science. Behind him could be seen a relief plan of a new canal for the city of Paris. Hébert then took the floor, and in an impromptu speech urged that the sanctuary of the arts be invested with the spirit of liberty. He further undertook to have Fourcroy's address printed at the expense of the Commune. A report was read on the culture of silk worms. Prizes were awarded for achievements in the mechanical and the agreeable arts. The session closed with a hymn to Apollo. The courses themselves followed nightly: On Monday, natural history, taught by Millin, who later became librarian of the *Bibliothèque nationale;* on Tuesday, amphibians, by Brongniart; on Wednesday, mineralogy by Tonnelier; on Thursday, vegetable physiology, by Fourcroy; on Friday, technology, by Hassenfratz; on Saturday, physiology, by Sue. The physical sciences had their due. In an evening spectacle open to ladies, citizen Val "a fait des tours de Physique amusante."[85]

Throughout the entire period, the spokesmen for science conducted its defense with neither dignity nor skill. They gave away the case for science as inquiry by themselves professing a vulgarly utilitarian valuation. One searches the sources for a single defense of science as a simple intellectual good, as a seeking for truth about nature, and one searches in vain. Instead, the Academy addressed Baconian platitudes about science and the arts to statesmen who were being told by the artisans themselves that the Academy was stifling creative industry, and who had been taught by the *Encyclopédie* that it is the artisan who knows, not the theorist. They drew up memoirs about the advantage and glory of the metric system for Deputies whose constituents were in the act of arresting Delambre on the not unreasonable suspicion that all his transepts and alidades and clambering about in steeples could only be the equipment and deportment of a spy.

Lavoisier was a very great scientist and a very bad politician. He was probably the one responsible for establishing the *Bureau de consultation* in order to divest the Academy of the increasingly vulnerable responsibility for technological supervision. Yet the Academy could not quite bear to let go, after all, for the initial constitution of the *Bureau* was so contrived that undoubted *savants* sat as representatives of the arts and trades, a Gallic maneuver which insured a firm majority of *savants,* deceived no one, and only served to exacerbate the suspicions of the *sociétés libres* for the academy, "don't les ramifications cachées et souterraines resistent encore aujourd'hui avec une force incalculable à ce qu'on arrache le tronc."[86] A plan of education was drawn in the name of the *Bureau de consultation* by Lavoisier, which preserved for science an independent organization in which the Academy might take refuge.[87] It is a fair surmise that this was an attempt to forestall the *Point central* in its favorite project of assimilating science to the arts. Lavoisier spared no efforts to draw off the menaces of that body. He and other academicians even joined it in sponsoring the *Lycée des arts.*[88] But Lavoisier was hopelessly out of his element. He drew up a report, for example, to the membership on the assaying of saltpetre. Unfortunately, so the journal informs us, its reading had to be abbreviated for fear of fatiguing the attention of the public by fixing it too lengthily on abstract subjects, although Lavoisier's resumé of the work in hand in the Academy was accepted courteously enough as "semblable au suc precieux que l'abeille va ramasser sur toutes les fleurs pour le deposer dans le ruche."[89]

As the Academy's days ran out, Lavoisier desperately wrote speeches to put in the mouth of Lakanal and ran about the streets trying to find his self-styled champion to send him to the Convention in time for some crucial vote. At the very moment of the Academy's dissolution, he warned that if France abused the devotion of her scientists, she would deprive herself of their services.[90] Nothing of the sort happened, of course. Reflecting on the behavior of the scientific community under the Revolutionary and Napoleonic regimes, one is tempted to apply to science the Schumpeter thesis about the *bourgeoisie's* needing a master,[91] or to recall Malraux's recent remark that an effective regime, by definition, finds its technicians.[92] As if to refute Lavoisier's warning, scientists were mobilized. They served brilliantly in the famous effort of war production celebrated in every text book.

France was not deprived of their services. Only science was.

This paper is a study of the consequences—not, be it emphasized, the "results," but the consequences—in actual events of the French Revolution of leading ideas of the Enlightenment. From it certain conclusions suggest themselves, which like the terms of the discussion, may best be stated categorically and explicitly, for they, too, will no doubt be controversial:

1. That the hostility of the revolutionary ideals of human nature to abstract physical science was not a passing irritation with the inhumane arrogance of mathematicians. It was profound and utter, rooted right down in the nature of the ideal itself. The Jacobins in their year of exaltation did not stop with their attempt to change human nature. They—I do not mean only the leaders, but especially the rank and file—proposed to substitute for the image of nature with which science confronts humanity a different one, one sympathetic to the ordinary man. Indeed, if the analogy with Marxism holds, the Jacobins ought to have had a philosophy of science, and out of their acts it is indeed possible to construct one, none the less real for being tacit. Theirs would be a science which, as to its technological aspect, would be a docile servant of humanity, and as to its conceptual, a simple extension of consciousness to nature, the seat of virtue, attainable by any instructed citizen through good will and moral insight: "Fut artiste et savant qui voulut," in Cassini's summary of the Revolution, "Heureuse liberté!"[93]

2. That the attack on the Academies in the Revolution was the political expression of a moral revolt against Newtonian science reaching back into the Enlightenment. But it traces back ultimately not, as I first expected (and, indeed, as I said in the earlier version of this paper) to Rousseau, but to Diderot. This is to dispute neither Rousseau's emotional hold over the revolutionary left, nor the attachment to him of the naturalists of

the *Société linnéene,* many of whom had actually botanized in his company. But his resentments were sporadic, unsystematic, and essentially trivial expressions of that petulance which loves nature and hates science. His actual writings on nature are surprisingly uninteresting and innocuous. Rousseau's chemical commentaries were sensible enough. His botany was a consoling hobby, making no pretense to science. On the question of the Linnaean system, hostility to which is the touchstone of romantic and metamorphic (or Stoicizing) tendencies in taxonomy as opposition to atomism is in physics, he was on the side of sobriety, analytical accuracy, and objective understanding. It would be ironical if his providentialism was what kept him in the objective camp, though that is possible, for providentialism permits the mind to sit loosely to necessity. But however that may be, Rousseau would at most diminish the importance of science. He would only attack it on occasion. He would not presume to alter its structure.[94]

For this reason, behind Rousseau, it was the influence of Diderot which was profound, reaching into the heart of science to turn it into moral philosophy. His was no feminine dislike of precision, no soulful sense of God in nature, but a philosophy of necessitarian and Spinozistic organism, a system according to which science could be, not just denounced, but altered. We must, therefore, review our interpretations of the *Encyclopédie.* In its theory of matter, in its enthusiasm for natural history, and in its sentimental humanitarianism about humility and truth in technology, it presaged in each essential element the events which engulfed the scientific community in the Revolution. Enlightenment Baconianism set the pattern in which those events transpired, and we must take this aspect of the Encyclopedic movement of thought, therefore, not as the expression of a developing scientific culture, but as strictly inimical to scientific culture in effect as in intent.

3. That this explains the curious *distance* which every student of the eighteenth century must feel between the work of the scientists themselves and the liberal ideology attributed to its influence. Practicing scientists did not participate in this revived Stoicism (unless one includes in their company Buffon and the early Lamarck). D'Alembert, Condorcet, and Lavoisier saw the importance of science in the light which the associationist psychology threw on the genesis of ideas. In their view, man is what he makes of his experience, and scientific explanation functions as a kind of cosmic education. Leading from Condillac to the *idéologues,* this is the

tradition which after Thermidor exerted the constructive influence of science in the Revolution—in the creation of the *Institut de France,* the *École polytechnique,* the *École normale,* etc.[95] But in view of its issuance in the religion of Saint-Simon and the authoritarianism of Comte, it is an open question whether this was a happier reading of science.

4. That, therefore, one has to go all the way back to Voltaire to find the *philosophe* who correctly understood the implications of Newtonian science for liberalism. For if I read him aright,[96] what drew Voltaire to Newton was precisely Newton's revelation of the absolute irrelevance of physics. This is what deprives dogma of any claim to draw authority from nature. After Newton, thought is really free, not just of the censor, but of the far more damaging tyranny of metaphysics. But science had only this one liberation to effect, and that, rather than the death of Madame du Châtelet, is why, once Voltaire has understood physics, he goes on about his business, which is not to explain nature, but to play on a world of men a critical intelligence. Nor is it inconsistent that science and theology should have co-existed loosely and easily enough in the breasts of Voltaire and before him Newton. The warfare between science and religion is relatively trivial: It is the conflict between science and any naturalistic moral philosophy which is profound—there is no co-existence in the breast of the thorough scientizing moralist, of a Diderot, a Comte, or a Marx. If science does not give them the nature they want, they will change it so that it does—or says it does.

5. That the interesting thing about the Jacobin philosophy of science is not that it was futile. The attempt to revive the Stoics' nature as the seat of civic virtue was, of course, bound to fail. Nature is not like that, and the interesting thing historically is rather that it happened at all. For it suggests how shallow was the penetration of culture by science at the end of the century which is always taken as the great century for that influence. Ultimately, the analogy with Marxism flags if it does not altogether fail. It is hardly conceivable that anything of the sort could happen today. The Lysenko controversy never went so deep. And, of course, it did not really matter to science that it happened then. It mattered only to scientists. One feature of the impersonality of science is the way it does survive untouched all such persecutions and all the divagations of thought and institutions which it inspires. In moments of discouragement, therefore, one is tempted to fear that the history of the influence of science on culture is bound to be the history of an unavoidable misunderstanding, in

which what changes is the way in which the import of science is misunderstood.

For men do not seem to be content to accept science for what it is intellectually: A great creation of the human mind, an inquiry about how nature works, the results of which are admirable and interesting in themselves, but empty of lessons, or promises, or comforts. They are bound to rummage about in it for liberal ideologies or social Darwinisms. They are bound to look for reassurance about free will in the unpredictability of the electron. Nor are they willing to accept as better evidence of what they seek their failure to find it there. So, perhaps, the question for a relevant sociology of science becomes this: What is the effect on men of living in a society whose most dynamic and characteristic activity moves ever further from their comprehension? Were the Jacobins (or the Marxists) right after all, and does it in practice come back to the fruits of technology? And perhaps, therefore, this paper is to be taken as a lengthy exemplification of the moral drawn for the author by a friend with whom he discussed it and who is a logician—one of those alarming friends who would embrace (and extinguish?) all one's interests in a single equation of linguistic analysis. It is impossible, he said, to move from a declarative to a normative statement, from an "is" to an "ought."

If you try, you only lose the "is."

References

1 Denis Diderot, *Oeuvres romanesques,* ed. Henri Bénac (Paris, 1951), p. 670.
2 On Stoic Physics, see S. Sambursky, *The Physical World of the Greeks* (London, 1956), and on its relation to ethics, E. Bréhier, *Chrysippe et l'ancien stoïcisme* (Paris, 1951). For influence of Stoic ideas in the sixteenth century, see Léontine Zanta, *La renaissance du stoïcisme* (Paris, 1914); and for a discussion of neo-Stoicism in seventeenth-century political philosophy, Gerhard Oestreich, "Justus Lipsius als Theoretiker des neuzeitlichen Machtstaates," *Historische Zeitschrift,* 161 (1956), 31–78.
3 J. L. Talmon, *The Rise of Totalitarian Democracy* (Boston 1952).
4 Jean Belin, *La Logique d'une idée-force: L'idée d'utilité sociale et la révolution française, 1798–1792* (Paris, 1939).
5 Printed documents in *Procès-verbaux du comité d'instruction publique,* ed., J. Guillaume (1894), II, 240–60; *Archives parlementaires de 1789 à 1860, fondé par MM. J. Mavidal et E. Laurent, Première série (1787 à 1799)* (82 vols.; Paris, 1867–1911).
6 For example Guillaume's introduction to the volume cited in note 5, pp. lxii–lxxii.
7 On the instructions of the Constituent Assembly, adopted on 20 August, 1790,

all academies were to draw up within one month revised statutes to bring their regimes into conformity with the new constitutional order (*Arch. parl.* XVIII, 173–74; *Archives nationales,* AD VIII, 11, pièce 1.) The Academy of Sciences conscientiously had anticipated this requirement, and began debating its own reform on 18 November, 1789, at the initiative of the Duc de la Rochefoucauld. On 21 August, 1790 it resolved to accelerate the discussions and hold four extra sessions a week, no doubt to conform with the will of the Constituent. On 13 September the new *règlement* was completed and adopted, one week under the deadline. In 1954 a copy of it was found by Professor Henry Guerlac in the *Archives de l'académie des sciences,* where too may be consulted the *procès-verbaux* of the meetings.

For a typical defense of the Academy by Lavoisier, see his "Observations sur l'académie des sciences," *Oeuvres de Lavoisier,* ed. J.-B. Dumas, (1868), IV, 616–23, where he points out (p. 618) that arts and letters do not need academies, and for the plan of education of the *Bureau de consultation, ibid.,* 650–68.

8   From Lavoisier's letters to Lakanal of 17 & 18 July, 1793 (*Oeuvres de Lavoisier,* IV 615, 623), it is evident that it was known at least by that date that measures were gathering against the academies. Indeed, the *registres* of the Academy are so slight for all of 1793 that the life of the body must have been increasingly paralysed as political uncertainty grew. But the measure itself was framed in great haste. Two days before it was adopted, Grégoire, who drew and presented it for the *Comité d'instruction publique,* wrote frantically to the Ministry to ask for information on what academies actually existed—Grégoire to Garat, 6 August, 1793, *Archives nationales,* F^{17} 1097, dossier 1, pièce 1.

9   For typical diatribes against scientists and academies, see Marat, *Les charlatans modernes* (1791); Vadier, *Le Montagnard Vadier à M. Caritat . . .* (1793), *Bibliothèque nat'le,* Fol. Lb^{41}.3196; Chabanon, *Adresse à l'Assemblée nationale . . .* (1791), British Museum, FR 450, No. 3; *Suppression de toutes les académies . . .* BM, FR 450, No. 4; *Suppression des Académies, Archives nationales,* D38 II 19.

10 The preceding five paragraphs paraphrase and quote from the article "Chimie" in the *Encyclopédie.*

11 Diderot, *Oeuvres philosophiques,* ed. Paul Vernière (Paris, 1956), "De l'interprétation de la nature" (xli), p. 217. For convenience, I cite in parentheses the paragraph numbers by which Diderot divided this essay.

12 *Ibid.,* (ii), pp. 178–79.

13 *Ibid.,* (iii), pp. 179–80. For the profound dissimilarity between Diderot's conception of probability, and that of modern physics, notice that he compares "ce que le sort met d'incertitude" in games with "ce que l'abstraction met d'inexactitude" in mathematical science.

14 "Principes philosophiques sur la matière et le mouvement, in *Oeuvres philosophiques,* pp. 393–400.

15 "Interprétation" (v) in *Oeuvres philosophiques,* pp. 181–82.

16 On his *Mémoires sur differens sujets de mathématiques* (1749), see L. G.

Krakeur and R. L. Krueger, "The Mathematical Writings of Diderot," *Isis,* XXXIII (1941), 219–32.

17 "Interprétation" (vi), in *Oeuvres philosophiques,* pp. 182–84.

18 "Le Rêve de d'Alembert," in *Oeuvres philosophiques,* 285–371.

19 *Ibid.,* p. 293.

20 "Interprétation" (xxiv), in *Oeuvres philosophiques,* p. 193.

21 *Ibid.,* prefatory apostrophe, p. 175.

22 *Ibid.* (vii and viii), pp. 184–85.

23 *Ibid.* (xl), p. 215.

24 *Ibid.* (xix), p. 191.

25 *Ibid.* (xxvi), p. 194.

26 *Ibid.* (xxx), pp. 196–97.

27 *Ibid.* (xxxi), pp. 197–98.

28. *Ibid.* (xi), p. 186.

29 *"Entretien entre d'Alembert et Diderot,"* in *Oeuvres philosophiques,* p. 277.

30 This is the tenor of the first five "conjectures," "Interprétation" (xxxii–xxxvi), in *Oeuvres philosophiques,* pp. 198–211. See, too, "Rêve," in *Oeuvres philosophiques,* pp. 263–64 and 289–90.

31 "Rêve" *Oeuvres philosophiques,* pp. 299–300.

32 *Ibid.,* pp. 296–303.

33 "Interprétation" (lviii), *Oeuvres philosophiques,* pp. 239–44. See, too, (l), pp. 224–30.

34 "Rêve" *Oeuvres philosophiques,* pp. 291–95.

35 By the way of the *fils naturel.* See Arthur Wilson's *Diderot* (New York, 1957), p. 255.

36 Delambre, "Notice sur la vie . . . de Lagrange," *Oeuvres de Lagrange,* (1867), I, xxxviii.

37 Maurice Daumas, *Lavoisier, théoricien et expérimentateur* (Paris, 1955), 157–78. Though it is only fair to say that M. Daumas does *not* treat the issue as one between objective concepts and qualitative principles, chemistry and anti-chemistry. He regards phlogiston as itself a principle, and oxygen as partaking of this eighteenth-century quality.

38 *Encyclopédie,* article "Chimie."

39 C. C. Gillispie, "The formation of Lamarck's Evolutionary Theory," *Archives internationales d'histoire des sciences* (Oct.–Dec. 1956), 323–38; see also my "Lamarck and Darwin in the History of Science," in Bentley Glass, ed. *The Forerunners of Darwin* (Baltimore, 1959).

40 See, for example, the serial attack on the new chemistry by J.-F. de Machy in the Masonic *Tribut de la Société nationale des neuf soeurs,* I–IV (1790–91). Machy was a significant figure among the enemies of the Academy, to which he was refused admission about the time of Lavoisier's election. He was a popular private teacher, a leader of the pharmacists, author of the volumes on distillation of the *Encyclopédie méthodique,* and one of the royal censors who refused a license to *Annales de chimie.* See L.-G. Toraude, *J.-F. de Machy* (Paris, 1907).

41 Daniel Mornet, *Les sciences de la nature en France au XVIIIe siècle* (Paris, 1911).

42 The essential documents were published by E.-T. Hamy, *Les derniers jours du jardin du roi et la fondation du muséum d'histoire naturelle* (Paris, 1893). See, too, *Procès-verbaux du comité d'instruction publique*, I, 479–87 and *passim*. Most of the archives of the Museum have been transferred to the *Archives nationales*, where they occupy sub-series AJ–15.

43 *Esquisse d'un tableau historique des progrès de l'esprit humain* (Paris, 1795), 27–28: "J'entends cette séparation de l'espèce humaine en deux portions; l'une destinée à enseigner, l'autre faite pour croire; l'une cachant orgueilleusement ce qu'elle se vante de savoir, l'autre recevant avec respect ce qu'on daigne lui révéler; l'une voulant s'élever au-dessus de la raison, et l'autre renonçant humblement à la sienne, et se rabaissant au-dessous de l'humanité, en reconnoisant dans d'autres hommes des prérogatives supérieures à leur commune nature."

44 The famous report on animal magnetism by Bailly in the name of a commission composed, in addition to himself, of Franklin, Le Roy, Bory, and Lavoisier, is in the *Histoire de l'Académie royale des sciences, année 1784* (Paris, 1787), pp. 6–15.

45 On the *Société linnéene* and its successor, the *Société d'histoire naturelle*, see its *Procès-verbaux, Bibliothèque mazarine*, MSS. 4, 441; together with fragmentary papers at the *Muséum d'histoire naturelle*, MSS. 298, 299, 300, 1998; and at the *Bibliothèque nationale*, MSS. FR, NA, 2760, f° 162–163 and 2762, f° 60. See, too, *Adresse des naturalistes*, BN Le[28]. 826. Publications were the *Actes de la Société d'histoire naturelle*, of which one folio volume appeared in 1792 (BN [Inv S 1333] and Academy of Natural Sciences, Philadelphia) and *Mémoires*, of which one volume appeared in 1799 (BN [Inv S 4649]). These are not to be confused with the *Journal d'histoire naturelle* (2 vols.; 1792), edited by Lamarck and others, which was bitterly hostile to Linnaean methods (BN S 11705 and S 11704). This, too, may be found at the Academy of Natural Sciences in Philadelphia—a collection, by the way, which is too little exploited by historians of science.

46 For Broussonet, Bosc, and Ramond, see documents in *Bibliothèque de l'Institut de France, Fonds Cuvier*, 186, 157, and 154. Bosc and Ramond have left autobiographies in manuscript. For the documents bearing on Broussonet's unsuccessful attempt to escape from France during the Terror, see *Archives nationales*, F[7] 4619. See, too, the relevant *éloges*, Georges Cuvier, *Recueil des éloges* (3 vols.; Strasbourg, 1819–27).

47 *Arch. parl.* XVIII (session of 16 August, 1790), p. 91. Creuzé's detailed objection to the Academy was submitted as an annex to the report of the session of 20 August, 1790, *ibid.*, pp. 182–84.

48 A.-C. Thibaudeau, *Mémoires sur la Convention et le Directoire* (2 vols.; Paris, 1824), "Parmi les chefs révolutionnaires d'alors, il y en avait qui regardaient les lumières comme des ennemis de la liberté, et la science comme une aristocratie; ils avaient leurs raisons pour cela. Si leur règne eût été plus long, ou s'ils l'eussent osé, ils eussent fait brûler les bibliothèques, égorgé les savans, et

replongé le monde dans les ténèbres. Ils répétaient contre les sciences les sophismes éloquens de quelques écrivains humoristes; elles étaient, disaient-ils, la source de toutes les erreurs, de tous les vices, de tous les maux de l'humanité: les plus grands hommes s'étaient formés d'eux-mêmes, et non dans les universités et les académies. Ces déclamations flattaient la multitude; les ignorans étaient ennemis des lumières pour la même raison que les pauvres le sont des richesses." Thibaudeau served on the *Comité d'instruction publique.*

49 Paul Dupuy, ed., *Centenaire de l'Ecole normale supérieure* (Paris, 1895), pp. 67–68. Dupuy reminds us that the *Comité d'instruction publique* was allowed little independence by Robespierre (p. 35).

50 C. Hippeau, *L'Instruction publique en France pendant la Révolution, débats législatifs* (Paris, 1883), pp. 65, 156.

51 J.-P. Brissot, *De la vérité* (Neuchatel, 1782), p. 335.

52 Brissot, *Mémoires* (4 vols.; Paris, 1830–32), I, 199.

53 As has been remarked by Paul Vernière in his edition of Diderot, *Oeuvres philosophiques,* p. 168.

54 Gillispie, "The Natural History of Industry," *Isis,* XLVIII (December, 1957).

55 See my introduction to *A Diderot Encyclopedia of Trades and Industry: Manufacturing and the Technical Arts in Plates from L'Encyclopédie* (2 vols; New York, 1959).

56 Diderot, *Encyclopédie,* article "Art."

57 Diderot, *ibid.,* article *"Encyclopédie."*

58 N° 1, 4 Sept. 1791, *B. N.* [8° Lc². 6381.

59 See, for example, *Adresse du Point central des arts et métiers,* 16 Oct., 1791, *Archives nationales,* AD VIII, 29; *Journal des sciences, arts, et métiers,* 22 Jan., 1792 and 29 Jan., 1792, *B. N.* V. 42735.

60 For example, the letters (in French) of Lexell and Marivetz to Euler, in 1781 and 1782, I. I. Liubimenko, ed., *Uchenaia Korrespondentsiia Akademii Nauk, XVIII veka,* Vol. II of Akademiia Nauk Soiuza Sovetskikh Sotsialisticheskikh Respublik, *Trudy Arkhiva,* ed. D. S. Rozhdestvensky (Moscow and Leningrad, 1937).

61 *Procès-verbaux de l'Académie des sciences,* 13 June 1789.

62 *Ibid.,* 3 April, 1791.

63 Compare, for example, in documents at the *Archives nationales,* F¹⁷. 1136, F¹⁷. 1137, AD VIII, 42 (for a typical sampling) the treatment of many *projets* already rejected by the Academy, in its *Procès-verbaux.*

64 Extracts from the *Procès-verbaux* of the *Bureau de consultation* were published by Charles Ballot, *Bulletin d'histoire économique de la Révolution* (1913). Three of the four volumes (the fourth has been lost) of the full minutes are preserved in the office of the Secretariat of the *Conservatoire des arts et métiers.* Some relevant documents were published in the *Oeuvres de Lavoisier,* and others in *Procès-Verbaux du Comité d'instruction publique.*

65 *Archives nationales,* AD VIII, 43; F¹⁷. 1097, dossier 4; F¹⁷. 1310, dossier 14; F¹⁷. 1350, dossier 2; *Journal de la société populaire . . . des arts* (B. N., V. 42711).

66 *Archives nationales,* C 686, pièce 2; F^7. 4239, dossier 2; and *Bibliothèque nationale,* MSS. FR, ancien supplément français, 8045, which *cote* is erroneously ascribed by the catalogue to the *Procès-verbaux* of the *Bureau de consultation.* In fact, the *procès-verbaux* are fragmentary remains of those of the *Société des inventions et découvertes.*

67 S.-J. de Boufflers, *Rapport . . . sur la propriété des auteurs de nouvelles découvertes et inventions* (30 Dec. 1790), B. N., Le^{29}. 1206, p. 12.

68 *Archives nationales,* AD VIII, 40, T. I, pièce 18, *Point central . . . à la Convention nationale* (26 Sept. 1793).

69 *Archives nationales,* AD VIII, 29, *Adresse du point central,* 16 Oct. 1791, p. 11.

70 *Archives nationales,* AD VIII, 29, *Establissement d'une Ecole athénienne, sous le nom de Lycée des Arts et Métiers.* Another version, B. N., Rz3007 & Rz3008.

71 *Archives nationales,* AD VIII, 40, T. I., pièce 18, *Point Central . . . à la Convention nationale.* The proposed constitution itself had been drawn up in March 1792 and sent to the Legislative Assembly with a "projet de décret." It was "redigé par la Société du Point Central des Arts et Métiers, en présence de MM. les Commissaires des Sociétés des Inventions et Découvertes et de la Commune des Arts" (B. N., Inv. Rz3001).

72 Lavoisier to Lakanal, 10 & 11 August 1793, *Procès-verbaux du comité d'instruction publique,* II, 314–17.

73 Lavoisier to Delambre, 8 August 1793: "Je ne sais si j'ai encore le droit de vous appeler mon confrère," he writes, assuring Delambre, however, that some way will be found to carry on the metric survey. This letter is not yet published, I believe. M. René Fric, who is editing Lavoisier's *Correspondance* (Paris, 1955—) very kindly and cordially allowed me to read through the copies of the letters which he has prepared for publication.

74 *Procès-verbaux du comité d'instruction publique* (14 August 1793), II, 319.

75 Lavoisier to *Comité d'instruction publique,* 17 August 1793, *ibid.,* II, 320.

76 *Ibid.,* II, 331–32.

77 The chairman was Grégoire. *Bibliothèque du muséum d'histoire naturelle,* MSS. 312 (I).

78 J.-D. Cassini, *Mémoires pour servir à l'histoire des sciences et à celle de l'Observatoire Royal de Paris* (Paris, 1810); J.-F.-S. Devic, *Histoire de la vie et des travaux scientifiques et littéraires de J.-D. Cassini IV* (Clermont, 1851); C. Wolf, *Histoire de l'Observatoire de Paris* (Paris, 1902); and for documents illustrative of the history of the observatory under the Terror, *Archives nationales,* F^{17}. 1065^A, dossiers 3 & 4; *Procès-verbaux du comité d'instruction publique,* II, 217–27.

79 J.-B. Delambre, *Grandeur et figure de la terre,* ed. G. Bigourdan (Paris, 1912); G. Bigourdan, *Le système métrique des poids et mesures* (Paris, 1901); Adrien Fabre, *Les origines du système métrique* (Paris, 1931); *Oeuvres de Lavoisier,* VI, 660–712; *Procès-verbaux de l'académie des sciences,* passim.

80 Georges Bouchard, *Prieur de la Côte-d'Or* (Paris, 1946).

81 F.-A. Aulard (ed.), *Recueil des actes du comité de salut public,* IX (1895), p. 600.

82 *Bibliothèque du muséum d'histoire naturelle,* MSS. 299. Article 2 states: "Le

gouvernement ne doit encourager les sciences que sous le rapport des arts, et toutes les fois qu'on demande de l'argent au peuple pour cet encouragement, il faut qu'il en puisse saisir facilement le but utile, autrement il croirait le produit . . . employé à satisfaire une vaine curiosité."

83 M. Berthelot, "Notice sur les origines et sur l'histoire de la Société philomatique," *Mémoires . . . a l'occasion du centenaire de sa fondation* (Paris, 1888), pp. i–xvii.

84 Its papers are conserved, though in a state of extreme disorder, at the *Bibliothèque de l'Université de Paris,* where in 1955 they were shelved in the hallway outside the office of the *Conservateur.* Whatever it became later, the *Société philomatique* in the 1790's remained the resort of young men and of mediocrities.

85 *Journal du Lycée des arts,* 15 April 1793 (B. N.: V. 28667).

86 *Journal des sciences, arts et métiers,* pp. 4–5 (B. N., V. 42735).

87 *Oeuvres de Lavoisier,* IV, 649–68.

88 In June, 1793, Lavoisier presided at the session of the *Lycée* (*Journal du Lycée des arts,* 8 July 1793).

89 *Journal du Lycée des arts,* 14 & 29 June 1793.

90 Lavoisier to Comité d'instruction publique, 10 August 1793, *Procès-verbaux du comité d'instruction publique,* II, 316–17, and "Observations sur l'Académie des sciences," *Oeuvres de Lavoisier,* IV, 616–23.

91 Joseph Schumpeter, *Capitalism, Socialism, and Democracy* (New York, 1942).

92 Interviewed in *L'Express* (29 Jan. 1955).

93 J.-D. Cassini, *Mémoires,* p. 93.

94 For Rousseau's botany, see *Lettres élémentaires sur la botanique* in *Oeuvres posthumes* (Geneva, 1782), and Albert Jansen, *Rousseau als Botaniker* (Berlin, 1885). For his chemistry, *Les institution chymiques,* ed. Maurice Gautier, *Annales de la société Jean-Jacques Rousseau,* XII–XIII (Geneva, 1918–21). For a discussion of his attitude to science, see F. C. Green, *Rousseau and the Idea of Progress,* The Zaharoff Lecture for 1950 (Oxford, 1950).

95 C. C. Gillispie, "Science in the French Revolution," *Behavioral Science,* IV (January, 1959), 67–73.

96 Particularly the *Eléments de la philosophie de Newton* (1741) where the scorn expressed for metaphysics is that on which classical physics in practice acted, and *Voltaire's Correspondence,* ed. Theodore Besterman (Geneva, 1953—), which is indispensable for his state of mind about physics in the 1730's.

Note

This essay is an elaboration of the first of two public lectures on "Science and the French Revolution" which I had the honor of delivering in Oxford University in April, 1955, and the draft of which I sent early in July, 1957, to my critics, Professors Henry Hill and Henry Guerlac, as a preliminary guide to my interpretations. Between then and the convening of the Institute in September, I developed the material into its present form. Upon my arrival in Madison, Professor Guerlac kindly showed me the typescript of a paper, "The Anatomy of Vandalism," which he had given before the History of Science Society, meeting in Washington, D. C. in December, 1954, at a time when I was working in Paris on a Guggenheim Fellowship (of which this essay is one result). Neither of us had wished to publish these papers, of which, therefore, we remained in ignorance, and although we have been aware of the convergence of our some- what differing interests on the period of the French Revolution, we were im- pressed at how closely the two papers agreed, even to the point that both of us hit upon the figure of a miniature French Revolution in the *Jardin du roi*. The gist of my own lecture occupies a certain portion of the essay published here- with. Our presentations were most strikingly parallel on the differential treat- ment accorded the *Jardin* and the *Académie des sciences*. Professor Guerlac went more deeply than I into the anti-scientific influence of Rousseau and Bernardin de St.-Pierre, whereas I developed the political activities of the volun- tary societies of naturalists, artisans, and inventors. I set this out since I should not wish it thought that the hostility of the Revolution to physical science and the Academy contrasted to its enthusiasm for natural history and the *Jardin* are my discoveries. Professor Guerlac expressed and communicated them before I did.

Since then I have not significantly revised my views on the events, but I have seen them from a new angle of importance as a result of three circumstances: First, an investigation into the origin of Lamarck's evolutionary theory (see above, n. 39); second, an interest in Stoic physics aroused by Professor S. Sam- bursky's *The Physical World of the Greeks* (London, 1956); and third, a study of the *Encyclopédie* of Diderot and d'Alembert in preparation for my edition of the technical plates (see note 55). This is the origin of the analysis which I now give the subject, treating the Jacobin attack on the Academy as the political expression of a half-Stoic, half-Baconian attempt to substitute organismic concepts and technology for Newtonian theoretical physics at the center of science. Professor Guerlac is not altogether persuaded by this analysis. He ex- presses some dissent and some elaboration in his critical remarks.

THE POLITICS OF SCIENCE

IN THE FRENCH REVOLUTION

L. Pearce Williams

In December of 1954, Henry E. Guerlac of Cornell University read a paper before a joint session of the History of Science Society and the American Historical Association entitled, "The Anatomy of Vandalism." Mr. Guerlac had been struck by the vast difference in the fates of two royal scientific institutions during the Revolution and his paper was an attempt to explain this difference. The *Académie royale des sciences* was suppressed; the *Jardin du roi* not only survived the revolutionary maelstrom but actually emerged from the Revolution, under the new name of the *Muséum d'histoire naturelle,* a larger and stronger institution with greater influence and prestige than it had ever enjoyed under royal patronage. In accounting for this difference, Mr. Guerlac stressed the political policies followed by the members of these two institutions and showed clearly that the personnel of the *Jardin du roi* were far more able in anticipating the shifts of public opinion which accompanied the onward rush of the Revolution than were their more conservative counterparts in the *Académie*. But Mr. Guerlac considered this factor, by itself, insufficient to explain satisfactorily the history of these two bodies in the Revolution. He introduced, therefore, a new and brilliant hypothesis which added a new dimension to the political account. Beginning with Rousseau and continuing through Bernardin de Saint-Pierre—who significantly became the first revolutionary director of the *Muséum*—Mr. Guerlac traced a current of anti-Newtonian science which arose in the second half of the eighteenth century and continued with increasing strength into the Revolution. The suppression of the *Académie* and the enlargement of the *Muséum,* according to this interpretation, involved not only the question of political agility but also a basic point of ideology. To the Jacobins, it was suggested, Newtonian science as represented by the *Académie* was dry, academic, and, above all, aristocratic, since only the few could hope to penetrate its mathematical mysteries. The science of Rousseau and of the *Muséum,* on the other hand, was a science of *sentiment,* of direct apprehension of the harmony of the world through contact with nature and, therefore, open to all. The triumph of the *Muséum* and the downfall of the *Académie* was, then, the result of the victory of the disciples of Rousseau—the Jacobins—and of this peculiarly Jacobin science.

This interpretation shed a new light on a problem with which I had been occupied for some time; namely, the failure of the Legislative Assembly and the Convention to reform the educational system of France. In a sense, there is seemingly no problem here at all. During the years 1789–94 the various governments of France were assailed from all sides by problems demanding immediate and effective action. The defection of the king, inflation, foreign invasion, civil war, and bitter party strife created an atmosphere in which concern over the proper education to be offered to future citizens had often to be relegated to the background. But if one examines the situation more closely, there is abundant evidence to show that the question of educational reform was not simply buried under the mass of more important problems.

Although symbolizing more the spirit of the Third Republic than that of the Revolution, there is some significance to the fact that Danton has been immortalized by the words, "Après le pain, l'éducation est le premier besoin d'un peuple," which are inscribed under his statue at the Place de l'Odéon in Paris. A more eloquent, although prosaic, confirmation of the ideal thus epigrammatically expressed is found in the hundreds of pages of debates and plans for educational reform in the *Archives parlementaires.*[1] In spite of these fine sentiments, sincerely felt, and the flood of words expended on this subject, the sorry fact is that from 1789 to 1794 no significant reform of education was forthcoming, and even the educational system of the *Ancien Régime,* with faults and virtues alike, was allowed to decay. By 1794, France was almost entirely without facilities for the training of the young.

This paper was originally conceived as an application of Mr. Guerlac's hypothesis to this impasse reached on education. Was the failure to reach agreement on education the result of an irreconcilable struggle between Newtonians and anti-Newtonians? As will become clear, this question is totally irrelevant in the area of education. In the almost endless debates on educational reform never once is the traditional science attacked by an anti-Newtonian rival. The origins of the deadlock must be sought elsewhere.

The major source of the impasse on educational reform lies deep within the history of the eighteenth century. The economic ruin resulting from Louis XIV's passionate search for glory and the political and moral decline which marked and marred the reign of Louis XV forced men to seek out new frames of reference within which hopes for the regeneration of France could be placed. Most thinking people of the Enlightenment clearly recognized the crying need for reform, but the bases of criticism of the *Ancien Régime* were not everywhere the same. More specifically, there developed throughout the eighteenth century three broad streams of reform which seemed merely to be different aspects of the same movement, as long as action was confined to the spoken and written word. When, during the Revolution, it became necessary to pass from words to deeds, these three movements were found to be incompatible with one another, particularly in the area of educational reform.

The oldest, and in terms of French tradition the most respectable of these reform movements, may be termed Enlightened Stoicism, for it drew both on the world of the Enlightenment and that of Republican Rome. Beginning with Fénélon and running through Montesquieu, d'Aguesseau, and Turgot, it was an outlook peculiarly suited to men whose thoughts had been influenced and shaped by public service to the *Ancien Régime*. Having viewed the corruption of French public life at close quarters, these men sought to reform the state by regenerating and reinstating those moral virtues without which liberty and justice could not exist. Although recognizing a fundamental and necessary order in the universe,[2] like their Stoic predecessors they concentrated their attention almost exclusively upon that aspect of this universal order which prescribed the proper rules of conduct necessary for the creation of public virtue. They were content to leave the physical order in the background to serve simply as a justification of the assumption of the moral law.[3]

The specific terms of the moral law were drawn directly from the Roman Stoics. Fénélon's *Télémaque,* prepared as a novelized and secular catechism for the Duke of Burgundy, stresses over and over the necessity of the prince's adherence to those values of honor, duty and self-sacrifice which alone can guarantee the common weal, a prosperous reign and eternal fame. Public virtue should not, however, reside only in the prince. If both liberty and justice are to exist, it must be present as well in the citizens composing the state. As Montesquieu pointed out in his *Considérations sur les causes de la grandeur des Romains et de leur décadence,* the general moral decay of the Romans caused by the twin evils of super-fluous wealth and the iniquitous philosophy of Epicurus so weakened the Roman state that it fell easy prey first to tyrants and then to the barbarian invaders.[4]

Public virtue was encountered only rarely in history, and governments could and did survive without it. The despot could rule by fear, the monarch by appealing to honor, but only the Republican form of govern-ment required such virtue for the continuance of its existence.[5] Montes-quieu leaves no doubt as to the place of public virtue in his hierarchy of values. In a passage certainly not without influence on a later generation, he indicates clearly that such virtue was not only the highest ideal to which man could aspire but also the only true guarantor of political liberty.

It is in the Republican form of government that all the power of education is needed. The fear of despotic governments is born automatically from its threats and punishments; honor in monarchies is stimulated by the passions and, in its turn, stimulates them: but political virtue involves self-renunciation which is always a difficult thing.

One can define this virtue as love of the laws and of one's country. This love, demanding a continual preference for the public interest to one's own, is the source of all particular virtues for they are nothing but this preference.[6]

This theme runs through much of the Enlightenment. The French must be schooled in virtue and the state must be reformed so that public virtue, just laws and a "patrie" deserving of love can bind citizen and government together into an invincible whole. Only then will France achieve that greatness for which her talents and resources had destined her.

The progress of science and its increasing prestige during the eight-eenth century gave rise to a second reform movement aimed at the modi-fication of the institutions of the *Ancien Régime*. The symmetry, order, and relative simplicity of the Newtonian world stood in vivid contrast to the chaotic hodge-podge of institutions, overlapping jurisdictions, irration-alities and sheer arbitrariness of the government of pre-revolutionary France. This contrast could not help but stimulate impatience and dis-content with the existing system. Even more important was the growth of relativism which science in the eighteenth century furthered. Conscious that man as a whole, in Newton's words, had only picked up a few pebbles on the beach of the sea of knowledge, the *philosophes* viewed the pretensions of Absolutism with skepticism.[7] Diderot, in his plan for a university in Russia, strongly urged the inclusion of the teaching of the theory of probabilities since, as he wrote, "Our whole life is only a game of chance; let us try to have the chances in our favor."[8] There must have been many people in France who concluded that the contemporary game of politics was played with loaded dice.

The influence of science in the Enlightenment, however, went far beyond challenging the intellectual justification of the *Ancien Régime*. It was on the model of Newtonian physics that a new "science" was created which had revolutionary implications. Just as Galileo, Descartes, and Newton had built the imposing structure of physics upon a foundation of basic physical principles, so too did Locke, Condillac, and their followers attempt to build a moral science upon certain fundamental properties of the human psyche. From the assumption that all knowledge comes from the senses and that all sensations are either pleasurable or painful, the sensationalists created their system for the moral and political regeneration of France.[9] In terms of reform, this system envisioned two necessary operations: First must come the destruction of superstition which since earliest times had held man in bondage, and then the true principles of social morality could be presented and accepted. The first step was uniquely the task of the sciences. When people could see for themselves what the system of the world was they would no longer tremble before either throne or altar and the supernatural sanctions which buttressed the power of both. With this fear removed, the natural morality of man could emerge to serve as the foundation of a society in which enlightened self-interest and social utility together produced political harmony and freedom.

The movement for reform, based on the writings of Jean-Jacques Rousseau, bears many striking resemblances to Enlightened Stoicism. Rousseau, like Fénélon and Montesquieu, found the source of contemporary evil in the corrupt morals of the eighteenth century. He, too, envisioned reform by means of a return to virtue and the establishment of a state in which perfect harmony existed between governors and governed. Where he differed from Enlightened Stoicism was in locating the source of virtue within the human heart. Virtue was not adherence to a stern moral law imposed from without by fiat, but the inner stirrings of one's intimate being. The origin of corruption was not simple forgetfulness of natural law, but an actual metamorphosis of the heart by which the vices of civilization have gained supremacy over the natural goodness inherent in every human breast.

In terms of criticism of the *Ancien Régime,* these three reform movements were perfectly compatible with one another and little difficulty would have been experienced in following all three at the same time. The arbitrariness of absolute rule, lack of concern for the public welfare, the selfish seeking for individual advantage at the expense of one's fellow man, oppression, privilege, corruption—all of these characteristics of the *Ancien Régime* were equally contemptible in the eyes of the proponents of each of these reform movements. Even the ultimate goals were without blatant contradictions. A society of free, moral men, following the path of duty, acting in the public welfare and conscious of the obligations binding them in a strong and free union, was a dream which could be willingly accepted by all those who desired the reform of France. The great area of incompatibility was, of course, that of the means for wiping out social evil and for creating the good society. Because the educational system of any society is necessarily concerned as much with means as with ends, this incompatibility of reform movements emerges with striking clarity in the debates over educational reform. And, as I hope to show, it is this three-way incompatibility which, without regard to party affiliation, led to the stalemate over the creation of a new system of public and national education in the most tumultuous years of the Revolution.

When the Estates-General met in May, 1789, there was little feeling of urgency about the question of educational reform. Not only were there much graver problems to be solved, but the whole atmosphere was one in which the nature of France's school system seemed of minor importance. The deputies came to Versailles to effect major changes in France's political institutions; they did not come to raze the very foundations of French society and erect a wholly new structure on the ruins. Specific major abuses of the *Ancien Régime* were the target of the National Assembly and it was felt that the correction of these evils would automatically restore the state to an even keel.

Both the *cahiers*[10] and the actions of the Constituent Assembly reflect the general feeling of apathy towards the problem of school reform. Many *cahiers* make no mention of this problem at all; others simply repeat a stock phrase calling for some action but leaving this action unspecified.

Only a few give any indication of the direction which reform should take and here the proposed changes are very minor. Almost without exception these *cahiers* call for the creation of chairs of law and of morality throughout France so that "les moeurs" will profit from the results of the activity of the deputies at Versailles.[11]

In the Assembly, there was almost complete silence on this question until 1791, when Talleyrand presented the first plan for a sweeping reform of French education.[12] His remarks were applauded and the Assembly ordered the printing of his discourse at public expense, but no action was taken since "we ought to leave something for our successors to do."[13] As long as the Revolution confined itself to reform within the broad framework of the monarchy, the problem of national education had few political implications and consequently was ignored.

By the spring of 1792 the situation had changed considerably. The hopes of turning France into a constitutional monarchy had been dashed by the flight of the King; the struggle between Left and Right was becoming increasingly bitter; Louis XVI's veto of the bill directed against the non-juring clergy had given heart to those who now saw the Revolution as a monster to be destroyed, and war with Austria and Prussia, threatening the very existence of France herself, loomed ever larger on the horizon.[14] The temper of the French and their legislators had altered. The Revolution was no longer a simple reform of the *Ancien Régime;* it was rapidly becoming a repudiation of that regime and the deputies were becoming increasingly conscious of the fact that if they and the Revolution were to survive, France must undergo a radical transformation. Where education before had been a relatively unimportant area for consideration, it now emerged as a powerful instrument for securing the fruits of the Revolution and for unifying and strengthening France.

On April 20, 1792, Condorcet presented his *Rapport et projet de décret sur l'organisation générale de l'instruction publique.*[15] The scope of this revolutionary educational system went far beyond the bounds of a simple reform of existing institutions. The schools of the *Ancien Régime* were already moribund. They should be allowed to die in peace while, in their place, a new system of education would be devised specifically intended to fit the coming generation for the New France so rapidly emerging from the Revolution. The old system, founded as it was on the prejudices and superstitions of the ancients and the Church had been suited more to enslave men than to free them.[16] The new France required a new education which could help to prepare the way for a future of liberty and equality.

To replace the classical education of the *Ancien Régime,* Condorcet had recourse to the Enlightenment and the truths which the eighteenth century had discovered. "You owe to the French Nation an education which is at the level of the spirit of the eighteenth century and of modern philosophy, which, by enlightening the present generation, already presages, prepares and anticipates that higher level of reason to which the necessary progress of the human race calls future generations."[17]

The method for achieving this "higher level of reason" lay almost entirely in the teaching of the physical and mathematical sciences.[18] For those who would never have the leisure nor the inclination to master any study thoroughly—and here Condorcet obviously is referring to the lower classes whose constant labor prevented them from pursuing a long course of studies—even the superficial study of the sciences was the surest method for developing their intellectual faculties and their ability to reason correctly. Moreover, the subjects of the sciences were immediately observable and the regularity of natural phenomena was so apparent that long years of purely verbal preparation were not necessary for the comprehension of scientific laws.

Even more important for the Revolutionary generation was the fact that the sciences combatted superstition and prejudices, those twin evils of the *philosophes,* and thus made the transition from an oppressed to a free society an easier one. While creating a society of free men, the sciences also prepared the way for the maintenance of this freedom in the material as well as the intellectual sphere. Already of immediate utility in the various professions, the time would come, Condorcet prophesied, when science would produce a revolution in the mechanical and agricultural arts and free the laborer from the routine and dependence upon others which, together with superstition and prejudice, had made him a slave.

To be enlightened, however, was not simply to commit a large body of truths to memory. Enlightenment was a never-ending process dependant upon the ever-expanding area of the sciences and constantly subject to revision and change. Man improves himself as he approaches truth and Condorcet viewed history as this process of improvement and not as the awesome leap from error to an immutable and monolithic perfection. It is not enough, then, to spread knowledge; new knowledge must constantly be discovered and worked into the social fabric. To accomplish this a hierarchy of studies was to be established. On the primary and secondary levels were to be taught those verities already discovered along with their social implications. The *lycées,* institutes, and National Academy of Sciences were to create and support the researchers whose task it would be to discover new facts and formulate new theories. In this way both the growth of enlightenment and the progress of the human race would be constant.[19] The only immutable elements in this system were love of truth and faith in reason. Data might change or theories undergo modification but the outlook of the enlightened citizen remained the same and provided a sure basis for a stable and free society.

Precisely because of this expansion of knowledge, education had to work two ways in the state. It not only had to reveal the justice of the laws to the citizen but, as knowledge increased, it had to serve as a guide to the legislators. Just as the scientist revised his concepts and theories, so too must the lawmaker revise the laws to keep them consonant with the growth of enlightenment. Education, therefore, must be free and "no public power should have the authority nor even the influence to prevent the development of new truths or the teaching of theories contrary to its particular policies or to its momentary interests."[20]

Discussion of Condorcet's project had to be deferred because of the growing crisis in France. With the declaration of war on Austria in April, 1792, Brunswick's Manifesto in August, the insurrection of August 10 and the September massacres, France was seriously threatened and all energies were directed toward the preservation of the Revolution. But as soon as the crisis had passed the Convention turned its attention back to the problem of educational reform. On October 2, 1792, the Convention selected twenty-four of its members to make up the Committee of Public Instruction and on the 12th of December this committee submitted a plan based closely on Condorcet's.[21]

The first attack came from the Right where Durand-Maillane, deputy from the Bouches-du-Rhône, objected strongly to the stress given to science in the proposed decree. What was needed was "less science than virtue"[22] and virtue was not the product of study but the simple love of country and obedience to her laws.[23] Love of country was a sentiment to be implanted in the hearts of all citizens, and as for the laws, "the whole duty of the citizen is to follow them and, to fulfill this duty, he has no need to sum them up or understand them."[24] Provided their forms are constitutional, the sole path the citizen should follow is that of obedience.

Science was not only irrelevant to the problem of public virtue, but a positive danger to it. Where Condorcet foresaw a rise in the standard of living as the result of the practical application of the sciences to industry and agriculture and a consequent lessening of social strife, Durand-Maillane, following both Rousseau and Montesquieu, saw the growth of luxury and a weakening of morality which would undermine the Republic. Where Condorcet foresaw the triumphant progress of science and enlightenment through the *lycées,* institutes, and the Academy of Sciences, Durand-Maillane saw the creation of a scientific guild both more powerful and more iniquitous than those abolished by the Revolution in the name of free enterprise.[25] But the real source of Durand-Maillane's opposition to Condorcet's reform lay in its disregard for the sanctions of religion. If the true moral virtues of duty and obedience were to flourish they must have the support of the Church in order to be effective. Enlightenment without the Church breeds anarchy whereas the Church, purged of its vices by the Revolution, could now serve as the effective guarantor of order, stability, and public rectitude so essential for the preservation of Republican France. The freedom demanded by Condorcet could only favor the spread of the cancer of subversion, and the entire school system, therefore, must be subjected to the strict control of a public censor to insure that enlightenment did not go too far and challenge the moral bases of the state.[26]

Although the Left had nothing but scorn for Durand-Maillane's position[27] there was soon to be an attack, from within its own ranks, levelled against the Committee of Public Instruction's plan. Invoking Rousseau as his beloved master, Citizen Petit called for a "School of Republicanism" in which neither Condorcet's sciences nor Durand-Maillane's Spartan virtue had a place. To be sure, the goal was the regeneration of morality but there was no need to seek this either in the physical universe or in the moral axioms of antiquity and the Christian religion. Virtue lies within the individual's heart and the only schoolmaster needed is Nature.[28] Before Nature could effect the efflorescence of the virtue inherent in all French citizens, the social evils which constricted and distorted the purest heart must first be eliminated. The twin evils of beggary and crushing labor were singled out as being particularly effective in corrupting the French. How could one expect virtue to flourish when many children were forced to become beggars to help their parents support the family? And how could parents allow their children to attend schools when every hand was needed to aid in the task of earning one's daily bread? First must come a reform of the state to wipe out these evils and then, almost automatically, virtue would flourish and the Republic would be saved. When the human heart is surrounded by virtue it will respond in kind and the whole problem of educational reform simply vanishes.[29]

For the next six months, these were the themes around which the debates in the Convention turned—the creation of enlightenment and virtue through the study of the sciences; the development of public virtue by the indoctrination of youth in the stern values of duty, patriotism, and self-sacrifice; the stimulation of natural virtue by surrounding the young with simplicity and justice so that the inclinations of their hearts could bear fruit in action.[30] No position lacked its ardent defenders but no compromise could be reached. As the battle between the Girondists and the Jacobins became increasingly bitter, the subject of educational reform was temporarily shelved.

Closely following the victory of the Jacobins at the beginning of June, 1793, a new Committee of Public Instruction, with Lakanal as Reporter, presented another plan for the Convention's consideration. The emphasis was shifted from enlightenment to sentiment, but unlike Rousseau's *Emile* this was mass sentiment guided by the state. Formal instruction was to be limited almost entirely to the teaching of reading and writing so that the natural goodness of the children's hearts would not be snuffed out by routine and pedantry. In place of these traditional pedagogical methods were to be substituted an appalling number of *fêtes* intended to stimulate specific feelings which, taken together, would guarantee the reign of virtue on this earth. Each administrative unit in France was to have its particular series of *fêtes*. On the cantonal level there would be *fêtes* of the beginning of work in the fields, the end of work in the fields, of the animals who were the companions of man, of youth, of marriage, of maternity, of the aged, of the perfecting of language, of the invention of writing, of the origin of commerce and the arts, of navigation and fishing, of the Rights of Man, of the social contract, the institution of primary assemblies and of the Sovereignty of the people, of popular elections and, finally, of the particular canton involved. In the districts there would be *fêtes* of the return of greenery, of the return of fruits, of the harvest, of the vintage, of the memory of one's ancestors, of equality, of liberty, of justice, of beneficence, and of the district. For the departments there were to be *fêtes* of spring at the vernal equinox, of summer at the summer solstice, of autumn at the autumnal equinox, of winter at the winter solstice, of poetry, letters, the sciences and the arts, of printing, of peace and war, of the destruction of the three orders in France, of the abolition of privileges, and of the department. For the *fêtes nationales* there was to be celebrated, visible nature, the fraternity of the human race, the French Revolution, the abolition of royalty and the establishment of the Republic, and the *fête* of the French people, one and indivisible.[31] Nor was this all! In every canton a national theater was to be established where the men could practice military maneuvers and both men and women could become proficient in music and the dance "in order to lend greater beauty and solemnity to the *fêtes nationales.*"[32]

Although momentarily stunned by this multitude of *fêtes* which, while doubtless stimulating virtue left very little time to do any work, the Convention soon rallied and, in the process of rejecting Lakanal's scheme, reverted back to the old deadlock between enlightenment, duty, and sentiment as the means for securing the Republic. Coupé de l'Oise and Lequinio, both good Jacobins, leapt to the defense of enlightenment and science while others once again raised the standards of Rousseau and Montesquieu.[33]

In an attempt to break this stalemate Robespierre himself ascended to the Tribune and presented a plan for educational reform conceived by the recently assassinated Michel Lepeletier. The fact that Robespierre utilized his growing influence in the Convention to plead for this project personally—not once but twice—seems adequate evidence from which to conclude that it reflected his own views on the subject of education.[34] As such, it is worthy of a rather detailed description.

Lepeletier began by distinguishing between education and instruction. The function of education is to create men; that of instruction is to propagate knowledge.[35] Education in a Republic must be universal, but instruction, although open to all, "by the very nature of things, will become the exclusive property of a small number of the members of society because of the differences in professions and of talents."[36] Lepeletier, and by extension, Robespierre, accepted Condorcet's system of higher instruction with its emphasis upon science but rejected the primary level, concerned with education. It was to this level—to the task of "creating a new people" through universal education that the whole of Lepeletier's plan for educational reform was devoted. The task was no easy one for the degradation of the French as the result of centuries of tyranny required draconian measures to correct. Could vice-ridden parents produce anything but vice-ridden children? And had not Montesquieu written that the young do not degenerate spontaneously but are lost only when the adult population is already corrupted?[37]

Lepeletier could see only one way to break this vicious circle. The Republic must take the young citizens of both sexes and, from the ages of five to puberty (12 for boys, 11 for girls) raise them as wards of the state away from their parents.[38] In the *maisons d'égalité* no distinction of birth, wealth or position would be recognized. All would live together in equality and thus learn that equality was one of the cornerstones of the Republic. In order to accustom the children to shun the luxury which saps the strength of Republics, the children were to live in Spartan simplicity. Hard beds, healthy but simple food, and rough clothing would root out the evil of the desire for the superfluous so characteristic of the *Ancien Régime*. The habit of hard work, too, was to be a result of the course of indoctrination. Not only would this habit aid the physical development of the youth of France, but it would form men truly free of dependence on others, further the ideal of equality, and place at the service of the state human energies with which prodigies could be accomplished.[39]

Actual instruction would be confined to the teaching of reading, writing, arithmetic, the principles of Republican morality as drawn from the Constitution and the emphasis of those virtues by which other Republics had risen to eminence.[40]

By banishing parental vices through the effective elimination of parental control, by indoctrinating French youth in Republican habits and virtues, and by providing a basic education in Republican morality the goal of the moral regeneration of France could be reached. The state had only to take the required action. Neither natural goodness nor enlightenment could serve to drag the French people up from the morass into which a millenium of degradation had plunged them. Politically the child belonged to the state and, if the state were to flourish, it must force the child into the proper mold.[41]

The Convention promptly voted the acceptance of Lepeletier's system but this action in no way terminated the debates on education. The followers of Rousseau, horrified at the whole concept of forcing both children and parents into a Republican mold, cried that "virtue cannot be taught by rote. It is not to be found in the intelligence but it is only in the human heart that it has its roots and grows."[42] And what blows this system struck at the human heart! To wrench a child of five away from the tenderness and love lavished upon him by his parents could produce only monsters and not citizens.[43]

The advocates of enlightenment, too, found this plan little to their liking in spite of the assurance that the higher degrees of instruction would pay ample attention to the sciences. With dogged persistence, Gilbert Romme submitted his project for educational reform based on the teaching of the sciences in October, 1793, but met with no success.

Here the matter rested until the time of the Directory when the proponents of enlightenment gained their long-sought victory and, with the *Ecoles centrales,* attempted to reform society through the medium of the physical sciences and the principles of morality deduced from them.

In spite of the fact that the debates on educational reform in the Legislative Assembly and the Convention bore no fruit during the Revolution, they are of considerable value for understanding the Revolution and lead to certain conclusions which help to clarify the whole question of the attitude of the revolutionaries towards learning in general and science in particular. In the first place, they illustrate with great clarity the fallacy of equating intellectual orientation with party affiliation. As I think has been conclusively shown, the Jacobins were hopelessly divided over the question of educational reform. That they could not agree on specific institutions for reforming educational practice might only illustrate differences in backgrounds and training of the deputies. But that they could not agree on the theoretical foundations of their new society is of much greater import. For this was no simple political debate. It involved basic ideological principles concerning the nature of man and of society. Indeed, in a sense, it is the *only* purely ideological debate in the entire Revolution for there was no pressure here to rationalize actions already taken which had to be made palatable by appealing to some higher justification than circumstances. On this basis, the debates on education show the folly of speaking of the Jacobins as more than adroit politicians. These debates reveal that Jacobinism was forged on the anvil of expediency and that he who seeks to make of it a coherent ideology must seek in vain.[44]

The deep divisions within the ranks of the Jacobins over educational reform should also make one wary of equating Jacobinism with a large-scale anti-intellectual movement. To be sure, there were Jacobins who felt that science was either a threat or irrelevant to their plans for the regeneration of France. But there were also men like Romme, Coupé de l'Oise, and many others who viewed science as the instrument for the liberation of their fellow citizens and the means of accomplishing their goal of a stable, orderly and enlightened Republic. Robespierre's acceptance of Lepeletier's ideas on educational reform would also seem to indicate that there was ample room for the sciences even in the breast of this man whom David depicted as having two hearts beating for liberty.

While there is ample evidence to show that some Jacobins did not wish to see the physical sciences become the core of the new education, nowhere in the debates on education, to return to Mr. Guerlac's hypothesis, is there to be found a suggestion that an anti-Newtonian science occupy the central place. As a matter of fact, there is no reference to such a science at all. The objections to Newtonian science were couched in moral terms which had nothing to do with any kind of science. That the anti-Newtonian school existed during the Revolution, I have no doubt. But there is not one shred of evidence to indicate that this anti-Newtonianism in any way played a part in the political decisions involving science or scientists during this period. Since the debates on education ran concurrently with the discussion of the fate of the *Académie des sciences,* surely this point of view would have revealed itself in the area of education where it would have been relevant. The execution of Lavoisier, the death of Bailly, the suicide of Condorcet, and the suppression of the Academy of Sciences would seem, then, to have been due to the political views of those concerned and not caused by either any widespread hostility to science per se or to the attacks of an anti-Newtonian group of Jacobins.

The debates on education reveal another point which, in my opinion, has not received proper attention.[45] This is the deep influence of the tradition of Fénélon and Montesquieu on the Revolution and the opposition of this tradition to that of Rousseau and his followers. One should not lose sight of the fact that, to Montesquieu, by definition, a republic was a Republic of Virtue, and that the means by which Robespierre sought to impose such a Republic upon France followed directly in the Stoic tradition. Certainly the means suggested by Robespierre and Lepeletier for the reform of education were unacceptable to Rousseau's disciples in the Convention. Perhaps it is permissible to suggest, on the basis of the evidence of the debates on education, that the roots of the Republic of Virtue of the Year II lie, not in Geneva but, paradoxically enough, in the Gironde.

Finally, one characteristic of the differences among the three reform movements noted in the debates should be underlined. Both the tradition of Montesquieu and Rousseau are absolute in the sense that they are impervious to the advance of knowledge. One draws its immutable moral dicta from historical example or revealed religion; the other, from the inherent and unchanging virtues of the human heart. To both, scientific knowledge can pose a threat, for science by appealing to the understanding can undermine confidence in the moral truths laid down by each system. Since both are ultimately political systems in which the demoralization of the citizen is not only an error but a crime, both must protect themselves by political means against such subversion. Only the scientific tradition, as enunciated by Condorcet, can and must leave itself room for development. The road to individual and political perfection runs parallel to that of scientific and general intellectual progress. The debates in the Legislative Assembly and the Convention, in which opposition to science came from both Left and Right are, in many ways, the first political manifestation in the modern world of the battle raging today between the open society and its enemies.

References

1 In a speech before the Convention, August 13, 1793. *Archives parlementaires de 1789 à 1860, Recueil complet des débats législatifs et politiques des chambres françaises, fondé par MM. J. Mavidal et E. Laurent, Première série (1787 à 1799)* (82 vols.; Paris, 1867–1911). LXXII, 126.

2 See Montesquieu, *De l'esprit des lois,* Book I, Chapter I. "Les lois, dans la signification la plus étendue, sont les rapports nécessaires qui dérivent de la nature des choses: et, dans ce sens, tous les êtres ont leurs lois; la Divinité a ses lois; le monde matériel a ses lois; les intélligences supérieures à l'homme ont leurs lois; les bêtes ont leur lois; l'homme a ses lois." All citations from Montesquieu are taken from the *Bibliothèque de la Pléiade* edition of Montesquieu's *Oeuvres complètes* (2 vols.; Paris, 1951). For the above quote see II, 232.

3 The eighteenth-century stoics had an advantage over their Roman counterparts for the Newtonian revelation of the physical order of the universe provided them with a much stronger foundation upon which to build their moral structure. But the specific findings of eighteenth century-science were irrelevant to the main problem of moral regeneration.

4 Montesquieu, *Considérations sur les causes de la grandeur des Romains et de leur décadence, Oeuvres complètes,* II, 120 ff.

5 Montesquieu, *De l'esprit des lois,* Book IV, Chap. I, *Oeuvres complètes,* II, 262.

6 *Ibid.,* p. 266.

7 This gulf between the royal will and what was true or good was widened enormously by the knowledge that, under Louis XV, first Mme. de Pompadour and then Mme. du Barry were, in reality, the makers of the royal will. To be-

lieve that God spoke to his vicar on earth through the medium of a Parisian streetwalker strained credulity to the breaking point.

8 Denis Diderot, "Plan of a University for the Russian Government," in F. de la Fontainerie (tr. and ed.), *French Liberalism and Education in the Eighteenth Century, The Writings of La Chalotais, Turgot, Diderot, and Condorcet on National Education* (New York and London, 1932), 231.

9 For a more detailed account of this moral science see F. Picavet, *Les Idéologues* (Paris, 1891), and L. Pearce Williams, "Science, Education and the French Revolution," *Isis,* XLIV (1953), 311 ff.

10 No attempt has been made to search the texts of all the published *cahiers.* The first six volumes of the *Archives parlementaires* contain what may be considered here as a representative sample and the conclusions drawn in the text of this paper are based on this sample.

11 See, for example, the following *cahiers* in the *Arch. parl.* Sénéchaussée d'Auch, Third Estate, II, 99; Bailliage de Saint-Lô, Third Estate, III, 61; Ville de Vienne, III, 86; Bailliage de Dôle, III, 162; Sénéchaussée de Forcalquier, Third Estate, III, 334; Bailliage de Nemours, Third Estate, IV, 164; Essones, Third Estate, IV, 532.

12 Talleyrand's *Rapport sur l'instruction publique* may be most conveniently consulted in C. Hippeau, *L'instruction publique en France pendant la Révolution: Rapports et Discours* (Paris, 1881). It also appears in *Arch. parl.,* XXX, 447.

13 *Arch. parl.,* XXXIII, 325. This remark was made by Prieur (de la Marne?) who objected to taking the time to consider Talleyrand's project when much work remained to be done before the Constituent Assembly dissolved and made way for the Legislative Assembly.

14 Condorcet's reading of his plan for educational reform was interrupted by the appearance of the King who, upon the advice of his Girondist ministers, urged and received a declaration of war against Austria.

15 *Arch. parl.,* XLII, 193.

16 Condorcet thus totally rejected the study of the writings of antiquity in the primary and secondary schools since, filled with errors as they were, they might do more harm than good. "Nous cherchons dans l'éducation à faire connaître des vérités, et ces livres sont remplis d'erreurs; nous cherchons à former la raison, et ces livres peuvent l'égarer. Nous sommes si éloignés des anciens, nous les avons tellement devancés dans la route de la vérité, qu'il faut avoir sa raison déjà toute armée pour que ces précieuses dépouilles puissent l'enrichir sans la corrompre." (*Arch. parl.,* XLII, 228.) Even for the higher degrees of instruction, it is clear that Condorcet sees little advantage in following the old classical education. *Ibid.,* p. 229.

17 *Ibid.*

18 *Ibid.,* p. 228.

19 *Ibid.,* p. 229 ff.

20 *Ibid.,* p. 193.

21 C. Hippeau, *L'Instruction publique en France pendant la Révolution; Débats législatifs* (Paris, 1881), 2.

22 *Arch. parl.*, LV, 27.

23 *Ibid.*, p. 28.

24 *Ibid.*

25 *Ibid.*, p. 29. Durand-Maillane was not the only member of the Right who felt this way. Roland, high in the councils of the Girondist party and then Minister of the Interior, agreed that the creation of a new corporation of scientists would be dangerous to the welfare of the state. *Arch. parl.*, LVI, 697.

26 *Arch. parl.*, LV, 30.

27 When Durand-Maillane finished speaking, a member of the Left replied sarcastically: "Oui, c'est une aristocratie de science; il ne faut qu'une seule école, qu'un seul enseignement public; il ne faut pas que l'on ait l'aristocratie d'être savant. (Rires ironiques et applaudissements à gauche.)" *Arch. parl.*, LV, 31.

28 *Ibid.*, p. 134.

29 *Ibid.*, p. 135.

30 See the following projects and discourses:
    Arch. parl., LV, 33 ff; 57 ff; 137 ff; 139 ff; 142 ff; 186 ff; 345 ff; 392 ff.

31 *Arch. parl.*, LXVIII, 507.

32 *Ibid.*

33 *Ibid.*, p. 103 ff; p. 109 ff; p. 149 ff; p. 179 ff; p. 194 ff; p. 209 ff.

34 Robespierre read the project to the Convention on July 13, 1793. Then, on July 29, it was Robespierre who acted as Reporter for the special *Commission d'Instruction Publique* and, with only minor modifications, submitted Lepeletier's plan in the form of a decree.

35 *Arch. parl.*, LXVIII, 662.

36 *Ibid.*

37 "Ce n'est point le peuple naissant qui dégénère; il ne se perd que lorsque les hommes faits sont déjà corrompus." (*De l'esprit des lois, Oeuvres complètes*, II, 267.)

38 *Arch. parl.*, LXVIII, 663.

39 *Ibid.*, p. 666.

40 *Ibid.*, p. 667.

41 *Ibid.*

42 *Arch. parl.*, LXIX, 661.

43 *Ibid.*

44 There is some relevance here, I feel, to the classic debate over "plot vs. circumstances" in the origins of the Terror. If one accepts the differences on educational reform as the sincere sentiments of the various protagonists, as I think one must, then one is tempted to extrapolate these differences into the area of the Terror itself. If this extrapolation is valid, then a "system of the Terror" becomes somewhat inconceivable.

45 The exception, Harold T. Parker, *The Cult of Antiquity and the French Revolutionaries* (Chicago, 1937), reinforces the conclusion here expressed.

HENRY BERTRAM HILL

on the Papers of Charles Coulston Gillispie and L. Pearce Williams

In his paper, Mr. Gillispie presented the thesis that the philosophical import of Newton, misunderstood by many of the men of the age of the encyclopedia, was obliquely attacked by those who wished to find moral purpose bedded in nature. Of these moralists Mr. Gillispie has found Diderot highly representative, and he used him to carry home his point in an impressive array of suggestive interpretations. I am rather at a loss as to which to select for comment, but it seems best to avoid a detailed debate and to concentrate instead upon my general reactions to the impact of the whole. Mr. Guerlac, who will be more precise than I, will be the bridge between.

As one long concerned with the French Revolution, these reactions are largely centered in the implications of the paper for the understanding of that event. By way of further preface I should add that I am no historian of science—quite the contrary. My own avenue of approach is via the history of institutions and of men in the aggregate, and I am interested in science and ideas only insofar as I can see their implications contemporaneously at work in the broad stream of events. It is from this quite different vantage point, therefore (no better or no worse), that I view Mr. Gillispie's paper.

To my way of thinking, the first conception about the French Revolution one should formulate as a basis of reference—what one should use to judge its every aspect—should be centered in what might be called its temper. I use this word because it can have any one of a number of related meanings: spirit, nature, tempo, feeling. That is, I believe it is historiographically risky to say (Mr. Gillispie does not say it) that the violence, both intellectual and physical, of the high revolution was largely the outcome of the natural development of the ideas of the day—that, to take a commonly assumed position, early neo-classical rationalism developed into cultish worship of reason and then by rebound into a veneration of emotion and totalitarian anti-intellectualism. Rather, I feel it to be much safer to approach the Revolution from the viewpoint of the total social situation and see the development of ideas as part of the full context.

Taken in that way, the years just before the Revolution, tested out so to speak by the opening months of that event, show a constantly increasing dessication of the elasticity in French society and the French mind. Year after year the frustrating stagnation in the social order, the lag in the institutional adjustment to changing conditions and insistent demands, led to a rigidification of relationships among classes and among those people who inevitably took intellectual stances of class as well as personal origin. I cannot see, under these circumstances, how the ruthlessness and dogmatism of the Terror can be attributed more than indirectly to the much maligned Rousseau or to Diderot or to any of the other formulators of ideas of the preceding era. To me, as a matter of fact, there is great wisdom in the view long held by Harold Lasswell that private motives are displaced on public objects and rationalized in terms of public interest. When you add to this that the rationalizations are drawn from whatever intellectual pabulum there is at hand of an appetizing and useful sort, you have the framework through which to view the process. Among the first impressions gained from looking at the Revolution this way is certainly of the nature of its inner forces, its drive, its spirit. Not only was life led at an accelerated pace, was history racing so to speak, but also, as is characteristic of such times, the sway of the irrational was greatly enlarged. Men were moved to action, not to thought, and the acceptable intellectual was not the far-reaching reasoner but the adept rationalizer. Thus it was that the cool moralizing of Montesquieu and the warmer moral exhortations of Rousseau and Diderot could become the hot blood-lettings of Marat and Saint-Just. But it is not possible to try to place the intellectual responsibility upon the immediate actors either. In other times Marat would have been ignored or locked up, and there is always an ample supply of Marats in the world. Nor can it be placed upon the inherent evolution of the implication of a set of ideas. The evolution of the implications, indeed, is always a product of the time and place, and may grow in any one of a number of directions men wish it to grow in the context of the given circumstances. For an example of this it is only necessary to look at the long and twisted casuistical history of the Christian church and the rivers of blood let supposedly in fulfilment of the words of Jesus Christ. Rather, popular ideas are directed toward a workable application to immediate issues, and the practical success of this at the hands of the Jacobins during the Reign of Terror is something of a measure of their astuteness as politicians and social manipulators. Unlike the Girondins on the one hand and the Enragées on the other, they were consciously aware of the temper of the times and how best to adjust both ideology and program to it. In the end, Robespierre was no longer able to maintain this adjustment, and he therefore went the way of his predecessors, but while things held together the dictates of adjustment called the turn in making each practical decision, whether it was to destroy something old or erect something new, and each time an appropriate rationale was provided.

Perhaps before now I should have said that I do not consider myself a cynic, but I had reserved that statement for inclusion at this point as a specific preface to what I am about to say. Given the nature of my audience it is especially necessary. Emile Durkheim in his famous book on suicide stated that free inquiry comes from the failure of old beliefs to continue to be satisfying. When men doubt their traditional beliefs they start looking for new ones. If they find them free inquiry ends. If they continue to fail to find them there is an age of tolerance and science. The French Revolution in microcosm reflects Durkheim's concept, just as does in a larger way the whole of modern history, and at times the open society did not prevail in either. Nor have all of the enemies been outsiders. For every conspicuous attack mounted from without, there has been an internal one in the cause of the worship of "reason" and "science"—the kind of unconscious attempt at self-destruction Mr. Gillispie alluded to in the case of the suggestion that a knowledge of geometry be required of all those presenting inventions to the Academy for approval. Whether the inventions themselves might be useful was almost beside the point. Such a sort of arrogance, when coupled with the pre-revolutionary social position of many of the academicians, can go a long way in explaining their fate during the Revolution—that time when black became blacker and white whiter and no man could stand on the thin edge of tolerance or even of deliberation.

There is, indeed, still another way in which rationalism brought trouble on the heads of its rigid proponents during the Revolution. Opposite the Scylla of arrogance was the Charybdis of the idea of progress. This feature, while not always accepted with the naïvete of Condorcet, was a characteristic piece of the apparatus, and in so far as ideas do make history must have played a major part as rationalization for much of the destruction of the Revolution, since the easiest way to build the future was seen as being on the ruins of the past. At one point Mr. Gillispie calls the Revolution "both purge and rebellion" and most rightly so. Just as long delay, old and deep hatreds, crystallized class lines, institutions so hopelessly outdated that renovation was well-nigh impossible, and whole classes so far evolved out of functional usefulness as to be beyond rehabilitation—just as all these social factors were to precipitate violent destruction, so the rationalist repudiation of tradition, adoption of the belief in the rationality of free men, and acceptance of the idea of progress were to provide a large portion of the justification of and rationalization for that destruction. This is not to say that inherent in rationalism were the seeds of violence, but only that in violent times the developing perimeter of the intellectual life of the day becomes a knife in the hands of the men of action. That is why at the beginning I said that the most important point from which to start, the thing to use as the basis of reference was the temper of the Revolution. The way in which ideas are developed and used depends upon what men wish to do with them, and from either an intellectual or a moral point of view the results may be good or bad, depending on circumstance—their prime utility is to spark action. That is the situation which makes me reject much of Talmon's *The Rise of Totalitarian Democracy*—to find it even misleading. I have no objection, as some have, to coupling the two words together—indeed at times they must be. What I do object to is the notion of the inevitability of the coupling, which I consider the result of the manipulation of ideas as abstractions outside the context of history. There are elements of totalitarianism in Rousseau, but they have been much exaggerated, and you can read totalitarianism into almost any text you wish to impose it on. In some ways Rousseau has become the victim of guilt by association, because a violent revolution took place in the country most readily identifiable with him. Critics rarely come down so hard on Hobbes, whose authoritarianism could certainly lead to totalitarianism. Why has he escaped? Mainly, I believe, because no violence followed him in England, where, indeed, he is venerated as one of the three main architects of the modern British constitution—his absolutism stiffening the reservation of rights by Locke and the preservation of tradition by Burke. Had Britain by some quirk of fate ultimately gone fascist I am sure that we would now point the finger of responsibility at Hobbes (let me skip Burke) and rightly so, for the legislative power that has made universal suffrage, social legislation, and the socialization of industries possible could, in a different atmosphere, legis-late in the apparatus of a fascist order.

I have here brought Rousseau in for a sort of whipping that Mr. Gillis-pie does not give him, because the men of the high revolution so laid him open to later attack. Mr. Gillispie has instead concerned himself with Diderot. This in itself is a good and necessary thing, and he has produced a fresh and very persuasive characterization of the place of the man. What most concerns me, however, is how this new interpretation of Diderot's influence can be placed in the historical context of the Revolution before the overtowering Rousseau is put in better focus. But enough of him for the present; I shall return to him when commenting on Mr. Williams' paper.

Mr. Gillispie says that the anti-science of the Revolution was "pat-terned rather than informed." To my way of thinking that is so close to being correct and with a little sharpening so important that it claims extra attention. The high revolution was a time of action—action along a forced and ever narrowing line—action fed and justified by rationalizations drawn from material both distant and close at hand. It was aimed at sweeping away an old and discredited order, and along with it anything tarred with the same brush. In some cases it involved institutions and concepts deeply imbedded in the past; in others it was no more than the vehicle of one man's private vengeance or sadistic impulse; but it swept in any ideas it could use and tried to destroy those it considered detrimen-tal, along with those men who propagated them. Mr. Gillispie stroke by stroke drew a poignant and touching portrait of Lavoisier. In some ways Bailly was an even more tragic symbol of the same thing. Also an acade-mician, Bailly was far from being the scientist that Lavoisier was, but he was not burdened with the social guilt of having been a tax farmer and he had taken a much more conspicuous part in the early Revolution. While he unwittingly bumbled his way to the guillotine, yet the cards were stacked against him, and he died one of the most horrible deaths of the Terror. But when M. Bailly's remains have been gathered up and disposed of, what is there to say? Only that he, like many other scientists had been tried in the fire of the Revolution and found wanting. What they lacked was the ability to divest themselves of enough of the traces of the old order, and that is what they died for. That such ruthlessness might injure the cause of science did not in the least bother the revolutionists—the Revolution came first. They were anti-everything else, and this was very definitely patterned—in the same way (so incomprehensible to our way of thinking) as with fascists and communists. And all of this, it seems to me, takes place in a complex far too intricate, far too crammed with the irrational, far too kaleidoscopic and hectic in its pace to be reduced to a formula. If, therefore, I have any quarrel with Mr. Gillispie (and I doubt that I really have) it would not be with his title: "The Encyclopedia and the Jacobin Philosophy of Science," nor with most of his subtitle: "A Study in Ideas and Consequences," but only with that last word: "Consequences." (This obviously is a small matter.) Mr. Gillispie at one time referred to his paper as a footnote to Mr. Williams' paper. Perhaps my remarks should be considered as a footnote to the footnote.

Mr. Williams' paper also dealt with the fate of science during the French Revolution, but while Mr. Gillispie made what might be called a line assault upon the problem Mr. Williams has made a frontal one. Tracing the fortunes of public education during the Revolution, he found that while the old system withered, no new structure was reared until the late and quiet days of the Directory. This long delay he found attributable to a stalemate among rival and deeply incompatible programs whose differences were rooted in basically hostile philosophical conceptions. These were three in number: enlightened stoicism, of which Montesquieu was perhaps best representative; scientific relativism, for which Mr. Williams first cites Diderot but later employs Condorcet; and what might be called moral reformism, of which Rousseau was the exponent—a sort of neo-Montesquieuism, to coin a real twister.

In an extremely compressed form this brings me to the last three paragraphs of Mr. Williams' excellent paper—and to the place where I wish, in a rather circuitous fashion, to cut in. Up to this point I felt no questions urging their presentation—with the one exception of his references to Rousseau. The same problem bothered me, but less directly, in the case of Mr. Gillispie's paper, but it seemed more appropriate to me to bring up the matter at this point.

I believe that by recognizing the full Rousseau it is possible to resolve some of the difficulties Rousseau himself created for the revolutionaries, and for us. Mr. Williams infers that Robespierre was not a good disciple of Rousseau. In some ways, on the contrary, he was perhaps too good a one. The educational system he adopted and propounded, as well as his other institutional proposals, were of that Spartan, thorough, and serious sort which revealed how much of this present world to him was bad and how much severe and searching discipline would be necessary to put man on the right road to Rousseau's utopia. This position, it is true, is quite contrary to the easy, simple, sentimental, permissive kind of Rousseauism that prevailed in many quarters during the Revolution and which, paradoxically, played so major a role in toppling Robespierre. Rousseau as prologue and mood producer was known and accepted; Rousseau as philosopher and social engineer was not. A contemporary example of this can be found in the case of the *Social Contract* itself. Everyone knew the opening sentence by heart: "Man is born free, yet everywhere he is in chains." Rather fewer knew that this was about the only violent expression in the entire book.

Although I would need more convincing before accepting Montesquieu as father of the Reign of Virtue, Mr. Williams quite briefly, but brilliantly, does partly equate Montesquieu with Rousseau and I think even more could be made of this. For one thing, Montesquieu too was misunderstood by most of the men of the Revolution. Hated in most respects as a reactionary after the first few weeks of the struggle, he was given repetitious lip service veneration only for his quite unimportant and unoriginal contribution of the doctrine of the separation and balance of powers—that notion referred to as a Frenchman's misconception of the English constitution which only Americans have been able to make work. The revolutionaries were, again as with Rousseau, generally quite unfamiliar with the significant part of Montesquieu, with his deep probing into the historical and organic structure of social institutions, particularly the state— the penetrating analysis which gave its title to his great work, *The Spirit of the Laws*. While he was a far better historian than Rousseau, and both more realistic and more limited in his reforming zeal, there nevertheless was much the two held in common. The moral order was the first premise of both, and the good society whether viewed functionally or ethically was based squarely on it. Montesquieu, because his approach was more analytical, was more of a relativist with respect to moral means, but not with respect to moral ends. Montesquieu and Rousseau also entertained similar views with regard to science, although in a slightly more complicated fashion than Mr. Williams was able to suggest in so limited a space. Both accepted science, Montesquieu most certainly and Rousseau too, as Mr. Gillispie implied near the end of his paper. Again, if I read Mr. Gillispie rightly, both would accept Newton as one who found science outside the realm of the moral order, although both would tip the scales of value far the other way. And so for them scientific knowledge posed no real threat, it was only scientists who did, for science as such, they would say, cannot really appeal to the understanding since the understanding is concerned with the interior life and moral values. Both, of course, were unfamiliar with the phraseology of twentieth-century social scientists, but if they had been they doubtless would have said that every going society rests upon an assumed but untestable system of values, and that the open society without such values would become the empty society.

But I am here trying to make Montesquieu and Rousseau answer something that Mr. Williams threw out to us as conjecture, as a stimulating and provocative salute at the end. Earlier he reached several conclusions that rested more directly upon the substance of his paper, and which are important enough to warrant re-emphasis. He rejected the idea that intellectual orientation could be equated with party affiliation, or Jacobinism with anti-intellectualism, and he also concluded that the Revolution was not hostile to science. These are not easy stands to defend—indeed much solid research points the other way—but I believe the balance is in Mr. Williams' favor, and I hope he keeps up the defense.

HENRY GUERLAC

on the Papers of Charles Coulston Gillispie and L. Pearce Williams

Approaching much the same subject from divergent points of view, and after examining different materials, Professor Williams and Professor Gillispie have each raised once again the important question of the role that science played in the thought of the men of the French Revolution.

Professor Williams has given us another of his well-documented studies of the educational proposals and experiments of the Revolutionary period. One of his points, I take it, is that the Revolutionary assemblies failed to reach a definite solution of the educational impasse, not because they were interrupted by pressing matters of greater urgency, but because of a fundamental cleavage on questions of educational principle and direction. Professor Williams distinguishes three main points of view, each of which he identifies with a well-defined current of eighteenth-century thought on matters of education; these he characterizes as the "enlightened Stoicism" of Montesquieu, Fénelon and Turgot; the Rousseauist tradition of natural virtue; and the Enlightenment's widespread faith in physical science and in the psychological doctrines of Locke and Condillac. These are helpful distinctions, but I am a little troubled by two small matters in Professor Williams' discussion of the newer Stoicism. I do not believe that the "enlightened" Stoics of the eighteenth century had—despite the dramatic example of Newton's physics—any deeper conviction of a regular physical order of nature than was held, on less cogent grounds, by their models in antiquity. Surely the regularity of astronomical phenomena and the belief in astrology combined in later antiquity to produce the overwhelming conviction that such an order must exist. In the second place, I wonder what evidence Professor Williams would adduce—he gives us none—for placing Turgot in this camp of "enlightened Stoics" rather than with the pro-scientific group, of which he makes Condorcet (a disciple of Turgot, it should not be forgotten) the spokesman.

But these are minor questions. Mr. Williams' chief message is that during the Revolutionary debates on educational reform these three points of view clearly emerge, and were soon shown to be fundamentally irreconcilable. Yet he believes it impossible to identify these views with particular party affiliations. More especially, he finds the Jacobins "hopelessly divided" over the question of educational reform, and not at all agreed as to the place that should be assigned to scientific training in a program of national education. Some Jacobin spokesmen were pure Rousseauists, objecting to the emphasis that others would put on the traditional science; others, like Gilbert Romme, were partisans of a scheme of education that gave a prominent place to scientific instruction; and finally the plan of Michel Le Pelletier offered a compromise that had elements of all three traditions. Though Professor Williams has considered only one kind of evidence—a most important kind, to be sure—and may, by this fact, have exaggerated the lack of correlation between educational ideology and party divisions, nevertheless he has warned us to be cautious about referring, as I once did and as Professor Gillispie has done, to a typically Jacobin attitude towards science.

Professor Gillispie's eloquent paper skips over a great deal of ground, and raises a number of fundamental questions about attitudes towards science during the Revolution. He believes he has discovered in an anti-Newtonian—indeed, from our point of view, an anti-scientific—current, a factor explaining the disaster that befell certain scientists and scientific institutions at the hands of the Jacobins. He finds the origin of this current in the *Encyclopédie* and the well-known rebellion of Diderot against an abstract, externalized and mathematized view of nature. But his stress is new, and his emphasis on the role that a resurgent interest in Stahlian chemistry played in this shift of scientific interest is a point of very considerable interest. Professor Gillispie, by the way, could perhaps equally well have cited the opinions of the Baron d'Holbach, who was also a pupil of the elder Rouelle, and whose views about matter, chemistry and Newtonian physics closely resemble Diderot's.

I do not doubt for a minute the existence of such an anti-Newtonian current as Professor Gillispie has described with such eloquence. I believe only that it had other, and perhaps more influential, manifestations in quarters about which he is silent. Perhaps we should examine more closely the role of Buffon in promoting this emancipation from a strictly mechanical world-view and in contributing to the organistic outlook. Above all, I think, we should emphasize the role of Rousseau.

Rousseau was the first to speak out bluntly and shockingly, so it seemed to his contemporaries, against the idol of Academic science. If we do not find Diderot translating his views directly into an attack on the Academies, this is not true of Rousseau. And unlike Rousseau—whose views were characteristically vague and contradictory—Diderot does not seem to have inspired a chain of disciples who carried his message, in the name of the master, down into the Revolutionary decade. But Rousseau did, and we can identify a group of his followers—Marat, Brissot de Warville, and Bernardin de Saint-Pierre—who clarified his thought, and kept up a sustained drum-fire against the Academic science of which Newton was the symbol, against the Academy of Sciences itself, and against Lavoisier as its dominating figure. There is ample evidence for this statement.

Some years ago I was fortunate enough to pick up from a Paris dealer a lost pamphlet of Brissot de Warville. This pamphlet, which biographers of Brissot had looked for in vain, was published anonymously about 1785 with the revealing title: *Un Mot à l'oreille des Académiciens de Paris.* Brissot, as a young man, was a fervent admirer of Rousseau; he was also a friend of the charlatanesque Marat, and in his *De la vérité* a few years earlier he had attacked the Academies and Academic science in the name of Rousseau, and accused the physical scientists of the Academy of Sciences of being unjust to Marat. In the anonymous pamphlet he goes further. He links together, with some others, those followers of Rousseau—the poet Roucher; the would-be opponent of Newton and Academic physics, Marat; and the author of the *Etudes de la Nature,* Bernardin de Saint-Pierre— whom he cites as victims of Academic intolerance. The pamphlet not only embodies a violent attack on the Academy of Sciences, but includes one of the earliest printed assaults on the person and integrity of Lavoisier. But Lavoisier is not singled out because he has perverted chemistry and failed to be the Paracelsus of whom poor Venel dreamed. He is attacked on the ground that he was supposed to be unscrupulous about acknowledging his scientific debt to others, and obviously because he was a power in the Academy of Sciences and a member of the hated *Ferme générale.* This did not end matters; we should not forget that both Brissot and Marat, as widely-read journalists of the Revolution, continued their attacks upon Lavoisier and the Academy of Sciences until almost the bitter end.

Bernardin de Saint-Pierre deserves a special word. Novelist, naturalist and biographer of Rousseau, who had botanized with the sage in his later years, Bernardin shared with Brissot the dream of a primitivistic sort of science, where man worked best in solitude and in intimate rapport with nature. His anti-Newtonian science drew its inspiration not from chemistry, but from natural history pursued in a new way. This is characteristic. Indeed the *Etudes de la Nature,* in which critics have found the first glimmerings of the ecological point of view, exemplified in its author's mind the fulfillment of Rousseau's dream of a new kind of study of nature.

If, as I believe, Professor Gillispie and I have each hit upon a separate, and complementary, phase of this anti-Newtonian movement, I concede at once that Diderot's thoughts are much more subtle and profound, and a deal more comprehensible, than Rousseau's. But I believe that the actual historical influence of Rousseau and his disciples, and of their particular form of an anti-Newtonian scientific view, emphasizing natural history rather than chemistry, was by far the greater. If there was a Jacobin philosophy of science, I feel it owed more to the scientific primitivism of Rousseau and his disciples than to Venel or Diderot. There is much to support this, and not least, perhaps, is the fact of Bernardin's selection as head of the *Museum* during the Revolution.

As far as the role of this anti-Newtonian science in the Revolution is concerned, I cannot agree completely with either Professor Williams or Professor Gillispie. Williams makes it clear that we must be cautious about identifying this attitude towards science with the Jacobin party. Yet I believe, with Professor Gillispie, that, as a group, men of the Jacobin persuasion were strongly attracted to some of the ideas of a Rousseauistic and more democratic science. But surely this ideological component does not completely explain the attack on the Academy or the execution of Lavoisier; political consideration and personal motives were powerful, and probably controlling. Least of all can I accept Professor Gillispie's view that Lavoisier was imprisoned and executed because he betrayed some dream of the nature of chemistry held by Diderot and Venel, and because his brand of New Chemistry was too physical and Newtonian. Diderot's (and Venel's) dream of a different chemistry had long-since been engulfed by events. Rouelle's more competent pupils, Macquer, Cadet, Darcet, Lavoisier and the rest, had drawn no such conclusions from his lectures and had gone on to found a chemistry of a valid sort. During the Revolution many of the most ardent Jacobins among the scientists—I need mention only Fourcroy, Adet, Hassenfratz and Monge among others—were believers in, and even propagandists for, the New Chemistry. This hardly supports the Gillispie theory. I can only conclude that this whole fascinating subject needs further, and extremely careful, investigation.

ENERGY CONSERVATION AS AN EXAMPLE
OF SIMULTANEOUS DISCOVERY

Thomas S. Kuhn

Between 1842 and 1847, the hypothesis of energy conservation was publicly announced by four widely scattered European scientists—Mayer, Joule, Colding, and Helmholtz—all but the last working in complete ignorance of the others.[1] The coincidence is conspicuous, yet these four announcements are unique only in combining generality of formulation with concrete quantitative applications. Sadi Carnot, before 1832, Marc Séguin in 1839, Karl Holtzmann in 1845, and G. A. Hirn in 1854, all recorded their independent convictions that heat and work are quantitatively interchangeable, and all computed a value for the conversion coefficient or an equivalent.[2] The convertibility of heat and work is, of course, only a special case of energy conservation, but the generality lacking in this second group of announcements occurs elsewhere in the literature of the period. Between 1837 and 1844, C. F. Mohr, William Grove, Faraday, and Liebig, all described the world of phenomena as manifesting but a single "force," one which could appear in electrical thermal, dynamical, and many other forms, but which could never, in all its transformations, be created or destroyed.[3] That so-called force is the one known to later scientists as energy. History of science offers no more striking instance of the phenomenon known as simultaneous discovery.

Already we have named twelve men who, within a short period of time, grasped for themselves essential parts of the concept of energy and its conservation. Their number could be increased, but not fruitfully.[4] The present multiplicity sufficiently suggests that in the two decades before 1850 the climate of European scientific thought included elements able to guide receptive scientists to a significant new view of nature. Isolating these elements within the works of the men affected by them may tell us something of the nature of simultaneous discovery. Conceivably, it may even give substance to those obvious yet totally unexpressive truisms: a scientific discovery must fit the times, or the time must be ripe. The problem is challenging. A preliminary identification of the sources of the phenomenon called simultaneous discovery is therefore the main objective of this paper.

Before proceeding towards that objective, however, we must briefly pause over the phrase simultaneous discovery itself. Does it sufficiently describe the phenomenon we are investigating? In the ideal case of simultaneous discovery two or more men would announce the same thing at the same time and in complete ignorance of each other's work, but nothing remotely like that happened during the development of energy conservation. The violations of simultaneity and mutual influence are secondary. But no two of our men even said the same thing. Until close to the end of the period of discovery, few of their papers have more than fragmentary resemblances retrievable in isolated sentences and paragraphs. Skillful excerpting is, for example, required to make Mohr's defense of the dynamical theory of heat resemble Liebig's discussion of the intrinsic limits of the electric motor. A diagram of the overlapping passages in the papers by the pioneers of energy conservation would resemble an unfinished crossword puzzle.

Fortunately no diagram is needed to grasp the most essential differences. Some pioneers, like Séguin and Carnot, discussed only a special case of energy conservation, and these two used very different approaches. Others, like Mohr and Grove, announced a universal conservation principle, but, as we shall see, their occasional attempts to quantify their imperishable "force" leave its concrete significance in doubt. Only in view of what happened later can we say that all these partial statements even deal with the same aspect of nature.[5] Nor is this problem of divergent discoveries restricted to those scientists whose formulations were obviously incomplete. Mayer, Colding, Joule, and Helmholtz were not saying the same things at the dates usually given for their discoveries of energy conservation. More than *amour propre* underlies Joule's subsequent claim that the discovery he had announced in 1843 was different from the one published by Mayer in 1842.[6] In these years their papers have important areas of overlap, but not until Mayer's book of 1845 and Joule's publications of 1844 and 1847 do their theories become substantially coextensive.[7]

In short, though the phrase "simultaneous discovery" points to the central problem of this paper, it does not, if taken at all literally, describe it. Even to the historian acquainted with the concepts of energy conservation, the pioneers do not all communicate the same thing. To each other, at the time, they often communicated nothing at all. What we see in their works is not really the simultaneous discovery of energy conservation. Rather it is the rapid and often disorderly emergence of the experimental and conceptual elements from which that theory was shortly to be compounded. It is these elements that concern us. We know why they were there: Energy *is* conserved; nature behaves that way. But we do not know why these elements suddenly became accessible and recognizable. That is the fundamental problem of this paper. Why, in the years 1830 to 1850, did so many of the experiments and concepts required for a full statement of energy conservation lie so close to the surface of scientific consciousness? [8]

This question could easily be taken as a request for a list of all those almost innumerable factors that caused the individual pioneers to make the particular discoveries that they did. Interpreted in this way, it has no answer, at least none that the historian can give. But the historian can attempt another sort of response. A contemplative immersion in the works of the pioneers and their contemporaries may reveal a subgroup of factors which seem more significant than the others, because of their frequent recurrence, their specificity to the period, and their decisive effect upon individual research.[9] The depth of my acquaintance with the literature permits, as yet, no definitive judgments. Nevertheless, I am already quite sure about two such factors, and I suspect the relevance of a third. Let me call them the "availability of conversion processes," the "concern with engines," and the "philosophy of nature." I shall consider them in order.

The availability of conversion processes resulted principally from the stream of discoveries that flowed from Volta's invention of the battery in 1800. According to the theory of galvanism most prevalent, at least, in France and England, the electric current was itself gained at the expense of forces of chemical affinity, and this conversion proved to be only the first step in a chain.[10] Electric current invariably produced heat and, under appropriate conditions, light as well. Or, by electrolysis, the current could vanquish forces of chemical affinity, bringing the chain of transformations full circle. These were the first fruits of Volta's work; other more striking conversion discoveries followed during the decade and a half after 1820.[11] In that year Oersted demonstrated the magnetic effects of a current; magnetism, in turn, could produce motion, and motion had long been known to produce electricity through friction. Another chain of conversions was closed. Then, in 1822, Seebeck showed that heat applied to a bimetallic junction would produce a current directly. Twelve years later Peltier reversed this striking example of conversion, demonstrating that the current could, on occasions, absorb heat, producing cold. Induced currents, discovered by Faraday in 1831, were only another, if particularly striking, member of a class of phenomena already characteristic of nineteenth-century science. In the decade after 1827, the progress of photography added yet another example, and Melloni's identification of light with radiant heat confirmed a long-standing suspicion about the fundamental connection between two other apparently disparate aspects of nature.[12]

Some conversion processes had, of course, been available before 1800. Motion had already produced electrostatic charges, and the resulting attractions and repulsions had produced motion. Static generators had occasionally engendered chemical reactions, including dissociations, and chemical reactions produced both light and heat.[13] Harnessed by the steam engine, heat could produce motion, and motion, in turn, engendered heat through friction and percussion. Yet in the eighteenth century these were isolated phenomena; few seemed of central importance to scientific research; and those few were studied by different groups. Only in the decade after 1830, when they were increasingly classified with the many other examples discovered in rapid succession by nineteenth-century scientists, did they begin to look like conversion processes at all.[14] By that time scientists were proceeding inevitably in the laboratory from a variety of chemical, thermal, electrical, magnetic, or dynamical phenomena to phenomena of any of the other types and to optical phenomena as well. Previously separate problems were gaining multiple interrelationships, and that is what Mary Sommerville had in mind when, in 1834, she gave her famous popularization of science the title, *On the Connexion of the Physical Sciences*. "The progress of modern science," she said in her preface, "especially within the last five years, has been remarkable for a tendency to . . . unite detached branches [of science, so that today] . . . there exists such a bond of union, that proficiency cannot be attained in any one branch without a knowledge of others."[15] Mrs. Sommerville's remark isolates the "new look" that physical science had acquired between 1800 and 1835. That new look, together with the discoveries that produced it, proved to be a major requisite for the emergence of energy conservation.

Yet, precisely because it produced a "look" rather than a single clearly defined laboratory phenomenon, the availability of conversion processes enters the development of energy conservation in an immense variety of ways. Faraday and Grove achieved an idea very close to conservation from a survey of the whole network of conversion processes taken together. For them conservation was quite literally a rationalization of the phenomenon Mrs. Sommerville described as the new "connexion." C. F. Mohr, on the other hand, took the idea of *conservation* from a quite different source, probably metaphysical.[16] But, as we shall see, it is only because he attempted to elucidate and defend this idea in terms of the new conversion processes that Mohr's initial conception came to look like conservation *of energy*. Mayer and Helmholtz present still another approach. They began by applying their concepts of conservation to well-known older phenomena. But until they extended their theories to embrace the new discoveries, they were not developing the same theory as men like Mohr and Grove. Still another group, consisting of Carnot, Séguin, Holtzmann, and Hirn, ignored the new conversion processes entirely. But they would not be discoverers of energy conservation if men like Joule, Helmholtz, and Colding had not shown that the thermal phenomena with which these steam engineers dealt were integral parts of the new network of conversions.

There is, I think, excellent reason for the complexity and variety of these relationships. In an important sense, though one which will demand later qualification, the conservation of energy is nothing less than the theoretical counterpart of the laboratory conversion processes discovered during the first four decades of the nineteenth century. Each laboratory conversion corresponds in the theory to a transformation in the form of energy. That is why, as we shall see, Grove and Faraday could derive conservation from the network of laboratory conversions itself. But the very homomorphism between the theory, energy conservation, and the earlier network of laboratory conversion processes indicates that one did not have to start by grasping the network whole. Liebig and Joule, for example, started from a single conversion process and were led by the "connexion" between the sciences through the entire network. Mohr and Colding started with a metaphysical idea and transformed it by application to the network. In short, just because the new nineteenth-century discoveries formed a network of "connexions" between previously distinct parts of science, they could be grasped either individually or whole in a large variety of ways and still lead to the same ultimate result. That, I think, explains why they could enter the pioneers' research in so many different ways. More important, it explains why the researches of the pioneers, despite the variety of their starting points, ultimately converged to a common outcome. What Mrs. Sommerville had called the new "connexions" between the sciences often proved to be the links that joined disparate approaches and enunciations into a single discovery.

The sequence of Joule's researches clearly illustrates the way in which the network of conversion processes actually marked out the experimental ground of energy conservation and thus provided the essential links between the various pioneers. When Joule first wrote in 1838, his exclusive concern with the design of improved electric motors effectively isolates him from all the other pioneers of energy conservation except Liebig. He was simply working on one of the many new problems born from nineteenth-century discovery. By 1840 his systematic evaluations of motors in terms of work and "duty" establishes a link to the researches of the steam engineers, Carnot, Séguin, Hirn, and Holtzmann.[17] But these "connexions" vanished in 1841 and 1842 when Joule's discouragement with motor design forced him to seek instead a fundamental improvement in the batteries that drove them. Now he was concerned with new discoveries in chemistry, and he absorbed entirely Faraday's view of the essential role of chemical processes in galvanism. In addition, his research in these years was concentrated upon what turned out to have been two of the numerous conversion processes selected by Grove and Mohr to illustrate their vague metaphysical hypothesis.[18] The "connexions" with the work of other pioneers are steadily increasing in number.

In 1843, prompted by the discovery of an error in his earlier work with batteries, Joule reintroduced the motor and the concept of mechanical work. Now the link to steam engineering is re-established, and simultaneously Joule's papers begin, for the first time, to read like investigations of energy relations.[19] But even in 1843 the resemblance to energy conservation is incomplete. Only as Joule traced still other new "connexions" during the years 1844 to 1847 does his theory really encompass the views of such disparate figures as Faraday, Mayer, and Helmholtz.[20] Starting from an isolated problem, Joule had involuntarily traced much of the connective tissue between the new nineteenth-century discoveries. As he did so, his work was linked increasingly to that of the other pioneers, and only when many such links had appeared did his discovery resemble energy conservation.

Joule's work shows that energy conservation could be discovered by starting from a single conversion process and tracing the network. But, as we have already indicated, that is not the only way in which conversion processes could effect the discovery of energy conservation. C. F. Mohr, for example, probably drew his initial concept of conservation from a source independent of the new conversion processes, but then used the new discoveries to clarify and elaborate his ideas. In 1839, close to the end of a long and often incoherent defense of the dynamical theory of heat, Mohr suddenly burst out: "Besides the known 54 chemical elements, there is, in the nature of things, only one other agent, and that is called force; it can appear under various circumstances as motion, chemical affinity, cohesion, electricity, light, heat, and magnetism, and from any one of these types of phenomena all the others can be called forth."[21] A knowledge of energy conservation makes the import of these sentences clear. But in the absence of such knowledge, they would have been almost meaningless except that Mohr proceeded immediately to two systematic pages of experimental examples. The experiments were, of course, just the new and old conversion processes listed above, the new ones in the lead, and they are essential to Mohr's argument. They alone specify his subject and show its close similarity to Joule's.

Mohr and Joule illustrate two of the ways in which conversion processes could affect the discoverers of energy conservation. But, as my final example from the works of Faraday and Grove will indicate, these are not the only ways. Though Faraday and Grove reached conclusions much like Mohr's, their route to the conclusions includes none of the same sudden leaps. Unlike Mohr, they seem to have derived energy conservation directly from the experimental conversion processes that they had already studied so fully in their own researches. Because their route is continuous, the homomorphism of energy conservation with the new conversion processes appears most clearly of all in their work.

In 1834, Faraday concluded five lectures on the new discoveries in chemistry and galvanism with a sixth on the "Relations of Chemical Affinity, Electricity, Heat, Magnetism, and other powers of Matter." His notes supply the gist of this last lecture in the words: "We cannot say that any one [of these powers] is the cause of the others, but only that all are connected and due to one common cause." To illustrate "the connection," Faraday then gave nine experimental demonstrations of "the production of any one [power] from another, or the conversion of one into another."[22] Grove's development seems parallel. In 1842 he included a remark almost identical with Faraday's in a lecture with the significant title, "On the Progress of Physical Science."[23] In the following year he expanded this isolated remark into his famous lecture series, *On the Correlation of Physical Forces*. "The position which I seek to establish in this Essay is," he said, "that [any one] of the various imponderable agencies . . . viz., Heat, Light, Electricity, Magnetism, Chemical Affinity, and Motion, . . . may, as a force, produce or be convertible into the other[s]; thus heat may mediately or immediately produce electricity, electricity may produce heat; and so of the rest."[24]

This is the concept of the universal convertibility of natural powers, and it is not, let us be clear, the same as the notion of conservation. But most of the remaining steps proved to be small and rather obvious.[25] All but one, to be discussed below, can be taken by applying to the concept of universal convertibility the perennially serviceable philosophic tags about the equality of cause and effect or the impossibility of perpetual motion. Since any power can produce any other *and be produced by it,* the equality of cause and effect demands a uniform quantitative equivalence between each pair of powers. If there is no such equivalence, then a properly chosen series of conversions will result in the creation of power, that is, in perpetual motion.[26] In all its manifestations and conversions, power must be conserved. This realization came neither all at once, nor fully to all, nor with complete logical rigor. But it did come.

Though he had no general conception of conversion processes, Peter Mark Roget, in 1829, opposed Volta's contact theory of galvanism because it implied a creation of power from nothing.[27] Faraday independently reproduced the argument in 1840 and immediately applied it to conversions in general. "We have," he said, "many processes by which the form of the power may be so changed that an apparent *conversion* of one into another takes place. . . . But in no cases . . . is there a pure creation of force; a production of power without a corresponding exhaustion of something to supply it."[28] In 1842 Grove devised the argument once more in order to prove the impossibility of inducing an electric current from static magnetism, and in the following year he generalized still further.[29] If it were true, he wrote, "that motion [could] be subdivided or changed in character, so as to become heat, electricity, etc.; it ought to follow, that when we collect the dissipated and changed forces, and reconvert them, the initial motion, affecting the same amount of matter with the same velocity, should be reproduced, and so of the change of matter produced by the other forces."[30] In the context of Grove's exhaustive discussion of the known conversion processes, this quotation is a full statement of all but the quantitative components of energy conservation. Furthermore, Grove knew what was missing. "The great problem that remains to be solved, in regard to the correlation of physical forces, is," he wrote, "the establishment of their equivalent of power, or their measurable relation to a given standard."[31] Conversion phenomena could carry scientists no further towards the enunciation of energy conservation.

Grove's case brings this discussion of conversion processes almost full circle. In his lectures energy conservation appears as the straightforward theoretical counterpart of nineteenth-century laboratory discoveries, and that was the suggestion from which I began. Only two of the pioneers, it is true, actually derived their versions of energy conservation from these new discoveries alone. But because such a derivation was possible, every one of the pioneers was decisively affected by the availability of conversion processes. Six of them dealt with the new discoveries from the start of their research. Without these discoveries, Joule, Mohr, Faraday, Grove, Liebig, and Colding would not be on our list at all.[32] The other six pioneers show the importance of conversion processes in a subtler but no less important way. Mayer and Helmholtz were late in turning to the new discoveries, but only when they did so, did they become candidates for the same list as the first six. Carnot, Séguin, Hirn, and Holtzmann are the most interesting of all. None of them even mentioned the new conversion processes. But their contributions, being uniformly obscure, would have vanished from history entirely if they had not been gathered into the larger network explored by the men we have already examined.[33] When conversion processes did not govern an individual's work, they often governed that work's reception. If they had not been available, the problem of simultaneous discovery might not exist at all. Certainly it would look very different.

Nevertheless, the view which Grove and Faraday derived from conversion processes is not identical with what scientists now call the conservation of energy, and we must not underestimate the importance of the missing element. Grove's *Physical Forces* contains the layman's view of energy conservation. In an expanded and revised form it proved to be one of the most effective and sought after popularizations of the new scientific law.[34] But this role was achieved only after the work of Joule, Mayer, Helmholtz, and their successors had provided a full quantitative substructure for the conception of force correlation. Anyone who has worked through a mathematical and numerical treatment of energy conservation may well wonder whether, in the absence of such substructure, Grove would have had anything to popularize. The "measurable relation to a given standard" of the various physical forces is an essential ingredient of energy conservation as we know it, and neither Grove, Faraday, Roget, nor Mohr was able even to approach it.

The quantification of energy conservation proved, in fact, insuperably difficult for those pioneers whose principal intellectual equipment consisted of concepts related to the new conversion processes. Grove thought he had found the clue to quantification in Dulong and Petit's law relating chemical affinity and heat.[35] Mohr believed he had produced the quantitative relationship when he equated the heat employed to raise the temperature of water 1° with the static force necessary to compress the same water to its original volume.[36] Mayer initially measured force by the momentum which it could produce.[37] These random leads were all totally unproductive, and of this group only Mayer succeeded in transcending them. To do so he had to use concepts belonging to a very different aspect of nineteenth-century science, an aspect to which I previously referred as the concern with engines, and whose existence I shall now take for granted as a well-known by-product of the Industrial Revolution. As we examine this aspect of science, we shall find the main source of the concepts—particularly of mechanical effect or work—required for the quantitative formulation of energy conservation. In addition, we shall find a multitude of experiments and of qualitative conceptions so closely related to energy conservation that they collectively provide something very like a second and independent route to it.

Let me begin by considering the concept of work. Its discussion will provide relevant background as well as opportunity for a few essential remarks on a more usual view about the sources of the quantitative concepts underlying energy conservation. Most histories or pre-histories of the conservation of energy imply that the model for quantifying conversion processes was the dynamical theorem known almost from the beginning of the eighteenth century as the conservation of *vis viva*.[38] That theorem has a distinguished role in the history of dynamics, and it also turns out to have been a special case of energy conservation. It could have provided a model. Yet I think the prevalent impression that it did so is misleading. The conservation of *vis viva* was important to Helmholtz's derivation of energy conservation, and a special case (free fall) of the same dynamical theorem was ultimately of great assistance to Mayer. But these men also drew significant elements from a second generally separate tradition—that of water, wind, and steam engineering—and that tradition is all important to the work of the other five pioneers who produced a quantitative version of energy conservation.

There is excellent reason why this should be so. *Vis viva* is mv^2, the product of mass by the square of velocity. But until a late date that quantity appears in the works of none of the pioneers except Carnot, Mayer, and Helmholtz. As a group the pioneers were scarcely interested in energy of motion, much less in using it as a basic quantitative measure. What they did use, at least those who were successful, was *f·s,* the product of force times distance, a quantity known variously under the names mechanical effect, mechanical power, and work. That quantity does not, however, occur as an independent conceptual entity in the dynamical literature. More precisely it scarcely occurs there until 1820 when the French (and only the French) literature was suddenly enriched by a series of theoretical works on such subjects as the theory of machines and of industrial mechanics. These new books did make work a significant independent conceptual entity, and they did relate it explicitly to *vis viva*. But the concept was not invented for these books. On the contrary it was borrowed from a century of engineering practice where its use had usually been quite independent of both *vis viva* and its conservation. That source within the engineering tradition is all that the pioneers of energy conservation required and as much as most of them used.

Another paper will be needed to document this conclusion, but let me illustrate the considerations from which it derives. Until 1743 the general dynamical significance of the conservation of *vis viva* must be recaptured from its application to two special sorts of problems: elastic impact and constrained fall.[39] Force times distance has no relevance to the former, since elastic impact numerically conserves *vis viva*. For other applications, e.g., the bachistochrone and isochronous pendulum, vertical displacement rather than force times distance appears in the conservation theorem. Huyghen's statement that the center of gravity of a system of masses can ascend no higher than its initial position of rest is typical.[40] Compare

Daniel Bernoulli's famous formulation of 1738: Conservation of *vis viva* is "the equality of actual descent with potential ascent."[41]

The more general formulations, inaugurated by d'Alembert's *Traité* in 1743, suppress even vertical displacement, which might conceivably be called an embryonic conception of work. D'Alembert states that the forces acting on a system of interconnected bodies will increase its *vis viva* by the amount $\Sigma m_i u_i^2$, where the u_i are the velocities that the masses m_i would have acquired if moved freely over the same paths by the same forces.[42] Here, as in Daniel Bernoulli's subsequent version of the general theorem, force times distance enters only in certain particular applications to permit the computation of individual u_i's; it has neither general significance nor a name; *vis viva* is the conceptual parameter.[43] The same parameter dominates the later analytic formulations. Euler's *Mechanica,* Lagrange's *Mécanique analytique,* and Laplace's *Mécanique céleste* give exclusive emphasis to central forces derivable from potential functions.[44] In these works the integral of force times differential path element occurs only in the derivation of the conservation law. The law itself equates *vis viva* with a function of position coördinates.

Not until 1782, in Lazare Carnot's *Essai sur les machines en général,* did force times distance begin to receive a name and a conceptual priority in dynamical theory.[45] Nor was this new dynamical view of the concept work really worked out or propagated until the years 1819 to 1839 when it received full expression in the works of Navier, Coriolis, Poncelet, and others.[46] All these works are concerned with the analysis of machines in motion. As a result, work—the integral of force with respect to distance— is their fundamental conceptual parameter. Among other significant and typical results of this reformulation were: the introduction of the term work and of units for its measure; the redefinition of *vis viva* as $\frac{1}{2}mv^2$ to preserve the conceptual priority of the measure work; and the explicit formulation of the conservation law in terms of the equality of work done and kinetic energy created.[47] Only when thus reformulated did the conservation of *vis viva* provide a convenient conceptual model for the quantification of conversion processes, and then almost none of the pioneers used it. Instead, they returned to the same older engineering tradition in which Lazare Carnot and his French successors had found the concepts needed for their new versions of the dynamical conservation theorem.

Sadi Carnot is the single complete exception. His manuscript notes proceed from the assertion that heat is motion to the conviction that it is molecular *vis viva* and that its increment must therefore be equal to work done. These steps imply an immediate command of the relation between work and *vis viva*. Mayer and Helmholtz might also have been exceptions, for both could have made good use of the French reformulation. But neither seems to have known it. Both began by taking work (or rather the product of weight times height) as the measure of "force," and each then rederived something very like the French reformulation for himself.[48] The other six pioneers who reached or came close to the quantification of conversion processes could not even have used the reformulation. Unlike Mayer and Helmholtz, they applied the concept work directly to a problem in which *vis viva* is constant from cycle to cycle and therefore does not enter. Joule and Liebig are typical. Both began by comparing the "duty" of the electric motor with that of the steam engine. How much weight, they both asked, can each of these engines raise through a fixed distance for a given expenditure of coal or zinc? That question is basic to their entire research programs as it is to the programs of Carnot, Seguin, Holtzmann, and Hirn. It is not, however, a question drawn from either the new or old dynamics.

But neither, except for its application to the electrical case, is it a novel question. The evaluation of engines in terms of the weight each could raise to a given height is implicit in Savery's engine descriptions of 1702 and explicit in Parent's discussion of water wheels in 1704.[49] Under a variety of names, particularly mechanical effect, weight times height provided the basic measure of engine achievement throughout the engineering works of Desagulier, Smeaton, and Watt.[50] Borda applied the same measure to hydraulic machines and Coulomb to wind and animal power.[51] These examples, drawn from all parts of the eighteenth century, but increasing in density towards its close, could be multiplied almost indefinitely. Yet even these few should prepare the way for a little noted but virtually decisive statistic. Of the nine pioneers who succeeded, partially or completely, in quantifying conversion processes, all but Mayer and Helmholtz were either trained as engineers or were working directly on engines when they made their contributions to energy conservation. Of the six who computed independent values of the conversion coefficient, all but Mayer were concerned with engines either in fact or by training.[52] To make the computation they needed the concept work, and the source of that concept was principally the engineering tradition.[53]

The concept work is the most decisive contribution to energy conservation made by the nineteenth-century concern with engines. That is why I have devoted so much space to it. But the concern with engines contributed to the emergence of energy conservation in a number of other ways besides, and we must consider at least a few of them. For example, long before the discovery of electro-chemical conversion processes, men interested in steam and water engines had occasionally seen them as devices for transforming the force latent in fuel or falling water to the mechanical force that raises weight. "I am persuaded," said Daniel Bernoulli in 1738, "that if all the *vis viva* hidden in a cubic foot of coal were called forth and usefully applied to the motion of a machine, more could be achieved than by the daily labor of eight or ten men."[54] Apparently that remark, made at the height of the controversy over metaphysical *vis viva*, had no later influence. Yet the same perception of engines recurs again and again, most explicitly in the French engineering writers. Lazare Carnot, for example, says that "the problem of turning a mill stone, whether by the impact of water, or by wind, or by animal power . . . is that of consuming the maximum possible [portion] of the work delivered by these agents."[55] With Coriolis, water, wind, steam, and animals are all simply sources of work, and machines become devices for putting this in useful form and transmitting it to the load.[56] Here, engines by themselves lead to a conception of conversion processes very close to that produced by the new discoveries of the nineteenth century. That aspect of the engine problem may well explain why the steam engineers—Hirn, Holtzmann, Séguin, and Sadi Carnot—were led to the same aspect of nature as men like Grove and Faraday.

The fact that engines could and occasionally did look like conversion devices may also explain something more. Is this not the reason why engineering concepts proved so readily transferable to the more abstract problems of energy conservation? The concept work is only the most important example of such a transfer. Joule and Liebig reached energy conservation by asking an old engineering question, "What is the 'duty'?" about the new conversion processes in the battery driven electric motor. But that question—how much work for how much fuel?—embraces the notion of a conversion process. In retrospect, it even sounds like the request for a conversion coefficient. Joule, at least, finally answered the question by producing one. Or consider the following more surprising transfer of engineering concepts. Though its fundamental conceptions are incompatible with energy conservation, Sadi Carnot's *Réflexion sur la puissance motrice du feu* was cited by both Helmholtz and Colding as the outstanding application of the impossibility of perpetual motion to a non-mechanical conversion process.[57] Helmholtz may well have borrowed from Carnot's memoir the analytic concept of a cyclic process that played so large a role in his own classic paper.[58] Holtzmann derived his value of the conversion coefficient by a minor modification of Carnot's analytic procedures, and Carnot's own discussion of energy conservation repeatedly employs data and concepts from his earlier and fundamentally incompatible memoir. These examples may give at least a hint of the ease and frequency with which engineering concepts were applied in deriving the abstract scientific conservation law.

My final example of the productiveness of the nineteenth-century concern with engines is less directly tied to engines. Yet it underscores the multiplicity and variety of the relationships that make the engineering factor bulk so large in this account of simultaneous discovery. I have shown elsewhere that many of the pioneers shared an important interest in the phenomenon known as adiabatic compression.[59] Qualitatively, the phenomenon provided an ideal demonstration of the conversion of work to heat; quantitatively, adiabatic compression yielded the only means of computing a conversion coefficient with existing data. The discovery of adiabatic compression has, of course, little or nothing to do with the interest in engines, but the nineteenth-century experiments which the pioneers used so heavily often seem related to just this practical concern. Dalton, and Clément and Désormes, who did important early work on adiabatic compression, also contributed early fundamental measurements on steam, and these measurements were used by many of the engineers.[60] Poisson, who developed an early theory of adiabatic compression, applied it, in the same article, to the steam engine, and his example was immediately followed by Sadi Carnot, Coriolis, Navier, and Poncelet.[61] Séguin, though he uses a different sort of data, seems a member of the same group. Dulong, to whose classic memoir on adiabatic compression many of the pioneers referred, was a close collaborator of Petit, and during the period of their collaboration Petit produced a quantitative account of the steam engine that antedates Carnot's by eight years.[62] There is even a hint of government interest. The prize offered by the French *Institut national* and won in 1812 by the classic research on gases of Delaroche and Bérard may well have grown in part from government interest in engines.[63] Certainly Regnault's later work on the same topic did. His famous investigations of the thermal characteristics of gas and steam bear the imposing title, "Experiments, undertaken by order of the Minister of Public Works and at the instigation of the Central Commission for Steam Engines, to determine the principal laws and the numerical data which enter into steam engine calculations."[64] One suspects that without these ties to the recognized problems of steam engineering, the important data on adiabatic compression would not have been so accessible to the pioneers of energy conservation. In this instance the concern with engines may not have been essential to the work of the pioneers, but it certainly facilitated their discoveries.

Because the concern with engines and the nineteenth-century conversion discoveries embrace most of the new technical concepts and experiments common to more than a few of the discoverers of energy conservation, this study of simultaneous discovery might well end here. But a last look at the papers of the pioneers generates an uncomfortable feeling that something is still missing, something that is not perhaps a substantive element at all. This feeling would not exist if all the pioneers had, like Carnot and Joule, begun with a straightforward technical problem and proceeded by stages to the concept of energy conservation. But in the cases of Colding, Helmholtz, Liebig, Mayer, Mohr, and Séguin, the notion of an underlying imperishable metaphysical force seems prior to research and almost unrelated to it. Put bluntly, these pioneers seem to have held an idea capable of becoming conservation of energy for some time before they found evidence for it. The factors previously discussed in this paper may explain why they were ultimately able to clothe the idea and thus to make sense of it. But the discussion does not yet sufficiently account for the idea's existence. One or two such cases among the twelve pioneers might not be troublesome. The sources of scientific inspiration are notoriously inscrutable. But the presence of major conceptual lacunae in six of our twelve cases is surprising. Though I cannot entirely resolve the problem it presents, I must at least touch upon it.

We have already noted a few of the lacunae. Mohr jumped without warning from a defense of the dynamical theory of heat to the statement that there is only one force in nature and that it is quantitatively unalterable.[65] Liebig made a similar leap from the duty of electric motors to the statement that the chemical equivalents of the elements determine the work retrievable from chemical processes by either electrical or thermal means.[66] Colding tells us that he got the idea of conservation in 1839, while still a student, but withheld announcement until 1843 so that he might gather evidence.[67] The biography of Helmholtz outlines a similar story.[68] Séguin confidently applied his concept of the convertibility of heat and motion to steam engine computations, even though his single attempt to confirm the idea had been totally fruitless.[69] Mayer's leap has repeatedly been noted, but its full size is not often remarked. From the light color of venous blood in the tropics, it is a small step to the conclusion that less internal oxidation is needed when the body loses less heat to the environment.[70] Crawford had drawn that conclusion from the same evidence in 1778.[71] Laplace and Lavoisier, in the 1780's, had balanced the same equation relating inspired oxygen to the body's heat losses.[72] A continuous line of research relates their work to the biochemical studies of respiration made by Liebig and Helmholtz in the early 1840's.[73] Though Mayer apparently did not know it, his observation of venous blood was simply a rediscovery of evidence for a well known, though controversial, biochemical theory. But that theory was not the one to which Mayer leaped. Instead Mayer insisted that internal oxidation must be balanced against *both* the body's heat loss *and* the manual labor the body performs. To this formulation, the light color of tropical venous blood is largely irrelevant. Mayer's extension of the theory calls for the discovery that lazy men, rather than hot men, have light venous blood.

The persistent occurrence of mental jumps like these suggests that many of the discoverers of energy conservation were deeply predisposed to see a single indestructible force at the root of all natural phenomena. The predisposition has been noted before, and a number of historians have at least implied that it is a residue of a similar metaphysic generated by the eighteenth-century controversy over the conservation of *vis viva*. Leibniz, Jean and Daniel Bernoulli, Hermann, and du Châtelet, all said things like, "*Vis* [*viva*] never perishes; it may in truth appear lost, but one can always discover it again in its effects if one can see them."[74] There are a multitude of such statements, and their authors do attempt, however crudely, to trace *vis viva* into and out of non-mechanical phenomena. The parallel to men like Mohr and Colding is very close. Yet eighteenth-century metaphysical sentiments of this sort seem an implausible source for the nineteenth-century predisposition we are examining. Though the technical *dynamical* conservation theorem has a continuous history from the early eighteenth century to the present, its metaphysical counterpart found few or no defenders after 1750.[75] To discover the *metaphysical* theorem, the pioneers of energy conservation would have had to return to books at least a century old. Neither their works nor their biographies suggest that they were significantly influenced by this particular bit of ancient intellectual history.[76]

Statements like those of both the eighteenth-century Leibnizians and the nineteenth-century pioneers of energy conservation can, however, be found repeatedly in the literature of a second philosophical movement, *Naturphilosophie*.[77] Positing organism as the fundamental metaphor of their universal science, the *Naturphilosophen* constantly sought a single unifying principle for all natural phenomena. Schelling, for example, maintained "that magnetic, electrical, chemical, and finally even organic phenomena would be interwoven into one great association . . . [which] extends over the whole of nature."[78] Even before the discovery of the battery he insisted that "without doubt only a single force in its various guises is manifest in [the phenomena of] light, electricity, and so forth."[79] These quotations point to an aspect of Schelling's thought fully documented by Brehier and more recently by Stauffer.[80] As a *Naturphilosoph*, Schelling constantly sought out conversion and transformation processes in the science of his day. At the beginning of his career chemistry seemed to him the basic physical science; from 1800 on he increasingly found in galvanism "the true border-phenomenon of both [organic and inorganic] natures."[81] Many of Schelling's followers, whose teaching dominated German and many neighboring universities during the first third of the nineteenth century, gave similar emphasis to the new conversion phenomena. Stauffer has shown that Oersted—a *Naturphilosoph* as well as a scientist—persisted in his long search for a relation between electricity and magnetism largely because of his prior philosophical conviction that one must exist. Once the interaction was discovered, electro-magnetism played a major role in Hebart's further elaboration of the scientific substructure of *Naturphilosophie*.[82] In short, many *Naturphilosophen* drew from their philosophy a view of physical processes very close to that which Faraday and Grove seem to have drawn from the new discoveries of the nineteenth century.[83]

Naturphilosophie could, therefore, have provided an appropriate philosophical background for the discovery of energy conservation. Furthermore, several of the pioneers were acquainted with at least its essentials. Colding was a protegé of Oersted's.[84] Liebig studied for two years with Schelling, and though he afterwards described these years as a waste, he never surrendered the vitalism he had then imbibed.[85] Hirn cited both Ocken and Kant.[86] Mayer did not study *Naturphilosophie,* but he had close student friends who did.[87] Helmholtz's father, an intimate of the younger Fichte's and a minor *Naturphilosoph* in his own right, constantly exhorted his son to desert strict mechanism.[88] Though Helmholtz himself felt forced to excise all philosophical discussion from his classic memoir, he was able by 1881 to recognize important Kantian residues that had escaped his earlier censorship.[89]

Biographical fragments of this sort do not, of course, prove intellectual indebtedness. They may, however, justify strong suspicion, and they surely provide leads for further research. At the moment I shall only insist that this research should be done and that there are excellent reasons to suppose it will be fruitful. Most of those reasons are given above, but the strongest has not yet been noticed. Though Germany in the 1840's had not yet achieved the scientific eminence of either Britain or France, five of our twelve pioneers were Germans, a sixth, Colding, was a Danish disciple of Oersted's, and a seventh, Hirn, was a self-educated Alsatian who read the *Naturphilosophen.*[90] Unless the *Naturphilosophie* indigenous to the educational environment of these seven men had a productive role in the researches of some, it is hard to see why more than fifty per cent of the pioneers should have been drawn from an area barely through its first generation of significant scientific productivity. Nor is this quite all. If proved, the influence of *Naturphilosophie* may also help explain why this particular group of five Germans, a Dane, and an Alsatian includes five of the six pioneers in whose approaches to energy conservation we have previously noted such marked conceptual lacunae.[91]

This preliminary discussion of simultaneous discovery must end here. Comparing it with the sources, primary and secondary, from which it derives, makes apparent its incompleteness. Almost nothing has been said, for example, about either the dynamical theory of heat or the conception of the impossibility of perpetual motion. Both bulk large in standard histories, and both would require discussion in a more extended treatment. But if I am right, these neglected factors and others like them would not enter a fuller discussion of simultaneous discovery with the urgency of the three discussed here. The impossibility of perpetual motion, for example, was an essential intellectual tool for most of the pioneers. The ways in which many of them arrived at the conservation of energy cannot be understood without it. Yet recognizing the intellectual tool scarcely contributes to an understanding of simultaneous discovery because the impossibility of perpetual motion had been endemic in scientific thought since antiquity.[92] Knowing the tool was there, our question has been: Why did it suddenly acquire a new significance and a new range of application? For us, that is the more significant question.

The same argument applies in part to my second example of neglected factors. Despite Rumford's deserved fame, the dynamical theory of heat had been close to the surface of scientific consciousness almost since the days of Francis Bacon.[93] Even at the end of the eighteenth century, when temporarily eclipsed by the work of Black and Lavoisier, the dynamical theory was often described in scientific discussions of heat, if only for the sake of refutation.[94] To the extent that the conception of heat as motion figured in the work of the pioneers, we must principally understand why that conception gained a significance after 1830 that it had seldom possessed before.[95] Besides, the dynamical theory did not figure very largely. Only Carnot used it as an essential stepping stone. Mohr leaped from the dynamical theory to conservation, but his paper indicates that other stimuli might have served as well. Grove and Joule adhered to the theory but show substantially no dependence on it.[96] Holtzmann, Mayer, and Séguin opposed it—Mayer vehemently and to the end of his life.[97] The apparently close connections between energy conservation and the dynamical theory are largely retrospective.[98]

Compare these two neglected factors with the three we have discussed. The rash of conversion discoveries dates from 1800. Technical discussions of dynamical engines were scarcely a recurrent ingredient of scientific literature before 1760 and their density increased steadily from that date.[99] *Naturphilosophie* reached its peak in the first two decades of the nineteenth century.[100] Furthermore, all three of these ingredients, except possibly the last, played important roles in the research of at least half the pioneers. That does not mean that these factors explain either the individual or collective discoveries of energy conservation. Many old discoveries and concepts were essential to the work of all the pioneers; many new ones played significant roles in the work of individuals. We have not and shall not reconstruct the causes of all that occurred. But the three factors discussed above may still provide the fundamental constellation, given the question from which we began: Why, in the years 1830 to 1850, did so many of the experiments and concepts required for a full statement of energy conservation lie so close to the surface of scientific consciousness?

References

1 J. R. Mayer, "Bemerkungen über die Kräfte der unbelebten Natur," *Ann. d. Chem. u. Pharm.*, XLII (1842). I have used the reprint in J. J. Weyrauch's excellent collection, *Die Mechanik der Wärme in gesammelten Schriften von Robert Mayer* (Stuttgart, 1893), pp. 23–30. This volume is cited below as Weyrauch, I. The same author's companion volume, *Kleinere Schriften und Briefe von Robert Mayer* (Stuttgart, 1893), is cited as Weyrauch, II.

   James P. Joule, "On the Calorific Effects of Magneto-Electricity, and on the Mechanical Value of Heat," *Phil. Mag.*, XXIII (1843). I have used the version in *The Scientific Papers of James Prescott Joule* (London, 1884), pp. 123–59. This volume is cited below as Joule, *Papers*.

   L. A. Colding, "Undersögelse on de almindelige Naturkraefter og deres gjensidige Afhaengighed og isaerdeleshed om den ved visse faste Legemers Gnidning udviklede Varme," *Dansk. Vid. Selsk.*, II (1851), 121–46. I am indebted to Miss Kirsten Emilie Hedebol for preparing a translation of this paper. It is, of course, far fuller than the unpublished original which Colding read to the Royal Society of Denmark in 1843, but it includes much information about that original. See also, L. A. Colding, "On the History of the Principle of the Conservation of Energy," *Phil. Mag.*, XXVII (1864), 56–64.

   H. von Helmholtz, *Ueber die Erhaltung der Kraft. Eine physikalische Abhandlung* (Berlin, 1847). I have used the annotated reprint in *Wissenschaftliche Abhandlungen von Hermann Helmholtz* (Leipzig, 1882), I, 12–75. This set is cited below as Helmholtz, *Abhandlungen*.

2 Carnot's version of the conservation hypothesis is scattered through a notebook written between the publication of his memoir in 1824 and his death in 1832. The most authoritative version of the notes is E. Picard, *Sadi Carnot, biographie et manuscript* (Paris, 1927); a more convenient source is the appendix to the recent reprint of Carnot's *Réflexions sur la puissance motrice du feu* (Paris, 1953). Notice that Carnot considered the material in these notes quite incompatible with the main thesis of his famous *Réflexions*. In fact, the essentials of his thesis proved to be salvageable, but a change in both its statement and its derivation was required.

   Marc Séguin, *De l'influence des chemins de fer et de l'art de les construire* (Paris, 1839), pp. xvi, 380–96.

   Karl Holtzmann, *Über die Wärme und Elasticität der Gase und Dämpfe* (Mannheim, 1845). I have used the translation by W. Francis in *Taylor's Scientific Memoirs*, IV (1846), 189–217. Since Holtzmann believed in the caloric theory of heat and used it in his monograph, he is a strange candidate for a list of discoverers of energy conservation. He also believed, however, that the same amount of work spent in compressing a gas isothermally must always produce the same increment of heat in the gas. As a result he made one of the early computations of Joule's coefficient and his work is therefore repeatedly cited by the early writers on thermodynamics as containing an important ingredient of their theory. Holtzmann can scarcely be said to have caught any part of energy conservation as we define that theory today. But for this investigation of simultaneous discovery the judgment of his contemporaries is more relevant than our own. To several of them Holtzmann seemed an active participant in the evolution of the conservation theory.

G. A. Hirn, "Études sur les principaux phénomènes que présentent les frottements médiats, et sur les diverses manières de déterminer la valeur mécanique des matières employées au graissage des machines," *Bulletin de la societé industrielle de Mulhouse*, XXVI (1854), 188–237; and "Notice sur les lois de la production du calorique par les frottements médiats," *ibid.*, pp. 238–77. It is hard to believe that Hirn was completely ignorant of the work of Mayer, Joule, Helmholtz, Clausius, and Kelvin when he wrote the "Études" in 1854. But after reading his paper, I find his claim to independent discovery (presented in the "Notice") entirely convincing. Since none of the standard histories cites these articles or even recognizes the existence of Hirn's claim, it seems appropriate to sketch its basis here.

Hirn's investigation deals with the relative effectiveness of various engine lubricants as a function of pressure at the bearing, applied torque, etc. Quite unexpectedly, or so he says, his measurements led to the conclusion that: "The absolute quantity of caloric developed by mediated friction [e.g., friction between two surfaces separated by a lubricant] is directly and uniquely proportional to the mechanical work absorbed by this friction. And if we express the work in kilograms raised to the height of one meter and the quantity of caloric in calories, we find that the ratio of these two numbers is very nearly 0.0027 [corresponding to 370 kg.m./cal.], whatever the velocity and the temperature and whatever the lubricating material" (p. 202). Until almost 1860 Hirn had doubts about the law's validity for impure lubricants or in the absence of lubrication (see particularly his *Récherches sur l'équivalent mécanique de la chaleur* [Paris, 1858], p. 83.) But despite these doubts, his work obviously displays one of the mid-nineteenth-century routes to an important part of energy conservation.

3   C. F. Mohr, "Ueber die Natur der Wärme," *Zeit. f. Phys.*, V (1837), 419–45; and "Ansichten über die Natur der Wärme," *Ann. d. Chem. u. Pharm.*, XXIV (1837), 141–47.

William R. Grove, *On the Correlation of Physical Forces: being the substance of a course of lectures delivered in the London Institution in the Year 1843* (London, 1846). Grove states that in this first edition he has introduced no new material since the lectures were delivered. The later and more accessible editions are greatly revised in the light of subsequent work.

Michael Faraday, *Experimental Researches in Electricity* (London, 1844), II, 101–104. The original "Seventeenth Series" of which this is a part was read to the Royal Society in March, 1840.

Justus Liebig, *Chemische Briefe* (Heidelberg, 1844), pp. 114–20. With this work, as with Grove's, one must beware of changes introduced in editions published after the conservation of energy was a recognized scientific law.

4 Since a few of my conclusions depend upon the particular list of names selected for study, a few words about the selection procedure seem essential. I have tried to include all the men who were thought by their contemporaries or immediate successors to have reached independently some significant part of energy conservation. To this group I have added Carnot and Hirn whose work would surely have been so regarded if it had been known. Their lack of actual influence is irrelevant from the viewpoint of this investigation.

This procedure has yielded the present list of twelve names, and I am aware of only four others for whom a place might be claimed. They are von Haller, Roget, Kaufmann, and Rumford. Despite P. S. Epstein's impassioned defense (*Textbook of Thermodynamics* [New York, 1937], pp. 27–34), von Haller has no place on the list. The notion that fluid friction in the arteries and veins contributes to body heat implies no part of the notion of energy conservation. Any theory that accounts for frictional generation of heat can embrace von Haller's conception. A better case can be made for Roget who did use the impossibility of perpetual motion to argue against the contact theory of galvanism (see note 27). I have omitted him only because he seems unaware of the possibility of extending the argument and because his own conceptions are duplicated in the work of Faraday who did extend them.

Hermann von Kaufmann probably should be included. According to Georg Helm his work is identical with Holtzmann's (*Die Energetik nach ihrer geschichtlichen Entwickelung* [Leipzig, 1898], p. 64). But I have been unable to see Kaufmann's writings, and Holtzmann's case is already somewhat doubtful, so that it has seemed better not to overload the list. As to Rumford, whose case is the most difficult of all, I shall point out below that before 1825 the dynamical theory of heat did not lead its adherents to energy conservation. Until the mid-century there was no necessary, or even likely, connection between the two sets of ideas. But Rumford was more than a dynamical theorist. He also said: "It would follow necessarily, from [the dynamical theory] . . . that the sum of the active forces in the universe must always remain constant" (*Complete Works* [London, 1876], III, 172), and this does sound like energy conservation. Perhaps it is. But if so, Rumford seems totally unaware of its significance. I cannot find the remark applied or even repeated elsewhere in his works. My inclination, therefore, is to regard the sentence as an easy echo, appropriate before a French audience, of the eighteenth-century theorem about the conservation of *vis viva*. Both Daniel Bernoulli and Lavoisier and Laplace had applied that theorem to the dynamical theory before (see note 95) without obtaining anything like energy conservation. I know of no reason to suppose that Rumford saw further than they.

5 This may well explain why the pioneers seem to have profited so little from each other's work, even when they read it. Our twelve men were not, in fact, strictly independent. Grove and Helmholtz knew Joule's work and cited it in their papers of 1843 and 1847 (Grove, *Physical Forces,* pp. 39, 52; Helmholtz, *Abhandlungen,* I, 33, 35, 37, 55). Joule, in turn, knew and cited the work of Faraday (*Papers,* p. 189). Liebig, though he did not cite Mohr and Mayer, must have known their work, for it was published in his own journal. (See also G. W. A. Kahlbaum, *Liebig und Friedrich Mohr, Briefe, 1834–1870* [Braunschweig, 1897], for Liebig's knowledge of Mohr's theory.) Very possibly more precise biographical information would disclose other interdependencies as well.

But these interdependencies, at least the identifiable ones, seem unimportant. In 1847 Helmholtz seems to have been unaware both of the generality of Joule's conclusions and of their large-scale overlap with his own. He cites only Joule's experimental findings, and these very selectively and critically. Not until the priority controversies of the second half-century, does Helmholtz seem to have recognized the extent to which he had been anticipated. Much the same holds for the relation between Joule and Faraday. From the latter Joule took illustrations, but not inspiration. Liebig's case may prove even more revealing. He could have neglected to cite Mohr and Mayer simply because they provided no relevant illustration and did not even seem to be dealing with the same subject matter. Apparently the men whom we call early exponents of energy conservation could occasionally read each other's works without quite recognizing that they were talking about the same things. For that matter, the fact that so many of them wrote from different professional and intellectual backgrounds may account for the infrequency with which they even saw each other's writings.

6 J. P. Joule, "Sur l'équivalent mécanique du calorique," *Comptes rendus,* XXVIII (1849), 132–35. I have used the reprint in Weyrauch, II, 276–80. This is only the first salvo in the priority controversy, but it already shows what the controversy is going to be about. Which of two (and later more than two) different statements is to be equated with *the* conservation of energy?

7 J. R. Mayer, *Die organische Bewegung in ihrem Zusammenhange mit dem Stoffwechsel* (Heilbronn, 1845) in Weyrauch, I, 45–128. Most of Joule's papers between 1843 and 1847 are relevant, but particularly: "On the Changes of Temperature produced by the Rarefaction and Condensation of Air" (1845) and "On Matter, Living Force, and Heat" (1847) in *Papers,* pp. 172–89, 265–81.

8 This formulation has at least one considerable advantage over the usual version. It does not imply or even permit the question, "Who *really* discovered conservation of energy first?" As a century of fruitless controversy has demonstrated, a suitable extension or restriction in the definition of energy conservation will award the crown to almost any one of the pioneers, an additional indication that they cannot have discovered the same thing.

The present formulation also bars a second impossible question, "Did

Faraday [or Séguin, or Mohr, or any one of the other pioneers, at will] really grasp the concept of energy conservation, even intuitively? Does he really belong on the list of pioneers?" Those questions have no conceivable answer, except in terms of the respondent's taste. But whatever answer taste may dictate, Faraday (or Séguin, etc.) provides useful evidence about the forces that led to the discovery of energy conservation.

9 These three criteria, particularly the second and third, determine the orientation of this study in a way that may not be immediately apparent. They direct attention away from the *prerequisites* to the discovery of energy conservation and towards what might be called the *trigger-factors* responsible for simultaneous discovery. For example, the following pages will show implicitly that all of the pioneers made significant use of the conceptual and experimental elements of calorimetry and that many of them also depended upon the new chemical conceptions derived from the work of Lavoisier and his contemporaries. These and many other developments within the sciences presumably had to occur before conservation of energy, as we know it, could be discovered. I have not, however, explicitly isolated elements like these below, because they do not seem to distinguish the pioneers from their predecessors. Since both calorimetry and the new chemistry had been the common property of all scientists for some years before the period of simultaneous discovery, they cannot have provided the immediate stimuli that triggered the work of the pioneers. As prerequisites for discovery, these elements have an interest and importance all their own. But their study is unlikely very much to illuminate the problem of simultaneous discovery to which this paper is directed. [This note has been added to the original manuscript in response to points raised during the discussion that followed the oral presentation.]

10 Faraday provides scarce and useful information about the progress of the significant controversy between the exponents of the chemical and contact theories of galvanism (*Experimental Researches*, II, 18–20). According to his account, the chemical theory was dominant in France and England from at least 1825, but the contact theory was still dominant in Germany and Italy when Faraday wrote in 1840. Does the dominance of the contact theory in Germany account for the rather surprising way in which both Mayer and Helmholtz neglect the battery in their accounts of energy transformations?

11 For the following discoveries see Sir Edmund Whittaker, *A History of the Theories of Aether and Electricity*, Vol. 1, *The Classical Theories* (London, 1951), pp. 81–84, 88–89, 170–71, 236–37. For Oersted's discovery see also, R. C. Stauffer, "Persistent Errors Regarding Oersted's Discovery of Electromagnetism," *Isis*, XLIV (1953), 307–10.

12 F. Cajori, *A History of Physics* (New York, 1922), pp. 158, 172–74. Grove makes a particular point of the early photographic processes (*Physical Forces*, pp. 27–32). Mohr gives great emphasis to Melloni's work (*Zeit. f. Phys.*, V [1837], 419).

13 For the chemical effects of static electricity see Whittaker, *Aether and Electricity*, p. 74, n. 2.

14 The single exception is significant and is discussed at some length below. During the eighteenth century steam engines were occasionally regarded as conversion devices.

15 Mary Sommerville, *On the Connexion of the Physical Sciences* (London, 1834), unpaginated "Preface."

16 Reasons for distinguishing Mohr's approach from that of Grove and Faraday will be examined below (note 83). The accompanying text will consider possible sources of Mohr's conviction about the conservation of "force."

17 The first eleven items in Joule's *Papers* (pp. 1–53) are exclusively concerned with improving first motors and then electromagnets, and these items cover the period 1838–41. The systematic evaluations of motors in terms of the engineering concepts, work and "duty," occur on pp. 21–25, 48. For Joule's earliest published use of the concept work or its equivalent, see p. 4.

18 Joule's concern with batteries and more particularly with the electrical production of heat by batteries dominates the five major contributions in *Papers*, pp. 53–123. My remark that Joule was led to batteries by his discouragement with motor design is a conjecture, but it seems extremely probable.

19 See note 1. This is the paper in which Joule is usually said to have announced energy conservation.

20 See note 7.

21 *Zeit. f. Phys.*, V (1837), 442.

22 Bence Jones, *The Life and Letters of Faraday* (London, 1870), II, 47.

23 *A Lecture on the Progress of Physical Science since the Opening of the London Institution* (London, 1842). Though the title page is dated 1842, the date is immediately followed by "[Not Published]." I do not know when the actual printing took place, but a prefatory remark of the author's indicates that the text itself was written very shortly after the lecture's delivery.

24 *Physical Forces*, p. 8.

25 Reasons for calling the remaining steps "obvious" are given in the closing paragraphs of this paper (see note 92).

26 Strictly speaking, this derivation is valid only if all the transformations of energy are reversible, which they are not. But that logical shortcoming completely escaped the notice of the pioneers.

27 P. M. Roget, *Treatise on Galvanism* (London, 1829). I have seen only the excerpt quoted by Faraday, *Experimental Researches*, II, 103, n. 2.

28 *Experimental Researches*, II, 103.

29 *Progress of Physical Science*, p. 20.

30 *Physical Forces*, p. 47.

31 *Ibid.*, p. 45.

32 I am not quite sure that this is true of Colding, particularly since I have not seen his unpublished paper of 1843. The early pages of his 1851 paper (note 1) contain many examples of conversion processes and are thus reminiscent of Mohr's approach. Also, Colding was a protegé of Oersted whose chief renown derived from his discovery of electromagnetic conversions. On the other hand, most of the conversion processes cited explicitly by Colding date from the

eighteenth century. In Colding's case, I suspect a prior tie between conversion processes and metaphysics (see note 83 and accompanying text). Very probably neither can be viewed as either logically or psychologically the more fundamental in the development of his thought.

33 Carnot's notes were not published until 1872 and then only because they contained anticipations of an important scientific law. Séguin had to call attention to the relevant passages in his book of 1839. Hirn did not bother to claim credit, but only attached a note denying plagiarism to his 1854 paper. That paper was published in an engineering journal that I have never seen cited by a scientist. Holtzmann's paper is the exception in that it was not obscure. But if other men had not discovered conservation of energy, Holtzmann's memoir would have continued to look like another one of the extensions of Carnot's memoir, for that is basically what it was (see note 2).

34 Between 1850 and 1875 Grove's book was reprinted at least six times in England, three times in America, twice in France, and once in Germany. The extensions were, of course, numerous, but I am aware of only two essential revisions. In the original discussion of heat (pp. 8–11), Grove suggested that macroscopic motion appears as heat only to the extent that it is *not* transformed to microscopic motion. In addition, of course, Grove's few attempts at quantification were quite off the track (see below).

35 *Physical Forces,* p. 46.

36 *Zeit. f. Phys.,* V (1837), 422–23.

37 Weyrauch, II, 102–105. This is in his first paper, "Ueber die quantitative und qualitative Bestimmung der Kräfte," sent to Poggendorf in 1841 but not published until after Mayer's death. Before he wrote his second paper, the first to be published, Mayer had learned a bit more physics.

38 It would be more precise to say that most pre-histories of energy conservation are principally lists of anticipations, and these occur particularly often in the early literature on *vis viva.*

39 The early eighteenth-century literature contains many general statements about the conservation of *vis viva* regarded as a metaphysical force. These formulations will be discussed briefly below. For the present notice only that none of them is suitable for application to the technical problems of dynamics, and it is with those formulations that we are here concerned. An excellent discussion of both the dynamical and metaphysical formulations is included in A. E. Haas, *Die Entwicklungsgeschichte des Satzes von der Erhaltung der Kraft* (Vienna, 1909), generally the fullest and most reliable pre-history of energy conservation. Other useful details can be found in Hans Schimank, "Die geschichtliche Entwicklung des Kraftbegriffs bis zum Aufkommen der Energetik," in *Robert Mayer und das Energieprinzip, 1842–1942,* ed. H. Schimank and E. Pietsch (Berlin, 1942). I am indebted to Professor Erwin Hiebert for calling these two useful and little known works to my attention.

40 Christian Huyghens, *Horologium oscillatorium* (Paris, 1673). I have used the German edition, *Die Penduluhr,* ed. A. Heckscher and A. V. Oettingen, Ostwald's Klassiker der Exakten Wissenschaften, No. 192 (Leipzig, 1913), p. 112.

41 D. Bernoulli, *Hydrodynamica, sive de viribus et motibus fluidorum, commentarii* (Basle, 1738), p. 12.

42 J. L. d'Alembert, *Traité de dynamique* (Paris, 1743). I have been able to see only the second edition (Paris, 1758) where the relevant material occurs on pp. 252–53. D'Alembert's discussion of the changes introduced since the first edition give no reason to suspect he has altered the original formulation at this point.

43 D. Bernoulli, "Remarques sur le principe de la conservation des forces vives pris dans un sens general," *Hist. Acad. de Berlin* (1748), pp. 356–64.

44 L. Euler, *Mechanica sive motus scientia analytice exposita,* in *Opera omnia* (Leipzig and Berlin, 1911–present), ser. 2, II, 74–77. The first edition was Petersburg, 1736.

J.-L. Lagrange, *Mécanique analytique* (Paris, 1788), pp. 206–9. I cite the first edition because the second, as reprinted in volumes 11 and 12 of Lagrange's *Oeuvres* (Paris, 1867–92), contains a very significant change. In the first edition, the conservation of *vis viva* is formulated only for time-independent constraints and for central or other integrable forces. It then takes the form $\Sigma m_i v_i^2 = 2H + 2\Sigma m_i \pi_i$, where H is a constant of integration and the π_i are functions of the position coordinates. In the second edition, Paris, 1811–15 (*Oeuvres,* XI, 306–10), Lagrange repeats the above but restricts it to a particular class of elastic bodies in order to take account of Lazare Carnot's engineering treatise (note 45) which he cites. For a fuller account of the engineering problem treated by Carnot, he refers his readers to his own *Théorie des fonctions analytiques* (Paris, 1797), pp. 399–410, where his version of Carnot's engineering problem is formulated more explicitly. That formulation makes the impact of the engineering tradition quite apparent, for the concept work now begins to appear. Lagrange states that the increment of *vis viva* between two dynamical states of the system is $2(P) + 2(Q) + \ldots$, where (P) — Lagrange calls it an "aire"—is $\Sigma_i \int P_i \, dp_i$, and P_i is the force on the i'th body in the direction of the position coordinates P_i. These "aires" are, of course, just work.

P. S. Laplace, *Traité de mécanique céleste* (Paris, 1798–1825). The relevant passages are more readily found in *Oeuvres complètes* (Paris, 1878–1904), I, 57–61. Mathematically, this treatment of 1798 actually resembles Lagrange's 1797 formulation rather than the earlier 1788 form. But, as in the pre-engineering formulations, the conservation law which includes a work integral is rapidly passed over in favor of the more restricted statement employing a potential function.

45 L. N. M. Carnot, *Essai sur les machines en général* (Dijon, 1782). I have consulted this work in Carnot's *Oeuvres mathematiques* (Basle, 1797) but rely principally on the expanded and more influential second edition, *Principes fondementaux de l'équilibre et du mouvement* (Paris, 1803). Carnot introduces several terms for what we call work, the most important being, "force vive latent" and "moment d'activité" (*ibid.,* pp. 38–43). Of these he says, "The kind of quantity to which I have given the name *moment of activity* plays a

very large role in the theory of machines in motion: for in general it is this quantity which one must economize as much as possible in order to derive from an agent [i.e., a source of power] all the [mechanical] effect of which it is capable" (*ibid.*, p. 257).

46 A useful survey of the early history of this important movement is C. L. M. H. Navier, "Details historiques sur l'emploi du principes des forces vives dans la theorie des machines et sur diverses roues hydraulique," *Ann. Chim. Phys.,* IX (1818), 146–59. I suspect that Navier's edition of B. de F. Belidor's *Architecture hydraulique* (Paris, 1819) contains the first developed presentation of the new engineering physics, but I have not yet seen this work. The standard treatises are: G. Coriolis, *Du calcul de l'effet des machines, ou considérations sur l'emploi des moteurs et sur leur évaluation pour servir d'introduction à l'étude speciale des machines* (Paris, 1829); C. L. M. H. Navier, *Résumé des leçons données a l'école des ponts et chaussées sur l'application de la mécanique à l'établissement des constructions et des machines* (Paris, 1838). Vol. 2; and J.-V. Poncelet, *Introduction à la mécanique industrielle,* ed. Kratz (3rd ed.; Paris, 1870). This work originally appeared in 1829 (part had appeared in lithoprint in 1827); the much enlarged and now standard edition from which the third is taken appeared in 1830–39.

47 The formal adoption of the term work (*travail*) is often credited to Poncelet (introduction, p. 64), though many others had used it casually before; Poncelet also (pp. 74–75) gives a useful account of the units (*dynamique, dyname, dynamie,* etc.) commonly used to measure this quantity. Coriolis *Du calcul de l'effect des machines,* p. iv) is the first to insist that *vis viva* be $\frac{1}{2}mv^2$, so that it will be numerically equal to the work it can produce; he also makes much use of the term *travail,* which Poncelet may have borrowed from him. The reformulation of the conservation law proceeds gradually from Lazare Carnot through all these later works.

48 As soon as he considers a quantitative problem in his first published paper, Mayer says: "A cause, which effects the raising of a weight, is a force; since this force brings about the fall of a body, we shall call it fall-force [Fallkraft]" (Weyrauch, I, 24). This is the engineering, not the theoretical dynamical, measure. By applying it to the problem of free fall, Mayer immediately derives $\frac{1}{2}mv^2$ (note the fraction) as the measure of energy of motion. The very crudeness of his derivation together with its lack of generality indicates his ignorance of the French engineering texts. The one French text he does mention in his writings (G. Lamé, *Cours de physique de l'école polytechnique* [2nd ed.; Paris, 1840]), does not deal with *vis viva* or its conservation at all.

Helmholtz uses the term *Arbeitskraft, bewegende Kraft, mechanische Arbeit,* and *Arbeit* for his fundamental measurable force (Helmholtz, *Abhandlungen,* I, 12, 17–18). I have not as yet been able to trace these terms in the earlier German literature, but their parallels in the French and English engineering traditions are obvious. Also, the term *bewegende Kraft* is used by the translator of Clapeyron's version of Sadi Carnot's memoir as equivalent to the French *puissance motrice* (*Pogg. Ann.,* LIX [1843], 446), and Helmholtz cites this translation (p. 17, n. 1). To this extent the tie to the engineering tradition is explicit.

Helmholtz was not, however, aware of the French theoretical engineering tradition. Like Mayer, he derives the factor of ½ in the definition of energy of motion and is unaware of any precedent for it (p. 18). More significant, he fails completely to identify $\int Pdp$ as work or *Arbeitskraft,* and instead calls it the "sum of the tensions" [*Summe der Spannkräfte*] over the space dimension of the motion.

49 The unit implicit in Savery's work is really the horsepower, but this includes weight times height as a part. See H. W. Dickinson and Rhys Jenkins, *James Watt and the Steam Engine* (Oxford, 1927), p. 353–54. Antoine Parent, "Sur le plus grande perfection possible des machines," *Hist. Acad. Roy.* (1704), pp. 323–38.

50 J. T. Desagulier, *A Course of Experimental Philosophy,* (3rd ed., 2 vols.; London, 1763), particularly I, 132, and II, 412. This posthumous edition is practically a reprint of the second edition (London, 1749).

John Smeaton, "An Experimental Inquiry concerning the Natural Powers of Water and Wind to turn Mills, and other Machines, depending on a Circular Motion," *Phil. Trans.,* LI (1759), 51. Here the measure is weight times height per unit time. The time dependence is, however, dropped in his "An Experimental Examination of the Quantity and Proportion of Mechanic Power necessary to be employed in giving different degrees of Velocity to Heavy Bodies," *Phil. Trans.,* LXVI (1776), 458.

For Watt see Dickinson and Jenkins, *James Watt,* pp. 353–56.

51 J. C. Borda, "Mémoires sur les rues hydrauliques," *Mem. l'Acad. Roy.* (1767), p. 272. Here the measure is weight times vertical speed. Height replaces speed in C. Coulomb, "Observation theorique et experimentales sur l'effet des moulins à vent, et sur la figure de leurs ailes," *ibid.* (1781), p. 68, and "Resultat de plusieurs expériences destinée à determiner la quantité d'action que les hommes peuvent fournir par leur travail journalier, suivant les differentes manières dont ils emploient leurs forces," *Mem. de l'Inst.,* II (1799), 381.

52 Mayer states that he loved to build model water wheels as a boy and that he learned the impossibility of perpetual motion in studying them (Weyrauch, II, p. 390). He could have learned simultaneously the proper measure of the product of machines.

53 Professor Hiebert asks if the concept of mechanical work may not have emerged from elementary statics and particularly from the formulation that derives statics from the principle of virtual velocities. The point needs further research, but my present response must be at least equivocally negative. The elements of statics were an important item in the equipment of all eighteenth-century engineers and the principle of virtual velocities, or an equivalent, therefore recurs in eighteenth-century writings on engineering problems. Quite possibly the engineers could not have evolved the concept work without the aid of the pre-existing static principle. But, as the preceding discussion may indicate, if the eighteenth-century concept work did emerge from the far older

principle of virtual velocities, it did so only when that principle was firmly embedded in the engineering tradition and only when that tradition turned its attention to the evaluation of power sources such as animals, falling water, wind, and steam. Therefore, reverting to the vocabulary of note 9, I suggest that the principle of virtual velocities may have been a prerequisite for the discovery of energy conservation but that it can scarcely have been a trigger. [This note added to original manuscript in response to points raised during discussion.]

54 *Hydrodynamica,* p. 231.

55 *De l'équilibre et du mouvement,* p. 258. Notice also that as soon as Lagrange turns to Carnot's problem (note 44), he speaks in the same way. In the *Fonctions analytiques,* he says that waterfalls, coal, gunpowder, animals, etc., all "contain a quantity of *vis viva,* which one can harness but which one cannot increase by any mechanical means. One may [therefore] always regard a machine as intended to destroy a given quantity of *vis viva* [in the load] by consuming some other given *vis viva* [from the source]." (*Oeuvres,* IX, 410.)

56 *Du calcul de l'effect des machines,* chapt. 1. For Coriolis the conservation theorem applied to a perfect machine becomes the "Principle of the Transmission of Work."

57 Helmholtz, *Abhandlungen,* I, 17. Colding, "Naturkraefter," *Dansk. Vid. Selsk.,* II (1851), 123–24. Particularly interesting evidence about the apparent similarities between the theory of energy conservation and Carnot's incompatible theory of the heat engine is provided by Carlo Matteucci. His paper, "De la relation qui existe entre la quantité de l'action chimique et la quantité de chaleur, d'électricité et de lumière qu'elle produit," *Bibliothèque universelle de Genève, Supplement,* IV (1847), 375–80, is an attack upon several of the early exponents of energy conservation. He describes his opponents as the group of physicists who "have tried to show that Carnot's celebrated principle about the motive force of heat can be applied to the other imponderable fluids."

58 Helmholtz, *Abhandlungen,* I, 18–19, gives Helmholtz initial abstract formulation of the cyclic process.

59 T. S. Kuhn, "The Caloric Theory of Adiabatic Compression," *Isis,* XLIX (1958), 132–40.

60 John Dalton, "Experimental Essays on the Constitution of Mixed Gases; on the Force of Steam or Vapour from Water and other Liquids in different temperatures, both in a Torricellian Vacuum and in Air; on Evaporation; and on the Expansion of Gases by Heat," *Manch. Mem.* V (1802), 535–602. The second essay, though it grew out of Dalton's meteorological interests, was immediately exploited by both British and French engineers.

Clément and Désormes, "Mémoires sur la théorie des machines à feu," *Bulletin des sciences par la societé philomatique,* VI (1819), 115–18; and "Tableau relatif à la théorie general de la puissance mécanique de la vapeur," *ibid.,* XIII (1826), 50–53. The second paper appears in full in Crelle's *Journal für die Baukunst,* VI (1833), 143–64. For the contributions of these men to adiabatic compression, see my paper, note 59.

61 S. D. Poisson, "Sur la chaleur des gaz et des vapeurs," *Ann. Chim. Phys.,* XXIII (1823), 337–52. For Navier, Coriolis, and Poncelet, all of whom devote chapters to steam engine computations, see note 46.

62 A. T. Petit, "Sur l'emploi du principe des forces vives dans le calcul de l'effet des machines," *Ann. Chim. Phys.,* VIII (1818), 287–305.

63 F. Delaroche and J. Bérard, "Mémoire sur la determination de la chaleur spe-cifique des differents gaz," *Ann. Chim. Phys.,* LXXXV (1813), 72–110, 113–82. I know of no direct evidence relating the prize won by this memoir to the problems of steam engineering, but the Academy did offer a prize for improve-ment in steam engines as early as 1793. See H. Guerlac, "Some Aspects of Science during the French Revolution," *The Scientific Monthly,* LXXX (1955), 96.

64 V. Regnault, *Mém. de l'Acad.,* XXI (1847), 1–767. The title of the work is quoted above.

65 See note 21 and accompanying text.

66 *Chemische Briefe,* pp. 115–17.

67 Colding, "History of Conservation," *Phil. Mag.,* XXVII (1864), 57–58.

68 Leo Koenigsberger, *Hermann von Helmholtz,* tr. F. A. Welby (Oxford, 1906), pp. 25–26, 31–33, implies that Helmholtz's ideas about conservation were com-plete as early as 1843, and he states that by 1845 the attempt at experimental proof motivated all of Helmholtz's research. But Koenigsberger gives no evi-dence, and he cannot be quite correct. In two articles on physiological heat written during 1845 and 1846 (*Abhandlungen,* I, 8–11; II, 680–725), Helm-holtz fails to notice that body heat may be expended in mechanical work (compare the discussion of Mayer, below). In the second of these papers he also gives the usual caloric explanation of adiabatic compression in terms of the change in heat capacity with pressure. In short, his ideas were by no means complete until 1847 or shortly before. But the papers of 1845 and 1846 do show that in these years Helmholtz was concerned to combat vitalism which he thought implied the creation of force from nothing. Also they show that he already knew the work of Clapeyron and of Holtzmann, which he thought relevant. To this extent, at least, Koenigsberger must be right.

69 *Chemins de fer,* p. 383. Séguin had tried unsuccessfully to measure the differ-ence in the quantities of heat abstracted from the boiler and delivered to the condenser of a steam engine.

70 Weyrauch, I, 12–14.

71 E. Farber, "The Color of Venous Blood," *Isis,* XLV (1954), 3–9.

72 A. Lavoisier and P. S. Laplace, "Mémoire sur la chaleur," *Hist. de l'Acad.* (1780), pp. 355–408.

73 Helmholtz touches on much of this research in his paper of 1845, "Wärme, physiologisch," for the *Encyclopädische Wörterbuch der medicinischen Wis-senschaften* (*Abhandlungen,* II, 680–725).

74 Haas, *Erhaltung,* p. 16, n. Quoted from *Institutions physiques de Madame la Marquise du Chastellet adressés à Mr. son Fils* (Amsterdam, 1742).

75 Haas, *Erhaltung,* p. 17.

76 None of the pioneers mention the eighteenth-century conservation literature in their original papers. Colding, however, says that he got his first glimpse of conservation while reading d'Alembert in 1839 (*Phil. Mag., XXVII* [1864], 58), and Koenigsberger says that Helmholtz read d'Alembert and Daniel Bernoulli early in 1843 (*von Helmholtz,* p. 26). These two counterexamples do not, however, really modify my thesis. D'Alembert omitted all mention of the metaphysical conservation theorem from the first edition of his *Traité,* and in the second he explicitly disowned the view ([2nd ed.; Paris, 1758], beginning of the "Avertissement" and pp. xvii–xxiv). In fact, d'Alembert was among the first to insist on freeing dynamics from what he considered to be mere metaphysical speculations. To take his ideas from this source Colding would still have required a strong predisposition. Bernoulli's *Hydrodynamica* is a more appropriate source (see, for example, the text which accompanies note 54), but Koenigsberger makes the very plausible point that Helmholtz consulted Bernoulli in order to work out his preexisting conception of conservation.

77 The roots of *Naturphilosophie* can, of course, be traced back through Kant and Wolff to Leibniz, and Leibniz was the author of the metaphysical conservation theorem about which both Kant and Wolff wrote (Haas, *Erhaltung,* pp. 15–18). The two movements are not, therefore, entirely independent.

78 Quoted by R. C. Stauffer, "Speculation and Experiment in the Background of Oersted's Discovery of Electromagnetism," *Isis,* XLVIII (1957), 37, from Schelling's *Einleitung zu seinem Entwurf eines Systems der Naturphilosophie* (1799).

79 Quoted by Haas, *Erhaltung,* p. 45, n. 61, from Schelling's *Erster Entwurf eines Systems der Naturphilosophie* (1799).

80 Émile Bréhier, *Schelling* (Paris, 1912). This is the most helpful discussion I have found and should certainly be added to Stauffer's list of useful aids for studying the complex relations of science and *Naturphilosophie* (*Isis,* XLVIII [1957], 37, n. 21).

81 *Ibid.,* p. 36, from Schelling's "Allgemeiner Deduktion des dynamischen Processes oder der Kategorien der Physik" (1800).

82 Haas, *Erhaltung,* p. 41.

83 It is, of course, impossible to distinguish sharply between the influence of *Naturphilosophie* and that of conversion processes. Brehier (*Schelling,* pp. 23–24) and Windelband (*History of Philosophy,* tr. J. H. Tufts [2nd ed.; New York, 1901], pp. 597–98) both emphasize that conversion processes were themselves a significant source of *Naturphilosophie,* so that the two were often grasped together. This fact must qualify some of the dichotomies set up in the first part of this paper, for the distinction between the two sources of the conservation concept is often equally hard to apply to individual pioneers. I have already pointed out the difficulty in Colding's case (note 32). With Mohr and Liebig I am still inclined to give *Naturphilosophie* the psychological priority, because neither had dealt much with the new conversion processes in their

own research and because both make such large leaps. Their cases appear in sharp contrast to those of Grove and Faraday who seem to proceed by a continuous path from conversion processes to conservation. But this continuity may be deceptive. Grove (*Physical Forces*, pp. 25–27) mentions Coleridge, and Coleridge was the principal British exponent of *Naturphilosophie*. Since the problem presented by these examples seems to me both real and unresolved, I had better point out that it affects only the organization, not the main thesis, of this paper. Perhaps conversion processes and *Naturphilosophie* should be considered in the same section. Nevertheless, they would both have to be considered.

84 Povl Vinding, "Colding, Ludwig August," *Dansk Biografisk Leksikon* (Copenhagen, 1933–44), pp. 377–82. I am grateful to Roy and Ann Lawrence for providing me with a précis of this useful biographical sketch.

85 E. von Meyer, *A History of Chemistry*, tr. G. McGowan, (3rd ed.; London, 1906), p. 274. J. T. Merz, *European Thought in the Nineteenth Century* (London, 1923–50), I, 178–218, particularly the last page.

86 G. A. Hirn, "Études sur les lois et sur les principes constituants de l'univers," *Revue d'Alsace*, I (1850), 24–41, 127–42, 183–201; *ibid.*, II (1851), 24–45. References to writings related to *Naturphilosophie* occur relatively often, though they are not very favorable. On the other hand, the very title of this piece suggests *Naturphilosophie*, and the title is appropriate to the contents.

87 B. Hell, "Robert Mayer," *Kantstudien*, XIX (1914), 222–48.

88 Koenigsberger, *von Helmholtz*, pp. 3–5, 30.

89 Helmholtz, *Abhandlungen*, I, 68.

90 Much biographical and bibliographical material for the study of Hirn's life and work can be found in the *Bulletin de la société d'histoire naturelle de Colmar*, I (1899), 183–335.

91 Séguin is the sixth, and the source of his idea remains a complete riddle. He attributes it (*Chemins de fer*, p. xvi) to his uncle Montgolfier about whom I have been able to get no relevant information.

The statistics above are not meant to imply that those exposed to *Naturphilosophie* were invariably affected by it; nor do I mean to argue that those whose work shows no conceptual lacunae were *ipso facto* not influenced by *Naturphilosophie* (see remarks on Grove in note 83). It is the *predominance* rather than the presence of pioneers from the area dominated by German intellectual traditions that constitutes the puzzle.

[The following paragraph was added to the original manuscript in response to points raised during the discussion.]

Professor Gillispie, in his paper, calls attention to a little known movement in eighteenth-century France that shows striking parallels to *Naturphilosophie*. If this movement had still been prevalent in nineteenth-century France, my contrast between the German scientific tradition and that prevalent elsewhere in Europe would be questionable. But I find nothing resembling *Naturphilosophie* in any of the nineteenth-century French sources I have examined, and Professor Gillispie assures me that, to the best of his knowledge, the movement to which his paper draws attention had disappeared (except perhaps from parts of biology) by the turn of the century. Notice, in addition, that this eighteenth-century movement, which was particularly prevalent among craftsmen and inventors, may provide a clue to the puzzle of Montgolfier (see above).

92 E. Mach, *History and Root of the Principle of the Conservation of Energy*, tr. Philip E. B. Jourdain (Chicago, 1911), pp. 19–41; and Haas, *Erhaltung*, chapt. IV. Remember also that in 1775 the French Academy formally resolved to consider no more purported designs of perpetual motion machines. Almost all of our pioneers make use of the impossibility of perpetual motion, and none feels the slightest necessity of arguing about its validity. In contrast, they do find it necessary to argue at length about the validity of the concept of universal conversions. Grove, for example, opens his *Physical Forces* (pp. 1–3) with a plea for a fair hearing of a radical idea. The idea turns out to be the concept of universal conversions developed at great length in the text (pp. 4–44). The impossibility of perpetual motion is casually applied to this idea without argument in the last seven pages (pp. 45–52). It is facts like these that led me to call the steps from universal conversions to an unquantified version of conservation "rather obvious."

93 For seventeenth-century theories of heat see, M. Boas, "The establishment of the mechanical philosophy," *Osiris*, X (1952), 412–541. Much information about eighteenth-century theories is scattered through: D. McKie and N. H. de V. Heathcote, *The Discovery of Specific and Latent Heat* (London, 1935), and H. Metzger, *Newton, Stahl, Boerhaave et la doctrine chimique* (Paris, 1930). Much other useful information will be found in G. Berthold, *Rumford und die Mechanische Wärmetheorie* (Heidelberg, 1875), though Berthold skips too rapidly from the seventeenth to the nineteenth century.

94 Since the caloric theory was scarcely presented in a developed form before the publication of Lavoisier's *Traité élémentaire de chimie* in 1789, it could hardly have eradicated the dynamical theory in the decade remaining before the publication of Rumford's work. For evidence that even the most pronounced caloricists continued to discuss it, see Armand Séguin, "Observations générales sur le calorique . . . reflexions sur la théorie de MM. Black, Crawford, Lavoisier, & Laplace," *Ann. de Chim.*, III (1789), 148–242, and V (1790), 191–271, particularly, III, 182–90. The material theory of heat has, of course, roots far older than Lavoisier, but Rumford, Davy, *et al.*, are really opposing a new theory, not an old one. Their work, particularly Rumford's, may have kept the dynamical theory alive after 1800, but Rumford did not create the theory. It had not died.

95 It is too seldom recognized that until almost the mid-nineteenth century, brilliant scientists could apply the dynamical conservation of *vis viva* to the theory that heat is motion without at all recognizing that heat and work should then be convertible. Consider the following three examples. Daniel Bernoulli, in the often quoted paragraphs from "Section X" of his *Hydrodynamica* equates heat with particulate *vis viva* and derives the gas laws.

Then, in paragraph 40, he applies this theory in computing the height from which a given weight must fall to compress a gas to a given fraction of its initial volume. His solution gives the energy of motion abstracted from the falling weight in order to compress the gas, but fails entirely to notice that this energy must be transferred to the gas particles and must therefore raise the gas's temperature. Lavoisier and Laplace, on pp. 357–59 of their classic memoir (note 72), apply the conservation of energy to the dynamical theory in order to show that for all experimental purposes the caloric and dynamical theories are precisely equivalent. J. B. Biot repeats the same argument, in his *Traité de physique expérimentale et mathematique* (Paris, 1816), I, 66–67, and elsewhere in the same chapter. Grove's mistake about heat (note 34) indicates that even the conception of conversion processes was sometimes insufficient to guide scientists away from this virtually universal mistake.

96 Grove, *Physical Forces,* pp. 7–8. Joule, *Papers,* pp. 121–23. Perhaps these two would not have developed their theories if they had not tended to regard heat as motion, but their published works indicate no such decisive connections.

97 Holtzmann's memoir is based on the caloric theory. For Mayer see Weyrauch, I, 265–72, and II, 320, n. 2. For Séguin see *Chemins de fer,* p. xvi.

98 The ease and immediacy with which the dynamical theory was identified with energy conservation is indicated by the contemporary misinterpretations of Mayer quoted in Weyrauch, II, pp. 320 and 428. The classic case, however, is Lord Kelvin's. Having employed the caloric theory in his research and writing until 1850, he opens his famous paper "On the Dynamical Theory of Heat" (*Mathematical and Physical Papers* [Cambridge, 1882], pp. 174–75) with a series of remarks on Davy's having "established" the dynamical theory fifty-three years before. Then he continues, "The recent discoveries made by Mayer and Joule . . . afford, *if required,* a perfect confirmation of Sir Humphry Davy's views" (italics mine). But if Davy established the dynamical theory in 1799 and if the rest of conservation follows from it, as Kelvin implies, what had Kelvin himself been doing before 1852?

99 The abstract theories of dynamical engines have no beginning in time. I pick 1760 because of its relation to the important and widely cited works of Smeaton and Borda (notes 50 and 51).

100 Merz, *European Thought,* I, 178, n. 1.

CONSERVATION AND THE CONCEPT OF ELECTRIC CHARGE: AN ASPECT OF PHILOSOPHY IN RELATION TO PHYSICS IN THE NINETEENTH CENTURY

I. Bernard Cohen

Among the egregious lacunae in the history of science are studies in the growth of science in the nineteenth century.[1] When, therefore, I was asked to give a paper that might make part of a session with Dr. Kuhn's study of conservation of energy, I wondered whether a less known aspect of conservation might possibly serve to illuminate the general role of conservation theories in nineteenth-century scientific thought and so perhaps help to provide a more ample perspective on this period. Furthermore, since I shall attempt to display some of the particular difficulties that have arisen from attempts to learn what it may be that is conserved in electrical phenomena, opportunity will be provided to witness the doubts as well as the certainties in nineteenth-century science.

To remind you of the importance that the study of electrical phenomena has had in recent times for general scientific ideas, let me quote the title of Albert Einstein's famous paper of 1905 on Special (or Restricted) Relativity. It was "Zur Elektrodynamik bewegter Körper," or "On the Electrodynamics of Moving Bodies."[2] At the time of writing it, Einstein did not yet know the equally fundamental paper of H. A. Lorentz, "Electromagnetic Phenomena in a System Moving with any Velocity less than that of Light,"[3] published a year earlier, which among many important steps in the pre-Einstein developments took cognizance of the publication of the experiments by W. Kaufmann,[4] which seemed to prove beyond any possible doubt that the "mass" of a "fundamental particle" like the electron is not constant but varies with speed. Hence electrical experiments and considerations of electrical phenomena were of the highest importance in the years 1902–1905 in challenging, undermining, and

357

eventually demanding revisions of the three primary concepts of physical science: space, time and mass.

Having mentioned the electron,[5] I should add that its discovery was startling enough in seeming to disclose the primary "atom" or "quantum" of negative electricity, the existence of which had been predicted by Johnston Stoney. But the electron had even more extraordinary implications when J. J. Thomson showed that this fundamental particle (at least its measurable value of e/m, the ratio of its charge to its mass) was the same no matter what type of glass was used to make the discharge tube, what kind of gas was employed, or what kind of metal the electrodes were made of.[6] It was a sub-atomic building block that had been found—a necessary constituent part of every kind of atom. What this conclusion eventually did to the chemist's conception of the indivisible atom is fairly well known. But to see how radical Thomson's conclusion was, it may be pointed out that he himself refused at first to recognize that this was the only inescapable conclusion. He later said that it was far easier to convince others of his discovery than it had been to convince himself. And recall that this statement was made with full memory that when he read his first paper on the subject, there was little discussion because his auditors did not believe he was serious; they thought that Thomson was playing some kind of practical joke.[7]

One final point about the discovery of the electron is that once its charge was determined the possibility arose of determining Avogadro's number, simply the number of electrons in Faraday's constant.[8] At once the actual mass (in grams) of the atoms of any chemical element could be computed. This striking result destroyed the doubts concerning the "atomic hypothesis" in so hardened a mind as that of Wilhelm Ostwald. Like Ernst Mach,[9] Ostwald held that the concept of atom was both useless scientifically and absurd philosophically. But others, while agreeing that the atom was not a sound philosophical concept, held it to be useful for the working scientist. Thus in 1867, Friedrich August Kekulé had made a clear distinction between the attitude of a philosopher and that of a chemist with respect to the existence of atoms:

The question whether atoms exist or not has but little significance from a chemical point of view: its discussion belongs rather to metaphysics. In chemistry we have only to decide whether the assumption of atoms is an hypothesis adapted to the explanation of chemical phenomena. More especially have we to consider the question whether a further development of the atomic hypothesis promises to advance our knowledge of the mechanism of chemical phenomena.

I have no hesitation in saying that, from a philosophical point of view, I do not believe in the actual existence of atoms, taking the word in its literal significance of indivisible particles of matter—I rather expect that we shall some day find for what we now call atoms a mathematico-mechanical explanation, which will render an account of atomic weight, of atomicity, and of numerous other properties of the so-called atoms. As a chemist, however, I regard the assumption of atoms, not only as advisable, but as absolutely necessary in chemistry. I will even go further, and declare my belief that *chemical atoms exist,* provided the term be understood to denote those particles of matter which undergo no further division in chemical metamorphoses. Should the progress of science lead to a theory of the constitution of chemical atoms— it would make but little alteration in chemistry itself. The chemical atoms will always remain the chemical unit; and for the specially chemical considerations we may always start from the constitution of atoms, and avail ourselves of the simplified expression thus obtained, that is to say, of the atomic hypothesis. We may, in fact, adopt the view of Dumas and of Faraday, that *whether matter be atomic or not, thus much is certain, that granting it to be atomic, it would appear as it now does.*[10]

Ostwald's point of view differed from Kekulé's, and from Mach's, in several noteworthy respects, as may be seen in the following extract:

For the representation of the simple and comprehensive laws to which the weight and volume ratios of chemical compounds are subject, a hypothetical conception has been in use since the time these laws were first discovered, which affords a very convenient picture of the actual relations, and processes, therefore of great value for the purposes of instruction and investigation. For this reason the above hypothesis has been made the basis of the language and modes of representation throughout the whole of chemistry, so that the results of chemical investigation are almost exclusively communicated in that language. For this reason alone a knowledge of the hypothesis is not necessary.

In general, an hypothesis is an *aid to representation.* Of the phenomena of the outer world, some are so familiar to us from repeated experience, that we know the relations which exist among them with great certainty. If now we find a new and unfamiliar class of phenomena, we unconsciously seek for similar ones among those that are known. If we succeed in discovering such a similarity we gain two advantages. In the first place, the fixing of the new facts in the memory is very greatly facilitated by the use of the similarity, and in the second place, the similarity affords us a means of making probable guesses concerning the behaviour of the new phenomena under conditions under which they have not yet been investigated. . . .

Ostwald then discussed the basic assumption of the atomic hypothesis: that all substances are composed of very small particles called atoms, the atoms of any one elementary substance being identical but different from atoms of other elementary substances. This hypothesis agrees well with the inconvertibility of the elements into one another and the combination of different elements to form compounds; furthermore, "the quantitative laws of combination are made intelligible." One can deduce the law of constant combining proportions. Since molecules of compounds are "congeries" of the atoms of the elements of which compounds are made, the atomic weight of each element multiplied by the number of atoms of that element in the molecule of the compound must give the observed combining proportions by weight. Hence, he said:

> Within the limits given here, the atomic hypothesis has proved to be an exceedingly useful aid to instruction and investigation, since it greatly facilitates the interpretation and the use of the general laws. One must not, however, be led astray by this agreement between picture and reality, and confound the two. So far as we have treated them, the chemical processes occurred in such a way as if the substances were composed of atoms in the sense explained. At best there follows from this the *possibility* that they are in reality so; not, however, the *certainty*. For it is impossible to prove that the laws of chemical combination cannot be deduced with the same completeness by means of a quite different assumption.
>
> One does not require, therefore, to give up the advantage of the atomic hypothesis if one bears in mind that it is an illustration of the actual relations in the form of a suitable and easily manipulated picture, but which may, on no account, be substituted for the actual relations. One must always be prepared for the fact that sooner or later the reality will be different from that which the picture leads one to expect.
>
> Especially, when any other well-founded speculation leads to a variance with the atomic hypothesis, one must not, on that account, regard the speculation as wrong. The blame can quite well attach to the hypothesis.[11]

Clearly, then, Ostwald's "conversion" was an event of note.

In the preface (1908) to the fourth edition of the *Outlines of General Chemistry,* Ostwald admitted: "I am now convinced that we have recently become possessed of experimental evidence of the discrete or grained nature of matter, which the atomic hypothesis sought in vain for hundreds and thousands of years." Primarily this new evidence came from the "isolation and counting of gas ions, . . . which have crowned with success the long and brilliant researches of J. J. Thomson." Together with the investigations of the Brownian movement (notably by J. Perrin), the work on electrons could "... justify the most cautious scientist in now speaking of the experimental proof of the atomic nature of matter. The atomic hypothesis is thus raised to the position of a scientifically well-founded theory, and can claim its place in a text-book intended as an introduction to the present state of our knowledge of General Chemistry."[12] Hence when Ostwald finally gave in and accepted the chemical atom, that atom was no longer the indivisible structural unit of matter that it had been thought to be during the previous century. In 1904 it could be said with confidence, "The chemist's atom is no longer the unit of the subdivision of matter, and the internal structure of the atom is now the object of experimental study."[13]

The electron proved to have yet further effects on the structure of physical thought. J. J. Thomson's discovery of the electron[14] occurred as a product of his investigations of cathode rays which—if of sufficient energy, as in Wilhelm Konrad Roentgen's experiments—produce X-rays, themselves rather sensational at the time of their discovery in 1895 and for many years afterwards. But, in many ways, the most significant by-product of the discovery of the electron came from studies of the interaction of light and electricity, notably the photo-electric effect. While the quantum theory originated in Max Planck's studies of classical radiation theory, there can be no question of the fact that it was Einstein's "heuristic" proposal of photons in 1905 that gave the concept of the quantum of electromagnetic energy its current form. Not until then had it been explicitly proposed that discrete units of such energy might maintain their identity while traveling through space. In many ways far more radical than the paper on relativity[15] published in the same year by Einstein in the same journal, the new views concerning the structure of light required decades for their acceptance.

Both directly and indirectly, then, the discovery of the unit of electric charge, of what it is that is conserved in electrical phenomena (by being transferred from one body to another, by being moved about on a given body, or by joining or leaving a body), was of paramount importance for the growth of physical science in the decade 1895–1905 and afterwards.[16] There is no want of evidence that the new discoveries in electricity—chiefly those relating to the nature of the electric charge (electron) and its interaction with light and with matter—were primarily responsible for the alteration in the basic concepts, theories, and general point of view of physical science. And we are therefore tempted to wonder whether this was a situation without precedent, or whether—during the previous century and a half—investigations of electrical phenomena had not also been of philosophical interest.

Let it be clearly understood that no attempt is being made here to minimize the importance of the concept of energy and the principle or law of its conservation for the physics of the nineteenth century and even for the philosophy of physics. Anyone who has read in the philosophy of science is well aware that energy was of primary concern to thoughtful men of science and to philosophers in the second half of the nineteenth century, and that the law of conservation of charge was of far less importance. Perhaps for that reason we are apt to overlook the nineteenth-century discussions of electricity; but to do so is to place the discovery of the law of conservation of energy in a unique position, rather than to see it as a part of a general trend in physical thought. The nineteenth-century scientists knew at least four major conservation laws or principles: the conservation, respectively, of momentum, charge, matter, and energy. Roughly we may say that the idea of conservation of momentum is the creation of the late seventeenth century, associated with what we generally call the physics of Newton,[17] and that the concept of conservation of charge is the creation of the mid-eighteenth century, associated with the name of Franklin. This pair of conservation principles differs from the second pair—matter, and energy—in that each of these principles is still considered valid as an independent statement about natural events. Whether we consider the behavior of gross objects, or discuss events on the atomic or sub-atomic level, we invoke these two principles without exception. No atom-model that violated either principle would be admitted into discussions of physics. On the other hand, the principle of conservation of matter—associated with the name of Lavoisier at the end of the eighteenth century—and the principle of conservation of energy (to which there are many claims of discovery) are valid in independence of one another only on the level of gross objects, or in macrophysics. In microphysical events, these two principles have validity only in terms of a sort of combined subject of matter and energy, linked together by the now famous equation of Einstein's restricted theory of relativity of 1905.

As to whether the independent status of the first two principles to be discovered has any special significance, I shall not say. It may merely be the case that the historical accident of my writing now rather than one or ten years hence enables me to mention their independence: No one knows how physics will progress. But what is significant, I believe, is that there have been found only four true conservation principles in physics, and that of these only two survive in independence so that perhaps we should say now there are only three such principles. Furthermore, all four were set forth in the first two centuries of the modern scientific era. This leads to the conclusion that this type of principle may not be very common. And so, if the notion of conservation in general was of any real importance to the growth of scientific thought, and to the nineteenth century in particular, the reason must be that these principles, though few in number, had wide applications or far-reaching results, or that their implications in the philosophy of science must have been very great, or that similar principles arose in other branches of science.[18]

Another aspect of this set of four conservation laws in physics is their essential unsimilarity. In the case of charge there is a polarity, two different types of charge which may combine to produce the effect of no charge at all. We shall return to this particular problem, but now let us merely observe that there is not in nineteenth-century physics and chemistry a kind of ordinary or positive matter, and a negative matter, which when combined give the effect of no matter at all. Hence the law of conservation of matter, or better of mass, says that in any closed system the total amount of mass is constant and hence that the total inertia remains unchanged. This mass may not, like charge, disappear and appear. Momentum, however, is in some ways—though by no means all—like charge, at least with respect to conservation. In a frame of reference or coördinate system, to each positive momentum there may be an equal but opposite momentum of negative sign; these two momenta—like the equal positive and negative charges—may add up to a null result, as in the collision of two inelastic objects of equal masses moving toward one another with speeds of equal magnitudes.

Two of the most original thinkers in the field of philosophy of science in the nineteenth century were J. B. Stallo and Ernst Mach. Stallo today has little or no reputation in his own country, although he is at the moment quite often mentioned by name by those who have not read him. Mach knew his worth and introduced him to the German public in his own books and in a translation of Stallo's *Modern Physics*.[19] The man who really knows most about Stallo is our colleague Stillman Drake, and I trust that before long he will see to it that we get a good edition of Stallo to go alongside his Galileo. Stallo devoted a considerable amount of his book to the weaknesses of the atomic theory, in the Machian positivist tradition; and like Mach he also studied the history and philosophical implications of the doctrine of conservation of energy. According to Stallo, in his consideration of what he called the "First Principles of the Mechanical Theory of the Universe," the mechanical theory undertakes to "account for all physical phenomena by describing them as variances in the structure or configuration of material systems." In other words, the aim is to "apprehend all phenomenal diversities in the material world as varieties in the grouping of primordial units of mass, to recognize all phenomenal changes as movements of unchangeable elements, and thus to exhibit all apparent qualitative heterogeneity as mere quantitative differences." The "ultimates of scientific analysis" are, therefore, *"mass* and *motion,* which are assumed to be essentially disparate."* Save for the substitute word of mass for matter, we recognize the two catholic principles of Boyle, or of Hooke and Galileo, Huyghens and Newton. Now the next point made by Stallo is that the "prime postulate of all science is that there is some constant amid all phenomenal variation." In fact, "Science is possible only on the hypothesis that all change is in its nature transformation." The constancy of mass and motion thus became for Stallo one of the three basic axioms. "Scientific analysis," he said, "yields mass and motion as its absolutely irreducible elementary terms" and "it follows that both are quantitatively invariable. Accordingly the mechanical theory of the universe postulates the conservation of both mass and motion."[20]

In the remainder of my paper I shall illustrate these ideas with respect to the law of conservation of charge in the eighteenth century. Then I shall show some of the difficulties faced by the nineteenth century in attempting to discover what produces charge, what is actually conserved—and the way in which these considerations are related to the philosophy of science.

In the theory of electricity which took form under Franklin's leadership in the 1740's, the fundamental concept was the existence of an electric matter contained inside bodies along with their ponderable matter. This electric matter was said to be composed of particles which repel one another or was, in eighteenth-century language, *elastic;* it was also *subtle* because it could penetrate ordinary matter. Electrification was conceived to be a process in which a body either lost or gained some electric matter or electric fluid. Since each body was postulated as having a natural quantity of electric fluid, there must be three possible states with respect to electrification: the *null* state, in which the body has its normal or natural quantity of electric matter; the *minus* state, in which the body has lost some of its normal quota of electric matter; and the *plus* state in which a body has gained some electric matter so as to have more than its natural or normal amount. If two bodies are brought into contact, say one rubbed against the other, and there is a transfer of electric matter, then the matter lost by one body must be gained by the other. That is to say, the negative charge acquired by one of the bodies must be quantitatively equal to the positive charge gained by the other. This is one example of the conservation of charge, the production of two charges of equal and opposite sign at the same time. The reverse example also occurs. Two insulated charged conductors with equal and opposite charges are brought together. One body returns to its neutral state by acquiring the excess fluid from the other, which simultaneously—of course—becomes neutral. Clearly this theory apprehends such phenomenal diversities as the difference between an electrified and a non-electrified glass rod, one attracting light objects and the other not, as varieties in the grouping of primordial units, the particles of electric fluid. Furthermore, this theory exhibits all apparent heterogeneity, as between a charged and an uncharged rod, or charges of different sign, as a mere quantitative difference—in the amount of electric fluid. Now in all cases of electrification, or changes in electrification, it is postulated that there is an alteration in the total amount of electric fluid in a body. The difference between a body's normal or natural quantity of electric fluid, and the amount after it has lost or gained some, is called the "charge," which is merely the net quantity of electric fluid gained or lost. Since whatever is lost by one body is gained by another, it follows that the electric fluid is conserved. In other words, it does not make much sense to speak separately of the concept of charge in terms of the electric fluid and the law of conservation of charge—as if one could exist without the other. This is in contrast to motion (defined by momentum) or to mass (defined by inertia). Yet all such principles need to be exhibited as valid by confrontation with experimental fact.[21]

I shall indicate only two of the many examples considered by Franklin. The first is the condenser or Leyden jar. In it there are two conductors separated by the glass of the bottle. One of the conductors is grounded and the other brought into contact with—usually—the positively charged prime conductor of an electric machine or turbo-electric generator. The result is eventually that the conductor which had been grounded is found to have acquired a negative charge, by influence or by electrostatic induction, while the other conductor has by contact acquired a positive charge. Furthermore, these two charges are of equal and opposite amounts, as is shown by a variety of experiments. Hence, in terms of the electric fluid a charged jar does not actually contain any more electric fluid than an uncharged jar. *Second example.* An insulated and *uncharged* conductor (a metal tube) hangs from a support by silk strings. At one end there is a bunch of moist, hence conducting, thread. Bring a charged glass rod near one end of the hanging conductor, but not in contact with it, and the strings will diverge. Why? One of the properties of the electric fluid is that its particles are mutually repellent. Hence the excess electric fluid on or around the positively charged or electrified glass rod will exert a force of repulsion on the electric fluid normally contained in the hanging conductor. This fluid will therefore be pushed away from the glass rod; since the insulated conductor *is* a conductor, charges may move on it, but they cannot move off of it because it is insulated. These charges thus collect in the threads, making them positively charged, and the threads diverge. The end of the tube near the charged rod must, by conservation, be negatively charged, or have lost some of its electric fluid. But note that the total amount of electric fluid on the hanging conductor has not been altered. What then will happen when the charged rod is withdrawn? Evidently the electric fluid will merely return to its original natural distribution and there will be no sign of charge on the insulated conductor. This is verified by experience.

Professor Kuhn has referred to the correlation of forces and the convertibility of one type of energy into another. In Franklin's case, we may observe his study of the relations between heat and electricity and between light and electricity as an example of the way in which conservation ideas which were developed in one part of science prove useful in another. First: light. In a series of experiments designed to test the possible ways of discharging bodies, Franklin set a cannon ball on a glass cup or insulator. He then found that the charged ball could be discharged by one's breath, by smoke, by a stream of sand grains, by candlelight, but not by sunlight. Candlelight, he decided, discharges the ball since "parts of the candle" separated by the flame carry off particles of electric matter, as do the particles of moist air, smoke, and sand. Hence, sunlight could not consist of particles, since it produced no discharge of this sort. Franklin disliked the corpuscular theory of light in any event. First of all, if the gross matter of the sun were being converted into particles of light, as Newtonians said, then by conservation principles, the mass of the sun must be constantly diminishing and hence the planets would suffer a diminution in their gravitational attraction by the sun. Furthermore, the speed of light is 186,000 miles per second. Conceive the particles of light as small as you wish; even so, the force will exceed that of a twenty-four pounder discharged from a cannon, and yet these particles will not move the smallest pieces of dust floating in air.[22] Observation shows neither a diminution in the force between the sun and the planets, nor any effect of a collision between "particles of light" and any tiny bodies.

In respect to heat, Franklin studied the relative conductivity of different substances, approaching the subject—as he said—like an electrician. Most notable are his explanations that the subtle fluid of heat is conserved in mixtures: heat being transferred from hot bodies to cold bodies. In such mixtures heat *is* conserved, but there is no such simple approach to the production of heat by friction: Where does *that* heat come from? And where does the *animal* heat come from? One of his friends once asked him in this regard whether he could explain why water pumped out of a spring was cool at first and then became warmer. Was it heated by being pumped? He wisely refrained from giving an explanation until he could have some assurance about the phenomenon. He would rather have expected, he said, that the person pumping would become warm and not the water. Friction had long been known to produce heat, he added, but all experiments to heat water by shaking or agitating it had failed.[23]

❧

Franklin's theory of electricity, and the principle of conservation of charge, had been premised on the concept of a single electric fluid in excess or defect or in a normal amount. It is well known that at the century's end there had arisen a rival theory based on the concept of two electric fluids. In this theory too, there was a conservation principle: In fact there were at least *two* conservation principles, because neither of the two fluids could ever be created or destroyed. All electrifications or changes in electrification were supposed due to the transfer of one, or the other, or both of these fluids. But, in every case of an electrification, there was in this theory too a prediction that the appearance and disappearance of charges in any system must represent a total net change that is null. This was a necessary condition because, as was said earlier, experiment shows the appearance and disappearance of simultaneous equal magnitudes of positive and negative charges.[24]

Now the situation with respect to these very different theories of electricity in the opening years of the nineteenth century may remind us of that holding with respect to the Ptolemaic versus the Copernican theories at the end of the sixteenth century, there being no experiment or observation to decide between them. Or, if you wish, between the Tychonic and the Copernican system after Galileo's telescope had shown that Venus's orbit encompasses the sun.

It would seem, to judge from the usual secondary accounts of science in the nineteenth century, that with respect to the two rival theories of electricity, men would have sought for some crucial experiment or observation to decide between them. Much has been made of the allegedly crucial experiment of measuring the speeds of light in free space and in water in the mid-nineteenth century as just such a test to decide between the wave and corpuscular theories of light. And from this point of view the almost contemporaneous determination of the annual parallax of a fixed star can be considered as a kind of crucial experiment to decide between the Copernican and the Tychonian theories of the universe. Neither of these statements is in any way more of an exaggeration than the other.[25]

In point of fact, the reaction of many nineteenth-century scientists to the problem of the theories of electricity was quite different. In a famous textbook of elementary physics the Abbé Haüy, founder of scientific crystallography, advocated the two-fluid theory of electricity even though ". . . the existence of these two fluids is not founded upon . . . the satisfactory reasons." He adopted this hypothesis, so he said, because it "conduces to a simple and plausible manner of representing the results of experience. . . ."[26] Another statement of this point of view was written by Olinthus Gregory, an advocate of the one-fluid theory: "It is of comparatively little importance," he wrote, "whether one fluid or two component fluids really exist, or be merely hypothetical." All that is necessary is that "the assumed hypothesis enables us faithfully and satisfactorily to exhibit and connect the results of experiment."[27] The same point of view was adopted by Thomas Young, who also favored the one-fluid theory. Young made it clear that there was no experimental basis for asserting the existence of a single electric fluid. Nor even on the assumption that there is such a fluid, is there any experiment to decide which of two electrified bodies is charged by a gain of something or a loss of something. Franklin had long before pointed to this major problem in his theory, that there was no basis in experiment for deciding whether the glass rubbed by silk is the positively charged member of the pair or the silk that rubbed the glass. Young was aware of this difficulty and stated his position as follows: "It is in fact of little consequence to the theory whether the terms positive or negative be correctly applied, provided that their sense remain determined; and that, like positive and negative quantities in mathematics, they always be understood of states which neutralise each other."[28] Even later in the century, Joseph Henry could maintain that it did not matter very much which of these two alternative hypotheses was adopted since they were mathematically equivalent. This attitude in the first half of the nineteenth century represents a degree of positivism that may prepare us for the later philosophical views of Stallo, Mach, Ostwald, and others.

The general preference of mathematical electricians in the early nineteenth century was for the two-fluid theory, which was much more symmetrical than the one-fluid theory which had to postulate two separate processes, gain and loss of electric fluid, for the two types of electrification. Furthermore, the phenomena of electrolysis, intensively studied at this period, led more naturally to a dualistic theory: the cations and anions respectively receiving a bit of negative or positive electricity at the two electrodes. In 1830, Thomas Thomson explained that it was "chiefly in consequence of the galvanic discoveries" that the hypothesis of two fluids was "gradually superseding the Franklinian hypothesis." Yet even Thomson adopted a somewhat positivist point of view when, having said of the hypothesis of two fluids that the "probability of its accord with truth is very great," he added that even though the concept of two electric fluids is only a mathematical hypothesis, it may be used in physics since it "enables us to reduce all the electrical facts into a simple and luminous system, and to foretell the result of any combination of electrical actions."[29]

Faraday did not adopt so sophisticated a point of view. He observed the differences that existed among the several advocates of the two-fluid hypothesis and, to use his own words, "sought amongst the various experiments for any which might be considered as sustaining the theory of two electricities rather than that of one." But he could not find "a single fact which could be brought forward for such a purpose." This does not mean that Faraday was a simple advocate of the one-fluid theory. He said on occasion that he gave "no opinion respecting the nature of the electric current." Although he spoke of the current "as proceeding from the parts which are positive to those which are negative," he said he did so "merely in accordance with the conventional, though in some degree tacit, agreement entered into by scientific men, that they may have a constant, certain, and definite means of referring to the direction of the forces of that current."[30]

Clerk Maxwell adopted a somewhat different point of view. He recognized that electricity behaved as if it were a substance. Like ordinary matter, therefore, its primary quality is to be conserved. But he warned his readers against assuming too hastily that electricity "belongs to any known category of physical quantities."[31] Conservation made it seem as if there were a *substance* of electricity, but what of the two electricities? Should one assume that there are two substances? We cannot, said Maxwell, "conceive of two substances annulling each other."[32] Although at first definitely favoring the one-fluid theory rather than the two-fluid theory,[33] Maxwell ended up by telling his readers he had introduced "the common phrase *electric fluid* only for the purpose of warning against it."[34] If one did not believe in a fluid or substance of electricity, or in two such fluids or substances, what was the significance of conservation of charge? What was being conserved?

Despite Maxwell's attempts to avoid the concept of electrical matter, the discussion of electrolysis led him to the idea of a "molecule of electricity." This phrase, though it was used by Maxwell, was said by him to be "gross" and "out of harmony with the rest of this treatise."[35] Such were the difficulties of nineteenth-century writers on electricity.[36]

It is small wonder that the physics of the nineteenth century gave rise to a positivist attitude, with the result that even the notion of conservation was said by Mach to be only an instance of "the economy of thought" for the reason that a "mere unrelated change, without fixed point of support, or reference, is not comprehensible, not mentally reconstructible."[37]

I believe that the history of nineteenth-century science makes it clear that the philosophical point of view of such men as Stallo, Mach, Duhem, Poincaré, and others, is only mistakenly considered in relation to the subsequent development of physics. Their philosophy was much rather a reaction to the problems that had been raised by the science of their own century. To illustrate this point, I will examine briefly the discussion of the law of conservation of charge and of the nature of electricity by certain major philosophers of science at the end of the nineteenth century and the beginning of the twentieth: Mach, Poincaré, and Duhem.

Ernst Mach's best known and most influential book is his *Science of Mechanics*. Also widely read are his *Conservation of Energy* and *Popular Scientific Lectures*. In his small book on the conservation of energy, electricity enters at two important stages of the argument. First Mach raises the question which, he says, bright students have sometimes put to him: "Is there a mechanical equivalent of electricity as there is a mechanical equivalent of heat?" The reply is, "There is no mechanical equivalent of *quantity* of electricity as there is an equivalent *quantity* of heat, because the same quantity of electricity has a very different capacity for work, according to the circumstances in which it is placed; but there *is* a mechanical equivalent of electrical energy."[38] Next he proceeds "to establish this assertion": that the "difference of view in our treatment of heat and of electricity" arises from a "reason [that] is purely historical, wholly conventional, and, what is still more important, is wholly indifferent." Hence electricity served Mach as an illustration of the arbitrary nature of scientific concepts.

The second use of electricity was in an important aside apropos of the arguments against the atomic-molecular theory. "Perhaps," Mach said, "the reason why, hitherto, people have not succeeded in establishing a satisfactory theory of electricity is because they wished to explain electrical phenomena by means of molecular events in a space of three dimensions."[39]

371

The same points are raised in the lecture "On the Principle of the Conservation of Energy" in Mach's *Popular Scientific Lectures*. Had the order of inventions been reversed, he said,

. . . what would be more natural than that the "quantity" of electricity contained in a jar should be measured by the heat produced in the thermometer? But then, this so-called quantity of electricity would decrease on the production of heat or on the performance of work, whereas it now remains unchanged; in that case, therefore, electricity would not be a *substance* but a *motion,* whereas now it is still a substance. The reason, therefore, why we have other notions of electricity than we have of heat, is purely historical, accidental, and conventional.[40]

Mach's concern with the nature of electricity appeared in two other lectures. In discussing "Comparison in physics," reference is made to the way in which "Ohm forms his conception of the electric current in imitation of Fourier's."[41] In order to present his views on "mental adaptation," Mach used the example of the old concept of electric fluid:

The electric current was conceived originally as the flow of a hypothetical fluid. To this conception was soon added the notion of a chemical current, the notion of an electric, magnetic, and an isotropic optical field, intimately connected with the path of the current. And the richer a conception becomes in following and keeping pace with facts, the better adapted it is to anticipate them. . . .

The ideas that have become most familiar through long experience, are the very ones that intrude themselves into the conception of every new fact observed. In every instance, thus, they become involved in a struggle for self-preservation, and it is just they that are seized by the inevitable process of transformation.

Upon this process rests substantially the method of explaining by hypothesis new and uncomprehended phenomena. Thus, . . . we imagine electrified bodies to be freighted with fluids that attract and repel, or we conceive the space between them to be in a state of elastic tension. In so doing, we substitute for new ideas distinct and more familiar notions of old experience—notions which to a great extent run unimpeded in their courses, although they too much suffer partial transformation.[42]

A whole lecture was devoted to "the fundamental concepts of electrostatics." It contains Mach's clearest statement on the usefulness and the limitations of "mental pictures" in the advance of physics. The "picture" is that of the two electric fluids postulated by Coulomb. Says Mach:

In the fact that this conception reproduces, lucidly and spontaneously, all the data which arduous research only slowly and gradually discovered, is contained its advantage and scientific value. With this, too, its value is exhausted. We must not seek in nature for the two hypothetical fluids which we have added as simple mental adjuncts, if we would not go astray. Coulomb's view may be replaced by a totally different one, for example, by that of Faraday, and the most proper course is always, after the general survey is obtained, to go back to the actual facts, to the electrical forces.[43]

I shall not attempt to summarize Mach's views on electricity, but shall conclude with a presentation of Mach's statement of the importance of the subject: "Particularly are electrical phenomena connected with all other physical events; and so intimate is this connexion that we might justly call the study of electricity the theory of the general connexion of physical processes."[44]

Henri Poincaré's major views on electricity are mainly to be found in his *Électricité et optique: la lumière et les théories électrodynamiques,*[45] of which the famous introduction begins:

La première fois qu'un lecteur français ouvre le livre de Maxwell, un sentiment de malaise, et souvent même de défiance se mêle d'abord à son admiration. Ce n'est qu'après un commerce prolongé et au prix de beaucoup d'efforts, que ce sentiment se dissipe. Quelques esprits éminents le conservent même toujours.

Pourquoi les idées du savant anglais ont-elles tant de peine à s'acclimater chez nous? C'est sans doute que l'éducation reçue par la plupart des Français éclairés les dispose à goûter la précision et la logique avant toute autre qualité.

That Poincaré recognized the fundamental character of conservation is shown by his many references to this principle. For example, in exhibiting the virtues of the theory of Lorentz, he said: "Cette théorie rend compte . . . du principe de la conservation de l'électricité, puisque l'hypothèse fondamentale n'est autre chose, après tout, qu'une traduction de ce principe lui-même."[46]

Portions of the introduction to this book, and of another on light, were combined to form chapter XII of Poincaré's famous philosophical work, *Science et hypothèse.*[47] The final chapter (XIII) is devoted to electrodynamics. Students of Poincaré's writings will recall that much of *Science et méthode* is devoted to theories of the electron, while chapter III ("The new mechanics and astronomy") opens with a perceptive comparison of Franklin's hypothesis and that of Lorentz. For Poincaré, as for Mach, the question of the nature of the electric charge was of major philosophical importance and he was concerned in a most significant way with the consequences of the discovery of the electron in relation to classical ideas of time, space, mass, atoms, and even force.[48]

One of Pierre Duhem's[49] major scientific works was entitled *Leçons sur l'électricité et le magnétisme*.[50] Although Duhem is usually conceived as a philosopher whose main scientific basis of thought was the new "energetics," he tells us in this work:

> En 1811, Poisson inaugura la théorie des phénomènes électriques; depuis ce temps, l'étude des lois auxquelles obéissent l'Électricité et le Magnétisme a suscité les efforts d'une foule de grands physiciens et de grands analystes. Leurs innombrables travaux se sont succédé sans relâche pendant toute la durée du siècle et dans tous les pays; leurs découvertes forment aujourd'hui l'un des plus vastes ensembles scientifiques qui soit au monde.
>
> Le moment semble venu de coordonner les résultats de tant d'efforts; de réunir en un faisceau unique ces recherches conçues d'après les idées les plus diverses, écrites dans toutes les langues, dispersées dans toutes les revues. Il semble que, si l'on parvenait à réaliser cette vaste synthèse, on se trouverait en présence du plus beau système de Philosophie naturelle qui ait jamais été engendré par l'esprit humain.
>
> Dans le présent Ouvrage, nous avons cherché, dans la limite de nos forces, à tracer une première ébauche de cette synthèse.[51]

The beginning of the first chapter is devoted to a dismissal of the older concept of the electric fluid,[52] followed by the introduction of an integral representing what is defined as "la quantité totale d'électricité que renferme le système." According to Duhem,

> Nous admettrons que cette quantité demeure invariable dans une modification quelconque du système.
>
> Cette hypothèse a été introduite en Physique par Franklin, lorsqu'il a admis que tout phénomène électrique pouvait être représenté par la formation de quantités égales de fluide positif et de fluide négatif aux dépens du fluide neutre. Depuis l'époque de Franklin, les physiciens n'ont cessé de l'admettre, explicitement ou implicitement, et d'en faire un fréquent usage dans leurs théories. C'est ainsi que, dans un Mémoire d'Ohm, paru en 1827, Mémoire dont nous aurons bientôt à signaler l'importance, nous trouvons ce principe très explicitement énoncé. "C'est une condition, dit Ohm, au cours de l'un de

ses raisonnements, qui résulte de ce principe fondamental, que les deux électricités se développent toujours dans le même temps en quantités égales." De nos jours, on a donné à cette hypothèse le titre de *Principe de la conservation de l'électricité.*

The writings of Duhem, like those of Poincaré, show us the need for a scholarly analysis of the implications of Maxwell's work in French scientific thought.[53]

Although, in this paper, the discussion has been limited to the problems associated with the concepts of electric charge and conservation of charge, electricity was of philosophical importance in the nineteenth century in relation to electromagnetism and particularly the electromagnetic theory of light. It has been mentioned above that the heat energy developed in current-carrying wires, as in the production of mechanical energy by electric motors, served to illustrate the general doctrine of conservation of energy.[54] But at the century's end, what captured men's minds was the concept of a unification of diverse types of physical phenomena in Maxwell's theory. Many would have agreed with John Trowbridge in 1896 when he said:

> Maxwell's theory that light and heat are phenomena of electro-magnetic waves which come to us from the sun is now the greatest generalization in physical science, and in stating it Maxwell lighted a torch which has illumined many hitherto dark regions. It is safe to affirm that the entire world is now working upon this great hypothesis. According to the electro-magnetic theory of light, the only difference between light, heat, and electricity consists in the length of waves in the ether of space. The sun is the source of electro-magnetic waves, and the earth is the scene of transformations of electric energy. A piece of coal burning in a grate has therefore a long electro-magnetic history. It owed its origin to electro-magnetic waves, and in burning it gives out again electro-magnetic waves, of which we can only detect the light and heat manifestations.[55]

According to Trowbridge, "the present views of scientific men" are that "the continuance of all life on the earth is due to the electrical energy which we receive from the sun; and physics, in general, can be defined as that subject which treats of the transformations of energy." He therefore "presented the varied phenomena of electricity in such a manner that the reader can perceive the physicist's reasons for supposing that all space is filled with a medium which transmits electro-magnetic waves to us from the sun."[56]

Electricity served in the nineteenth century as a basis for attempts to understand the forces of chemical combination and the transmission of nervous impulses. Apart from the practical applications in technological devices, the development of electricity as a science was of major importance in so many different areas of scientific thought that it may fairly be called *the* science of the nineteenth century. The ultimate failure of the electromagnetic theory of radiation led to quantum theory, which was seen to be "a complete departure from the old Newtonian system of mechanics."[57] The mid-twentieth-century reader may wonder, therefore, at the enthusiasm of the men of the 1890's for a theory that was about to collapse under its failures and that had already exhibited dangerous weaknesses and insoluble puzzles such as the very nature of the aether that vibrated with electromagnetic radiation. But, in the last decade of the nineteenth century the triumphs of the electromagnetic theory were great enough to overshadow most doubts and perplexities. The physics of the nineteenth century was apparently ending at its moment of greatest triumph.

References

1 Perhaps one of the major reasons for the situation with respect to nineteenth-century science is that such historical studies are apt to require so great a technical competence in science that many historians of science do not attempt them, abandoning the field to the scientist, whose approach to history often suffers from the consequences of a purely scientific training. In many ways, the best attempt to comprehend the science of the nineteenth century is still John Theodore Merz, *A History of European Thought in the Nineteenth Century* (3rd ed., 2 vols.; Edinburgh, London: William Blackwood & Sons, 1907). Also useful are Albert Bordeaux, *Histoire des sciences physiques, chimiques et géologiques au XIX* siècle* (Paris et Liége: Librairie Polytechnique ch. Béranger, 1920), Franz Carl Müller, *Geschichte der organischen Naturwissenschaften im Neunzehnten Jahrhundert: Medizin und deren Hilfsswissenschaften, Zoologie und Botanik* (Berlin: Georg Bondi, 1902), Ernst Cassirer, *The Problem of Knowledge: Philosophy, Science, and History since Hegel,* tr. William H. Woglom and Charles W. Hendel (New Haven: Yale University Press, 1950), Alfred Russel Wallace, William Ramsay, William Mathew Flinders-Petrie, *et. al., The Progress of the Century* (New York, London: Harper & Brothers, 1901), Alfred Russel Wallace, *The Wonderful Century: Its Successes and its Failures* (New York: Dodd, Mead & Company, 1899).
2 *Annalen der Physik,* XVII (1905), 891–921.

3 Lorentz's paper is reprinted from the English version in the *Proceedings* of the Academy of Sciences of Amsterdam, 1904, in H. A. Lorentz, A. Einstein, H. Minkowski, H. Weyl, *The Principle of Relativity: A Collection of Original Memoirs on the Special and General Theory of Relativity,* with notes by A. Sommerfeld, tr. W. Perrett and G. B. Jeffery (London: Methuen & Co., 1923). That Einstein was ignorant of Lorentz's paper when writing his own is stated in a footnote on p. 38 of the above-mentioned book.

4 W. Kaufmann, "Die elektromagnetische Masse des Elcktrons," *Physikalische Zeitschrift,* IV (1902), 54–57; "Ueber die 'Elektromagnetische Masse' der Elektronen," *Nachrichten von der Königl. Gesellschaft der Wissenschaften zu Göttingen* (Math.-phys. Kl.) (1903), pp. 90–103.

5 For an account of the development of physical thought, and of science in general, during this period, see Robert Andrews Millikan, *Electrons (+ and −), Protons, Photons, Neutrons, Mestrons, and Cosmic Rays* (Chicago: University of Chicago Press, 1947—or the earlier editions, beginning in 1917 under the simpler title, *The Electron*); Sir Arthur Schuster, *The Progress of Physics During 33 Years (1875–1908)* (Cambridge, Eng.: The University Press, 1911); William Wilson, *A Hundred Years of Physics* (London: Gerald Duckworth & Co., 1950); Herbert Dingle (ed.), *A Century of Science, 1851–1951* (London, New York: Hutchinson's Scientific and Technical Publications, 1951); A. E. Heath (ed.), *Scientific Thought in the Twentieth Century* (New York: Frederick Ungar Publishing Co., [1951] 1954).

6 See Thomson's paper, "On the existence of masses smaller than atoms," read at the Dover meeting of the British Association for the Advancement of Science. Of his many early statements, the following expresses Thomson's views as clearly as any: "The explanation which seems to me to account in the most simple and straightforward way for the facts is founded on a view of the chemical elements which has been favourably entertained by many chemists: this view is that the atoms of the different chemical elements are different aggregations of atoms of the same kind. In the form in which this hypothesis was enunciated by Prout, the atoms of the different elements were hydrogen atoms; in this precise form the hypothesis is not tenable, but if we substitute for hydrogen some unknown primordial substance X, there is nothing known which is inconsistent with this hypothesis. . . . Thus on this view we have in the cathode rays matter in a new state, a state in which the subdivision of matter is carried very much further than in the ordinary gaseous state: a state in which all matter—that is, matter derived from different sources such as hydrogen, oxygen, etc.—is one and the same kind; this matter being the substance from which all the chemical elements are built up" (Quoted from Lord Rayleigh, *The Life of Sir J. J. Thomson* [Cambridge, Eng.: The University Press, 1942], p. 91).

7 See Thomson's *Recollections and Reflections* (London: G. Bell & Sons, 1936), p. 341: "At first there were very few who believed in the existence of these bodies smaller than atoms."

8 If the charge on a single monovalent ion (e.g., H^+, Na^+, Cl^-) is one electron

charge $\pm e$, then—since 96,500 coulombs [Faraday's constant F] deposits or liberates M grams of any monovalent element, where M is the atomic weight—$\frac{M}{N}$ grams should give the mass of an atom of a monovalent element; N is Avogadro's [or Loschmidt's] number, $N = \pm F/e$. Once the mass of a single atom of any element was known, the table of "atomic weights" served to give the mass of an atom of any other element. Since that time, of course, the above statements require amendment for the existence of isotopes.

9 See Ernst Mach, *History and Root of the Principle of the Conservation of Energy,* tr. Philip E. B. Jourdain (Chicago, London: The Open Court Publishing Co., 1911), pp. 86–87; *Popular Scientific Lectures,* tr. Thomas J. McCormack (La Salle, Illinois: The Open Court Publishing Co., 1910), p. 207; *The Science of Mechanics,* tr. Thomas J. McCormack (Chicago, London: The Open Court Publishing Co., 1907), p. 492. For an account of Albert Einstein's discussion with Mach of the concept of atom, see I. Bernard Cohen, "An interview with Einstein," *Scientific American,* 193 (1955), 72–73.

Typical also of this point of view is the statement of Sir William Thomson (later Lord Kelvin) in an article on "contact electricity" (originally published in the *Proceedings of the Literary and Philosophical Society of Manchester* in 1862, reprinted in Thomson's *Reprint of Papers on Electrostatics and Magnetism* [2nd ed.; London: Macmillan & Co., 1884], §400): "I can now tell the amount of the force, and calculate how great a proportion of chemical affinity is used up electrolytically, before two such discs come within 1/1000th of an inch of one another, or any less distance down to a limit within which molecular heterogeneousness becomes sensible. This, of course, will give a definite limit for the sizes of atoms, or rather, as I do not believe in atoms, for the dimensions of molecular structures."

10 Quoted in Ida Freund, *The Study of Chemical Composition: An Account of its Method and Historical Development, With Illustrative Quotations* (Cambridge, Eng.: The University Press, 1904), p. 624.

11 Wilhelm Ostwald, *Principles of Inorganic Chemistry,* tr. by Alexander Findlay (London, New York: Macmillan & Co., 1902), pp. 146–48.

12 Wilhelm Ostwald, *Outlines of General Chemistry,* tr. W. W. Taylor (3rd ed.; London: Macmillan & Co., 1912), p. vi. Ostwald added, "That I have considered it expedient to give it [the atomic hypothesis] this place in the above-mentioned new chapters, in which the proofs are to be found, is scientifically justifiable. From the point of view of stoichemistry the atomic theory is merely a convenient mode of representation, for the facts, as is well known, can be equally well, and perhaps better, represented without the aid of the atomic conception as usually advanced. In this part of the book, then, I have made little use of the atomic theory. . . ." In the text proper (p. 540), Ostwald said, "The old chemical hypothesis that all the elements are compounds of a primal matter has been revived very recently in consequence of the transformations of the radioactive elements . . . , the objection based on the irrational ratios of the atomic weights has lost its force, since the mass of the electron is so small that the lightest of

all atoms, that of hydrogen, must, according to this theory, consist of about 1000 electrons."

13　A statement of Frederick Soddy, quoted by Freund, *Study of Chemical Composition*, p. 613. More information on this topic may be found in John Howe Scott, "The Nineteenth-Century Atom: Undivided or Indivisible," *Journal of Chemical Education*, XXXVI (1959), 64–67.

14　For the exact sense in which it may be said that Thomson "discovered" the electron, see Lord Rayleigh, *Life of Sir J. J. Thomson*, pp. 77 ff.

15　To see how truly radical Einstein's proposal was, recall that he presented his new concept of photon from only a "heuristic" point of view—the word "heuristic," be it noted, is not even found commonly in scientific papers of the period. A second gauge of the radical character of Einstein's new concept is that there is no claimant to simultaneous independent discovery. Some would give the credit for special relativity to Lorentz or to Poincaré (e.g., Sir Edmund Whittaker, *A History of the Theories of Aether and Electricity*, [London: Thomas Nelson, 1953], Vol. 2), and Einstein himself said that if he had not discovered special relativity, the French physicist Paul Langevin would have.

16　One further aspect of the importance of the science of electricity for the growth of general ideas in physical science is the study made during the closing years of the nineteenth century of the possible electromagnetic "nature of mass."

17　This principle should be credited to John Wallis.

18　I purposely omit from this discussion the variational principles, such as the law of least constraint, Le Châtelier's principle, the principle of virtual work, the law of least action, and also the older law of conservation of *vis viva,* and such principles as Lenz's law.

19　See the writer's discussion of Stallo and Mach in "Ethan Allen Hitchcock," *Proceedings,* American Antiquarian Society, 1951, LXI (1952), 29–136; especially p. 32 for details of Mach's learning about Stallo. Stallo's book appeared in German as *Die Begriffe und Theorien der Modernen Physik*. Nach der 3. Auflage d. englischen Originals übers. u. herausg. v. Dr. Hans Kleinpeter. Mit einem Vorwort von Ernst Mach. Mit einem Porträt d. Verfassers (Leipzig: Verlag von Johann Ambrosius Barth, 1901). The French edition is entitled *La Matière et la Physique Moderne.* Avec une préface sur la théorie atomique par C. Friedel (Paris: Félix Alcan, Editeur, 1884). Mach included an autobiographical letter of Stallo's in his foreword. A second edition of the German translation appeared in 1911. Mr. Drake's study of Stallo will be included in the forthcoming *Essays in the History of Science* (University of Washington Press).

20　J. B. Stallo, *Concepts and Theories of Modern Physics* (New York: D. Appleton, 1882), p. 25. A new edition of Stallo's *Concepts of Modern Physics,* edited by P. W. Bridgman, has been announced by the "John Harvard Library," to be published by the Belknap Press of Harvard University Press.

21　See *Benjamin Franklin's Experiments, a New Edition of Franklin's "Experiments and Observations on Electricity,"* ed. I. Bernard Cohen (Cambridge: Harvard University Press, 1941), and I. B. Cohen, *Franklin and Newton, an Inquiry into Speculative Newtonian Experimental Science and Franklin's Work in Electricity as an Example Thereof* (Philadelphia: American Philosophical Society, 1956), Ch. 8, 10.

22 For details, see the writer's *Franklin and Newton,* pp. 320ff.

23 *Ibid.,* pp. 323ff., 326ff., 334ff.

24 The name of "Law of conservation of charge" was apparently introduced late in the nineteenth century. For an account of the rise of the two-fluid concept of electricity, see *Franklin and Newton* ch. 12. The best historical study of the concept of "electric charge" is Duane Roller and Duane H. D. Roller, *The Development of the Concept of Electric Charge* (Cambridge: Harvard University Press, 1954). Of great interest for any student of the electrical theories of the nineteenth century are: Edmund Hoppe, *Geschichte der Elektrizität* (Leipzig: Johann Ambrosius Barth, 1884); Ferd. Rosenberger, *Die moderne Entwicklung der elektrischen Principien. Fünf Vorträge* (Leipzig: Verlag von Johann Ambrosius Barth, 1898); Sir Edmund Whittaker, *A History of the Theories of Aether and Electricity,* Vol. 1, *The Classical Theories* (New York: Philosophical Library; Edinburgh, London: Thomas Nelson & Sons, 1951); Paul F. Mottelay, *Bibliographical History of Electricity and Magnetism, Chronologically Arranged* (London: Charles Griffin; Philadelphia: J. B. Lippincott, 1922).

25 This topic will be discussed at greater length in a forth-coming study of optical theories in the last three centuries.

26 René-Just Haüy, *Traité élémentaire de physique* (Paris: Delance et Lesueur, 1803—a second edition, Paris: Chez Courcier, 1806), English translation by O. Gregory, *An Elementary Treatise on Natural Philosophy* (London: George Kearsley, 1807).

27 Translator's note in English edition of Haüy's *Treatise,* I, 347.

28 Thomas Young, *A Course of Lectures on Natural Philosophy* (London: Joseph Johnson, 1807), Lecture XIV.

29 Thomas Thomson, *An Outline of the Sciences of Heat and Electricity* (London: Baldwin & Cradock; Edinburgh: William Blackwood, 1830), pp. 347–48.

30 Michael Faraday, *Experimental Researches in Electricity,* (3 vols.; London: [1 & 2] Richard and John Edward Taylor; [3] Richard Taylor and William Francis, 1838, 1844, 1855), pp. 511, 516, 667.

31 James Clerk Maxwell, *A Treatise on Electricity and Magnetism* (Oxford: Clarendon Press, 1873), art. 35.

32 *Ibid.,* art. 36.

33 That is, Maxwell in his *Treatise* (art. 37) showed that the two-fluid theory had major faults and that the critics of the one-fluid theory had been in error.

34 This point of view appears in the text of his lectures on electricity given at the Cavendish Laboratory, published posthumously, *An Elementary Treatise on Electricity* (2nd ed.; Oxford: Clarendon Press, 1888), art. 10.

35 Maxwell had held that "electrolysis appears the most likely to furnish us with a real insight into the true nature of the electric current, because we find currents of ordinary matter and currents of electricity forming essential parts of the same phenomenon." As to "the fact of the constant value of the molecular charge," one was tempted to call this "constant molecular charge, for con-

venience in description, *one molecule of electricity."* This phrase, he said, "will enable us at least to state clearly what is known about electrolysis, and to appreciate the outstanding difficulties." See Maxwell, *Treatise* Pt. 2, Ch. 4, arts. 255–60. Typical of the bewilderment of men of science of this period is Hermann von Helmholtz, "On the modern development of Faraday's conception of electricity," *Journal of the Chemical Society,* XXXIX (1881), 277–304. An analysis of the conflict between "atomic" or "molecular" theories of electricity and the Faraday-Maxwell views on electricity, as seen in the writings of the second half of the nineteenth century, may be found in *Franklin and Newton,* ch. 12.

36 McCormack (tr.), *Science of Mechanics,* p. 504.

37 See footnote 9 above. *The Analysis of Sensations and the Relation of the Physical to the Psychical,* tr. C. M. Williams, revised and supplemented by Sidney Waterlow (Chicago, London: The Open Court Publishing Co., 1914) and *Space and Geometry,* tr. Thomas J. McCormack (Chicago: The Open Court Publishing Co.; London: Kegan Paul, Trench, Trübner & Co., 1907) have been less widely read. *The Principles of Physical Optics: An Historical and Philosophical Treatment,* tr. John S. Anderson and A. F. A. Young (London: Methuen & Co., 1926; paper-backed reprint now available from Dover Publications, New York) is a strange book since it omits ". . . radiation, the decline of the emission theory of light, Maxwell's theory, together with relativity," which subjects are said to be treated in a "subsequent part" which was never completed. In the preface, Mach said: "The questions and doubts arising from the study of these chapters formed the subject of tedious researches undertaken conjointly with my son, who has been my colleague for many years. It would have been desirable for the collaborated second part to have been published almost immediately, but I am compelled, in what may be my last opportunity, to cancel my contemplation of the relativity theory. I gather from the publications which have reached me, and especially from my correspondence, that I am gradually becoming regarded as the forerunner of relativity. I am able even now to picture approximately what new expositions and interpretations many of the ideas expressed in my book on Mechanics will receive in the future from the point of view of relativity. It was to be expected that philosophers and physicists should carry on a crusade against me, for, as I have repeatedly observed, I was merely an unprejudiced rambler, endowed with original ideas, in varied fields of knowledge. I must, however, as assuredly disclaim to be a forerunner of the relativists as I withhold from the atomistic belief of the present day. The reason why, and the extent to which, I discredit the present-day relativity theory, which I find to be growing more and more dogmatical, together with the particular reasons which have led me to such a view—the considerations based on, the physiology of the senses, the theoretical ideas, and above all the conceptions resulting from my experiments—must remain to be treated in the sequel. The ever-increasing amount of thought devoted to the study of relativity will not, indeed, be lost; it has already been both fruitful and of permanent value to mathematics. Will it, however, be able to maintain its

position in the physical conception of the universe of some future period as a theory which has to find a place in a universe enlarged by a multitude of new ideas? Will it prove to be more than a transitory inspiration in the history of this science?" Mach's other two books, *Erkenntnis und Irrtum* (Leipzig: Verlag von Johann Ambrosius Barth, 1905) and *Die Principien der Wärmelehre.* 4. Auflage (Leipzig: Verlag von Johann Ambrosius Barth, 1923), are both of sufficient interest to warrant translation into English.

A complete bibliography of Mach's writings may be found in Hans Henning, *Ernst Mach als Philosoph, Physiker, und Psycholog* (Leipzig: Verlag von Johann Ambrosius Barth, 1915).

38 Mach, *History and Root of the Principle of the Conservation of Energy,* pp. 44, 46.

39 *Ibid.,* p. 54.

40 McCormack (tr.), *Popular scientific lectures,* pp. 169–70.

41 *Ibid.,* pp. 249–50.

42 *Ibid.,* pp. 226–27, 228–29.

43 *Ibid.,* pp. 113–14.

44 *Ibid.,* p. 131.

45 Deuxième édition, revue et complétée par Jules Blondin et Eugène Néculcéa (Paris: Gauthier-Villars, 1901). Poincaré advocated tentatively a theory based on two electric fluids, thus departing from "la théorie du *fluide unique,* à laquelle se rattache la théorie de Maxwell."

46 *Ibid.,* p. 599.

47 Poincaré's three major works on the philosophy of science—the two mentioned above and *La valeur de la science*—have been translated into English by George Bruce Halsted and issued in a single volume under the general title *The Foundations of Science* (Lancaster, Pa.: The Science Press, 1913).

48 I ignore here the many attempts to make of Poincaré the founder of relativity theory. See Whittaker (footnote 15 above) and appendix 2 of René Dugas, *A History of Mechanics,* English version by J. R. Maddox (New York: Central Book Co., n. d.).

49 See Armand Lowinger, *The Methodology of Pierre Duhem* (New York: Columbia University Press, 1941).

50 Tome I: "Les corps conducteurs à l'état permanent" (Paris: Gauthier-Villars, 1891).

51 *Ibid.,* introduction, p. v.

52 *Ibid.,* p. 1: "Symmer et Franklin ont apporté le premier contingent à l'ensemble des hypothèses qui représentent les propriétés des corps électrisés. Ces hypothèses étaient, dans leur origine, intimement liées à des suppositions sur la nature de l'électricité; mais il est aisé aujourd'hui de briser ce lien, de laisser de côté ces suppositions sur la nature de l'électricité, suppositions si étrangères au véritable objet de la Physique, que cette science n'a même pas le droit d'en montrer la vanité; de ne laisser, enfin, aux hypothèses fondamentales que le caractère de définitions de paramètres analytiques qui est essentiellement le leur."

53 Only a glance at the index is necessary to convince any reader that electricity (notably the nature of the electric charge) is an important subject in Duhem's major philosophical work, *The Aim and Structure of Physical Theory,* tr. Philip P. Wiener (Princeton: Princeton University Press, 1954); on page 119, apropos of Franklin, there is a presentation and discussion of the law of conservation of charge.

54 The subject of electricity, to take one example, entered Balfour Stewart's *The Conservation of Energy* ("International Scientific Series," New York: D. Appleton & Co., 1890) to illustrate "the forces and energies of nature," special attention being called to the conservation of charge, p. 63, § 82.

55 John Trowbridge: *What is Electricity?* ("International Scientific Series," New York: D. Appleton & Co., 1896), p. 5.

56 *Ibid.,* p. v.

57 J. H. Jeans: *Report on Radiation and the Quantum-Theory* (London: "The Electrician" Printing & Publishing Co., 1914), p. 1.

CARL B. BOYER

on the Papers of Thomas S. Kuhn and I. Bernard Cohen

⬛ The two very thought-provoking papers presented by Professor Kuhn and Professor Cohen are actually quite different in character: The one presented a clear-cut thesis argued systematically about a single conservation law; the other called attention to some of the broader aspects of the general notion of conservation. But the two papers are so closely related that we shall discuss them together. Inasmuch as the framework of Professor Kuhn's paper is essential to his thesis, I shall make it the basis of my comment, with interpolations as we go along referring to points in Professor Cohen's lecture. Moreover, while I shall agree with virtually everything that Professor Kuhn says with respect to detail, I shall take issue in a fundamental way with the thesis as a whole; and the nature of this disagreement will be made clear by incorporating some of the material from Professor Cohen's paper into the framework of the paper of Dr. Kuhn.

Let us recall first that Professor Kuhn recognized a dozen discoverers of the conservation of energy, grouped into three categories which we designate as (A) Mayer, Joule, Colding, and Helmholtz; (B) Carnot, Séguin, Holtzmann, and Hirn, and (C) Mohr, Grove, Faraday, and Liebig. He also recognized three significant factors leading to the discovery which we shall list in order as (I) the availability of conversion processes; (II) the concern with engines, and (III) the philosophy of nature. I shall argue that material in the two papers suggests the inclusion of two additional significant factors in the discovery.

Before presenting the argument for two additional factors, I wish to call attention to the fact that the statement by Professor Kuhn of the meaning of the phrase "simultaneous discovery" is none too sharp. I have no quarrel with the use of the word simultaneous, for I agree that this shall mean simply "at about the same time." I do not insist on a Dedekind Cut in the time scale; and the simultaneity of discovery in our case is striking whether one's conception of "about the same time" allows for an interval of a year or a decade. My complaint is not with the word "simultaneous," but with the word "discovery." It is explained that the discovery meant different things to different men (which I take to be a truism), but what is not explained is what the discovery is to mean in this paper. Just what qualifies a man to be a discoverer? If you will allow me to analyze our discovery into three elements, I think it will be clear why I do not follow part of Professor Kuhn's thesis. The discovery of the Law of the Conservation of Energy (abbreviated as L. C. E., or "Elsie"), in perhaps over-simplified form, consists of three main steps: (1) the formation of the concept of energy, (2) the statement that energy is neither created nor destroyed, and (3) a reasonable justification of step 2. Now it must be granted immediately that not one of these three steps is without ambiguity; but I think that we need not worry about the ambiguity, for it is not likely to affect our thesis. For example, the word energy has not meant the same thing throughout the past century or more, but the differences are not sufficiently great to change our problem. Professor Kuhn's statement that "energy *is* conserved; nature behaves that way," seems to imply a notion of energy as something actually *in* nature which we discover, rather than as a concept which we invent as one appropriate means of describing nature. But it is enough that we agree to regard energy as real only in the sense that we can ascribe properties to it; and I believe that the relevant properties would pretty closely resemble each other in either case. I take it that all of the twelve "discoverers" listed above meet the discovery criterion which I have listed first, even though there are recognizable differences in conception as expressed in their works. Moreover, I feel that we certainly can accept Professor Kuhn's first factor as perhaps the chief one leading to our striking case of simultaneity. A concept as general as that of energy does not arise from one or two cases, but rather from a multiplicity of instances such as Professor Kuhn has well described. Nevertheless, there are some cases in which I don't quite like the phrase "conversion process," for it seems to put the cart before the horse. When, for example, Rumford's experiments with the boring of cannon showed that the motion was accompanied by heat, did the motion *become* heat or did it merely act as a catalytic agent to release the heat? Or when a wire moves in a magnetic field, does the motion necessarily *become* electricity? We say now that the energy of motion becomes the energy of heat, or that the energy of motion becomes electrical energy; but this confirms the notion that to speak of these as conversion processes before 1842 is an anachronism, for not until the concept of energy appeared could they really be said to refer to conversions of something into something else. But this is not important to our theses. (Incidentally, this question of conversion points up a fact upon which we seem to agree, but which is often misunderstood. The kinetic theory of heat was not a necessary factor in the development of "Elsie," for this law is independent of any particular view as to the nature of heat. Mayer himself rejected the view that heat could be a form of motion.)

Step 2 as given above has some degree of ambivalence, but this is easily recognized. If one insists on the interconvertibility of all forms of energy, including our modern atomic energy, then I suspect that no one has been or can be a discoverer of "Elsie." On the other hand, if one limits himself to some special case, such as the equivalence of heat and mechanical energy, then all twelve men listed above were discoverers. For my part, I should take an intermediate position and hold that the men in category B above failed to express the generality implied in what we now speak of as the conservation of energy, whereas those in categories A and C did. Even such a limitation, however, still leaves us with a very striking case of simultaneity of discovery.

The emphasis to be placed upon step 3 above represents the crux of the problem I should like to raise with Professor Kuhn. The phrase "reasonable justification" can mean many things, but I shall take it to mean some clear-cut mathematical or experimental verification. Inasmuch as the law in question involves a quantitative invariance, what would seem to be called for is a proof or an experiment showing a strict equivalence between various manifestations of energy. If one insists that the equivalence must be demonstrated in all cases of the conversion processes, then again no one of the dozen men above was a discoverer; but I expect that we can agree that we will be satisfied with at least one acceptable quantitative case going beyond work and motion. This would include the men in category A above but would exclude those in category C, and I believe that there is good historical tradition behind such an exclusion. Both Professor Kuhn and Professor Cohen seem to have underestimated the importance of step 3. Cohen quotes with approval Frege's "admiration of that unknown great *discoverer* who first *suggested* that the sun which rose this morning might be the same object that set last night"* (italics are mine). There is, of course, quite a difference between a suggestion and a

* In the first draft of Mr. Cohen's paper; now deleted.

discovery, and most of this is represented by step 3 above, as the history of science shows. Yet Kuhn has undervalued this. When his "discoverers" had completed step 2, he says, "To this point the notion of a universal convertibility of natural powers does not at all imply their conservation. But as Roget, Faraday, and Grove all showed, the remaining steps are *small and rather obvious.*"* (Italics are mine.) Kuhn cites the maxim *causa aequat effectum,* almost as though it settles the problem; but in reality one does not know whether or not it applies here until one has shown the quantitative equivalence between pairs of powers. Grove saw that "The great problem that remains to be solved, in regard to the correlation of physical forces, is the establishment of their equivalent of power, or their measurable relation to a given standard." That is, Grove saw that the law *ought* to follow. But does it? In view of their inability to verify the conjecture, one wonders to what extent the men in category C deserve to be called discoverers. As Kuhn himself later admitted, "The quantification of energy proved, in fact, insuperably difficult for those pioneers whose principal intellectual equipment consisted of concepts related to the new conversion processes."

The essential nature of step 3 in our discovery becomes clear if we consider another conservation principle, the Law of the Conservation of Matter or Mass ("Maggie," for short). This law was clearly implied by the ancient atomists; but credit for its discovery generally is reserved for Lavoisier, who first supplied satisfactory quantitative verification. If one does not insist on this last step, it is easy to multiply conservation laws. I shall suggest the Law of the Conservation of the Soul ("Sophie," for short), for this word soul appears in various forms, such as charity, sympathy, brotherly love, and it is not unreasonable to assume that these forms are interconvertible. Should I insist on the Conservation of the Soul, however, you would immediately ask me for quantitative verification. I shall therefore abandon "Sophie" and suggest another conservation principle proposed by Descartes—the Law of the Conservation of Motion ("Mollie," for short). Descartes did not clearly define the word motion, being satisfied with an easy metaphysical intuition which contrasts sharply with the laborious mathematical achievements of Newton. Quantitative verification of "Mollie" seemed at first to fail, as did such verification of the vortices of Descartes; but "Mollie's" reputation was rehabilitated when it was found

* A quotation drawn from the first draft of Mr. Kuhn's paper; now somewhat modified.

that the quantity mv is to be treated as a vector. Here the importance of polarity reminds one of another conservation principle stressed by Cohen, the Law of the Conservation of Charge in electricity ("Carrie," for short), a product of the eighteenth century.

While men were verifying "Mollie" quantitatively, it was discovered by Huyghens and others that there was another type of motion, now called kinetic energy but then known as *vis viva,* which was conserved under elastic impact. This led to the Law of the Conservation of *Vis Viva* ("Vivian," for short), and it was not long before Huyghens saw that this was related to the older principle of work. From Huyghens' law $v^2 = 2as$ and Newton's law $f = ma,$ one sees immediately that $fs = \frac{1}{2}mv^2.$ In the language of Leibniz, *vis viva* is the continuous sum (integral) of infinitely many *vires mortuae*—that is, $\frac{1}{2}mv^2 = \int f \, ds,$ as is easily verified by differentiating with respect to s to obtain $mv \dfrac{dv}{ds} = f.$ Since $\dfrac{dv}{dt} \cdot \dfrac{dt}{ds} = \dfrac{a}{v}$ we have $f = ma.$

Leibniz, like Descartes, was fond of metaphysical generalizations, and he suggested elevating "Vivian" to the status of a universal principle valid also for inelastic impact. This would make "Vivian" virtually equivalent to "Elsie," and would justify placing Leibniz among the discoverers above in category C. Leibniz was confident that in inelastic impact the *vis viva* which appeared to have been dissipated had not vanished but had taken on a new and latent form. This was not unlike Black's use of latent heats; but, unfortunately, Leibniz could not supply quantitative verification. Newton was sceptical of the principle, and he suggested that there must be some activating principle which was able to restore the lost *vis viva*. In the early eighteenth century there were a number of supporters of "Vivian," and Kuhn has quoted from Daniel Bernoulli's *Hydrodynamica* of 1738 an acute passage on the equivalence of work and heat. But on the whole "Vivian" went into eclipse for just about a century after 1742. Just what part, if any, "Vivian" had in the simultaneous discovery of "Elsie" is not clear. Kuhn feels that Bernoulli's remark apparently "had no later influence." In fact, Kuhn would see little connection between the scholarly tradition of *vis viva* and the rise of the conservation of energy, citing two arguments with neither of which I can agree. In the first place, Kuhn argues that "Elsie" arose from an engineering tradition in which work rather than *vis viva* was the standard unit of measure; but this seems to me to carry little weight in view of the fact that work and *vis viva* were known to be readily converted into each other, as we have seen above in the work of Huyghens and Leibniz. In the second place, Kuhn holds that our discoverers are not likely to have known about "Vivian" because "To discover the *metaphysical* theorem, the pioneers of energy conservation would have had to return to books at least a century old." Were all of our discoverers avoiders of libraries? And if not, is it likely that they would have found the works of Leibniz and the Bernoullis covered with dust? The *Hydrodynamica* in particular was a classic to which Joule would be expected to turn in connection with his work on the velocity of hydrogen molecules, if not for other themes. It seems to me that Kuhn has dismissed the written scholarly tradition rather too casually. Even if the "metaphysical" form of "Vivian" was not directly influential, it is likely that the more restricted mechanical form was indeed a factor, and one which I would place high on my list.

Professor Kuhn, a scholar and a scientist, would appear to have leaned over backward and to have undervalued not only the library factor but also the laboratory factor. While it undoubtedly is true that a concern with engines played a part, the material of Kuhn's paper suggests strongly another factor which has not been singled out for inclusion in his trivium —the laboratory work in physical and chemical thermotics. Chemistry before 1800 had been a largely unquantitative study of heat and mixture; physics had been mainly a quantitative study of forces and motions. Toward the middle of the nineteenth century, chemistry and physics were less sharply divided, and the study of heat was less distinctly a part of chemistry. The quantitative study of heat in laboratory research included, among other things, the adiabatic compression of gases, and data compiled in this connection became invaluable to some of the discoverers above. It may well be that it was the study of engines which first suggested "Elsie" to Mayer, for he said that the heat beneath the boiler of a locomotive is converted into the motion of the train, and is again deposited as heat in the axles and wheels. But the efficiency of steam engines is incredibly low—so low that this set of conversion processes could never have served in a quantitative verification of our law. For this Mayer fell back on laboratory data on the adiabatic compression of gases, which had, as Kuhn has expressed it, "nothing to do with the interest in engines." In the case of Joule the part played by laboratory experiments is everywhere recognized; and it may be no accident that Joule discovered "Elsie" shortly after he had temporarily deserted physics for chemistry.

The possible role played in our discovery by *Naturphilosophie* has been well described by Kuhn, and I see no reason for questioning its importance. I shall therefore sum up by listing the factors which I should recognize as significant in our problem of simultaneity of discovery:

1. The availability of conversion processes
2. The tradition of *vis viva*
3. Laboratory work in quantitative thermotics
4. The concern with engines
5. The philosophy of nature

In substituting this quintet for Kuhn's trivium, it will be noticed that there is no serious disagreement, except perhaps as concerns factor 2. The discrepancy in our points of view is largely a matter of the relative emphasis to be placed on the third step in the discovery—the quantitative verification.

It may be well to point out in closing that of the four universal principles referred to as "Elsie," "Maggie," "Mollie," and "Carrie," the first two recently have been combined into a broader law which we may call "Mellie" (the Law of the Conservation of Mass-Energy). Is there any chance that the last two, which incidentally are those which have polarity in common, will some day be subsumed under a broader universal principle?

ERWIN HIEBERT

on the Papers of Thomas S. Kuhn and I. Bernard Cohen

My attention will be directed first to Mr. Kuhn's paper on "Energy Conservation as an Example of Simultaneous Discovery." Mr. Kuhn has presented us with some of the conceptual and experimental elements which went into the formulation of the law of conservation of energy in the two decades between 1830 and 1850. Specifically, he has considered aspects of the activity of twelve individuals from whose writings the notion of energy conservation was compounded. I think that he has clarified the issue enormously by dividing them up into three groups: Mayer, Joule, Colding, and Helmholtz (1842–47), the first three working in independence of each other, and each providing generality of formulation as well as concrete quantitative application; Sadi Carnot, Seguin, Holtzmann, and Hirn (1832–55), providing quantitative computations for the interchange of heat and work (something like a conversion coefficient), but lacking in generality of formulation; and Mohr, Grove, Faraday, and Liebig (1839–44) supplying generality of statement without any computed quantitative application. Only the first-named group of four individuals provided anything akin to both generality and concrete application in their statement of the law of conservation of energy.

Mr. Kuhn's explicit qualifications concerning the meaning of "simultaneous discovery" seem rather important to me as a prelude to the launching of his historical investigation. In brief, that no two of these men wrote anything like the same thing; that, as a matter of fact, most of them disagreed, at that time, on the interpretation of what was being stated. Conceding this, the central concern boils down to an examination of very different approaches to one central problem—the object of the examination being to analyze a number of disorderly emergences, composed of experimental and conceptual elements, in order to discover how the notion of energy conservation could have been compounded from them.

The historian is concerned here, among other things, with locating a rationale for this great surge of interest in conservation over a relatively short period of time. Why was the idea of conservation accessible to so many? And why was the idea of conservation recognized by so many? Factors are allegedly innumerable. Mr. Kuhn singles out three: (1) The availability of conversion processes. (2) Concern with engines. (3) *Naturphilosophie.* I shall comment on each of them in turn.

Concerning the first of these I should merely underline, but with a somewhat different emphasis, Mr. Kuhn's own point; *viz.,* that a number of newly discovered physical facts connected with electricity, chemistry, heat and light, and the qualitative observation of their interconversions, led scientists to suspect that seemingly disparate research efforts might converge in some manner, or at least, that certain of these facts provided physical examples of earlier intuitions of the conservation of a metaphysical force. I should not wish to place so great an emphasis upon the new discoveries in electricity, except perhaps in the work of Joule, and certainly not in the work of Mayer and Colding. That leaves only Helmholtz in the first of Mr. Kuhn's group of four, and his work, more than that of any of the twelve individuals named, depended upon the prior activities of the rest of them. It would be rash to claim that the new experiments in electricity did not produce a favorable climate of thought for consideration of energy conversions, yet it would be valuable to see this influence more clearly documented from the writings of some of the pioneers employing energy conservation notions. That is a lot to ask for. I would suggest as an alternative interpretation, that what was needed more than, and in addition to, the collection of new facts unknown to the eighteenth century, was a workable agreement, on the conceptual level, of what may be referred to as a general notion of "completed action" (energy—thermal, chemical, etc.), over and against the then-already clear notion of "force." Closely related to this question was the need for establishing a usable terminology for "completed action," and of setting up intercomparable physical units for the capacity and the intensity factors whose product is conserved.

The principles of the impossibility of a *perpetuum mobile* and of the philosophical notion of equality between cause and effect, seem to me, very far from being the "remaining small and obvious steps" applied to the universal convertibility of natural powers, as Mr. Kuhn implies. I should accept them rather as the most universally held and agreed upon principles pointing men's minds in the direction of exploring the universal convertibility of natural powers. The principle of the impossibility of a *perpetuum mobile* must certainly have been based upon experience, before any thought was given to notions of conservation.

Concerning Mr. Kuhn's second factor, I am in full agreement as to the fruitful consequences resulting for conservation theory from the nineteenth-century concern with engines. I mean with reference to the establishment of a physical measure of the power of heat in terms of a mechanical equivalent. This study, which still needs to be made by historians of science, is of paramount importance. Yet, I was more than a little worried by Mr. Kuhn's cavalier dismissal of the dynamic tradition in mechanics in terms of its possibility of having provided a background or model for the concept of mechanical work. What bothers me is the tremendous amount of seventeenth- and eighteenth-century dynamical literature concerned with problems which make an attempt to account for both the *vis viva* and the *vis mortua* in mechanical systems; examples being: free fall; restricted motion as in the pendulum; the pulley and the inclined plane; and elastic and inelastic collisions. In addition we find numerous cases in this same literature where problems in statics are being attacked, at least implicitly, through an appeal to essentially dynamic arguments through the use of the principle of virtual work. The expression "virtual work" is, I admit, somewhat of a misnomer and should certainly not be equated with "mechanical work." The principle is more legitimately looked upon as a corollary of the principle of "virtual velocity"—the *"vitesse virtuelle"* of John Bernoulli which first appeared in print in Varignon's *Nouvelle mécanique ou statique* of 1725. I would hope to maintain that neither classical dynamics nor the dynamic tradition of treating problems in statics is as far removed from the nineteenth-century principle of "mechanical work" as Mr. Kuhn supposes. Eighteenth-century texts in theoretical mechanics supply the evidence for this connection.

What I should like to insist on is that the coinage of a host of new expressions for "mechanical work" in the writings of the French mechanicians, does not necessarily preclude the prior mathematical usage of combinations (products) of terms amounting to the same thing. After all, in most if not all of the problems posed by the French mechanicians in their applications of "mechanical work" to the action of machines, the use of "mechanical work" is restricted to the unsophisticated meaning of force times distance, i.e., not in the sense of the product of mass, acceleration and distance, but simply as the product of weight and distance. Now the usage of the mathematical product of weight and distance for explaining the "completed action" of the so-called five simple machines, and fairly complicated combinations of them, has an almost continuing tradition from the time of Hero of Alexandria's *Mechanica* through the eighteenth century. The question I would really like to ask Mr. Kuhn is to what extent the use of the principle of "mechanical work" in the writings of the French mechanicians differs from this long earlier tradition.

Neither Mr. Kuhn nor I know of any historical attempt to connect the engineering tradition in the use of the concept of "mechanical work" with that of the eighteenth-century theoretical mechanics treatises. It may well be that a careful historical analysis would reveal that dynamics (either in the *vis viva* problem or in the dynamical approach to statics via the principle of virtual work) did not provide a model. Mr. Kuhn says, "It could have provided a model. Yet I think the prevalent impression that it did so is misleading." I am not convinced, in the first place, that this is a prevalent impression, except in the writings of Mach, and I can only suggest that it would be worthwhile immersing oneself in the documents in an attempt to show up a connection where there could have been one. But that too, is an enormous assignment. A point, for example, that would merit investigation would be to search for the reason why Colding says in his writings that the first leading idea of the perpetuity of energy came to him by considering d'Alembert's principle of lost forces.

The last of the three factors mentioned by Mr. Kuhn is that the metaphysical notion of an imperishable force was stimulated by *Naturphilosophie*. Of his three factors this one appears to me the least difficult to document, at least superficially in the case of Mayer, Joule, and Colding; and we might add Liebig, Mohr, and Séguin as well. Mayer cannot possibly conceal his Schelling influence. Colding dwells with pride upon Oersted's influence. But then, the fact that many of these men were steeped in *Naturphilosophie* still does not answer the question which of its elements were congenial to notions of energy conservation, and why.

Mr. Kuhn limited himself to three major factors. Do these three provide the fundamental constellation? He recognizes other factors at work— *perpetuum mobile* notions, and the argument based on the equivalence between cause and effect. May I suggest others. There is, first of all, the question tied to concerns about "quantity of motion." In particular, there were two kinds of explanations offered for the apparent loss of quantity of motion in the special case of inelastic collision. The first had its origin in the Epicurean doctrine according to which the motion of a compound body was simply the outward expression of the internal atom collisions, the state of greater motion in one body than in another being attributed to more atoms moving in the direction of the compound body in one than in the other, and the state of rest, or external equilibrium, being maintained when the sum total of internal atom movements were balanced. Motion was simply transferred from the macroscopic body to the motion of its invisible internal parts. Gassendi used this argument against Descartes in his *Animadversiones* of 1647, when he taught that the inherent force (*vis*) of atoms does not perish when complex substances pass into the state of rest. Nor, he thought, was force generated when things began to move; force merely thereby acquired its freedom, so that as much *impetus* remained constant in objects as was there originally. This same argument was invoked many times later by natural philosophers grappling with the problem of the mechanical losses of inelastic collision. The argument was used by Borelli in the *De vi percussionis* of 1667. Leibniz' arguments in the *Dynamica de potentia et legibus naturae corporeae* are close to this view except that for him the *vis,* apparently lost in inelastic collision, is transferred to the *impedimenta;* and in his *Essay de dynamique* he speaks of masses being pressed together without returning completely to their original form, though without loss of *vis.*

A second more subtle explanation for the loss of *vis* or quantity of motion on inelastic collision contained the germ of the general concept of potential energy which has been so fruitful in modern thermodynamics. It might profitably be examined within the context of this discussion. It was based upon the argument that when the *vis viva* of a body decreased as a result of inelastic collision, that was simply a case of the conversion of *vis viva* into a potential *vis,* or of "virtual" (taken here in the sense of "potential") work. This "virtual" work then took the form not of the elevation of a weight in a gravitational field; it existed rather in the form of constraint, or of unreleased compression, or of temporary deformation—an internal constraint stored by the body in a manner similar to the way in which "virtual" work is stored in a raised weight. This view was expressed by John Bernoulli in his *De vera notione,* and by 's Gravesande in the *Mathematical Elements of Natural Philosophy* (1719), and by Christian Wolf who threw out the interesting suggestion in his *In conflictu corporum* that the *vis viva* was not lost just because it did not reappear immediately. It was simply a matter of time over which the constraint might be released. The most convincing speculation however was provided by Daniel Bernoulli in his *Hydrodynamica* of 1738 where he speaks of the transfer of *vis* and of *ascensus potentialis* being used to compress bodies; it was not, in this case, given back to the body but went over into a *materiae subtili* where it remained. It is regrettable that Bernoulli did not expand this idea since he also held to the view of the mechanical theory of heat, and may reasonably have been led to the possibility of treating heat as a form of *vis,* equivalent in its actions to the *ascensus potentialis* of a weight in a gravitational field. Mr. Kuhn quotes Bernoulli, "I cherish the convictions. . . ." But the full meaning of Bernoulli's conception is obtained from the two preceding sentences in that same text where he says, "It is clear, that from the relation between the conservation of *vis viva* in compressed air and that residing in a body which has fallen from a given height, no advantage is to be anticipated for machines from the principle of the compression of air. However, since air is compressed in many ways other than by the application of force, and achieves by nature an extraordinary degree of expansive force, it is to be hoped that from this natural state of things great advantage for the motion of machines can be devised, since D. Amontons has already taught a mode of moving machines by the force of heat."

Related to this same notion of internal constraint, or "potential energy" waiting to be released, is the rather common eighteenth-century concern to explain why a small spark can provide the moment of release of very great quantities of "energy" in inflammable substances and especially in explosive materials like gunpowder.

A second important possible factor at work in opening up men's minds to the idea of energy conservation should be sought in the activities of physicians and physiologists who were concerned, between 1820 and 1840, with the mechanical origin of animal heat. It would take too long to go into this problem here and I might only mention that some pertinent remarks on this subject appear on pages 27–34 of Paul Epstein's *Textbook of Thermodynamics,* although that statement suffers from a number of misconceptions regarding the work of Albrecht von Haller. It will suffice to mention that Haller's theory of friction of the blood was used by physicians and quoted in a number of handbooks to explain the fact that combustion experiments alone seemed to indicate that the oxidation of carbon from the carbohydrates in the blood was insufficient to account for the heat developed in animals, being short by a factor of about 20 per cent. Mayer and Helmholtz may reasonably have been motivated by this discrepancy which was not removed until after the publication of Dulong's work which showed that the heat of combustion of water had been estimated too low.

As a final comment to Mr. Kuhn's paper let me suggest a type of thinking about the conservation of energy problem which begins with a somewhat maturer statement of the law of conservation of energy than any which were submitted before 1850. I am referring to the work of Clausius and Kelvin by about 1885. That of course goes beyond Mr. Kuhn's time limits of investigation and these considerations are introduced here merely for the light that they may shed upon the earlier investigation. In this maturer formulation energy is defined as a thermodynamic property. A thermodynamic property is one which has the characteristic of being an exact differential, such that the summation of the product of all capacity factors α_i and all intensity factors X_i, i.e., $\Sigma_i \, \alpha_i \, X_i$, have the same value regardless of how one goes from a given initial state to a given final state. In other words these are processes in which the change of energy is independent of the path of the process, independent of the mechanism of proceeding from the initial to the final state, and independent of all working substances involved. That is all that we can say. It exhausts the general statement. Many useful examples of such processes were formulated in the last quarter of the nineteenth century. Thus mechanical work (W), force (f), and distance (s) are related so that $W = \int f \cdot ds$. For gravitational work (W_g), mass (m), acceleration of gravity (g), and height (h) we have the relation that $W_g = \int m \cdot g \cdot dh$. Where pressure is ($P$), volume ($V$), volts ($v$), coulombs ($c$), field strength ($H$), magnetization ($X$), surface tension ($\sigma$), increase of surface area (a), temperature (T), entropy (S) we may formulate the expressions: Work of expansion $= \int P \cdot dV$; Electrical energy $= \int v \cdot dc$; Magnetic energy $= \int H \cdot dX$; Surface tension $= \int \sigma \cdot da$; and Reversible heat $= \int T \cdot dS$. There are no limits of application to this general form of the equation provided that one considers reversible processes, since all forms of potential energy can be thus formulated as the product of a capacity factor and an intensity differential.

Thus informed about the mature energy conservation formulation let us go back to an analysis of earlier conservation notions. And if we restrict our attention to mechanical systems isolated from all non-mechanical interactions (which is after all what seventeenth- and eighteenth-century physicists were doing in their theoretical treatises on mechanics) then

I would like to maintain that one will find beautiful models of each of these conceptual elements of conservation theory in mechanics before 1750 —specifically in the mechanical works (dynamics and statics) of Huyghens, Descartes, Galileo, John Bernoulli, and Daniel Bernoulli as well as in the hydrostatics of Pascal and Torricelli. As a word of caution let me say, definitely, that the explicit mathematical formulation of the product of weight and distance does not enter into these works. But the conjoint action of weight and distance are constantly being considered in the solution of various problems. But beyond that, and this is my point, the authors are quite explicit about the basic features of what I have referred to as the mature notion of energy conservation, *viz.* dependence only on initial and final states, independence of path (vertical displacement or constrained fall as in the pulley, the pendulum and the inclined plane), and total independence of working substance. That this perfect mechanical model of energy conservation should have provided a major stimulus to nineteenth-century energy conservation notions seems rather plausible to me. But lest I commit the error, which that kind of unhistorical reasoning taken by itself may lead to, I hasten to add again, that this question can only be answered by total immersion in the documents of history.

Mr. Cohen mentioned the importance of the principle of conservation of weight in the development of nineteenth-century chemistry, and spoke at greater length of the concept of electric charge in relation to conservation principles in the nineteenth century. I should like to consider some other examples drawn from both mechanics and chemistry which demonstrate that energy conservation provided a really convincing unifying principle for nineteenth-century science.

In view of the successes of the eighteenth-century energy mechanics, or energetics, it is understandable that nineteenth-century scientists would consider setting aside the Newtonian force mechanics in order to attempt a grand unification of nineteenth-century physical theory based upon the law of conservation of energy and the division of all of mechanics in two classes of systems: conservative and non-conservative. We should be reminded that Newtonian dynamics was able in principle to solve all of the problems in mechanics which were tackled by the energetic principles. In practice the mode of attack was ultimately dictated by the ease of solution of the mathematical problem, and in general by convenience and economy of thought. We may say that while the force mechanics and the energy mechanics were both very successful in the nineteenth century, yet it was somewhat of a surprise to discover that energy notions were able to compete with the classical force mechanics of Newton at all. But how about the physical sciences other than mechanics?

In some areas of nineteenth-century physical science we discover that conservation theory provided greater overall unity than had ever been known before. Take chemistry as a case in point. And let us ask the question: How would one specifically apply the concept of force to chemistry? It was certainly attempted by nineteenth-century theoretical chemists in terms of affinity notions and in works on chemical statics. An impossible task! There were no obvious capacity factors analogous to the concept of force.

One of the central concerns of nineteenth-century physical chemistry was to locate the criteria for chemical spontaneity, i.e., to find the basis of calculating whether a given chemical reaction might proceed or not. And if so, in what direction and to what extent? And where in terms of the quantities of initial reactants and final products would equilibrium prevail? Here the energy model of the nineteenth century gave the only answer—and that, for a wide variety of experimental conditions. The criteria for spontaneity turned out to be a number of so-called thermodynamic properties having the dimensions of energy: the work content, or free energy (A), of Helmholtz who first introduced the term "free energy" in 1882; the free energy (F), a function whose value was first indicated by Massieu in 1869, and then effectively and comprehensively employed by Gibbs beginning in 1873; the entropy (S), first defined by Clausius as the capacity factor for isothermally unavailable energy; internal energy (E); and heat content (H). The criteria for spontaneity were: that all chemical reactions would proceed from left to right as written, (i.e. were not at equilibrium) when, $A_{V,T} < 0$, or $F_{P,T} < 0$, or $S_{E,V} > 0$, or $E_{S,V} < 0$, or $H_{P,S} < 0$; where the subscripts V for volume, T for temperature, P for pressure, etc. indicate isochoric conditions (constancy of volume), isothermal conditions (constancy of temperature), isobaric conditions (constancy of pressure), etc. The important point is that the question of chemical spontaneity was answered for many experimental conditions by calculating different energy functions.

Thus while the model of the force mechanics was well nigh useless to chemistry, the model of the energy mechanics, extended to include the physical manifestations of heat as in the principle of the conservation of energy, answered chemistry's crucial question. The successful extension of the energy concept to include chemistry in this manner was surely one of the major achievements of nineteenth-century science and an outstanding example in which the principle of conservation of energy became a unifying principle for nineteenth-century science.

Such successes of the conservation theory in chemistry, as elsewhere in electricity, magnetism and optics, led late nineteenth-century and early twentieth-century scientists to contend with some right that "energy," being the one aspect of matter conserved in all of nature's manifestations, was the key concept which would provide a new unity for all of the physical sciences. In fact we know that energetic notions were incorporated into economics, sociology, religion, etc., and soon became the basis for metaphysical cults and technocratic organizations dedicated to the task of re-evaluating the world's resources and the output of man and machines in terms of energy certificates.

In addition, we should note that philosophers of science, writing at the very end of the nineteenth century and at the beginning of the twentieth century, had a heyday trying to bolster their own peculiar view of the theoretical foundations of science by appealing above all to examples drawn from the unitary picture of the physical world which had resulted from the application of the principle of energy conservation. The writings, between 1872 and 1910, of a very prominent group of philosophers of science—Mach, Duhem, Enriques, Poincaré, Meyerson, and Cassirer—all bear upon the question of conservation of energy as a unifying principle in nineteenth-century science. In each of their works historical examples, drawn from nineteenth-century energy conservation theory, perform a very central role in the exposition of a special philosophical position. And this is again an indication of the tremendous impact which the unity of nineteenth-century conservation theory had exerted upon a group of men who were looking back at the turn of the century over the developments of nineteenth-century physical science. And they were individuals, let us note, who had achieved eminence as scientists, philosophers, and historians. Each of them has a claim to all three of those titles.

BIOLOGY ATTAINS MATURITY
IN THE NINETEENTH CENTURY

J. Walter Wilson

In the decade of the 1850's three notable ideas were published which led biology to the attainment of maturity as a science. Virchow, in his *Die Cellularpathologie,*[1] stated that all cells come from preceding cells. Pasteur,[2] in disproving spontaneous generation of bacteria, proved finally that all organisms come from preceding organisms. Darwin,[3] in his *Origin of Species,* proved that all species are derived from pre-existing species. These three great investigators were, in fact, saying virtually the same thing in different language. Working in entirely different fields and from entirely different points of view, they were, in effect, saying that living substance arises only from living substance; this is the doctrine of the continuity of living matter, of protoplasm. This was declared by Whitman,[4] as the century drew to a close, to be the equivalent in biology of the law of the conservation of energy in physics. He called it the Law of Genetic Continuity or the Law of Homogenesis.

It had been suggested that my topic be "Evolution as a Unifying Principle in the Biological Sciences of the Nineteenth Century." From the point of view of biology, it would be difficult, indeed, to say anything about evolution that has not already been said. I have, therefore, elected to deal with the maturation of biology on a somewhat broader scale. I do this because, although evolution did play an important and perhaps the most important role, it was only one factor and it would only have done so in a very elaborate setting in which many factors were involved. This paper is, then, an attempt to present the major factors in the setting from which biology as a modern science emerged.

It should be emphasized that not only biology but all science, if not, indeed, all scholarship, came to maturity in the nineteenth century. In biology the most important achievement was its liberation from the various psychic entities that had been called upon as the controlling and directing agents in explaining biological phenomena: the will of the Diety, His ideas, and His beneficence and omnipotence, Anima, the Vis Creatrix, Entelechies, all that we would call supernatural in that they do not have any of the characteristics of matter or energy and are not amenable to control in the laboratory. We have still with us, it is true, the remains of such thinking, in various forms of vitalism, but however important they may be in the "Philosophy of Science" and whatever their role in the directive process of nature may eventually prove to be, we cannot too emphatically state that in the basic phenomena that the biologist studies today, they play no role. The digestion and assimilation of food, the fertilization of the egg and its development, the contraction of a muscle, and the detection of light by the eye, to cite a few examples, are natural phenomena to be studied directly in the laboratory and involve nothing more than the matter and energy that the physicist uses. At least this is the "mature" point of view in biology. I propose here to review the events by which the concept of protoplasm came to take the place of all vitalistic or animistic concepts in the explanation of biological phenomena.

The basic problem of the life sciences is to explain the specific forms of organisms and the relation of their form to the functions they perform. Living matter is always found organized in the specific form of some organism. In this it is unique. It is possible to conceive of other matter, a lump of coal, for example, in any size and in many different forms. Some inorganic matter does display specific form in the shape of crystals. That this type of inorganic form might be somehow related to the form of organisms has long been recognized and the idea has an interesting history. However, even if the phenomena are related, they have one essential difference which is a critical one; both crystals and organisms grow but the crystal does not change its shape as it grows. Little crystals have the same shape as big ones. Only the organism develops its form as it grows and it is by the form developed that we recognize and identify the myriads of kinds of animals and plants.

The early naturalists, down through the eighteenth century, had studied and described the forms of different kinds of organisms and had recognized that, in general, there are differences and similarities between them. Such observations suggested categories into which the organisms could be classified so that a system of classification naturally grew out of them. However, as the number of known organisms increased and different naturalists used different criteria in classification, a condition verging on chaos developed. In the mid-eighteenth century Linnaeus[5] made two important contributions. He proposed that the simple binomial nomenclature be adopted and that the description of a species be reduced to a terse statement made up of clearly defined terms. These suggestions were adopted with enthusiasm. The consequence need not be reviewed here in detail. Chaos was rapidly reduced to a semblance of order so that by the early nineteenth century a fairly satisfactory, comprehensive view of the forms of animals and plants had been attained. While from the point of view of the technician in taxonomy, many important developments have taken place since then, the fundamental nature of his work has not changed. I believe this statement will hold water in spite of the fact that evolution with its phylogenetic trees, together with geographical distribution, have given us a different idea of relationships between the categories of the system and now modern cytogenetics is giving us an idea of the essential basis of these relationships and the origin of the differences. The primary work of the taxonomist today is the same as that of Linneus and his immediate followers.

The study of the relation between form and function has a somewhat different history. This has always been the province of the medical man. His problems have to do, for the most part, with aberrations of functions, departures from normal activities. From the days of antiquity he has taken organisms apart, as a boy would a watch, to find out how they work. The study of gross anatomy had reached its height in the Renaissance with the publication of the great work of Vesalius, *De humani corporis fabrica*.[6] John Fulton, I believe, interpreted this as "the works of the human body" like "the works of a watch." Harvey's discovery of the circulation of the blood was one of the last great contributions of this method of study. The general plan of operation of the body and how it works was pretty clear but it was evident that what makes it work was not at all revealed.

With the advent of the microscope in the late seventeenth century, anatomical description received a new impetus at an entirely different level. When the microscope revealed the intricate and exquisite structure beyond the limits of visibility to the naked eye, it was hoped that the failure of the older gross anatomy to explain function was speedily to be rectified. This hope gradually subsided. In spite of the fact that a much more precise knowledge of how the parts work was developed as the microscope was progressively improved, why they work continued to elude the keenest worker. By the end of the nineteenth century it was generally concluded that the secret of the "works" must lie between the structure of the molecule and the structures revealed by the microscope. To designate this structure, the cytologists[7] around the close of the nineteenth century coined the word *metastructure*. This is not a vitalistic concept, as has been suggested, but merely a formal expression of ignorance. It is perhaps of interest that in our own generation with the enormously increased magnification of the electron microscope, the same hope was rekindled and we are still in search of a working system whose operation we can understand in terms of the simple mechanisms of physics and chemistry. We are substituting for "metastructure" a concept of "ultrastructure," something that at this new level we can actually take apart and study.

Two ideas derived from the study of form were in vogue at the beginning of the nineteenth century; the realization of the importance of the comparative point of view, comparative anatomy, and the concept of "general anatomy." The comparative point of view in the study of anatomy was no novelty. It must have been obvious to the earliest hunters, butchers, and even fishermen that the "works" of the bodies of different animals and even fish had important features in common. This is amply testified to in Aristotle's *Parts of Animals*. Because of difficulties in the way of dissection of the human body, this was taken advantage of by the student of medicine. He studied monkeys, pigs, and dogs extensively. During the latter half of the eighteenth-century studies of the anatomy of many species of animals were carried on, particularly by an active group in Paris. Here the basis was laid for the later work of Cuvier and the other comparative anatomists of the nineteenth century. It should be emphasized that this type of study of anatomy was essentially basic to physiology, to the understanding of function, and in no way implied "morphology," which is a much more sophisticated study of form for its own sake, including a theory about it. This did grow in part out of the comparative point of view but more especially out of the Nature Philosophy of the early nineteenth century which will be discussed below. Comparative anatomy may be considered in some ways the culmination of gross anatomy. It contributed much to taxonomy, especially to the relationships within the higher categories, orders, classes, and phyla, that is not revealed by a superficial examination of animals.

"General anatomy" or what we today call histology was the culmination of the search for an explanation of activity of the parts described in gross anatomy. The attempt to analyze the activities of animals into "faculties" reached a point of absurdity in the late Renaissance so that it obviously had no value whatsoever in explanation. The introduction of a new concept, function, by Haller in the mid-eighteenth century led to a very different type of analysis, irritability, contractility, etc., and to a search for components of organs that could account for the functions. These components were found in the "tissues" of which the organs are composed. The success of the analysis of activities into functions, and organs into tissues, was immediate and lead to the publication in 1801 of Bichat's *Anatomie Générale*.[8] The importance of this concept in our present day thinking is hardly realized, it has become such an essential part of our science. But Bichat found nothing in the tissues to explain the phenomena of life. He therefore again fell back on vitalism.

Biology entered the nineteenth century, then, with a fairly satisfactory grasp of taxonomy, gross and comparative anatomy, and general anatomy, the relation between activity and visible structure. The facts of taxonomy, especially the existence of many forms, was satisfactorily accounted for by the concept of creation; and the obvious fact of adaptation, the fitness of the organism, and its parts, for the activities it displayed, was equally well accounted for by the wisdom and foresight of the Creator. The facts of general anatomy, the exhibition of functions by parts, was accounted for by Bichat himself by a resort to vitalism; the parts were operated by some vitalistic entity. All such explanations would be unsatisfactory to the modern scientist. Furthermore, two problems were so far from explanation that they could hardly be considered scientifically. One, the reason for, or "cause of," the relation between species revealed by taxonomy and comparative anatomy and, the other, the origin of the individual and the development of its form, that is, the phenomena of reproduction.

The relationships between the different forms of life, revealed by the study of structure, led to the formulation in antiquity of the concept of the Scala Naturae, the Great Chain of Being, so interestingly discussed by Lovejoy.[9] This recognized not only similarities, but the fact that organisms could be arranged in a progressive series from the "lowest" to the "highest," which, of course, must be man. That there were gaps was recognized. This gave rise to the concept of "the Missing Links" sought for even before the publication of *The Origin of Species* and the acceptance of the concept of evolution. It is to be emphasized that the Scala

Naturae did not necessarily involve any implication of descent. In fact, it was not originally in any sense an "historical" idea.

Lovejoy speaks of the "Temporalization" of the Scala Naturae during the eighteenth century. It did not occur abruptly and did not receive immediate universal acceptance. It produced, however, the first doctrine of descent. This differed from the Darwinian concept in that it was considered as a simple unidirectional process, the ascent of a ladder, whereas Darwin's concept was one of diversification, the ascent of a tree, a phylogenetic tree. Another difference was in the mechanism. The components of the Scala were presumed to be created but, in the "Temporalized" form, to have been created in sequence with the Creator progressively "changing his mind as he went along." The "progress" displayed in the Scala is the expression of progress in the mind of the Creator.

The change from the inert scale to the temporal, progressive scale is perhaps related to a new view of history generally. It has been pointed out that until recently history was little more than an account of events that occurred in sequence and receded negatively into the past (Merz[10] and Miller[11]). As told, it was for the most part the story of the past exploits of a family or of a nation. There was little attempt at "interpretation" beyond the general idea that somewhere in the past were "the good old days," undoubtedly derived from the disillusioning experiences of all men as they grow older. The view of history as a progressive process seems to be a sophisticated one, especially when we seek in the past for causes of the present and future trend of events. This would appear to be the maturing factor in the history of history. It is related to the idea of evolution. Miller would seem to imply that it was the result of the Darwinian point of view, but Merz has attributed it to the philosophers, especially Hegel, and such historians as Carlyle. It would seem, then, rather to be a general point of view that contributed to the acceptability of Darwinism. This point might well stand further study. Applied to the Scala Naturae in the eighteenth century, it transformed it from a taxonomic device comparable to a set of postage stamps mounted according to ascending values, to a doctrine of descent.

The historical point of view was introduced into science in two ways in the eighteenth century. The attempt to explain the universe as a mechanism led to the formulation of the nebular hypothesis of its origin, and the study of the earth's surface and its fossils led to the development of geology and the concept of the history of the earth. Neither needs detailed discussion here, but the bearing of the latter on biology is particularly important. Geology became an historical science as it became clear that the successive strata of the surface of the earth must have been laid down in temporal sequence. With a more intense study of fossils in the strata, it became clear that they are not accidents of nature nor cast off molds or imperfect inventions of the Creator, but rather the actual remains of animals and plants some of which have become extinct. Furthermore, it was also obvious that the fossils of the older strata tend to be limited to animals lower in the scale which helped to give the Scala Naturae its historical significance, made it, in fact, a pattern of evolution.

The idea of organic evolution as we understand it became clearly formulated in the late eighteenth century in the writings of Erasmus Darwin and Lamarck. Lamarck's statement was the more significant from the point of view of theory. Although Lamarck pictured it in terms of the Scala, he saw the process very much as Charles Darwin later did, except that his idea of the mechanism of transformation was entirely different. Lamarck believed that, for each individual, the development of form was the work of a psychic entity and that if this entity had a strong desire for a change to a more suitable form for a particular purpose, a transformation might gradually occur in successive generations and this new form thus become the standard of the species. This would account not only for transformation but for adaptation. The stock example is the stretching of the giraffe's neck in response to a desire on the part of the giraffe to brouse on the leaves of trees. It was this psychological element, not the inheritance of acquired characters, that Darwin characterized as "Lamarckian Nonsense." Darwin's great contribution was to substitute for the psychic element, the purely mechanical processes of chance variation and natural selection.

Lyell's[12] book, with its exposition of the historical idea of geology, made a profound impression on Darwin and contributed substantially to the acceptance of the principal of evolution. But evolution, with the transformation of species, was not a necessary consequence of the geological story. Two great naturalists, students of paleontology and of anatomy and taxonomy, were ardent anti-evolutionists. Cuvier contended violently against Lamarckianism and Agassiz equally refused to accept Darwinism. Both arrived at a satisfactory explanation of the temporalization of the Scala Naturae that did not involve transformation of species or evolution. Their explanations were related; in fact, they both could find no satisfactory unifying factor except in the Creator. Agassiz's important work, "An Essay on Classification," was published the same year as Darwin's *Origin*. To me there is no more striking document in the history of biology than the chapter in this work that discusses the parallelism between the sequence of form in the taxonomic scale, in the fossil record, and in the stages of development of the individual. Our textbooks present this parallelism no less convincingly than Agassiz does, but in support of evolution. Agassiz' chapter ends with the following statement: "... enough has already been said to show that the leading thought which runs through the succession of all organized beings in past ages is manifested again in new combinations in the phases of the development of the living representatives of these different types. It exhibits everywhere the working of the same creative Mind, through all times, and upon the whole surface of the globe."[13]

The point of view of Cuvier had been much the same. He had felt that the concept of a Creator continually in action was preferable to the individual psychic components that Lamarck depended on and, therefore, he rejected evolution. The paleontological sequence he explained by accepting one of the current theories of the geological sequence, the cataclysmic theory. This was that the strata are evidences of a succession of great catastrophies in the history of the earth. The paleontological sequence, with its ascent of the Scala, was accounted for by a new creation following each catastrophe. Obviously, the Flood of the Bible story was the latest of the series of catastrophies.

To understand the point of view of these men, it is necessary to review the background of their thinking. Both had been exposed to the Nature Philosophy, which has been described as a disease emanating from Germany in the late eighteenth and early nineteenth centuries. This is one of the outstanding examples of "arm chair" science in history. It was an attempt to construct a complete picture of the universe and its operation by sheer intellectual effort. The only purpose of observation was to discover evidence for its conclusions.

For our present discussions the most important product of this kind of thinking was morphology, a philosophy of form. As pointed out above, this led to a study of form for its own sake whereas the older anatomy merely sought a mechanism for function. In the analysis of organic form, it was considered that there were only a few basic types, or perhaps only one. These were the archetypes from which all others were derived. The classic example for student consumption is the Oken-Goethe theory that the mammalian skull is a series of transformed vertebrae. This type of exercise became so attractive to anatomists that it became the main occupation of many; for example, Owen, who became one of Darwin's bitter opponents and a target of Huxley's polemics.

Cuvier was a profound student of comparative anatomy and carried its theory far beyond that of his predecessors. He developed the doctrine of "the correlation of parts," which is that the parts of an organism are interrelated and correlated to suit its activities. Thus, the teeth of an herbivorous mammal must have wide grinding surfaces. The jaw must, therefore, be large to carry them, and it must, therefore, have large surfaces to carry big muscles to handle it. These big muscles must also have large surfaces on the skull for the attachment of the other end. The big skull would require a large ligament and tendon to hold it up and this with the necessary muscles would require large dorsal processes of the vertebrae. As this sort of reasoning was carried out, it became clear that it might be possible to reconstruct an extinct animal from a single fossil tooth.

To explain this remarkable correlation, Cuvier fell back on something like the Platonic idea, now in the mind of the Creator. In spite of the great diversity of animal forms, there were only a few fundamental ones, the rest being more or less extensive variants of these. The Creator had started with these basic ideas and, in the production of the different variants, had adapted the details to the kind of life the animal was intended to lead. This accounted for the correlation.

Cuvier's "correlation of parts" was an example of the fitness of things that had been long recognized in organisms generally as adaptation. Agassiz adopted it and when he came to America found little difficulty in grafting it onto the Natural Theology that he found entrenched here. This had been a prevalent point of view in England and America in the seventeenth and eighteenth centuries. It was expounded in Paley's *Natural Theology*,[14] which became a standard textbook throughout England and America. It was part of the academic curriculum to study nature in order to discover examples of the fitness of things, adaptation, to illustrate the wisdom and beneficence of the Creator. Darwin testifies that he had studied Paley with great interest and profit.

Although, as has been frequently pointed out, Darwin was not a morphologist, his theory profited by morphology and also contributed

409

substantially to it. Russell summarizes this as follows: "The morphology of the 'fifties lent itself readily to evolutionary interpretation. Darwin found it easy to give a formal solution of all the main problems which pre-evolutionary morphology had set—he was able to interpret the natural system of classification as being in reality genealogical, systematic relationship as being really blood-relationship; he was able to interpret homology and analogy in terms of heredity and adaptation; he was able to explain the unity of plan by descent from a common ancestor, and for the concept of 'archetype' to substitute that of 'ancestral form.' "

As mentioned above a striking feature of organic form is its development. The problem of the reproduction and development of form, and the remarkable specificity of the whole process is within the experience of all mankind and is unquestionably one of the great mysteries. Each new generation is a new creation. The origin of each individual is, indeed, as great a mystery as the origin of the first one, except that it has an antecedent specifically like itself, to which in some unknown way it owes its existence and form. The possibility that at some levels individuals develop spontaneously was, of course, always present. But the hen and the egg, and the phenomena of pregnancy and birth speak eloquently of some type of continuum. The microscope led to the discovery of the minute eggs of lower forms and the experiments of Redi, Spallanzani, and others in the eighteenth century century cast a serious doubt on the idea of spontaneous generation except for the very lowest and simplest animals. Finally the discovery and the description of the mammalian egg by Von Baer[15] in the 1820's brought even mankind within the general concept of reproduction by egg formation. What, however, is an egg? When Harvey formulated the dictum "Omne vivum ex ovo," the word ovum had very little of the clear cut significance it has today. It meant little more than "a specific product of the mother." Our idea is that the egg is a cell, derived from a maternal cell by mitosis, and that it is made up of some of the maternal protoplasm. To Harvey it was an unformed product, like a secretion. To bring it into shape, to induce its development, some guiding influence akin to thought and thought transference must be postulated. This general idea is found in Aristotle and was the best theoretical explanation of reproduction, heredity and development. It made the Lamarckian concept of evolution a reasonable one. If a psychic agent brings about development, it might by desire to do so make it different from the parent. To the hard-headed scientist, however, it was not an acceptable theory. This type of psychic influence was too elusive for manipulation and really explained nothing to his satisfaction.

The only available alternative to this "epigenetic" idea was a negation of the whole problem in the doctrine of "evolution," that is, of the evolution of the individual rather than the evolution of species. This was adopted by the great leaders of the eighteenth century, Haller and Bonnet. This was simply that the development and heredity that we see is an illusion. Really nothing occurs but the unfolding of an individual already completely present in the egg. In its complete form, which to us would seem a *reductio ad absurdum* but which did not to those who formulated it, it was conceived that all individuals of a species that were or ever will be were created at the same time at the original creation. The first female carried, then, all individuals of the species, one generation within another, in infinitely small dimensions and yet completely formed. All that was necessary was that at each successive reproduction, under the proper stimulus, the existing form would unfold as the individual grows. No limit of subdivision of matter comparable to our concept of atoms and molecules stood in the way of this, to us, amazing flight of the imagination. It, too, however, was unsatisfactory to the hard-headed scientist. It did not solve the problem. It merely placed it in the lap of the Creator who already presented problems to scientific explanation. Furthermore, as microscopes were improved, observations indicated that the development of the embryo is not an unfolding of a pre-existing form, but the development of a new form through a sequence of very different ones. Wolff,[16] in the late eighteenth century, placed this type of observation on a firm footing with his studies on the development of the chick. Hence, awaiting the basic concepts of the cell theory and the continuity of protoplasm, there was no satisfactory scientific theory about this, the greatest of mysteries. Today we know that in their reasoning both schools of thought, the epigeneticists and the evolutionists, were in part right, but for the most part, wrong.

The background of Virchow's dictum that all cells come from cells has been told in detail and considerably discussed.[17] The publication of the cell theory by Schleiden and Schwann[18] in 1839 was one of the important events in the century. But their cell theory was a far cry from that of the present day biologist. Schwann's concept was that all organisms are composed of cells or cell products. His idea of the cell was, however, very different from ours, although the fact that there must be a nucleus was recognized. The wall of the cell was still to him important as it had been to the student of plant cells since Hooke[19] first used the term in 1665. The two important components of our concept of the cell that were missing in Schwann's concept are the material, protoplasm, and the fact that new cells arise by division of parent cells. Schwann's idea of the origin of new cells was indeed a crude one and entirely erroneous. He believed that new cells arise in an intercellular fluid, the cytoblastema, by a process of crystallization. He says: "Disregarding all that is specially peculiar to the formation of cells, in order to find a more general definition in which it may be included with a process occurring in inorganic nature, we may view it as a process in which a solid body of definite and regular shape is formed in a fluid at the expense of a substance held in solution by that fluid. The process of crystallization in inorganic nature comes also within this definition, and is, therefore, the nearest analogue to the formation of cells."[20]

Actually, this concept was little different from that generally current concerning the origin of infusoria by spontaneous generation. It was, in fact, a theory of cell origin by spontaneous generation. That some cells divide was, however, already known and new examples were rapidly discovered. Division is easily observed in some egg cells, protozoa, and blood cells. But where a large number of cells arise quickly, as in pus, this is not so easily observed. In such a case, spontaneous generation from a special fluid would seem quite probable. However, before Virchow's statement was published twenty years later there were enough instances of division known to make many suspicious that it is the rule, although it is noteworthy that in the first edition of Strasburger's[21] monograph on mitotic cell division in 1875, "free cell formation" was still seriously discussed. Virchow's conclusion was based on no new striking discovery but seems, rather, to have been related to his feeling against spontaneous generation generally. He himself contributed significantly to the studies of the life history of parasites. The complete statement of his concept reads:

Even in pathology we can now go so far as to establish, as a general principle, *that no development of any kind begins* de novo, *and consequently as to reject the theory of equivocal* [spontaneous] *generation just as much in the history of the development of individual parts as we do in that of entire organisms.* Just as little as we can now admit that a taenia can arise out of saburral mucus, or that out of the residue of the decomposition of animal or vegetable matter an infusorial animalcule, a fungus, or an alga, can be formed, equally little are we disposed to concede either in physiological or pathological histology, that a new cell can build itself up out of any non-cellular substance. Where a cell arises, there a cell must have previously existed (*omnis cellula e cellula*), just as an animal can spring only from an animal, a plant only from a plant. In this matter, although there are still a few spots in the body where absolute demonstration has not yet been afforded, the principle is nevertheless established, that in the whole series of living things, whether they be entire plants or animal organisms, or essential constituents of the same, an eternal law of *continuous development* prevails.[22]

The conclusive evidence in favor of the universality of division came only in the 1870's when with the new techniques of sectioning and staining and still better microscopes, the facts of mitosis were discovered. It rapidly appeared that mitosis occurs everywhere. It was easier to prove this than to disprove free cell formation which, like many other concepts, simply disappeared by neglect.

Until the possibility of spontaneous generation was disposed of, evolution in the modern sense would be meaningless. If flies are generated in a decaying carcass, frogs in mud, mice in grain, and deer in old tree stumps, individuals do not have to come from preceding individuals, so neither would species have to come from antecedents. It is customary to say that spontaneous generation by large free-living animals was definitely disproven by the experiments of Redi. In these classical experiments he demonstrated that if decaying meat is screened so that flies cannot reach it, no flies are developed in it. However, while this does prove that the flies are necessary for the reproduction of flies, and to the sceptical scientist this may be tantamount to the disproof of spontaneous generation, such a concept can never be completely disproven.

As long as the doctrine of creation was prevalent, the possibility of spontaneous generation was also prevalent. Until it was shown that the egg is a cell derived from the parent and carrying its protoplasm (Gegenbaur, 1861),[23] the origin of every individual seems in a sense a spontaneous generation. The organization of unorganized matter into an individual was likened to either coagulation or crystallization, but to account for the specific form assumed by the individual, some sort of animistic force must be called upon. Abnormal and monstrous births could be accounted for on this basis. Even fossils could be explained as due to activity of such forces in unprofitable places or materials. Von Baer, even after his discovery of the mammalian egg, believed that it was necessary to consider spontaneous generation a possibility even in the origin of higher forms.

There were, furthermore, two types of phenomena not easily dismissed that even in the mid-nineteenth century still seemed explainable only by spontaneous generation. First, the microscope revealed hosts of very small organisms that appear suddenly in water to which all sorts of nonliving materials have been added; hay, tea, pepper, meat, and even dried mud. Some of these microscopic organisms are almost structureless, others quite elaborate. When a little dried dust from an eaves-trough is wet with the purest rainwater, rotifers may appear almost immediately. This is because they are able to withstand dessication for long periods of time. In this condition their form has collapsed so that in dried dust they would never have been recognized. They can, in this form, be carried long distances by the wind, and even longer perhaps by birds. Before this was demonstrated, the simplest explanation of their appearance was spontaneous generation.

The origin of the simpler forms of life that appear to be almost form-less, or at least with very little form and that superficial, presented a more difficult problem. These organisms become so numerous in a very short time in a broth or tea of any kind that their origin by reproduction is almost inconceivable. Furthermore, there arose a concept developed by both Buffon and Oken that they are elementary unit particles of "life" lib-erated in decay. This was the more plausible because their occurrence in large numbers is usually associated with the decomposition of once-living material. Such particles had been postulated since the days of the phi-losophers of ancient Greece. As soon as they were found with the micro-scope, it was felt that the problem of life would be solved by their study.

The role of Pasteur in the spontaneous generation controversy is gen-erally known. He proved by simple and direct experiments that even with free access to pure air, broths that had been boiled remained free of organisms. While his proof was satisfactory to most scientists, the fact that for many years afterwards attempts were made to prove spontaneous generation shows how difficult final disproof must be. It was indeed fortu-nate for Pasteur's experiments that he did not have to do with a spore-bearing organism, whose spores could withstand boiling temperature.

Even more puzzling is the appearance of internal parasites. They seem to appear from nowhere and the complicated life histories of some of them, with alternate generations in quite a different form, make their source very difficult to detect. Here, also, spontaneous generation where they are found was the simplest explanation of their presence. With the knowledge that reproduction must involve eggs, the life history of one after another was worked out. This occurred particularly between 1840 and 1860. Early in the 1850's enough was already known to make it clear that it only required detailed work on each species to make spontaneous generation unnecessary. Cole has summarized this problem by stating:

It would be literally correct to state that Küchenmeister was the first to establish experimentally (in 1853) the complete life cycle of a Cestode, but to go no further would be to take the difficulties and merit of that achievement for granted, and to ignore the element of time. If, however, the statement is amplified, and it is added that the discovery was the result of the efforts of six distinguished naturalists spread over a period of sixteen years, we begin to realize that the investigation must have been an exacting one, and we can only completely realize its perplexities, and appraise the genius of the time, by including in our survey the labours of that much larger company of ineffective workers and critics, whose activities in the main served only to complicate and postpone the solution of the problem.[24]

It is of interest, as was mentioned above, that some of the work of Virchow had to do with this problem. He contributed to the proof that internal parasites do not arise spontaneously.

If we adopt the entire concept of evolution, however, spontaneous generation must, at some time, have occurred. Hence, it might occur today. Spontaneous generation at the origin of life is still a live question. Jacques Loeb used to contend that the biologist who teaches that life can only come from preceding life is really leading his students astray. He should, rather, teach them that the primary job of the biologist is to find out how life arose from the non-living and proceed, himself, to create it. This is cited here to emphasize that this is not what was meant by spontaneous generation. The old controversy had really nothing to do with the modern, quite sophisticated question of the ultimate origin of life but rather with the production of known specific forms of organisms.

This, then, completes the review of the background of the three ideas stated in the introduction. It should be emphasized that the basic ideas were not completely novel. The basic idea of evolution, that species are derived from parent species, was not new. The doubts about the occurrence of spontaneous generation, with the belief that all organisms come from parents, was also not new, and the concept that organisms are made up of cells was not entirely new. The really new item was the nature of the cell and its reproduction. The concept of protoplasm as a universal substrate of life was a complete novelty. It had its inception in the observation that eggs and protozoa seem to be composed of a clear, homogeneous jelly.

The improved microscopes of the 1830's were used assiduously, almost as a fad. Years later Henle wrote: "Those were the great times. Any day a bit of tissue shaved off with a scalpel or picked to pieces with a pair of needles might lead to important and fundamental discoveries." Since this involved the study of fresh cells, the jelly-like substance would be constantly observed. It came, soon, to be associated with movement, partly because of the activity of protozoa and partly in the cyclosis of the plant cell. Since movement is a recognized attribute of life, it gradually came to be considered the important living part of the cell. Observations by Cohn, von Mohl and others lead finally to the studies of Max Schultze[25] which, published in 1862, established the generalization that, essentially, a cell is a mass of protoplasm with a nucleus. With the establishment of the fact that cells come only from parent cells, protoplasm became a self-perpetuating material and when Gegenbaur in 1861 established finally that the egg itself is a cell derived from a maternal cell, protoplasm became the material bridge between generations. Thus, quietly, and with no fanfare, for it was not the work of a single great mind, the Law of Genetic Continuity was established. It was at first felt that protoplasm might be something fairly simple in structure, but Brucke,[26] in an essay, the importance of which was immediately recognized, emphasized that no simple jelly could perform all the tasks ascribed to it. Hence, it must have some kind of organization which is perpetuated along with the substance. While we still are not sure of the nature of this organization, Brucke's concept is the modern concept of protoplasm. Huxley's[27] famous essay, "The Physical Basis of Life" (1868), indicated the importance of the concept for all biology. We have to seek in protoplasm the mechanism of all activities including not only those studied by the physiologist but those of development, heredity, and evolution.

The necessity of calling on vitalistic and animistic entities to explain biological phenomena began to disappear in the early nineteenth century. Woehler[28] in 1828 synthesized urea, an organic substance up to that time only produced in living organisms, presumably through the action of vital forces. One of the features of Schwann's theory of cell origin was that crystallization, a mechanical process rather than a vital force, was involved. Darwin's theory was conspicuously successful for it substituted a purely mechanical device for the mind of the Diety in the origin of new species and their adaptations. With the elimination of spontaneous generation a whole array of vital forces became unnecessary. It was the elimination of these non-physical or supermechanical entities from biological theory that placed biology on a firm scientific basis. Its maturation was completed with the establishment of the unifying idea of protoplasm, that is, with the Law of Genetic Continuity.

It would not be difficult to find and quote many statements, some of them elaborate and almost extravagant, to the effect that the great transformation of biology was due to evolution, particularly to Darwin's theory, as a unifying principle. This is in part, at least, because of its great popular appeal and its controversial aspects. There can be no doubt but that, from the point of view of the advancement of the intellectual side of our culture, Darwin's contribution was very great. It was one of the most effective impacts that the biological sciences had made on the development of thought. This is itself a thesis worthy of discussion but it is a different story from the one I have told here. The Theory of Genetic Continuity, the continuity of protoplasm, while it implies evolution, is of more far-reaching significance in biology as a science.

To the man in the street this theory has a meaning he can hardly grasp. It was presented dramatically by my great teacher, Dr. Mead, as follows: "When we wash our hands we remove a multitude of dead cells. They float away down the drain. The death of each of those cells is the first death that has occurred in a continuous line of living protoplasm extending back through our whole lives to the fertilized egg, through this to our parents, grandparents, and distant ancestors; to primitive man and his prehuman ancestors; to the primitive mammal and thus to the lower vertebrates; to the ancestors of these vertebrates and finally to the earliest form of life, the original life that appeared on earth." This represents an extent of time and of experience that is almost beyond comprehension. During this time, the organization of matter which we call protoplasm and which manifests itself as life has endured as a very sensitive equilibrium in the midst of hazardous vicissitudes, and in it has occurred the majestic process of evolution, a progressive process leading to almost infinite transformations, in which the old "Scala Naturae" has disappeared in a welter of adaptations, some "higher" and some "lower." This is the unifying idea that completed the maturation of biology as a science.

J. Walter Wilson

References

1 R. Virchow, *Die Cellularpathologie* (Berlin, 1858).

2 L. Pasteur, *Ann. Chim. Phys.,* Ser. 3, LII (1858), 404; *Comptes rendu de l'académie des sciences,* XLVII (1858), 224.

3 C. Darwin, *On the Origin of Species* (London, 1859).

4 C. O. Whitman, "Evolution and Epigenesis," in *Biological Lectures delivered at Marine Biological Laboratory, Woods Hole, in 1894* (Boston, 1896).

5 Carl von Linnaeus, *Systema naturae* (1st ed.; Leyden, 1735).

6 Andreas Vesalius, *De humani corporis fabrica, libri septem* (Basel, 1543).

7 Martin Heidenhain, *Plasma und Zelle* (Jena, 1907–11).

8 X. Bichat, *Anatomie Générale* (Paris, 1801).

9 A. O. Lovejoy, *The Great Chain of Being* (Cambridge, Mass., 1936).

10 J. T. Merz, *A History of European Thought in the Nineteenth Century* (Edinburgh and London, 1896). Volume I.

11 H. Miller, *History and Science* (Berkeley, Cal., 1939).

12 C. Lyell, *Elements of Geology* (London, 1839).

13 L. Agassiz, *An Essay on Classification* (London, 1859), p. 175.

14 William Paley, *Natural Theology* (London, 1802).

15 K. E. von Baer, *De ovi mammalium et hominis genesi* (Leipsig, 1827).

16 K. F. Wolff, *Theoria Generationis* (1759); *De formatione intestinorum* (1768–69.

17 *The Cell Theory, its Past, Present and Future,* ed. Joseph Mayer (Biological Symposia, Lancaster, Pa., 1940). Volume I. J. W. Wilson, *Isis,* XXXV (1944), 168; *Isis,* XXXVII (1947), 14. John R. Baker, "The Cell Theory, History and Critique," *Quart. J. Micros. Sci.,* LXXXIX, 103; XC, 87; XCIII, 157; XCIV, 407; XCVI, 449.

18 T. Schwann, *Mikroskopische Untersuchungen über die Uebereinstimmung in der Struktur und dem Wachsthum der Thiere und Pflanzen* (Berlin, 1839); M. J. Schleiden, "Beitrage zur Phytogenesis," *Arch. Anat. Physiol. Wiss. Med.* (1938), pp. 137–76.

19 Robert Hooke, *Micrographia* (London, 1665).

20 Schwann, *Mikroskopische,* Smith, translator (London, 1847), p. 2.

21 E. Strasburger, *Zellbildung und Zellteilung* (Jena, 1875).

22 Virchow, *Die Cellularpathologie,* quoted from translation by Chance (London, 1860), p. 54.

23 K. Gegenbaur, cited in A. W. Meyer, *The Rise of Embryology* (Stanford University, Calif., 1939).

24 F. J. Cole, *Early Theories of Sexual Generation* (Oxford, 1930), p. 197.

25 M. Schultze, *Arch. Anat. u. Phys.* (1861), p. 11.

26 E. Brucke, "Die Elementarorganism," *Sitzungsber. Akad. Wiss.,* Vienna, Math.-Natur. (1861, Klasse. 44, abt. 2; (1862), Klasse. 46, abt. 2, 35, 629.

27 T. H. Huxley, "On the Physical Basis of Life" (1869).

28 Woehler, "Ueber Künstliche Bildung des Harnstoffs," Poggendorff's *Annalen der Physik* (1828), XII, 253.

BIOLOGY AND SOCIAL THEORY

IN THE NINETEENTH CENTURY:

AUGUSTE COMTE AND HERBERT SPENCER

John C. Greene

The simultaneous emergence of evolutionary theories in biology and sociology in the nineteenth century presents an interesting problem in historical interpretation. How were these two types of evolutionary theory related to each other? Was one derived from the other? Did they develop independently of each other? Was there a continuing interaction between the two? If so, what was the nature of the interaction? The present essay attempts to take a first step toward solving this problem by analyzing the relation of biological to sociological theory in the writings of Auguste Comte and Herbert Spencer.

The first truly evolutionary speculations in modern social theory appeared at approximately the same time as the first transformist ideas in biology. The same mid-eighteenth-century years which produced the speculations of Maupertuis and Diderot also produced Rousseau's *Discourse on the Origin and Foundations of Inequality Among Men*.[1] In this famous essay Rousseau developed a theory of human evolution in a purely speculative manner, by thinking away all of man's acquired characteristics so as to discover original human nature. He arrived at the conclusion that man had begun his career on earth as a brute-like creature distinguished from other animals only by his ability to perfect himself, that is, to invent ways and means of ameliorating his condition. Given this ability and the pressure of circumstances, the development of society, language, and the arts and sciences followed necessarily. Rousseau was uncertain whether to regard the progress of civilization as a blessing or a curse to mankind, but as the century wore on, developments in commerce and industry, combined with the revelations of geology, paleontology, and ethnology, produced a growing conviction that both nature and history were inherently progressive. Then came the French Revolution, posing its great question as to the future of Western civilization and awakening hope that reason might supplant custom as the molder of social institutions. On every hand there were proposals for creating a social science, a science which would reorder society, banish superstition, and guide mankind along the path of perpetual progress.

Nineteenth-century social science took its general character from these events and aspirations. It was highly normative in orientation, claiming to disclose man's duty and destiny as well as the solution of his immediate political and economic problems. Except for its "dismal" branch, political economy, it took progress for granted and set out to discover the laws of historical development. Theories of social evolution proliferated, as theories of the earth had in the eighteenth century. The striking resemblance between these two types of speculation was more than coincidental, for, given the conviction of gradual progress from a primitive condition whether of man or of the earth, the problem was to explain the assumed development by means of a few judiciously selected principles supported by an assortment of judiciously selected facts. In both cases attempts at scientific explanation ranged all the way from mere speculation to profound analysis based on careful investigation. A broad comparative study of these evolutionary theories would throw much light on the history of thought in the eighteenth and nineteenth centuries. The present paper, limited as it is to the social theories of Comte and Spencer, can do little more than suggest the possibilities inherent in this field of research.

The genesis of nineteenth-century social science is nowhere better illustrated than in the life and works of Auguste Comte, the founder of positivism. Born at Montpellier in 1798, he received an excellent training in the exact sciences at the École Polytechnique in Paris, where he soon distinguished himself by his mathematical talents and his republican fervor. There followed several years of intensive study and political agitation, 1817–25, during which Comte collaborated with Count Henri de Saint-Simon in the latter's various literary projects, meanwhile developing his own ideas. These took definite shape in Comte's first and briefer *System of Positive Polity,* published under Saint-Simon's auspices in 1824. Saint-Simon died in the following year, and Comte soon after began a course of private lectures, the famous *Cours de philosophie positive,* which, published in six fat volumes between 1830 and 1842, laid the theoretical groundwork for his system of practical polity and established his claim to be one of the founders of the science which he named *sociology.*[2]

How did Comte arrive at his conception of sociology as a science concerned primarily with the laws of social evolution? In his earliest writings, dating from the period of his collaboration with Saint-Simon, Comte took as his starting point the moral and spiritual anarchy of post-Revolutionary Europe, an anarchy reflected in political instability, social unrest, and intellectual chaos. In Comte's opinion, the old order had been demolished not by the French Revolution but by the steady growth of science, undermining in its progress the system of beliefs and sentiments which held the aristocratic order together and gave it moral prestige. God was dead, and little could be done to restore social harmony and political stability until a new system of positive beliefs was erected on scientific foundations. "The period has . . . at last arrived," wrote Comte, "when the human mind, as the final result of all its previous labours, can complete the ensemble of Natural Philosophy by reducing Social phenomena, as all others have been reduced, to Positive theories. . . . Such is the great philosophic effort reserved for the Nineteenth Century by the natural progress of our intellectual development."[3]

As he set himself to this task, Comte examined one by one the efforts which had already been made toward constituting a science of society. First, there were those like Condorcet who held out the hope that mathematics, especially the calculus of probabilities, could be applied to the analysis of human behavior. But Comte, mathematician though he was, regarded this as a false lead. He was convinced that the phenomena of life and of society were too complex for mathematical treatment. Secondly, there were the political economists, the champions and analysts of the new commercial and industrial order, who claimed to have demonstrated the self-regulating character of the economy and thereby to have established a scientific basis for a policy of *laissez-faire*. But, although Comte had great respect for Adam Smith and Jean Baptiste Say, he found their conception of social science too narrow for his purposes. They explained everything in terms of the rational pursuit of self-interest, assuming a kind of pre-ordained harmony between individual interests and the welfare of society. It was simply not true, said Comte, that men acted only, or even mainly, from calculation or that they were capable of calculating wisely when they did. Doubtless there was a tendency toward equilibrium in economic affairs, but this equilibrium was by no means automatic, and it presupposed a moral order capable of moderating class antagonisms and international rivalries and of mitigating the divisive effects of the ever-increasing division of labor. Political economy was at best a partial and one-sided social science, incapable of creating or sustaining the moral order which it presupposed or of providing an adequate basis for scientific prevision.[4]

Finally, there were the bio-sociologists, men like George Cabanis, Antoine Destutt de Tracy, and Francis Joseph Gall, who envisaged the new social science as a branch of zoology. In Comte's opinion, however, these writers overlooked the crucial characteristic of human society— "the progressive influence of the successive generations upon each other." Biology, he declared, was relevant only insofar as it disclosed man's natural endowment and insofar as it could "throw light on the formation of primitive aggregations of men, and deduce the history of the childhood of our race down to the period when the first impulse of Civilisation was given by the creation of Language."[5] From that point on, it was the province of social physics to discover "by what necessary chain of successive transformations the human race, starting from a condition barely superior to that of a society of great apes, has been gradually led up to the present stage of European civilisation."[6] Thus, social physics was the study of the progress of civilization. That civilization had progressed seemed to Comte too obvious to require proof. Its progress was to be seen in the course of human history as a whole and in the recurrence of an identical pattern of historical development in civilizations isolated from each other. The recurrence of this pattern seemed to prove beyond doubt that social evolution was not haphazard but issued inevitably from "the fundamental laws of human organization" and was governed by "a natural law of progress, independent of all combinations, and dominating them."[7]

What were the sources of this evolutionary conception, embodied in Comte's famous law of the three successive stages of civilization: the theological, the metaphysical, and the positive? It was not from biology that his inspiration was drawn; his writings and letters in the formative period sing the praises of Bichat and Gall but not of Lamarck. His intellectual debt in social theory lay in a different direction—to Condorcet's *Sketch of an Historical Picture of the Progress of the Human Mind,* to the historical writings of Hume and Robertson, and to the ideas of Saint-Simon. Above all, his doctrine emerged from his studies in the history and philosophy of science, studies begun at the École Polytechnique and subsequently broadened to include the life sciences. The history of science and technology was the example *par excellence* of cumulative growth, of the progressive transformation of society by the progressive development of positive knowledge. It seemed to show that each of the sciences had passed through a theological and a metaphysical stage before becoming truly scientific, and the philosophy of science established the necessary sequence in which the various sciences underwent this transformation. Thus, the growth of science and the scientific attitude indicated the general direction of historical development, and the discovery of the pre-ordained sequence in which the sciences attained a positive status made it possible to locate Comte's own age in the stream of history and to predict the outcome of the entire development. For the science of society was in its very nature a science of the whole historical sequence. With the establishment of sociology, all fields of thought would have become positive in orientation, and historical development would have reached its final phase. Emancipated from theological and metaphysical conceptions, civilized by the growth of science and industry, unified by a common allegiance to science and the scientific method, men would abandon their attempts to control each other and would unite in efforts to increase human happiness by controlling nature.[8]

Comte's biological philosophy is set forth in the third volume of his *Cours de philosophie positive,* his sociology in the three following volumes. It will be interesting to treat the two together in order to discover how they were related to each other. As has been indicated, Comte's social evolutionism was not derived from his biological studies. He specifically rejected Lamarck's development hypothesis, but not because it implied the transmissibility of acquired characters. Comte regarded it as an "incontestable principle" that there was a tendency "to fix in races, by hereditary transmission, . . . modifications at first direct and individual, so as to augment them gradually in each generation if the action of the environment continues unaltered."[9] Lamarck's error, he maintained, lay in assuming that nature had unlimited time at her disposal and that organisms were indefinitely modifiable. The assumption of unlimited time he considered a defect "so glaring that there is no need to examine it specially." The indefinite mutability of species, on the other hand, had been disproved by the "luminous argumentation" of Cuvier. The species observed by Aristotle had not changed since his day, the Egyptian mummies exhibited a physical type still preserved among their descendants, and the fossil record provided numerous examples of species which had remained unaltered throughout geologic time. Plants and animals which had been

domesticated and transported to new environments had not been transmuted into new species. The same was true of the human species despite
the variety of climates to which it had been exposed. Finally, the cases
were legion in which organisms had failed to adapt themselves to changed
circumstances. Many mammals had been rendered extinct by the spread
.of human civilization, and some primitive races of men had all but disappeared from the earth. Extinction, not progressive modification, was
the consequence of alterations which destroyed the equilibrium between
organism and environment. "If," said Comte, "one conceives that all possible organisms were successively placed during a suitable time in all
imaginable environments, most of these organisms would necessarily end
by disappearing, thus leaving alive only those which could satisfy the general laws of that fundamental equilibrium: it is probably through a succession of analogous eliminations that the biological harmony was established little by little on our planet, where indeed we still see it modifying
itself unceasingly in a similar manner."[10] Thus, Comte invoked the principle of the survival of the fit to account for the gradual establishment of
the present equilibrium of life and its terrestrial environment, but he made
no provision for the emergence of new species to take the place of those
extinguished in the course of time. Only his limited time scheme prevented this omission from becoming a serious embarrassment. If he had
thought the matter through, he would have had to admit either the idea
of special creations, a conception totally inconsistent with his positivistic
outlook, or some kind of development hypothesis, unless he was willing
to envisage the progressive extinction of all forms of life.

The inadequacy of Comte's discussion of the development hypothesis
is explained in large measure by his rather restricted view of the field of
biology. His main interest was in anatomy and physiology, which comprised for him *la biologie proprement dite*. Natural history in Buffon's
broad sense he recognized only as a "concrete, particular, and hence secondary science" subordinate to the more abstract, general, and basic disciplines of pure biology. Geology and paleontology, with their strong
historical orientation, scarcely entered into his calculations. In Comte's
view, biology began by assuming a certain equilibrium between the organism and its environment and then sought to discover the laws of their
interaction: "in a word, to connect constantly, in a manner both general
and special, the twofold idea of organ and of milieu by a means of the idea
of function."[11] Viewed statically, this was a problem in comparative anat-

omy; viewed dynamically, it was a problem in physiology. Comparative anatomy, grounding itself in Bichat's work on the tissues of the body, aimed first of all at deriving all the various elementary tissues from a single primitive tissue—"the essential term of every organism."[12] On the other hand, it moved toward a natural classification of living forms, assigning each its distinctive place in the organic hierarchy, which Comte conceived as descending from the adult male human being as the primordial and most perfect type to the lowest of the low in the vegetable kingdom. It was in this connection, in showing how the concept of the scale of nature would assist in perfecting the natural system of classification, that Comte felt called upon to discuss Lamarck's development hypothesis, which, though it did not impugn the basic conception of the scale of being, placed enormous difficulties in the way of assigning species and genera their proper places in that scale. Lamarck's hypothesis disposed of, Comte found it possible to "regard as demonstrated the necessary discontinuity of the great biological series." "The various transitions," he conceded, "can, doubtless, eventually become more gradual either by the discovery of intermediate organisms or by a more careful study of those already known. But the essential fixity of the species guarantees that the series will always be composed of terms clearly distinct, separated by unbreakable intervals."[13]

Although Comte did not believe in organic evolution, biology and biological analogy played a large role in his social theory. In the first place, he drew from biology, especially from the writings of Gall, his theory of basic human nature, that nature which made possible and inevitable man's social evolution, determined its general direction of movement, and prescribed the limits of its variability. According to Comte, Gall had demonstrated scientifically the preponderance of the affective over the intellectual side of human nature and had proved the innate sociability of man, "in virtue of an instinctive tendency toward common life independently of all personal calculation and often contrary to the most energetic individual interests."[14] Without this preponderance of the emotions, without this natural sociability, the formation of society would have been impossible, the sentiments being necessary to stimulate intellectual activity and to give it a definite goal, as well as to restrain man's egotistic impulses in behalf of social order. On this "double opposition" of man's affective and intellectual nature and his social and egotistic impulses society was founded. These propositions Comte believed had been given

425

a firm scientific basis by the phrenological researches of Franz Joseph Gall.

Comte also drew from biology his distinction between the statical and the dynamical approach to the study of organisms. Social statics he defined as "the fundamental study of the conditions of existence of society."[15] Every system, he declared, requires and presupposes a certain solidarity or internal harmony, the subjection of the parts to the functioning of the whole. In biological organisms the activities of the various tissues and organs are coördinated more or less automatically by means of the nervous system. In human society the coördination of actions is rooted in the sentiments and is moral in character. Its non-intellectual, predominantly affective, character is best exemplified in the basic social unit, which is not the individual but the family. In the family is to be found the primitive division of labor, based on differences of sex and age. In the family is to be found that social solidarity, that *union véritable,* which is "essentially moral and very incidentally intellectual," and which is "destined to satisfy directly the ensemble of our sympathetic instincts independently of any thought of active and continued coöperation toward any end whatever."[16] In the later stages of social evolution, Comte conceded, voluntary coöperation based on rational appreciation of self-interest plays a larger and larger role in social coördination, but even then social solidarity is predominantly affective in character. Voluntary coöperation, far from having given birth to society, presupposes it. As the division of labor progresses, the moral ties uniting society are loosened, and increased social control is required to supply the deficiencies of the ever-less-automatic social harmony. Thus, a true social science, far from assuming the natural identity of interests in modern society, perceives the tendency of industrial evolution to dissolve the bonds of social solidarity in the acid of self-interest and class interest and moves to prescribe the necessary social remedies, warned by its knowledge of social statics that the elements of the social organism cannot safely be viewed in abstraction from the functioning of the whole.[17]

So much for Comte's social statics, notable for its discussion of the basis of social order, or what, by analogy to biology, might be called the internal conditions of existence of the social organism. One might suppose that Comte would carry the analogy farther and consider the external conditions of existence of particular societies, exhibiting the functions of the various social organs and their modifications in response to changes in the environment, as Spencer was later to do. Such was not the case, however. For Comte was less interested in the evolution of particular societies than in the progress of the human race, more especially the white race. The pecularity of human society, he insisted, lay in the influence of each generation on succeeding generations, a peculiarity which tended more and more "to transform artificially the species into a single individual, immense and eternal, endowed with a constantly progressive action on exterior nature."[18] Through the cumulative developments in each particular society and the ever-increasing contact and borrowing among different societies there was set in motion an irreversible development of the human race as a whole, a development which could terminate only in the universal brotherhood of man. The direction and general character of this historical process was dictated by man's biological endowment, his original complement of abilities, instincts, and impulses. Social evolution could not alter basic human nature, but it could and did alter the relative influence of its various components, gradually bringing about the ascendancy of the social, peaceable, and intellectual side of man's nature over the egotistic, combative, and sensual side. In this respect, said Comte, it was necessary to invoke the "undeniable principle of the illustrious Lamarck" concerning "the necessary influence of a homogeneous and continued exercise in producing, in every animal organism, and especially in man, an improvement susceptible of being gradually fixed in the race after a persistence sufficiently prolonged."[19] The evolution of humanity was best observed, however, in its intellectual development, more especially in the development of science and the scientific spirit, which, though intimately connected with social and economic changes, was nevertheless the prime mover in history. To grasp the laws of this development it was necessary to apply the comparative method historically, exhibiting the progressive complication of social organization and cultural interdependence in the progress of mankind. Thus, whereas in biology progressive gradation was conceived as being given once and for all in the organic hierarchy or great chain of being, in sociology it was regarded as evolved in the course of time, as moving gradually toward full realization. And just as in biology the progressive gradation of the organic hierarchy was thought to be recapitulated in the embryological development of the individual human organism, so the *échelle sociale,* or progressive societal development, was conceived as mirrored in the mental life history of each individual in civilized society.[20]

The cause of human progress Comte found in the very nature of man.

It was implicit in "that fundamental instinct, a complex resultant of the combination of all our natural tendencies, which directly pushes man to ameliorate his condition incessantly in every way, . . . to develop always in every regard the whole of his physical, moral, and intellectual life as much as the circumstances in which he is placed permit."[21] Thus, for Comte, circumstances were not so much the cause as the occasion of human development. Climate, geography, racial mixture, social conditions, political policies, and the like might accelerate or retard it, but they could never alter its direction or halt its onward progress. For in every situation, favorable or unfavorable, men would seek to understand their world and control it, and so seeking they would be led by slow but inevitable steps from a theological to a metaphysical to a scientific way of thinking and from a militaristic and predatory to an industrial and peaceful way of living, a transition necessarily accompanied by a corresponding development of the higher brain centers. It was in this "exact harmony" between the findings of historical analysis and the expectations derived from the biological theory of human nature that sociological science found its ultimate verification.

Herbert Spencer provides an interesting contrast to Auguste Comte. Like Comte, Spencer ranged the whole gamut of the sciences and sought to bring human nature and society within the scope of scientific method. Like Comte, he rejected supernaturalism and the search for ultimate reality, confining himself to investigating the laws governing phenomena. Like Comte, he believed that social science, since it aimed at discovering the laws of historical development, held the key to man's duty and destiny and afforded a scientific basis for individual ethics and social policy. But whereas Comte rejected the competitive society of the nineteenth century as anarchic in both theory and practice, regarding it as a mere transition between the community of the medieval period and the coming community of the scientific age, Spencer glorified the individualism of his day and modelled his society of the future on its pattern. Far from rejecting the social atomism of the eighteenth century, Spencer sought to ground it firmly in the science of life and in a general view of the cosmos. Whereas for Comte progressive development was the peculiar characteristic of human society, for Spencer it was the general attribute of existence, the universal law of nature. In Comte's view biology was relevant to social theory chiefly as it threw light on man's original nature; in Spencer's it provided a model of social theory in both its statical and its dynamical aspects.

In its earliest manifestations Spencer's thought exhibited two features which were to characterize it to the end: (1) a firm belief in the *laissez-faire* policy in social and political matters, and (2) adherence to the development hypothesis. It was about 1840 that Spencer first read Lyell's *Principles of Geology.* The arguments there advanced against Lamarck's zoological philosophy, far from inducing Spencer to reject it, inclined him to accept it. "Its congruity with the course of procedure throughout things at large, gave it an irresistible attraction," Spencer wrote later, "and my belief in it never afterwards wavered, much as I was, in after years, ridiculed for entertaining it."[22] No less ardent was his attachment to the *laissez-faire* principles of classical political economy. His first book, *Social Statics,* published in 1850, attempted to strengthen those principles by deriving them from a broader social science rooted in an evolutionary conception of nature, human nature, and society. The strongly normative character of the inquiry was indicated by the author's definitions of social statics and social dynamics. Social statics, Spencer declared, concerns itself with "the equilibrium of a perfect society." "It seeks to determine what laws we must obey for the obtainment of complete happiness."[23] Social dynamics, on the other hand, studies "the forces by which society is advanced toward perfection," that is, the influences which gradually dispose human beings to obey the laws conditioning their happiness. The perfect society is one in which every individual is free to exercise his natural faculties to the limit and in which individuals have lost all inclination to engage in activities harmful to others.

By defining social statics as the study of the conditions of equilibrium in a *perfect* society, Spencer introduced a dynamic point of view immediately. Obviously the society of his own day was not the perfect society; it did not give each individual equal freedom to exercise his faculties, nor was it composed of individuals whose desires were completely compatible with the rights of others. These imperfections in European society were explained, said Spencer, by the fact that human nature had not yet become adapted to the conditions of life in the modern industrial order. All imperfection consisting in "unfitness to the conditions of existence," this unfitness must consist "in having a faculty or faculties in excess; or in having a faculty or faculties deficient; or in both."[24] In what respect, then, was modern man maladapted to the conditions of modern life? In the sense, answered Spencer, that his original nature was adapted to a primitive condition of life, a condition in which survival depended on a fierce struggle with nature and with other men. For, "the aboriginal man must have a constitution adapted to the work he has to perform, joined with a dormant capability of developing into the ultimate man when the conditions of existence permit. To the end that he may prepare the earth for its future inhabitants . . . he must possess a character fitting him to clear it of races endangering his life, and races occupying the space required by mankind. . . . In other words, he must be what we call a savage, and must be left to acquire fitness for social life as fast as the conquest of the earth renders social life possible."[25] Thus, man began as a brute-like creature hard pressed to survive. The pressure of necessity drove him into the social state and led him to invent techniques whereby the circumstances of life were ameliorated. Man's inventiveness transformed the conditions of his existence, and the new conditions of existence operated in turn to transform human nature. In primitive society, men could be held together only by the crude forces of fear and hero worship. But as civilization progressed, the moral sense, based on sympathy and the instinct of personal rights, grew in strength, and voluntary coöperation began to supplant coercive arrangements. When fully developed, the moral sense would render government both unnecessary and impossible. This whole progress of events was predetermined from the beginning: "given an unsubdued earth; given the being man, appointed to overspread and occupy it; given the laws of life what they are; and no other series of changes than that which has taken place, could have taken place. . . . Progress, therefore, is not an accident, but a necessity. Instead of civilization being artificial, it is a part of nature; all of a piece with the development of the embryo or the unfolding of a flower. The modifications mankind have undergone, and are still undergoing, result from a law underlying the whole organic creation; and provided the human race continues, and the constitution of things remains the same, those modifications must end in completeness."[26]

Given the fact and the general pattern of societal development, what were the implications for social policy? Like Comte, Spencer maintained that social policy, though it might accelerate or retard the predetermined course of social evolution, could never halt it or deflect it from its main line of march. But whereas Comte envisaged an ever-increasing interposition of social control to make up for the gradual dissolution of social solidarity, Spencer anticipated the gradual withering away of government as the individuals composing society acquired the natures requisite for the successful operation of a free society. This process of adaptation could best be hastened, he thought, by throwing the individual ever more on his own, thus compelling him to develop industry, thrift, foresight, moral independence, and capacity for voluntary coöperation. "It is not by humanly-devised agencies, good as these may be in their way," he declared, "but it is by the never-ceasing action of circumstances upon men—by the constant pressure of their new conditions upon them—that the required change is mainly effected."[27] This action, he asserted, is akin to "the stern discipline of Nature" in the biological realm. Just as in nature the struggle for existence operates to remove the sickly, the malformed, the unfit, so in society competition eliminates the ignorant, the improvident, and the lazy. "Partly by weeding out those of lowest development, and partly by subjecting those who remain to the never-ceasing discipline of experience, nature secures the growth of a race who shall both understand the conditions of existence, and be able to act up to them. It is impossible in any degree to suspend this discipline by stepping in between ignorance and its consequences, without, to a corresponding degree, suspending progress."[28] Such an intervention, moreover, would interfere with the automatic processes of internal adjustment whereby the energies of the social organism are directed where they are most needed.

Despite its title, *Social Statics* was concerned more with the progress than with the structure of society. Spencer's treatment of social structure and of the basis of social order was very sketchy, and it appealed to physical and chemical rather than to biological analogies. Society was conceived as an aggregate of individuals, and the proposition was advanced that the characters of the individuals comprising the social aggregate determine the nature of the aggregate. Stability in the social system was conceived to arise from the properties of the constituent units. "Sympathy and the instinct of rights," Spencer wrote, "do not always coexist in equal strength any more than other faculties do. Either of them may be present in normal amount, whilst the other is almost wanting. . . . The instinct of rights, being of itself entirely selfish, merely impels its possessor to maintain his own privileges. Only by the sympathetic excitement of it, is a desire to behave equitably to others awakened; and when sympathy is absent such a desire is impossible. Nevertheless this does not affect the general proposition, that where there exists the usual amount of sympathy, respect for the rights of others will be great or small, according as the amount of the instinct of personal rights is great or small."[29] Thus, "the first principle of a code for the right ruling of humanity in its state of *multitude,* is to be found in humanity in its state of *unitude*— . . . the moral forces upon which social equilibrium depends, are resident in the social atom—man; and . . . if we would understand the nature of those forces, and the laws of that equilibrium, we must look for them in the human constitution."[30]

Spencer's treatment of social dynamics also left a good deal to be desired. He asserted again and again that the historical change in the circumstances of man's existence had produced and continued to produce a change in the characters of the individuals composing society. But as to what had brought about the change of circumstances he had little to say, except for the suggestion that a change in character, once accomplished, operated to produce new changes in circumstances. The reason for Spencer's vagueness on this point is not hard to guess. The question of the causes underlying these changes was an historical one, and Spencer was not interested in history. "You can draw no inference from the facts and alleged facts of history," he wrote his friend Edward Lott, "without your conceptions of human nature entering into that inference: and unless your conceptions of human nature are true your inference will be vicious. But if your conceptions of human nature be true you need none of the inferences drawn from history for your guidance. If you ask how is one to get a true theory of humanity, I reply—study it in the facts you see around you and in the general laws of life. For myself, looking as I do at humanity as the highest result yet of the evolution of life on the earth, I prefer to take in the whole series of phenomena from the beginning as far as they are ascertainable. I, too, am a lover of history; but it is the history of the Cosmos as a whole."[31] In view of this statement, it is not surprising that when Spencer did hazard a speculation concerning the causes of historical change, it was quasi-biological in character. In his "Theory of Population," published in 1852, he found the "proximate cause of progress" in the pressure of population on resources. Starting from two propositions: (1) that the power of self-preservation varies inversely as the power of reproduction, and (2) that the degree of fertility varies inversely as the development of the nervous system, he endeavored to show that population pressure in any species would operate to produce complication of the nervous system and a decline in fertility, thus reestablishing an equilibrium between population and the means of subsistence. Since the human race was rapidly multiplying in numbers, it followed on this hypothesis that mankind must also be undergoing a development in nervous organization which must eventually diminish the supply of vital energy available for the reproduction of the species. That such a development was actually taking place was apparent, said Spencer, from the difference in cranial capacity between the savages of Australia and Africa and the civilized peoples of Europe, a difference of thirty per cent from lowest to highest. The reasons for the development were equally apparent. For, "this inevitable redundancy of numbers—this constant increase of people beyond the means of subsistence—involving as it does an increasing stimulus to better the modes of producing food and other necessaries—involves also an increasing demand for skill, intelligence, and self-control—involves, therefore, a constant exercise of these, that is—involves a gradual growth of them. Every improvement is at once the product of a higher form of humanity, and demands that higher form of humanity to carry it into practice."[32] It should not be thought, Spencer warned, that every people hard-pressed to subsist inevitably undergo a progressive improvement in character. Although all mankind are subjected sooner or later to Nature's stern discipline, not all peoples benefit by it. "For, necessarily, families and races whom this increasing difficulty of getting a living which excess of fertility entails, does not stimulate to improvements in production—that is, to greater mental activity—are on the high road to extinction; and must ultimately be supplanted by those whom the pressure does so stimulate."[33] Those who survive the ordeal, whether as individuals or as races, are nature's elect, and their survival insures the onward progress of mankind and the decrease in human fertility required to adjust population to resources. Thus, population pressure is the great engine of progress whether in nature or in society.

It produced the original diffusion of the race. It compelled men to abandon predatory habits and take to agriculture. It led to the clearing of the earth's surface. It forced men into the social state; made social organization inevitable; and has developed the social sentiments. It has stimulated to progressive improvements in production, and to increased skill and intelligence. It is daily pressing us into closer contact and more mutually-dependent relationships. And after having caused, as it ultimately must, the due peopling of the globe, and the bringing of all its habitable parts into the highest state of culture—after having brought all processes for the satisfaction of human wants to the greatest perfection—after having, at the same time, developed the intellect into complete competency for its work, and the feelings into complete fitness for social life—after having done all this, we see that the pressure of population, as it gradually finishes its work, must gradually bring itself to an end.[34]

As the decade of the fifties progressed, Spencer applied the development hypothesis to an ever-widening range of phenomena. In his *Principles of Psychology,* first published in 1855, he endeavored to establish psychology as a branch of evolutionary biology by showing that mental processes, like physiological processes, are modes of adaptation of the organism to its environment and that in both cases nature displays a "progressive evolution of the correspondence between organism and environment." Having established the general similarity of mental processes and life processes in general, Spencer proceeded to trace the various gradations of these modes of awareness of the environment from the simplest organic responses to environmental stimuli to the highest thought processes of human beings, exhibiting the progressive extension of the correspondence of internal and external relations in space and time and in generality and speciality, and indicating the necessary connection between this progressive correspondence and the progressive complication in organic structure, particularly in the nervous system. At this point in the argument, Spencer tacitly introduced the development hypothesis, arguing that the progressive complication in nervous organization and in the accompanying correspondences of internal and external conditions had slowly evolved in response to changes in the environment. On this assumption he was able to show that not only reflex and instinctive sequences but also the very forms of thought, the Kantian categories, were products of mental evolution. "The doctrine that the connections among our ideas are determined by experience," he wrote, "must, in consistency, be extended not only to all the connections established by the accumulated experiences of every individual, but to all those established by the accumulated experiences of every race. The abstract law of Intelligence being, that the strength of the tendency which the antecedent of any psychical change has to be followed by its consequent, is proportionate to the persistency of the union between the external things they symbolize; it becomes the resulting law of all concrete intelligences, that the strength of the tendency for such consequent to follow its antecedent, is, other things being equal, proportionate to the number of times it has thus followed in experience. The harmony of the inner tendencies and the outer persistencies, is, in all its complications, explicable on the single principle that the outer persistencies produce

434

the inner tendencies."[35] Thus, associations of ideas may be inherited.

Throughout the course of this argument Spencer appealed frequently to the facts or alleged facts of human history in support of his contentions. On the one hand, he cited the progress of science and technology as evidence of the ever-widening correspondence between man and his environment. On the other hand, he viewed the progress of humanity-in-general as a product of gradual biological improvement in the human stock, involving the superseding and displacement of inferior varieties of human beings by superior types, these superior types being evolved in response to the more complicated physiographical and social environments in which some races of men found themselves. In Spencer's mind the difference between the "small-brained savage" and the "large-brained European" was the strongest kind of evidence favoring the hypothesis that both brain structure and mental processes were the cumulative product of the life habits of the race, life habits dictated primarily by the conditions of existence. "We know," he declared, "that there are warlike, peaceful, nomadic, maritime, hunting, commercial races—races that are independent or slavish, active or slothful,—races that display great varieties of disposition; we know that many of these, if not all, have a common origin; and hence there can be no question that these varieties of disposition, which have a more or less evident relation to habits of life, have been gradually induced and established in successive generations, and have become organic."[36] The logical implication of Spencer's argument was that the progress of humanity had been accomplished by a competition of races in which superior types produced by more favorable combinations of circumstances had pushed aside those types whose life circumstances had been less conducive to mental growth.

Spencer produced no general work on sociology in this period, but the trend of his thought concerning society was made clear in a series of essays bearing such titles as "The Genesis of Science," "Manners and Fashions," "The Social Organism," "Overlegislation," and the like.[37] In general, the themes of these essays were those of his earlier writings. There was a firm insistence on the genetic approach to the study of social institutions. Society was again compared to a biological organism, and the comparison was worked out in greater detail. Social institutions were held to develop "under the pressure of wants and necessities," their particular forms being dictated by the characters of the peoples involved, these characters, in turn, being the cumulative and slowly-changing product of varying life circumstances. The poles of this evolutionary process were primitive society, represented by the "almost structureless" aggregations of Australian savages, and the highly complex and interdependent society of Europe. The sequence of institutional development was viewed as dictated by the relative importance of the various kinds of social institutions for the survival of the race. First came arrangements for defense, then political institutions, then economic organization and the division of labor, then science, literature, and the arts, each undergoing a progressive complication in its turn. This institutional growth was conceived as self-regulating and beneficent so long as it continued unhampered by well-intended but ill-conceived government regulation.

There was little new in all of this. The main novelty of these years, apart from Spencer's venture into psychology, was his steady progress toward a philosophical synthesis in which the concept of evolution played the central role. As Spencer himself pointed out, the time was ripe for a philosophical reconstruction which would tie together with a few leading concepts and broad generalizations the growing accumulation of empirical knowledge in the various fields of scientific inquiry. In his essay entitled "Progress: Its Law and Cause," published in 1857, he called attention to the fact that developmental concepts had invaded many fields of science, especially anatomy and physiology, but also geology, paleontology, astronomy, linguistics, and social theory. It was in the social sphere that he found his most conclusive evidence of the universal tendency toward progress. The nebular hypothesis had not yet been accepted by astronomers, and although Von Baer's researches in embryology had introduced developmental conceptions into anatomy and physiology, the evidence for and against a general evolution of organic forms was too evenly balanced to permit a clear conclusion. In human history, however, Spencer found overwhelming evidence of a progressive trend from homogeneity to heterogeneity. It was to be seen, he declared, in the multiplication of human races, in the transition from the relatively undeveloped brain of the barbarian to the superior mental capacity of the civilized European, in the growing complexity of political, social and economic institutions, in the evolution of languages, in the growth of science, in every field of human endeavor. Progress was written into the constitution of the universe. It was "not an accident, not a thing within human control, but a beneficent necessity."[38]

A decade and a half of public discussion of Darwin's theory of evolution intervened between the publication of Spencer's essay on the law and cause of progress and the appearance of his major sociological treatises: *The Study of Sociology, Descriptive Sociology,* and *Principles of Sociology*. On the surface it seemed that his general point of view had changed but little. The description of the task of sociological science in *The Study of Sociology* had a familiar ring:

> Setting out, then, with this general principle, that the properties of its members determine the properties of the mass, we conclude that there must be a Social Science expressing the relations between the two. Beginning with types of men who form but small and incoherent social aggregates, such a science has to show in what way the individual qualities, intellectual and emotional, negative further aggregation. It has to explain how modifications of the individual nature, arising under modified conditions of life, make larger aggregates possible. It has to trace, in societies of some size, the genesis of the social relations, regulative and operative, into which the members fall. It has to exhibit the stronger and more prolonged influences which, by further modifying the characters of citizens, facilitate wider and closer unions with consequent further complexities of structure. . . . In every case its object is to interpret the growth, development, structure, and functions, of the social aggregate, as brought about by the mutual actions of individuals whose natures are partly like those of all men, partly like those of kindred races, partly distinctive.
>
> These phenomena have, of course, to be explained with due reference to the conditions each society is placed in—the conditions furnished by its locality and by its relation to neighboring societies. Noting this merely to prevent misapprehensions, the general fact which here concerns us, is that, given men having certain properties, and an aggregate of such men must have certain derivative properties which form the subject-matter of a science.[39]

In the working out of these general principles, however, there were significant changes. The old polarity between primitive society, held together by force and fear, and modern society, based increasingly on voluntary coöperation, was now elaborated in a correlative distinction between militant and industrial societies, the former characterized by compulsory coöperation and centralized regulatory systems, the latter by voluntary coöperation, free exchange, and the diminution of centralized controls. Spencer was no longer certain, however, that the trend from the militant to the industrial type of society was inevitable and irreversible. He listed a whole series of factors which tended to delay and disfigure the transition and conceded the possibility that a society of the industrial type might

437

regress to militancy "if international conflicts recur."[40] Indeed, he saw that very process of reversion taking place before his eyes in Western Europe.

This recognition of the contemporary regression to militancy and centralized controls was but one manifestation of a fundamental reorientation in Spencer's thinking about evolution. In his essay entitled "Progress: Its Law and Cause" Spencer had tended to identify change and development, that is, to think of change as developmental in its very nature, progress from homogeneity to heterogeneity following ineluctably from a basic law of change, namely, that "every active force produces more than one change." In social matters this progress was conceived as characterizing the development of each particular society as well as the development of humanity at large.[41] By the time of the sociological treatises, however, Spencer was less inclined to identify change and progress.

Evolution [he wrote] is commonly conceived to imply in everything an intrinsic tendency to become something higher; but this is an erroneous conception of it. In all cases it is determined by the co-operation of inner and outer factors. This co-operation works changes until there is reached an equilibrium between the environing actions and the actions which the aggregate opposes to them—a complete equilibrium if the aggregate is without life, a moving equilibrium if the aggregate is living. Thereupon evolution, continuing to show itself only in the progressing integration that ends in rigidity, practically ceases. If in the case of the living aggregates forming a species, the environing actions remain constant from generation to generation, the species remains constant. If the environing actions change, the species changes until it re-equilibrates itself with them. But it by no means follows that this change in the species constitutes a step in evolution. Usually neither advance nor recession results; and often, certain previously-acquired structures being rendered superfluous, there results a simpler form. Only now and then does the environing change initiate in the organism a new complication, and so produce a somewhat higher type. Hence the truth that while for immeasurable periods some types have neither advanced nor receded, and while in other types there has been further evolution, there are many types in which retrogression has happened.

As with organic evolution, so with super-organic evolution. Though, taking the entire assemblage of societies, evolution may be held inevitable as an ultimate effect of the co-operating factors, intrinsic and extrinsic, acting on them all through indefinite periods of time; yet it cannot be held inevitable in each particular society, or even probable. A social organism, like an individual organism, undergoes modifications until it comes into equilibrium with environing conditions; and thereupon continues without further change of structure. When the conditions are changed meteorologically, or geologically, or by alterations in the Flora and Fauna, or by migration consequent on pressure of population, or by flight before usurping races, some change of social structure is entailed. But this change does not necessarily imply advance. Often it is towards neither a higher nor a lower structure. Where the habitat entails modes of life that are inferior, some degradation results. Only occasionally is the new combination of factors such as to cause a change constituting a step in social evolution, and initiating a social type which spreads and supplants inferior social types. For with these super-organic aggregates, as with the organic aggregates, progression in some produces retrogression in others: the more-evolved societies drive the less-evolved societies into unfavourable habitats; and so entail on them decrease of size, or decay of structure.[42]

This revised version of evolution afforded a convenient explanation of the failure of some organisms, whether biological or social, to progress. It explained the anomalous fact that the Australian savages, notwithstanding their primitive social condition, had a complicated kinship structure and system of marriage rules—presumably a holdover from a higher stage from which they had regressed. In other respects, however, the new point of view had disturbing implications. It envisaged the progress of humanity as the outcome of race conflict, the progress being made by those races which happened to develop in such a way that they were able to "spread and supplant inferior social types." Such, according to Spencer, had been the history of human progress "throughout long periods," a history marked by "a continuous over-running of the less powerful or less adapted by the more powerful or more adapted, a driving of inferior varieties into undesirable habitats, and, occasionally, an extermination of inferior varieties."[43] But if progress had occurred in this way in the past, why should it not be generated in the same way in the future? And if the militant, highly-centralized form of society was better adapted to survival in the competition of races, what reason was there to regard the peaceable, industrial type as somehow higher or more evolved? Was not survival in the competitive conflict the test of superiority?

Spencer seems not to have grasped the full implication of his revised view of evolution. Having admitted that progress is neither necessary nor probable in any particular society, he nevertheless continued to invoke the analogy of the individual life cycle in opposing programs of radical social change, arguing that such programs served to retard "the normal course of evolution" in societies which adopted them.[44] Again, in his reply to James Martineau's attack on Darwinian evolution he argued that survival of the fittest was not necessarily or even probably survival of the best,[45] ignoring the fact that he had represented and continued to represent social progress as a product of race warfare in which "superior races" supplanted and even exterminated "inferior races." As if to confound confusion, he then proceeded to assert in his *Principles of Sociology* that the time had come when no further progress could be expected to ensue from conflicts between peoples. In the past, he declared, war had been a means whereby the earth was peopled with "the more powerful and intelligent races"; in the future, however, progress would result from "the quiet pressure of a spreading industrial civilization on a barbarism which slowly dwindles."[46] The integration of simple groups into compound and doubly compound ones by military conquest had been carried "as far as seems either practicable or desirable." But what assurance was there that it would not be carried farther by natural causes? Spencer could offer none. He could only contemplate the possibility with growing alarm.[47] And if some nation, claiming to be the master race and the bearer of a higher civilization, should throw down the gauntlet to bourgeois industrial nations, would this not be perfectly natural, and would not the victory of such a nation in war validate its claim to cultural and racial superiority?

To these questions Spencer had no satisfactory answer. The truth of the matter is that his social ideal had never really been grounded in biological science, much as he liked to pretend that it was. His youthful optimism had deluded him into thinking that the historical process was moving steadily toward a free, competitive society, and his interest in the development hypothesis, combined with the omnipresent influence of the natural theology of his day, had suggested the possibility of viewing the progress of history as a simple extension of the progress of nature. But both history and biology betrayed him. The revival of militarism cast a dark shadow over the future of the free society, and Darwinian biology placed the progress of nature in an equally sombre perspective. Biological evolution no longer appeared as a steady progress of organic nature onward and upward in response to the demands of the environment but rather as a haphazard, zigzag course twisting through the whitened remains of those creatures whom the chances of heredity and environment had doomed to extinction. Social evolution took on an equally grisly and uncertain aspect. Faced with this depressing turn of events and ideas, Spencer stood his ground undaunted, somewhat disconcerted and disillusioned but little suspecting that he had helped to fashion an ideology which would be used to justify a barbarous onslaught on the free society.

🔣

The foregoing analysis of two theories of social progress does not go very far toward solving the problem of the relations between biological and social evolutionism in the nineteenth century. A few tentative conclusions would seem warranted, however, some of them drawn directly from the materials presented above, the others from more general considerations.

1. There was no necessary or inevitable connection between theories of biological evolution and theories of sociological evolution in the nineteenth century. Comte's rejection of Lamarck's development hypothesis should make this clear, and the same point could be made with respect to the social evolutionism of Marx and Engels.

2. Even when, as in the case of Comte, the development hypothesis was specifically disavowed, biological concepts and analogies often played an important role in social theory. Thus, Comte conceived the progress of the human race as a kind of temporalization of the great chain of being and invoked Lamarck's law of use and disuse to explain the evolution of human nature. Oddly enough, Comte, though less inclined than Spencer to assimilate social to biological theory, had a much firmer grasp of the concept of organism. Spencer never got beyond conceiving society as an aggregate of individuals.

3. The primary stimulus to speculation concerning social evolution was the growing conviction that the early condition of the human race had been a bestial one, scarcely distinguishable from that of the higher animals. This conviction sprang partly from increasing knowledge concerning the anthropoid apes but even more from growing contact with primitive cultures. It was related to biological inquiry but not specifically to evolutionary biology.

4. Given the conviction that human history was a gradual ascent from brute-like beginnings and the concomitant conviction that all natural and historical events are subject to law, it was inevitable that there should have been attempts to formulate the laws of historical development. It was inevitable, too, that these attempts should have involved heavy borrowing of concepts and principles from older fields of inquiry. In this respect

Comte was much more cautious than Spencer, realizing that sociology must develop concepts appropriate to its own peculiar subject matter.

5. Finally, it was natural that the social evolutionists of the nineteenth century should have attempted to validate their programs of political and social action by claiming the sanction of science for their philosophies of history. As supernatural sanctions were discredited and the prestige of science grew, social prophets assumed the pose of the scientist. But the imposture could not long deceive. Science went on her way, a prolific but cruel mother, forever spawning scientisms and forever abandoning her illegitimate offspring.

References

1 See, for example, Arthur O. Lovejoy, "Some Eighteenth Century Evolutionists," *Popular Science Monthly*, LXV (1904), 238–51, 323–40; "The Supposed Primitivism of Rousseau's Discourse on Inequality," *Modern Philology*, XXI (1923), 165–86; "Monboddo and Rousseau," *Modern Philology*, XXX (1933), 275–96. Also, Gladys Bryson, *Man and Society: The Scottish Inquiry of the Eighteenth Century* (Princeton, 1945); Howard Becker and Harry Elmer Barnes, *Social Thought from Lore to Science* . . . (2nd ed.; Washington, D. C., 1952). Volume I.

2 An exhaustive account of Comte's early years, with comprehensive bibliography, is to be found in Henri Gouhier's *La Jeunesse d'Auguste Comte et la Formation du Positivisme* (Bibliothèque d'Histoire de la Philosophie), (3 vols.; Paris, 1933). See also Franck Alengry, *Essai historique et critique sur la sociologie chez Auguste Comte* (Paris, 1900).

3 Auguste Comte, *Early Essays on Social Philosophy,* tr. Henry D. Hutton, (London, 1911), p. 247.

4 *Ibid.*, p. 193, 324. See also Gouhier, *La Jeunesse d'Auguste Comte*, I, 222.

5 *Ibid.*, pp. 205–206.

6 *Ibid.*, pp. 237–38.

7 *Ibid.*, pp. 149–50. On p. 147–48: "All men who possess a certain knowledge of the leading facts of history, be their historical views what they may, will agree in this, that the cultivated portion of the human race, considered as a whole, has made uninterrupted progress in Civilisation, from the most remote periods of history to our own day." In another passage (p. 308, n.) Comte declares that the words *improvement* and *development* are not intended to suggest "absolute excellence and indefinite amelioration" but rather "a certain succession of states which the human race reaches in accordance with fixed laws." But Comte's disclaimer of a teleological implication, like Spencer's later on, runs counter to the optimistic tenor of his writing and thinking. Neither man shared Rousseau's doubts concerning the ultimate character of human progress. Concerning human nature Comte wrote (*Early Essays,* 61–62): "The majority of men desire power, when placed within their reach, not as an end but as a

means. They value it, less from love of authority, than because their idleness and incapacity disposes them to employ others in procuring enjoyments instead of themselves joining in their labour. The dominant aim of almost all persons is not to act upon Man, but upon Nature."

8 *Early Essays,* 131 ff., 153 ff. See also Comte's *Cours de philosophie positive* (Paris, 1908), IV, 123–24, where he refers to "the evident origin of our fundamental notion of human progress, which, spontaneously issuing from the gradual development of the various positive sciences, still finds there today its most unshakable foundations."

9 Auguste Comte, *Cours,* III, 296.

10 *Ibid.,* III, 296–97. Although Comte rejected the development hypothesis, he had great admiration for Lamarck's other contributions, especially his classification of invertebrate animals. See also Alengry, *Essai historique,* 142 ff.

11 *Ibid.,* III, 158. For Comte's remarks on the relation of natural history to biology, see pages 185–86, 247–48. His conception of biology was profoundly influenced by the writings of Marie Francois Xavier Bichat and Henri Marie Ducrotay de Blainville. See in this connection Henri Gouhier, "La Philosophie 'Positiviste' et 'Chretienne' de D. de Blainville," *Revue philosophique,* CXXXI (1941), 38–69. See also *Cours,* III, 151 ff.

12 *Ibid.,* III, 278–279. Comte rejected the newly-proposed cell theory on the ground that it involved an attempt to reduce biology to physics and chemistry. See p. 279.

13 *Ibid.,* III, 301.

14 *Ibid.,* IV, 285. See also volume III, 325, 403–4, 422.

15 *Ibid.,* IV, 167. See also pp. 170, 183–85.

16 *Ibid.,* IV, 310.

17 "It is clear, indeed, that not only political institutions properly so-called but social customs on the one hand and the *mores* and ideas on the other are reciprocally solidary and that this entire ensemble is connected constantly by its very nature to the corresponding state of integral development of humanity considered in all its various modes of activity whatever, intellectual, moral, and physical, concerning which no political system, whether temporal or spiritual, could ever have any other real aim than to regularize suitably the spontaneous activity in order to direct it better toward a more perfect accomplishment of its natural, previously determined end." *Ibid.,* IV, 176.

18 *Ibid.,* III, 156.

19 *Ibid.,* IV, 201. In Comte's opinion, man was the most modifiable of all animals. One of his objections to Lamarck's development hypothesis was that it seemed to imply a higher degree of variability in the lower animals than in man. See *Cours,* III, 297–98.

20 *Ibid.,* IV, 329, 341, 374; VI, 506, 511–12, 515–16. On pp. 511–12: "In every degree on the sociological scale and in all its statical and dynamical relations, biology necessarily furnishes concerning human nature, insofar as it can be known by considering the individual alone, the fundamental notions which must always control the direct findings of sociological exploration and even

correct and perfect them. Moreover, in the lower part of the series, without descending to the initial state in which biological deductions alone can guide us, it is clear that biology, although always dominated . . . by the sociological spirit, must especially make known that elementary association, spontaneous intermediary between purely individual existence and a full social existence, which results from domestic existence properly so called, more or less common to all the higher animals, and which constitutes in our species the true primordial base of the larger collective organism. However, the original elaboration of this new science [sociology] must be essentially dynamical in view of the fact that the laws of harmony have there almost always been considered implicitly as laws of succession, the clear appreciation of which could alone constitute social physics today. Also, its highest scientific connection with biology consists now in the fundamental link which I have established between the sociological series and the biological series and which permits envisaging philosophically the first as a simple, gradual prolongation of the second, although the terms of the one are coexistent and those of the other successive. Except for this unique general difference, which does not prevent connecting the two series, we have, in effect, recognized that the essential character of human evolution results necessarily from the ever-increasing preponderance of the same higher attributes which place man at the head of the animal hierarchy. . . . One comes thus to conceive the immense organic system as actually linking the least vegetative existence to the noblest social existence by a long intermediate progression of higher and higher modes of existence whose succession, although necessarily discontinuous, is nevertheless essentially homogeneous."

21 *Ibid.*, IV, 191. For Comte's views on the action of climate, race, etc. in accelerating or retarding the normal course of social evolution, see *ibid.*, IV, 208–10, 332 ff.

22 Herbert Spencer, *Autobiography,* I, 201. See also David Duncan, *Life and Letters of Herbert Spencer* (New York, 1906), II, 156 ff.

23 Herbert Spencer, *Social Statics: Or, the Conditions Essential to Human Happiness Specified, and the First of Them Developed* (New York, 1882), 447. This edition is a reprint of the original edition of 1850. The views set forth in *Social Statics* were adumbrated in Spencer's first extensive political essay, "On the Proper Sphere of Government," published as a series of letters in *The Nonconformist* in 1842. During the rest of the decade of the forties Spencer broadened the range of his inquiries, making excursions into biology and psychology. He was much interested in Robert Chambers' *Vestiges of the Natural History of Creation* (1844) and in the phrenological doctrines of Gall and Spurzheim. In an article in the *Philosophical Magazine* in 1844, "On the Theory of Reciprocal Dependence in the Animal and Vegetable Creations as Regards Its Bearing upon Palaeontology," he propounded the theory that the amount of oxygen in the earth's atmosphere had been steadily increasing and that there must consequently have been "a gradual change in the character of the animate creation" involving "a continual increase of the hot-blooded tribes and an apparent diminution of the cold-blooded ones."

24 *Ibid.,* p. 79.

25 *Ibid.,* pp. 448–49.

26 *Ibid.,* pp. 80, 447–48, 467. Like Comte, Spencer regarded the progressive complication of social institutions as analogous to the progressive gradation of living forms in nature. Unlike Comte, however, he viewed both types of complication as products of temporal evolution. See *Social Statics,* pp. 480–81, 493.

27 *Ibid.,* p. 384.

28 *Ibid.,* pp. 413, 426–27.

29 *Ibid.,* p. 119.

30 *Ibid.,* p. 29.

31 Letter from Herbert Spencer to Edward Lott, dated April 23, 1852, as quoted in Duncan, *Life and Letters,* I, 80–81.

32 Herbert Spencer, "A Theory of Population, Deduced from the General Law of Animal Fertility," *The Westminster Review,* LVII (1852), 498–99.

33 *Ibid.,* pp. 499–500.

34 *Ibid.,* p. 501.

35 Herbert Spencer, *The Principles of Psychology* (London, 1855), p. 529. Spencer first openly espoused the development hypothesis in a brief article in *The Leader* in March, 1852. Shortly thereafter he came across Von Baer's description of embryological development as a passage from homogeneity to heterogeneity and seized upon it as a clue to the character of organic development generally. In 1858 he attacked Richard Owen's theory of ideal types and defended the nebular hypothesis. In the following year he launched into the field of botany with an essay on the modification of leaf forms by the action of environmental forces, adducing evidence suggesting "that the forms of all organisms are dependent on their relations to incident forces," the hereditary type changing slowly in response to the continued action of these forces. A few months later appeared his review of Hugh Miller's geological works, devoted to proving that the discontinuities in the paleontological record were not conclusive against the development hypothesis. In private conversations with Huxley, too, he pressed his favorite conception, but without success. See the *Autobiography,* I, 445 ff.; II, 321–22. Also, Herbert Spencer, *Illustrations of Universal Progress; A Series of Discussions* (New York, 1878), in which many of the articles referred to above are reprinted.

36 *Ibid.,* pp. 526–27. Again, page 581: "Every one of the countless connections among the fibres of the cerebral masses, answers to some permanent connection of phenomena in the experiences of the race."

37 See Spencer's *Illustrations of Universal Progress,* cited above.

38 Herbert Spencer, "Progress: Its Law and Cause," *ibid.,* p. 58.

39 Herbert Spencer, *The Study of Sociology* (New York, 1881), pp. 52–53.

40 Herbert Spencer, *The Principles of Sociology* (New York, 1877), I, 600.

41 "The change from the homogeneous to the heterogeneous is displayed equally in the progress of civilization as a whole, and in the progress of every tribe or nation; and is still going on with increasing rapidity." See "Progress: It's Law and Cause," *Illustrations of Universal Progress,* 12. In this same essay Spencer

attempted to strip the idea of progress of its teleological implications, but his description of progress as a "beneficent necessity" indicated that he still thought of the world process as fraught with good for mankind. See p. 58.

42 Herbert Spencer, *Principles of Sociology,* I, 106–8.

43 *Ibid.,* I, 43.

44 "Doubtless, from all that has been said it follows that, if surrounding conditions remain the same, the evolution of a society cannot be in any essential way diverted from its course; though it also follows . . . that the beliefs and actions of individuals, being natural factors that arise in the course of the evolution itself, and aid in further advancing it, must be severally valued as increments of the aggregate force producing change. . . . For though the process of social evolution is so far pre-determined that no teaching or discipline can advance it beyond a certain normal rate, which is limited by the rate of organic modification in human beings; yet it is quite possible to perturb, to retard, or to disorder the process. The analogy of individual development again serves us. . . . Growth and development may be, and often are, hindered or deranged, though they cannot be artificially bettered. Similarly with the social organism. Though, by maintaining the required conditions, there cannot be more good done than that of letting social progress go on unhindered; yet immense mischief may be done in the way of disturbing and distorting and repressing, by policies carried out in pursuance of erroneous conceptions." *Study of Sociology,* p. 396.

45 Herbert Spencer, "Mr. Martineau on Evolution," *Various Fragments* (New York, 1908), p. 379.

46 Herbert Spencer, *Principles of Sociology,* II, 663–64.

47 *Ibid.,* II, 590 ff.

RICHARD H. SHRYOCK

on the Papers of J. Walter Wilson and John C. Greene

In Professor Wilson's paper we have a most effective study, which presents the simultaneous establishment of continuity in cells (Virchow), in organisms (Pasteur), and in species (Darwin). I doubt if many of us, even if we were familiar with the facts, had heretofore brought these three into such juxtaposition in our own minds. Each of these achievements, of course, nicely illustrates the penchant of science for bringing order into apparently disparate phenomena.

Professor Cohen has already commented on the similarity of these developments in biology with concomitant work in the physical sciences on the conservation of energy. There may be distinctions between "continuity" in the one area and "conservation" in the other, but there is at least a close analogy between them—on the basis of which we might take off on a flight of speculation about the *Zeitgeist* of the period. While resisting this temptation, I will only make the obvious point that each of the developments discussed by Dr. Wilson brought order into a particular field as well as into biology as a whole. Virchow's cellular ideas, for example, provided a unifying concept in pathology, which that discipline had sadly lacked since the abandonment of the speculative theories of preceding centuries.

Since I have only admiration for Dr. Wilson's general synthesis, there would be no purpose in recalling each section of his paper in turn. Nor would there be any point in quibbling about isolated terms or statements; as in asking whether Darwin fully "proved" his theory of evolution; or in inquiring, again, whether preformation theories in embryology required a more "amazing" flight of imagination than is now needed in order to conceive of the potentialities of genes. It seems preferable, rather, to explore the implications of certain terms or assumptions in the paper, in order to see what larger questions these may raise.

Dr. Wilson offers, for example, a structural differentiation between living and non-living materials. Implicit here, apparently, is the old problem of a definition of "life"; the answer being that living things develop their own forms, while the non-living do not. Incidentally, I am in some confusion about the facts. I would have thought, as a layman, that protozoa multiplying by fission do not determine their own forms—that two baby parameciums would bear a striking family resemblance to mama-papa paramecium, even as little crystals take the same shape as big crystals. Assuming, however, that this is not true and that all organisms *do* develop their own forms in a distinctive manner, have we here anything more than a striking coincidence which has no bearing on the basic nature of living things? Perhaps Dr. Wilson was not implying that it did. But, in any case, I am not clear from the context whether he does or does not retain a teleologic concept of "life."

In discussing the background to Virchow, Pasteur, and Darwin—and especially that of the latter—Dr. Wilson recalls the development of historical geology, of animal and plant breeding, and of economic theory. He omits but could just as well have included also the individual precursors of Darwin; that is, those who actually formulated Darwin's thesis or came so close to it that the distinctions are subtle. Perhaps Dr. Wilson does not consider these men significant; but in this audience, biologists should assert their claim to just such precursors—perhaps even to bigger and better precursors—than those exhibited by the physical scientists. In Darwin's case, the problem is that of the so-called discoverer who flatly denies any knowledge of those who anticipated him. One can, of course, take this as final and assume that all earlier theses were still-born; if so, this fact in itself raises other problems. But so simple an answer presents difficulties, when we recall that the number (or "density") of anticipations increases as we approach Darwin's date.

One may revert at this point to some earlier comments made by Father Clark and by Professor Hall. The former injected a subjective note in viewing "precursoritis" as an affliction of historians; and Dr. Hall, if I recall correctly, referred to the historians' bent for revealing the new or unexpected—which we do when we uncover anticipations.

One may add to these motivations another which seems more respectable; namely, the desire of serious historians to do just what the mid-nineteenth century biologists did—to show continuity in development. Historians, indeed, often distinguish themselves from antiquarians, by claiming that they discover continuities in the past while antiquarians see only contrasts. Many historians, I think, would view the very commonness of precursor phenomena as significant in itself; for this suggests that basic ideas must be presented a number of times before final "discovery"— whether or not each intermediate advance or connecting link can be demonstrated.

The historian who is over-anxious to find precursors may, no doubt, "read them into" his data; and this, I presume, was what Father Clark had in mind. But is there not another subjective factor which counteracts this among historians at large; namely, the lure of the "great man" concept or of what Carlyle once called "heroes and hero worship?" In extreme form, this may also approach a morbid state—a sort of elephantiasis which is projected outward into historic figures, swelling them up out of all proportion to their contemporaries. The scholar in this state of mind will not tolerate anticipations of his hero.

One of the most arresting statements in Dr. Wilson's paper is the assertion that biology attained "maturity" by about 1860. This figure of speech is appropriately biological and seems justified by the context. Yet it also suggests some intriguing questions. If biology became mature after 1860, can it in some sense become "more mature" in the future; let us say, by analyzing its problems in molecular terms and so revealing continuity between organic phenomena and the inorganic? Or does the term "maturity" imply the inevitability of future senescence?

Let us not, however, quibble about figures of speech. What Professor Wilson is probably implying here, is what most of us would assume; namely, that there is observable progress or advance in sciences toward levels which are defined by our contemporary standards or understanding. We imply this more definitely if we speak of the "modernization" of a science.

Assuming that there is such progress toward maturity or modernization, note that Dr. Wilson recognizes or identifies it—in the case of biology —with the formulation and general acceptance of certain concepts anent cells, organisms, and species. The significance of conceptualizations in science is certainly profound and Dr. Wilson has given us an excellent illustration of this. There is, however, another way of recognizing major advances in science; that is, in terms of basic improvements in logic or methods. To what extent, for example, can we say that seventeenth-century dynamics advanced beyond late-medieval dynamics—despite the latter's promising concepts—primarily because the former actually did become an "experimental philosophy"?

It may be said, of course, that we have here only a logical distinction between aspects of science which are really inseparable—concepts and methods. But have there not actually been times, places, and disciplines in which concepts seem to have been more promising than methods, or vice versa? The first case, as noted, may be illustrated by medieval dynamics; the second, by eighteenth-century pathology. In the latter case, the Newtonian model was available, experimental and quantitative procedures or devices were well known, and medical men often asserted that their problems should be reduced to those "of matter and motion." Yet experimental and quantitative procedures were rarely employed and the discipline exhibited a sort of confused adolescence rather than anything approaching maturity. This confusion may be ascribed in part to the simple need for data; but even more, I believe, to the ineffectiveness of prevailing concepts. The latter were such that they simply did not call for any resort to the methods available, however promising the latter may have been in themselves.

Much the same thing may be said of the introspective psychology of the same period and of certain other disciplines, then or later. Indeed, one may picture the history of science as a whole, after 1700, as a continuing effort within each discipline to attain such concepts as would permit the use of effective methods. The Newtonian model, with its conjunction of effective concepts and methods, was visible to all; but it required time to implement the scientific revolution in one science after another.

The problem of why and how this revolution finally occurred, in turn and in varying degrees, in chemistry, in clinical medicine, in psychology, and in sociology, is almost as fascinating as is the problem of the primary revolution in dynamics. Few of these secondary transformations have been studied as carefully as has the first, but they make one thing clear. The Galileo-Newtonian epic might conceivably be interpreted in terms of the social environment and intellectual climate of the seventeenth century. *Any* revolution occurring within a single century *could* be so interpreted. But the time interval between, let us say, the transformation of dynamics in the 1600's and that of psychology in the 1800's, is such that these occurred against quite different social and cultural backgrounds. The implication here seems to be either that (1) the social and cultural milieu of the first period was especially *en rapport* with dynamics, and that of the second, with psychology; or (2) that sciences at large exhibit considerable independence of their social and cultural settings.

450

If I have wandered somewhat afield in these remarks from Dr. Wilson's paper, he must at least take the responsibility for having provided a starting point. In the language of our opening discussions, indeed, I would say that he provided me with considerable impetus.

⚜

Professor Greene, in his stimulating paper, provides the only analysis of the history of social thought which has been presented in these discussions. Hence the first question which poses itself, is whether we have here a sudden jump between science and another order of thought—a discontinuity in our discourse. True, Dr. Greene's subjects for analysis—Comte and Spencer—were pioneers in elaborating a claim to having established a "science of society." Yet the contrast between the whole nature of, let us say, Galileo's concepts, methods or results and those of Spencer, was so obvious that it gave rise to the scorn which natural scientists have exhibited toward the very idea of "social science."

Professor Greene's penetrating analysis of both Comte and Spencer demonstrates, nevertheless, that we cannot dismiss social science in so cavalier a fashion. True, both men are shown to have owed much to earlier historians, and the latter hardly then made any pretense to being scientists—whatever they may have held since. But Spencer derived his ideas about social evolution, and even of the evolution of the cosmos, from such sciences as geology and biology. And Comte, although he curiously rejected evolution in organisms while affirming it in society, took some of his ideas about the latter from the medical literature.

Up to this point, one can only say that science influenced social thought, even as it did literature or the arts. But both Comte and Spencer were more than just "influenced" by science. Each was convinced that he could actually demonstrate laws of social evolution, even as laws could be demonstrated in dynamics. That they failed in this seems implicit in Dr. Greene's paper. Why this failure? Perhaps it resulted from their normative outlook; each was anxious to "teach lessons" to society, even as had the early-medieval naturalists. And is not such a moral purpose inconsistent with the objectivity long since demanded by natural scientists?

More serious, I think, was the fact that Comte's and Spencer's concepts were so general and sweeping that they did not lend themselves to verification except by debate. How much direct observation, to say nothing of experimentation or quantification, was involved in the work of either man? I would therefore suggest that we have here a special case of the situation, in which interesting concepts were presented but in such a form that they were not susceptible to verification by effective methods.

Far from saying that the thought of Comte and Spencer was not scientific, however, I would say that it was *potentially* scientific in much the sense that late-medieval dynamics may be said to have been potentially scientific. They provided general conceptions which still seem to influence us as historians; for example, Comte's view that science is somewhat autonomous from society and therefore a prime mover in history. Or Marx's concomitant and reverse opinion, mentioned by Dr. Greene in an earlier version, that science is not autonomous but emerges from the social environment. More than this, I assume that Quetelet and his colleagues were influenced by Comte—though Dr. Greene may correct me on this—and, if so, Quetelet illustrates how quickly a "social physics" could and did shrink its concepts to a form susceptible to effective (quantitative) treatment.

Dr. Greene, in his earlier version, raised the interesting question whether theories of biological and of social evolution arose independently, were derived from one another, or continuously interacted? He examines the problem in terms of the influence exerted by biologic ideas on sociology, rather than vice versa; and makes it clear that Spencer was so influenced, while Comte was not. Hence we have only one case each way; though, in the earlier version, Marx is also presented as owing little to biology. I wonder whether Dr. Greene thinks that social evolutionists of the later nineteenth-century can be viewed, in a general way, as reflecting pre- or post-Darwinian, evolutionary viewpoints in biology? Or do the examples of Comte and Marx suggest that social evolutionary ideas would have evolved, shall we say, no matter what had occurred in biology?

Meantime, can one reverse the order here and find evidence of an influence exerted by social evolutionary concepts upon the biologic? Darwin was indebted to Malthus, but hardly for an evolutionary concept. Finally, what are we to make of the fact which Dr. Greene notes; namely, that both the biologic and the social concepts begin to emerge in the same period (late eighteenth and early nineteenth centuries), regardless of what connections can be shown between them in particular cases? Does this imply his alternative thesis of interaction—of a give-and-take rather than of a one-way process? Or can we conceive of still another alternative, in which both biologic and social evolutionary concepts arose independently out of a common background? And, if so, can this background be made more definite than that which is implied in a mystic-whole termed the *Zeitgeist?*

CONWAY ZIRKLE

on the Papers of J. Walter Wilson and John C. Greene

I do not believe that it is either possible or desirable to comment on the whole of Professor Wilson's most informative paper. If I did, I could only say that he has marshalled most of the pertinent facts, organized them logically, and reached conclusions that most biologists will accept. However, in one or two instances, I believe, his remarks could be extended a bit further, provided of course that he had the time. Nineteenth-century biology not only attained its maturity as Professor Wilson indicated, it also made a major impact on nineteenth-century thought. Indeed, it modified all of the really serious thinking of the period, so much so, in fact, that few of the contemporary sciences escaped its influence. The social sciences were especially affected, and some of the social scientists found it necessary to keep themselves informed of every new and major biological advance. A few even sought to discover a compensatory flow of ideas from the social to the biological sciences, particularly as they influenced the personal development of Charles Darwin.

Today the view is widely held that Darwin was stimulated to discover that individuals in nature competed with each other for their very existence, through his familiarity with the general Victorian idea of the respectability of competition in business. This is a view I shall challenge for it has led, I believe, to an inaccurate interpretation of the factors that stimulated the great nineteenth-century growth of biology.

First, I would like to point out that two of the three scientists described by Professor Wilson, the scientists who did so much for biology a hundred years ago (Virchow, Pasteur, and Darwin), made contributions that were essentially negative. (1) Virchow, like his predecessors, held that cells originated through the division of pre-existing cells, but, in addition, he emphasized that they were *not* formed in any other way— they were not produced through free cell formation. (2) Pasteur, for his part, showed that bacteria were *not* generated spontaneously. Even Darwin demonstrated that it was *not* necessary for each species to be created by a special act of the Deity. Thus these contributions that added so much to the science of biology illustrate very vividly a point that is frequently missed, and that is that a large part of scientific advance consists in discarding errors. Every time we eliminate one of two competing hypotheses, we take a definite step forward.

The work of Virchow, Pasteur, and Darwin cleared away a lot of rubbish, and it allowed biology to concentrate on ideas that were exceptionally fertile. Now, the contributions of these three great men are major contributions which fit together well, and today we see that they were all part of the same picture. Professor Wilson puts them together in his paper and, to us at least, they belong together. They supplement each other beautifully. There is little doubt, however, that, to Virchow, Pasteur, and Darwin, the unity underlying their contributions was not at all obvious. They certainly did not look upon themselves as members of a team. Virchow, for instance, was exceptionally conservative; he bitterly opposed the teaching of evolution in the German schools. Pasteur was a believing Christian, and his disproof of the biogenesis of bacteria did not fit into the current, mechanistic picture of a unified cosmos, subject to invariant law, and run by mechanical means. And Pasteur apparently was quite happy with this implication of his work. If life could not be generated equivocally, it would have to be called into being univocally—it would have to be created, supposedly by the Deity. Darwinian evolution thus could not be extended to include the origin of life itself; something extra was needed. As Professor Wilson pointed out, the spontaneous generation of the higher forms of life made evolution unnecessary (Aristotle, for example, discussed the question as to whether an animal, produced spontaneously, belonged to the same species as one born in the usual manner), but if evolution were to be a unifying philosophical force, it needed spontaneous generation as a starting point. Darwin himself did not attempt to extend evolution this far. But today the origin of life is a live and lively issue, and today we feel that evolution, as a unifying principle, should explain how living carbon compounds evolved from those that were not living.

Now to the possible role of the prevailing Victorian climate of opinion upon the biological discoveries of Charles Darwin! Darwin *was* influenced by Malthus, as he himself stated very frankly. Alfred Russell Wallace was also led to his discovery of natural selection by reading Malthus' famous *Essay on the Principles of Population*. Obviously Malthus, a clergyman and an economist, did play an important part in the development of the theory of evolution. Thus there does seem to be a *prima facie* case for the influence of the nineteenth-century social thought on the growth of biology. Let us examine some of this evidence first; I shall attempt to refute it later.

The analogy between the struggle for existence in nature, and business competition in a *laissez faire* economy is so clear that almost as soon as Darwin brought natural selection to the attention of the scholarly world, it was recognized and commented upon. As early as January 16, 1861, Karl Marx stated in a letter to Ferdinand Lassalle, "Darwin's book is very important, and serves me as a basis in natural selection for the class struggle in history," and a little later on he wrote to Friedrich Engels on June 18, 1862, "It is splendid that Darwin again discovers among plants and animals his English society with its division of labour, competition, opening up of new markets, 'inventions' and Malthusian 'struggle for existence.' This is Hobbes' *bellum omnium contra omnes,* and reminds one of Hegel in his *Phenomenology* in which civic society is expressed as the 'spiritual animal kingdom' whereas with Darwin the animal kingdom represents civic society."

On March 29, 1865, Engels wrote to F. A. Lange, "I too was struck, the very first time I read Darwin, with the remarkable likeness between his account of plant and animal life and the Malthusian theory. Only I came to a different conclusion from yours: namely, that nothing discredits modern bourgeoise development so much as the fact that it has not succeeded in getting beyond the economic forces of the animal world." Later, in the *Dialectics of Nature,* Engels stated, "The whole Darwinian theory of the struggle for life is simply the transformation from society to organic nature of Hobbes' theory of *bellum omnium contra omnes,* and the bourgeoise economic theory of competition . . ."

The Marxians were not the only ones who saw the analogy between natural selection and *laissez faire* business competition. Walter Bagehot also saw the connection, and he went so far as to extend natural selection from the biological to the social sciences, but in this Herbert Spencer had anticipated him. Spencer had even anticipated Darwin's use of competition between different types as a selecting agent. In 1852, seven years before the *Origin of Species* was published, Spencer claimed that competition served as a selecting agent in human society, and thus he adumbrated the doctrine that was later to be called "social Darwinism." Soon this doctrine dominated orthodox economics where it served to integrate the newer social sciences into the overall biological picture. Both biologists and social scientists were concerned professionally with a species that had emerged from and dominated all other living forms. Before long, social Darwinism permeated the thinking of the Victorian Age, and it is only recently that we have recovered from it.

Now Charles Darwin lived in Victorian society. Moreover, he stated definitely, just as Alfred Russell Wallace did, that he had gotten his idea of natural selection from reading Malthus' famous *Essay* on population. In this connection he saw that the over-production of young, with the consequent enormous death rate, could cause species to evolve, i.e., if the young had heritable variations and if the death rate were different for different types—if the different types competed for existence. Thus on the surface, Darwin would seem to be a real child of his age, and it would only be reasonable for us to assume that he was led to his explanation of evolution by the ideas that he got from the age in which he lived.

However, if we examine the historical evidence in detail, and give a rigid logical analysis to the ideas involved in Darwin's contribution—the ideas implied in natural selection—the matter seems far from settled. Direct conclusive evidence to the effect that *laissez faire* influenced Darwin is really lacking, but of course we can never establish a negative conclusion from a mere lack of positive evidence. But when we examine and evaluate the ideas that did impinge on Darwin, we can show that he was influenced by sources other than those derived from the business ideals of Victorian society. And these sources were sufficient in themselves to suggest natural selection. Indeed, they led others to the concept of natural selection and in circumstances where *laissez faire* did not enter the picture at all. It is possible that the current Victorian belief in the *rightness,* naturalness, and efficiency of business competition had no influence at all on Darwin, and certainly did not lead him to his epoch-making contribution.

This point is not trivial. It is relevant to one of the most important problems that faces the historians of science, and it has recently been the subject of an important controversy. It is very pertinent to the problem of planning science—to the problem of how to support science adequately in an age that needs great and continuing technicological advances. Very recently a whole school of thought existed that insisted that science was stimulated to advance by the needs of society, and that the direction of its progress was socially controlled. Hence, this school propagated the doctrine that the prevailing climate of opinion not only directed the scientists as to how they should work, but it also gave the scientists their ideas and motivation. In contrast to this view, many of the scientists themselves believe that the progress of science is conditioned by its own internal logic and not by the desires of a dominant social class, or even by the technical needs of a whole culture. It is in the light of these conflicting claims that the personal and scientific development of Charles Darwin becomes so important. The question at issue can be stated as follows: Was Darwin led to his explanation that the evolution of new species came about through natural selection primarily by what he absorbed from his social environment or was he led to it by his own work in biology?

Let us assemble the facts and examine their logical connection. Let us develop them also according to their own internal logic. I shall propound eight numbered propositions.

1. The concept of natural selection does not depend on that of *laissez faire*. We know this because natural selection is by far the older of the two. In classical times, natural selection was described by Empedocles, Epicurus, and in great detail by Lucretius.

2. A belief in natural selection does not necessarily lead to a belief in evolution. The classical philosophers named above used it to explain the existence of adaptation (through the natural death of the unadapted), and hence as an hypothesis to compete with teleology. Aristotle preferred teleology to natural selection and rejected natural selection explicitly and completely. Lactantius attacked natural selection violently, but the concept persisted, although it lived in great philosophical disrepute. In the eighteenth century Denis Diderot preferred natural selection to teleology, and showed how it could explain adaptation.

3. Natural selection can explain the existence of adaptation even in the total absence of any form of competition. Logically, it is independent of *laissez faire*. It can operate where there is no "struggle for existence" at all, and it can even explain some evolution without resorting to population pressure. In fact, it was first used to explain evolution in 1813 by W. C. Wells, and Wells did not utilize any form of competition as a selecting agent. In all marginal environments, the better adapted live while the unadapted are destroyed by the environment itself.

4. The struggle for existence induced by population pressure does not necessarily lead to natural selection or to evolution. In a homozygous

population nature cannot select, no matter how great the struggle. If all of the individuals of a species are equal in their biological inheritance, that species cannot evolve.

5. Darwin acknowledged his indebtedness to Malthus, but Malthus was an anti-evolutionist. His opposition to "perfectability" as described by Condorcet and Godwin was so intense that he rejected the possibility of any unlimited alteration of living forms. He stated definitely that a rat would never become as large as a sheep. Darwin could not have gotten his idea of evolution from Malthus. Malthus had the concept of natural selection right in his hands but he muffed it. He did not get the idea that the enormous death rate in nature could be differential.

6. Evolution grew up in Darwin's mind by an organization of the following concepts and observations: (a) species are unstable and changing units and old species produce new ones, as is shown by the regularities in their distribution; (b) varieties, if altered sufficiently, develop into species; (c) varieties can be altered by artificial selection, as was shown by the work of Roger Bakewell, and by the eighteenth- and nineteenth-century animal breeders; (d) given heritable variation to start with, population pressure in nature may alter the wild forms just as artificial selection changes the forms of our domestic breeds. Malthus gave Darwin the idea of population pressure *ca.* 1838, and twenty years later he gave the same idea to Alfred Russell Wallace.

7. The struggle for existence in nature did not seem to be analogous to competition in business until after Darwin published his *Origin of Species*. Before this, the two concepts seemed to be quite distinct, except to Herbert Spencer. For example, the struggle in nature had been described by Schopenhauer and by Tennyson as a brutal, amoral and vicious affair, one that no civilized man could contemplate without pain. On the other hand, to the Victorians in general, business competition seemed to be a very different affair; to them it was merely an effective device for securing a free market. The Victorian business man probably did not look upon himself as the Marxians looked upon him—he did not consider himself to be a man motivated by jungle ethics. He may not even have looked upon himself as a reactionary and inhuman creature, who was motivated primarily by greed. It is true that he did not have the highest social position in Victorian society, but in general he was considered a worthy and virtuous citizen. The resemblance between natural selection and business competition thus is not immediately obvious to the average citizen.

8. Charles Darwin himself offers us a problem: (a) He had a real talent for ignoring ideas outside of his own specialty; (b) but he also did not investigate the history of his specialty. He did not know that natural selection had been used to explain evolution by Wells (1813) or Matthews (1831), until after he had published the *Origin of Species*. As a rule, his historical notes would be published in the second editions of his works, and he stated very honestly that they had been called to his attention by his friends. His limited interests would indicate that he got his ideas from his specialty, but his lack of historical research does not allow us to ascribe his ideas to his predecessors in biology. Darwin certainly *thought* that the concept of natural selection was his own original creation, as he showed by his statements concerning Wallace when the latter discovered natural selection independently. I believe that Darwin was right. He stated very frankly just where he got the component parts of the theory, but the synthesis of its parts he held to be his own.

This discussion is focussed on a single passing comment of Professor Wilson. It is not, of course, an adequate discussion of his whole paper.

To me the most striking feature of Professor Greene's paper is his demonstration of the remarkable parallelism in the development of the nineteenth-century biological and social sciences. This parallelism is especially clear as the sciences approach the concept of uniformitarianism, in their attempt to explain the successive stages in the development of man and his society. Both the biological and social sciences were groping toward a theory of evolution and both groped in the same way. It seems remarkable that their explanations of evolution fell into the same basic patterns.

There are really innumerable theories of evolution but all of the theories seem to fall into one of three groups. Granting the *fact* of evolution, the causes assigned to it are not hard to classify. For my own convenience I have placed them under the following three heads: (1) some form or other of mysticism, (2) the inherited effects of acquired modification, and (3) the preservation of certain variations by nature and the elimination by nature of others. While the first of these categories is little more than a lumber room, whose contents were soon discarded by most scientists, the second and third categories remained in good standing throughout most of the nineteenth century. As the three groups are not mutually exclusive, many thinkers built hypotheses that include concepts taken from two or even from all three, but every thinker, to the best of my knowledge, tended to emphasize the concepts taken from a single group and to reduce the concepts from the other groups to ancillary roles.

1. Under the rubric "mysticism" we can group all of the teleological explanations of evolution, all concepts of natural tendencies toward perfectability, all doctrines of orthogenesis or notions as to the inevitability of progress, all beliefs in the "science" of historical development, and, of course, all of the doctrines that held that the human species was the ultimate end and aim of some Deity (i.e., all forms of deism) or that mankind was the true apex of the Cosmos itself (i.e., all forms of humanism).

Professor Greene has shown excellently, I believe, how one or another of these concepts influenced the thinking of the early nineteenth-century evolutionists. August Comte, for example, had basically a mystical conception of evolution although he did rely in part on the inheritance of induced modifications.

2. That acquired characters are inherited is an extremely ancient hypothesis. The myth of Phaethon shows that it existed in the Bronze Age. It was endorsed by Hippocrates, Aristotle, Antigonus, Strabo, Pliny, Plutarch, Suetonius, etc. I have recorded some hundred descriptions of it before that of Lamarck. It had been a generally accepted view for about 2500 years. Erasmus Darwin (1794) and Lamarck (1802) used it to account for the origin of new species. The humorless Lamarck, however, brought the doctrine into a half-century of disrepute but it was later adopted as a subordinate hypothesis by Charles Darwin and thus was revived until the decade 1890–1900. Now it has no scientific standing whatever, although it is still an official doctrine in the Communist world and it is still an integral part of Marxian biology.

As Professor Greene showed, Comte accepted the inheritance of acquired characters as did Herbert Spencer. I would like to call attention here to a point made by Professor Greene that might be misunderstood. He says, "He [Spencer] was first attracted to Lamarck's hypothesis on reading Lyell's refutation of it in his Principles of Geology." Now today Lamarck's hypothesis means the inheritance of acquired characters, and Professor Greene's statement might be interpreted to mean that Lyell did not believe in the inheritance of such characters. Lyell did believe in this inheritance and to this extent he was a Lamarckian. When Lyell wrote the *Principles of Geology,* however, he just did not believe in organic evolution and to this extent he rejected Lamarck. Herbert Spencer remained a Lamarckian his entire life and endorsed the doctrine as late as 1890—even after Weismann had brought it into some deserved disrepute.

Both Marx and Engels accepted the inheritance of acquired characters because it was basic to their scheme of social betterment. In fact, Lamarckism is so much a part of Marxism that the Communists in the Soviet Union have made it canonical. It has also been incorporated frequently in the doctrines of others who wanted to make the world over in a hurry, e.g., it was endorsed by Condorcet and by Godwin and many others. Even today it lives a sort of underground life in a number of disciplines that have little or no contact with biology.

3. The theory of natural selection and its use to explain evolution has had a most complex history. In classical times, natural selection was an hypothesis which competed with teleology. Early in the nineteenth century (Wells, 1813; Matthews, 1831) it was used to explain evolution. Today natural selection plays *the* major role in the theory of evolution. In fact, we know now that evolution depends upon it, that evolution could not take place without it, and that it operates on three different levels. (a) On the first level it explains the evolution of fitness or adaptation to the environment, e.g., in a xerophytic environment, the plants that survive are those best adapted to drought; in the northern sector of a species' range those individuals survive which are winter hardy, etc. On this level, natural selection can operate in the total absence of population pressure or of any form of intra-species competition. (b) On the second level natural selection utilizes the enormous over-production of young and the resulting differential death rate to change the norm of the species. The recognition of this is Darwin's first great contribution to biology and he was stimulated into developing this idea by reading Malthus. Darwin, of course, recognized that natural selection operated on both of these levels. Herbert Spencer also accepted these two functionings of natural selection. Karl Marx, however, accepted only the first operation of natural selection and denied the second violently. The utilization of population pressure as a selecting agent would not be compatible with his social ideals. Marx never judged any biological principle on the available evidence but always accepted or rejected a principle according to whether it fit or did not fit into his proposed social reforms. The Communists of

462

today have followed Marx. Lysenko, for example, denies the existence of any intra-species competition. (c) The third level on which natural selection works was unknown to Darwin. Indeed it could not be recognized until after genes and gene mutations were discovered. As we know, genes mutate constantly and, in some instances, at a rate that can be measured. This mutation, when described quantitatively, is called mutation pressure. Now the overall effects of mutation pressure are deleterious. Mutation pressure, unless checked, would inevitably cause degeneration and ultimate extinction, because all large mutations are destructive as are most of the smaller ones. The third level on which natural selection operates is that of the checking of overall mutation pressure, the elimination of the harmful mutations, and the taking advantage of the rare mutation that is beneficial. Natural selection has been described as a mechanism that achieves a high degree of improbability. Without natural selection operating on all three of these levels we could not have evolution. The Marxians reject the functioning of natural selection on the second and third levels, and rely instead on the inheritance of acquired characters.

At this point it might be well to point out some of the social implications of the factors that cause evolution. To do this I need two terms that I will have to use in a special and limited sense. These terms are "aristocratic" and "democratic." A "democratic" type of evolution would be one in which the whole population participated. With this type of evolution the whole group would evolve together and would advance as a unit. There would be no hindmost for the devil to take. On the other hand, an "aristocratic" evolution would be secured through the segregation of an elite, who would take advantage of their superior endowment and would, in due course, supercede the mediocre majority, and be superceded in turn by a new and super-elite. And this process would continue indefinitely.

Now the doctrine of the inheritance of acquired characters can fit into either a "democratic" or an "aristocratic" evolutionary process. If the characteristics that an individual acquires during his life can be transmitted to his progeny, then his experiences and the effects of environment upon him assume a genetic importance. All living conditions which improve him as an individual would also improve his progeny, hence also his species. In addition, the transmission of acquired characters would furnish a technique for securing a real biological equality of all individuals. That is, an altruistic concern by the exceptionally able for the welfare of their less fortunate fellows, their giving every possible advantage to the backward and the stupid fraction of mankind, would, in time, make these depressed human specimens equal to the best.

Once equality were reached, the whole population could move forward as a unit and everyone would evolve in the same direction and, with very little social adjustment, at the same rate. And this mode of progress still seems to be ideal to some very powerful political philosophies. "From each according to his ability, to each according to his needs" could, under these conditions, be the slogan of a rapidly evolving and improving species. Thus it is not remarkable that the present Communists, as well as those others who get their intellectual directives from Marx and Engels, accept the inheritance of acquired characters as an article of faith.

But another and antithetical application of the doctrine can also be made and the two applications are so far apart that men, as philosophically and ethically antagonistic as Karl Marx and Herbert Spencer, could both incorporate the doctrine into their systems of thought. According to this second view, the successful social-Darwinian (or rather Spencerian) competitor, by grabbing the best of everything and retaining a disproportionate share, could assure his children having "the most of the best" and, strengthened by their superior environment, they would be in a better position to grab for themselves and for their own children and so on, as long as evolution lasted. In such a system, "he should take who has the power and he should keep who can" and this taking and keeping would ensure evolutionary progress.

On the other hand evolution by natural selection is exclusively an "aristocratic" process—at least it was as it was understood in the nineteenth century. Natural selection means that the fit survive. The fit are the better adapted for whatever conditions exist at the time, and they survive or leave the greater number of offspring while the unfit—the unadapted —perish or leave the fewer offspring. The discovery of Mendelian segregation, however, introduced a most puzzling complication. Mendelian genes do not follow any laws of primogeniture. The ablest fraction of mankind are heterozygous for the genes that make them able, and they do not breed true. Their children are only seedlings. This throws the problem of aristocratic selection into some of the complex mathematics of population genetics and surprisingly it revives Galton's old "law" of ancestral regression.

This does not mean, of course, that the children of our exceptionally able minority will have the same gene frequency as the children of the mediocre. Far from it! The able will produce more children of higher ability than will the average; otherwise human ability could never have evolved in the first place—otherwise our intelligence could never have risen above the simian. But some of the mediocre, even some of the sub-mediocre, will produce some very able children through the chance combinations of Mendelian genes. This means that the elite of every generation will be recruited from many different groups within the whole population but, of course, in very different ratios from the different groups. Thus it follows that, unless the opportunity for an individual to develop into a member of elite is widely distributed throughout the whole population, the elite itself will suffer. This introduces a "democratic" element into evolution by natural selection.

Thus it follows that the ideal society for an evolutionary improvement by means of natural selection is an "open" society, a society where vertical migration is both easy and of common occurrence, a society where the able rise and the dimwits sink. It must also be a society in which the able reproduce copiously and in which the bums do not breed to excess.

This picture that I have given is, of course, much too simple. I have left out all complications caused by selective mating, by panmixia, by the environmental inhibitions of biological potentialities, by the lack of reproduction of the institutionalized nitwits at the lower end of the distribution curve, and by the breeding habits of the 6 per cent of our race that has an I.Q. of 69 or less. In fact, I have left out all complications. I am merely pointing out the social implications of evolution by natural selection and of evolution through the inheritance of acquired characters.

To me it is very significant that most of those who have very definite beliefs in regard to the social implications of human evolution accept or reject the several biological principles not on the basis of the biological evidence available but on the way the several principles fit into their firmly established social ideals.

I would like to call attention finally to the very clear, precise, and eloquent way that Professor Greene has presented the ideas and conclusions of Herbert Spencer. In reading an advance copy of his paper, I found myself unconsciously entering into a debate with Spencer, and checking in the margins of the pages the points Spencer made that I wanted to correct. This, of course, was very silly of me. I do not believe that I or any one else could, in a normal lifetime, correct all of the mistakes that Herbert Spencer made. I can admire Spencer's lack of sentimentality but he seems to me to be so misinformed factually that, on the whole, his conclusions are about as erroneous as they could be. But this is not necessarily his fault, as he could build only on the knowledge that he had.

To give one example of his inadequate data: His time scale of organic evolution is off by a factor of about one hundred. Then too, his account of social evolution, as described by Professor Greene, is simplified to a point where it has no validity whatever. Primitive man did not lead a life that was solitary, brutal, and short. We know now that our species became gregarious long before it became human. And once our ancestors became gregarious, nature selected not only individual men but also human groups. Thus nature selected in us those traits that preserve both ourselves and our sodalities, i.e., both our egotistic and our altruistic characteristics.

Natural selection not only leads to evolutionary progress, it also leads species into blind alleys and to extinction. The individualistic type of competition, that Spencer liked so well, favors only the competitor who is best suited to the immediate conditions in which the competition is held. The winners may, of course, advance their species along the paths of progress and improvement, but also they may not. The winners may lead their species into a trap and, as often happens, to extinction. Spencer did not really understand these complexities of natural selection and we should be grateful to Professor Greene for telling us so clearly just where Spencer's thought led him.

THE DEVELOPMENT OF IDEAS

ON THE STRUCTURE OF METALS

Cyril Stanley Smith

Metallurgists have taken inspiration from many parts of the purer sciences and have contact with wide areas of practical knowledge. Though metallurgy lacks the singlemindedness of physics and is less logical in the development of its theory, the very diversity of the source material for its history serves to emphasize the importance of interdisciplinary interaction. The science of metals and alloys became recognizably modern only toward the very end of the nineteenth century, when it arose from the putting in order of a host of empirical observations on their microstructure by the application of the phase rule, and by observations of the relationship between structure and composition with each other and with the mechanical and physical properties that had been determined by physicists and engineers throughout the century.

The present paper traces the development of the idea that metals are composed of a host of microcrystals—a seeming simple idea, particularly in view of the widespread occurrence of visibly polycrystalline rocks, but one which was very late in being correctly formulated.

PRESENT CONCEPTS OF THE STRUCTURE OF METALS AND ALLOYS

Metals are universally crystalline, but it is very rarely that they occur in the crystallographers' polyhedron with plane faces and sharp edges. An ordinary piece of metal consists of a large number of microcrystals, arranged usually at random orientations. It seems to have been very difficult for man to realize that such materials are crystalline rather than granular (in the sense of a grain of wheat with an individual envelope) and that crystallinity lies in internal order rather than external appearance. When single microcrystalline grains are separated from the mass, they have a shape that is similar to that of a cell in a froth of soap bubbles, with no plane or parallel surfaces, for the structure is determined by the two-dimensional interface between the crystals, where the atoms are essentially disordered, not by the planes of order in the crystals themselves. There is a substructure (of considerable interest today), but to a first approximation the orientation is substantially the same through each grain and changes abruptly at the boundary. At appropriate temperatures and compositions many alloys develop structures with two or more crystalline phases in equilibrium, which may either be randomly oriented or may bear some special relation to each other in shape and orientation. The minimizing of interfacial energy is the principal factor determining shape.

These structures can be best seen by examining a carefully polished and etched section of the metal under the microscope at magnifications from 10 to 2000 or more. The techniques are relatively simple but they were not developed until 1863 and did not become at all widespread until after 1880. Until then structure had to be inferred from fracture, which frequently distorted the specimen so much that the crystalline aspects were rendered invisible. It is the purpose of this paper to outline both the practical observations and the theoretical ideas of the development of structure on this scale. The determination of actual atomic positions in the crystal lattice, made possible by the discovery of x-ray diffraction in 1912, gave a second great impetus to structural studies, but this will not be discussed here.

꛷ DECORATIVE ETCHING AND FRACTURE TESTS

Before structure can be seen on the surface of a metal piece something must be done to the surface to distinguish between parts which differ in composition, crystal structure, or orientation. This is most simply done by treatment with an appropriate chemical reagent to produce differential staining or corrosion to show crystallographic facets.

The first use of etching is extremely old.[1] Some type of chemical attack was used by the Egyptians to develop uniform color on various alloys and it is likely that chemical pickling was used much as now to remove scale after annealing; however, the first suggestion that a *differential* chemical attack was used seems to lie in the existence of Merovingian pattern-welded sword blades. These blades, perhaps made only by the Franks, appear first in archaeological sites of the end of the second century A.D. They were carried far by Roman mercenary and Viking hands and were widespread throughout all of Europe until the tenth century.[2] They are characterized by a special twisted pattern of welded iron and steel forged integrally into the center of the blade. All surviving examples are far too deeply rusted to show the original surface. The pattern was undoubtedly intended to be seen, perhaps partly as a kind of trademark to designate a superior blade, and partly just for aesthetic reasons. Its preparation was certainly a step toward metallography, regardless of whether it was brought out by a chemical etch on a uniformly smooth surface (as seems most probable) or by some technique such as relief polishing.

In Europe these blades were replaced after the tenth century by more carefully heat-treated blades which were visually homogenous. Textured metals continue to this day to be of interest in the Orient. The true Damascus blade (a high carbon steel with a visible structure directly related to the solidification mechanism) and the forged so-called Damascus gun barrel (a highly contorted forged mixture of iron and steel) both inspired attempts at duplication in the eighteenth century which had important effects on European metallurgy.

Etching reappeared in Europe in the fifteenth century,[3] at first for the etched decoration of armor and later for the production of incised lines for intaglio printing. The designs came from a locally protective varnish and the etching did not develop any visible structure of the metal. Although unintentional variations in the metal must often have annoyed an artisan, there is no indication that the effect was intentionally utilized to reveal heterogeneities until the end of the eighteenth century, as will be discussed below.

The structure of metals can be revealed by their fracture, although the fact that the structure is distorted before breaking, and that cracks may sometimes pass through the grains and sometimes between them has made interpretation difficult. There can be little doubt that the earliest metallurgists, interested as they were in the ductiliy of their material (its principal virtue over stone) must have noticed the appearance of the broken surfaces of defective pieces, and must soon have related specific details of the fracture to variables in manufacturing procedures. The first written comment, however, seems to have been in 1540, by Biringuccio[4] who remarks that fracture is used specifically to follow the completion of the conversion of iron into steel and to adjust the tin content in a bronze.

Louis Savot in 1627[5] describes the fracture test explicity: "Foundrymen judge the quantity of tin that they should put [in bellmetal] by breaking a piece of the material before they cast and make a bell out of it. If they find the grain too large they put in more tin; if it is too fine they augment the copper." In the same year Mathurin Jousse[6] described the selection of iron and steel for fine blacksmith work on the basis of its fracture:

Soft iron can be recognized by the color of its fracture. When this is black over the entire cross-section of the bar, it is a sure sign that you have good iron which can be formed cold and worked with the file; for the blacker the fracture, the softer will be the iron to the file and the more ductile.

. . . .

Other iron bars have a gray fracture, that is, black mixed with white. Iron of this color is much harder and stiffer on bending than the previous kind.

. . . .

Still another kind of iron has a mixed fracture—part of it being white and the rest black or gray—and a somewhat coarser grain than the iron previously described. This is often the better iron. It forges more easily, is not likely to have cindery spots, has no [imbedded hard] grains, and polishes more easily. I believe that it is the best kind for forging, working with the file, or easy polishing; for it is refined by forging and acquires an entirely black fracture when it is being worked.

. . . .

Still other bars have a very small grain, like steel, and are ductile when cold.

Similar comments appear thereafter in most treatises for the practical worker. Although these fracture tests presuppose some kind of a structure and the relation of this to serviceability, they do not by themselves contribute much to the understanding of the nature of the component parts of the metal. It was Robert Boyle in 1672 who first performed scientifically-motivated experiments on fracture.[7] As can be seen from the following quotation from his *Origin and Virtue of Gems,* he has a remarkably clear idea of the ordering of the parts that occurs during solidification, and the relation of this to temperature gradients. He clearly distinguishes between the random orientation of a slowly cooled ingot and the directional columnar grains in the chilled ones with rapid withdrawal of heat. Bismuth when broken "will discover a great many smooth and bright planes, larger or lesser according to the bigness of the lump; which sometimes meet and sometimes cross over at different angles." A casting made in a one-inch diameter bullet mold "being warily broken, did seem to be, as it were, made up of a multitude of little shining planes, so shaped and placed that they seemed orderly to decrease more and more as they were further and further removed from the superfices of the globe . . . almost like so many radiuses of a sphere from the center or middle part." This structure results from the fact that after the coagulation has begun, "the parts of the remaining fluid, as they happened to pass by this already cooled matter, with a motion which . . . was now slackened, they were easily fastened upon the already stable parts." "If the corpuscles of a body be so shaped, as to be fitted by their coalition, to constitute (and if I may so speak) glossy planes though they be variously shuffled and discomposed as to their pristine order, yet if they be but a little kept in a state of fluidity . . . they will presently be brought to convene into smooth and shining planes, and the situation of those planes in reference to one another will be more uniform and regular than almost any one would expect in a concretion so hastily made; notwithstanding which their internal contexture will be much diversified by circumstances as particularly the figure of the vessel or mould wherein the fluid matter concretes." Though this model would allow for many crystals within its diversity, their separateness is not discussed. Similar experiments were done by Réaumur early in the eighteenth century, which will be separately discussed below.

PHILOSOPHY AND METALS
IN THE SEVENTEENTH CENTURY

Despite the visible granularity of many materials structural ideas were virtually ignored from the time of the Greek atomists until the revival of atomism and the development of corpuscular philosophy in the seventeenth century. Aristotle and his followers regarded metal as essentially structureless and homogeneous. Paracelsus, when he selected salt, sulfur, and mercury as his principles, was unconsciously making an important classification of different classes of inter-atomic bonding, but he seems to have had no structural concepts, although, like Aristotle, he knew that many natural objects were of mixed character with parts that could be easily distinguished.

The obvious malleability of metals, their general fusibility, and their reversible conversion into calces, were properties so central to the rise of phlogiston theory and to its displacement by modern chemistry that metals have always had a central place in chemical theory, but until recently, chemists have neither assumed nor explained any structure beyond the molecule. Moreover, since the time of Dalton they seem rather to have avoided the study of metals, because so many intermetallic phases ignore the rules of simple combining proportions.

The revival of Greek atomistic philosophy (see E. C. Millington[8] and particularly the excellent discussion by Marie Boas[9]) deeply influenced men like Boyle, Hooke and Newton, but its promise was not fulfilled, for it was qualitative in nature and did not lend itself to exact and conclusive experiment at the time. The various properties which the philosophers chose to explain in terms of their atoms or corpuscles were actually not due to the ultimate atom but rather frequently were properties of aggregates of many atoms on higher scales of organization. From our present point of view it is this hierarchy of structures which is of greatest interest.

Sennert,[10] in developing his atomic philosophy, gives a number of examples taken from metallurgical operations, but these all involve the persistence of metallic atoms through solution, precipitation, melting, oxidation, reduction, and amalgamation. Like the older atomists he was relatively unconcerned with specific shapes of particles or the manner of their organization, but Gassendi,[11] a representative seventeenth-century atomist, explained many properties directly on the basis of the specific shapes and interrelationships of the fundamental particles. In this he was probably acting under the stimulus of Descartes (*vide infra*) although he rejected this philosopher's basic premises. He considered the hardening of iron (steel) as due to the fact that heat opened its structure by the ingression of the atoms of fire, while on quenching the water particles remained "imprisoned in the small incontinguities, . . . making the body of the iron somewhat more solid or hard than otherwise it would have been." Annealing by the smith relaxes the structure and evaporates the water. (This concept of sticking particles together by filling in the spaces between them with a kind of cement, obviously derived from the behavior of wet sand, occurs in the earliest atomistic writings and remains even in papers by Roberts-Austen and Osmond at the end of the nineteenth century.) Soft bodies generally have parts "so separated from each other in many points, that more and larger inane spaces be intercepted among them," while hard bodies are reduced to a closer order. The great ductility of gold is due to the compactness of the structure and to the great tenuity of the component particles, conveniently provided with numerous hooks and claws whereby they reciprocally implicate each other and maintain the continuity of the whole mass. "No sooner does one particle dissociate from its neighbour but instantly it lays hold of and fasteneth upon another, and as firmly coheareth thereunto as to its former hold, so that mutual cohesion is maintained even above the highest degree of extension or attenuation which any imaginable art can promise." Brittleness, on the other hand, Gassendi thought to be due to the fact that the percussion or pressure is propagated from point to point successively throughout the structure.

The explanation of properties of bodies in terms of the shapes and interaction of the particles was developed much more extensively by Descartes,[12] and particularly by the Cartesian physicists who flourished toward the end of the seventeenth century. Although he believed in no *ultimate* particle, he had to invoke some structure and in practice his ideas are quite compatible with those of the atomists. An essential part of Descartes' theory of matter was the existence of three different forms of it. These were essentially stages of comminution, rather than states of aggregation, but between them they completely filled all space. Though he was wrong in insisting on the impossibility of empty space and on the possibility of infinite subdivisibility, he was quite right in trying to explain many properties of substances in terms of aggregates of matter on different scales. To a modern physical metallurgist the writings of the Cartesians are almost prescient, and it is regrettable that when Cartesian cosmology was rejected his beginning interest in solid state physics also disappeared. In some degree it was the success of the Newtonian mathematical method which displaced interest in the real structure of matter for this is too complicated to be treated rigorously in many aspects even today. Theoretical mechanics became more mathematical than physical and it developed only by paying the price of completely ignoring real polycrystalline matter with all its interesting departures from uniform ideal elasticity. To be sure, corpuscular ideas crystallized into mathematical crystallography on one hand and the chemical atomic theory on the other, but, basically important though they are, these extremities of the scale of organization fail to explain a good many of the properties of matter—indeed some of the very properties with which both the philosopher and the practicing metallurgist had been concerned. A. R. Hall[13] has remarked, "Cartesian mechanism, the dinosaur of seventeenth-century scientific thought, could not adapt itself to a new intellectual en-

473

vironment. It was committed dogmatically in too many points of detail where it proved to be false." Yet, at least from the viewpoint of the present paper, the dinosaur was interested in the right things. The surviving forms of life were excessively preoccupied with things that could be mathematically considered, which inevitably necessitated over-simplification and resulted in many parts of a wide and beautiful landscape remaining unappreciated.

It seems probable that Descartes had spent some time in a smith's forge, for in a letter[14] to Mersenne on January 9, 1639, he discusses the hardening of steel (he also discusses the hardening of *iron,* seemingly regarding it as more interesting, which indeed it would be were it not due merely to incidental carbon content). The description of the making of steel in his *Principia Philosophiae* (1644) could only have been written by someone who had himself observed the granular nature of wrought iron coming to nature in the bath of molten cast iron in the hearth of the forge fire.

The entire bath of molten metal becomes divided into many small lumps or drops, the surface of which become smooth. For all the particles of metal somehow joined together make up one of these drops, and because it is pressed upon on all sides by other drops which surround it and which move in different direction from its own, none of the points or branches of the particles could come out of its surface ever so little more than the others without being continuously pushed back towards its core by the other drops. This causes the surface to become smooth and also causes the particles that compose each drop to draw together and be even more closely joined.[15]

Descartes seems to be groping toward the concept of surface tension, and certainly sensed the aggregation of the metal into individual grains, which, in the presence of liquid, could not stick to each other. Steel, he says, when melted, is divided into small drops, which incessantly fall apart and reassemble when the metal is liquid, but if it is cooled rapidly it becomes very hard, stiff and brittle, almost like glass. (So slowly did ideas change that the famous paper by Vandermonde, Berthollet, and Monge in 1786 still explained the hardening of steel in almost identical terms. To illustrate the clumping they use the analogy of sandstone, with the "molecules" bonded closely together into grains which have few contacts among themselves and are loosely bonded. This, however, characterizes only the high temperature state). Descartes regarded steel as hard because its parts are joined very tightly; stiff and springy because, by bend-

474

ing it, the arrangement of its parts cannot be changed but only the shape of the pores; brittle "because the small drops of which it is composed are joined only by the contact of their surfaces, which touch directly only in a very few small spots."

The word translated above as drops is *gouttes* in French, but *guttulas sive grumulos* in the original Latin. They are unmistakably modern grains, aggregrates of much smaller particles, more tightly bound to each other than to particles in adjacent grains. Descartes even realized that it was some kind of orientation of the component parts which distinguished one "drop" of the solid from another. He supposed that the particles of iron had grooves in them, so that a channel would be formed if, moving at random, two particles came together with their grooves matching. This allowed freer circulation of his most tenuous particles, the pressure of which held all matter together. With further properly-oriented additions, the channels would be kept open even in large aggregates; the particles composing the aggregates would be held together while their surfaces would reject particles, either individually or clumped, which were not correctly aligned.

The followers of Descartes developed his particle-fitting concepts, but lost the idea of oriented grain. Cartesianism, at least in its solid-state aspects, became quite flexible. Not only did it come to utilize the iatrochemical principles, but a great deal of misspent imagination was applied to the dreaming up of special shapes for particles which would tend to account for observed properties. Iron, for instance, according to LeGrande[16] was composed of

... thick and branching particles whose surfaces lie close together and whose pores are penetrated by striate matter [i.e., salt particles which had been beaten into little swords by being dashed against the sides of the pores in the earth]. The reason for the ductility of metals is because the particles of metals are of a longish figure and are so disposed that they lie upon one another according to their whole surface; which makes that when they are prest under the hammer or in drawing, they fall down sideling, and joyn side to side without any separation. Thus it comes to pass that metals when under the hammer may be extended into length and breadth, still retaining firm cohaesion of their parts.

The French Cartesian physicists carried to the extreme the explanation of matter in terms of the fitting together of shaped parts. Perrault, for example,[17] believed that hardness depended on there being many perfectly flat faces of the parts, directly in contact with each other and held together by the pressure of subtile "air." Forging or cold working most metals hardens them because the strong compression causes a great number of the faces to join together, and presses out slippery particles which had been interposed, but there are some metals like lead and tin which are softened by cold work because this mixes up the particles of a fluid nature with the flat-faced ones. It is hardly surprising to find that the author of this treatise was connected with building: He was, indeed, none other than the famous architect of the Louvre.

Hartsoeker, in a book which he entitles *Principes de Physique* (1696) but which is composed principally of uninhibited philosophic speculations, developed even more specific mechanical models depending upon shape.[18] Difficultly fusible metals, he supposed, had more nearly cubic parts, capable of being packed in close order with large areas of flat faces in contact, while easily melted ones were dodecahedral. (It is conceivable that this distinction is based upon a cubic cleavage fracture often observed in a metal like iron and the intergranular fracture which can be seen in metals like lead or zinc under certain conditions, and in which the pentagonal dodecahedron is observed more frequently than any other regular body.) By also introducing the concept of the roughness of surfaces of the particles and by varying the areas of contact he was naturally able to explain everything. Iron he believed to be made of particles in the form of triangular prisms with a rough surface and a hole down in the center to allow for easy corrosion. In common with most Cartesians he believed that the steel particles were put in motion and disarranged by the fire, quenching giving a hard metal because the parts did not have time to rearrange on cooling. Some of his molecules were amazing contraptions—corrosive sublimate is a ball of mercy with particles of salt-like needles sticking out of it, and air is a hollow ball made of wire rings. After speculations on the various ways in which salt particles could be shaped like knives with cutting and pointed edges in various combinations, he desists for he did not wish "to deprive the reader of the pleasure of making investigations for himself following the principles that have been established." The last Cartesian whom we will quote is better disciplined, and displays better physical intuition. He is Jacques Rohault, whose *Traité de Physique*[19] was published in Paris in 1671. Rohault was the son-in-law of Claude Clerselier, who had been a friend of Descartes and devoted most of his life to promoting the philosopher's ideas. With true

476

Cartesian breadth, Rohault covers all material science in his book, including astronomy, medicine, and meteorology. It had a great influence on the development of French science, and, via a critically annotated Latin translation by Samuel Clark, was instrumental in popularizing its rival Newtonian physics in England.

Following Descartes, Rohault believed that all properties of matter depend upon the shape, arrangement, and motion of its particles. The particles were essentially all of the same composition although chemical differences are sometimes explained in terms of a kind of segregation. In its motion, the elementary matter gets broken up into three kinds of particles, a very fine dust produced by the attrition, particles which are rounded in form, and polyhedral particles which are more or less the first fragments. The particles, forming the Cartesian elements, are in principle transformable into each other. Although in their origin they are not assumed to be uniform, yet in their use to explain the properties of matter they seem always to be essentially of a single size and shape in any given substance.

Although Rohault is much concerned with the manner of packing the particles, he is not specific as to the type of force that holds them together, whether their own attraction or under the influence of the external pressure of subtile matter.

There are pores between the third kind of particles, through some of which the second matter can move and through others only the first. The pores will be enlarged under a tensile stress, contracted under compression: In a bent spring the subtile matter in circulating through the pores may either wear the particles or restore them to the state they were in before bending, thus causing springback.

Cold working a metal increases its power of springing back, because the beating does nothing else but make the parts approach nearer to one another and so straightening the pores. Bodies that bend plastically without breaking are conceived as being made of parts with complicated textures intermixed with each other and hooked together like the rings of a chain or the threads of a cord, while brittle bodies are of simple texture with particles touching one another at only a few places, so that they cannot be separated even a little without the whole continuity being destroyed.

In Part III Rohault talks of the engendering of metals in restricted cavities in the earth from rising particles of water, salt, and oily substances mixed together combined to form little hard bodies which are supposed to be the component parts of metals. Such a model would allow rearranging the small component parts to produce transmutation of the metals, though it is incredibly improbable that it can ever be done by man.

The portion of Rohault dealing with ductility and hardness is particularly revealing:

But however this be, we cannot but think that the component Parts of Metals are long; otherwise we cannot understand how Metals should be so ductile as they are, whether they be forged upon an Anvil, or drawn through a wire-drawing Iron; whereas, if we suppose them to be somewhat long, it is easy to conceive, that when they are pressed on one Side, they will slip sideways of each other without quite separating.

Further, it is not possible to conceive, that when a Piece of Metal is continually pressed upon one Way the Parts of it should be able to lye cross; on the contrary, we cannot but think, that they must necessarily so order themselves as to place themselves by each other's Side, and correspond Length-ways to the Length of the whole Piece, which will make it easier to bend that Way than any other; And this agrees with Experience; for Metals which are beaten into Rods upon an Anvil, or drawn into Wire through a Wire-Iron, are very strong Length-ways, but Breadth-ways they are many Times easier to break than Workmen would have them. And we observe Strings in them, as in the Slip of an Ozier.

These Strings ought not to be in Metal that is cast and has not been forged: And so we find that cast Metal is as easy to break one Way as another.

Steel, which is nothing else but fine Iron, is capable of being made the hardest of all Metals: The Way of making it so is this; only to heat it red-hot in the Fire, and then throw it all at once into cold Water; and this manner of hardning is what they call *tempering* it, and this makes it capable of cutting or at least of breaking all Sorts of Bodies without Exception, even Diamonds themselves: For it is certain they will break in Pieces with a small Stroke with a Hammer if it hits right.

In order to account for this Effect (which perhaps is one of the most admirable, and doubtless one of the most useful Properties that we know) we must suppose that the Heat of the Fire, which makes the Steel almost ready to melt, puts the small Particles, which each component Part is made up of, into Motion, and thereby causes the Particles of the two nearest component Parts, (whose Distance from each other was very small, though far enough) to approach a little nearer one another, so that the Metal becomes more uniform than it was before; after this, being cast on a sudden into the cold Water, the metallick Parts lose the Motion they were in, before they have Time to gather together again into gross component Parts, with considerable Intervals between them: Whence it follows, that the Points or Edges of Gravers and the Teeth of Files can only slip over them without entering into them.[20]

It is probably worth remarking that the concept that steel is nothing but refined iron (see above) was almost universally believed prior to Réaumur. It can be traced to Aristotle (*Meteorologica,* IV, 6) and is perhaps a reasonable supposition since steel resulted from the further heating of iron in the fire, and fire is known to purify. Robert Hooke took the opposite view, believing that vitreous matter was present in steel and responsible for its properties.

On the hardening of steel Rohault had earlier remarked that "Thus we can perceive by our Senses that the Parts of Steel which are not tempered, are larger, and consequently the Pores wider, than those of tempered steel."[21] By this he was supposedly referring to the visible fracture, which becomes finer on quenching, and it seems certain therefore that the "parts" he is talking about verge on the visible—in other words, are commensurate with the modern metallurgist's grains.

In discussing the formation of little cubes of salt on evaporating a concentrated solution of salt, Rohault remarks on the ordering of the particles as they lie against each other during crystallization, but he is farther than was Descartes from the concept that crystallinity consists just in order. This is the more remarkable since Robert Hooke had already suggested that spheres packed in ordered array would give rise to all of the faces of simple crystals, and the Cartesians lay frequent emphasis on the geometric contact of plane faces. Nevertheless, Rohault's "gross component parts" (French *gros grumeaux,* the common word for clot or clump but obviously descended from Descartes' *gouttes* or *grumulos*) carry clearly the idea of an aggregation of particles and were analogous to the microcrystalline grains except for the emphasis on crystalline order.

It is interesting to compare these attempts to discuss properties of matter of practical importance with the much more restrained statements of Isaac Newton. To quote once more a much-quoted passage:

The Parts of all homogeneal hard Bodies which fully touch one another, stick together very strongly. And for explaining how this may be, some have invented hooked Atoms, which is begging the Question; and others tell us that Bodies are glued together by rest, that is, by an occult Quality, or rather by nothing; and others, that they stick together by conspiring Motions, that is, by relative rest amongst themselves. I had rather infer from their Cohesion, that their Particles attract one another by some Force, which in immediate Contact is exceeding strong, at small distances performs the chymical Operations above-mention'd, and reaches not far from the Particles with any sensible Effect.

He believed, moreover, in aggregates of particles on successive scales:

Now if we conceive these Particles of Bodies to be so disposed amongst themselves, that the Intervals or empty Spaces between them may be equal in magnitude to them all; and that these Particles may be composed of other Particles much smaller, which have as much empty Space between them as equals all the Magnitudes of these smaller Particles: And that in like manner these smaller Particles are again composed of others much smaller, all which together are equal to all the Pores or empty Spaces between them; and so on perpetually till you come to solid Particles, such as have no Pores or empty Spaces within them.

And later, more specifically:

There are therefore Agents in Nature able to make the Particles of Bodies stick together by very strong Attractions. And it is the Business of experimental Philosophy to find them out.

Now the smallest Particles of Matter may cohere by the strongest Attractions, and compose bigger Particles of weaker Virtue; and many of these may cohere and compose bigger Particles whose Virtue is still weaker, and so on for divers Successions, until the Progression end in the biggest Particles on which the Operations in Chymistry, and the Colours of natural Bodies depend, and which by cohering compose Bodies of a sensible Magnitude. If the Body is compact, and bends or yields inward to Pression without any sliding of its Parts, it is hard and elastick, returning to its Figure with a Force rising from the mutual Attraction of its Parts. If the Parts slide upon one another, the Body is malleable or soft. If they slip easily, and are of a fit Size to be agitated by Heat, and the Heat is big enough to keep them in Agitation, the Body is fluid.[22]

These statements of Newton's are clearer than those of the Cartesians, they extend little beyond what is warranted by the facts, they conform much better to accepted standards of modern scientific writing, and are in every way admirable; yet there is nothing to indicate that they served to inspire and provide the basis for further experiments by more practical men in the way that Rohault's ideas clearly influenced Réaumur, as will be discussed later. Indeed, as science became mathematical under the influence of Newton (the *Principia* more than the *Opticks*), speculation on the nature of matter was excluded, and even interest in it vanished. Following the *Principia* and Galileo's *Due Nuove Scienze*—both superb books—mechanics became purely theoretical, and theory for a time was *only* mathematics. The most important tool of science came to be regarded as its body. Although practical engineers tested things and built fine structures, that strange field called "Elasticity and Strength of Materials" decayed into exercises in applied math, with virtually complete disregard of the intricate inhomogeneities of real materials.

⚎ SEVENTEENTH-CENTURY MICROSCOPY

The microscope became available about 1600 and was used to examine the fine structure of all kinds of things shortly thereafter. Many inorganic crystals and chemical precipitates were examined by the early microscopists, and it was not until the nineteenth century that the center of gravity of microscopic observation became so exclusively biological. There is, however, remarkably little on the structure of metals. Henry Power[23] seems to be the first, in 1664, to record the appearance of a piece of metal, but he saw little enough. "Look at a polish'd piece of any of these Metals [gold, silver, steel, copper, mercury, tin, and lead] and you shall see them all full of fissures, cavities, and asperities, and irregularities; but least of all in Lead which is the closest and most compact solid Body probably in the world." Mercury resembles a little globular looking-glass reflecting the surroundings in fine detail, and "whereas in most other metals you may perceive holes, flaws and cavities, yet in quicksilver none at all are discoverable." Robert Hooke, in his famous *Micrographia*,[24] gives as his first object the point of the needle, which he regards as very gross compared to the points made by nature, and next he examines a razor which showed nothing but scratches resulting from particles of dust on the hone or flinty parts in it. He has some interesting remarks on the vitrification of the sparks struck by flint from steel (whose fused nature had been observed by Power) and suggests that steel is partly vitreous, the vitreous material exuding during tempering to produce the temper colors which he had correctly identified as being due to thin lamellae of transparent material. Hooke observed the regular branching dendritic structure on the surface of lead containing arsenic, but he seems nowhere to have examined the fracture of a piece of steel or brass, much less that of a metal subject to proper chemical attack to reveal its structure.

Rohault commented on the rugged and uneven appearance of gold burnished with a bloodstone.[25] Leeuwenhoek reproduced a sketch of the dendritic pattern of a silver "tree"[26] and in a later paper records his observations on the fine particles in metals, which he says are inconceivably small.

One may indeed by the help of a good microscope just discover the exceeding small particles of gold and silver but one cannot perceive what figure they are; and who can tell of what multitude of parts these little particles which we see by the help of a microscope are again composed. . . . [In steel] we can only discover the broken gaps or notches of a razor, for instance, and the coarser the parts are of which these metals are composed, as we may see in cast iron, the less valuable are the said metals; but the finer the particles are, the more valuable in my opinion will be the steel and iron which they compose. [Forged metals] consolidated by strokes and pressure of the smith's hammer that they seem to us to be but one body, though they do consist of a great many small particles, the coarsest of which are always obvious when we come to break the metals: and how often so ever you melt any of these metals and break them again after they are cold, you will always be able to discover the grainy particles thereof; but you will find them so strongly joined and riveted to one another that they appear to be one body.[27]

Thereafter, though microscopists grew in number and microscopy became an important field for the dilettante scientist, metals ceased to be of interest and the only observations are the incidental ones made by physicists, chemists and others. This is perhaps not surprising, since the surfaces available to them would show no structure except accidental features due to the rather brutal methods of polishing. Had only European microscopists had access to the surfaces of Japanese swords and sword furniture, with the wonderful detail of *niye,* and the shimmering eutectic network structure of *shibuichi,* both finished by methods which clearly reveal the structure without distortion (see note 1), metallography could well have developed in the seventeenth century instead of the nineteenth, and its observations would have helped to place corpuscular philosophy on a firmer experimental basis.

THE EIGHTEENTH CENTURY:
OBSERVATIONS ON CRYSTALS AND CRYSTALLIZATION

The beginning of the eighteenth century saw the remarkable structural observations by Réaumur, and the end of it the birth of mathematical crystallography under Häuy and the "chemical revolution" in which metals played no small part. Most of the century, however, is devoid of scientific study of structure, and, indeed, relatively few observations that were not mere repetition of the earlier ones.

Réaumur's work is, however, really outstanding.[28] He was essentially Cartesian in his approach and was strongly influenced by Rohault, yet unlike anyone before him or for long after, he related his ideas on the structure of steel to the practical problems of making and treating it. Conversely he used his experimental observations on both plant and laboratory scale to develop his theoretical ideas, and unlike most of his contemporaries he combined the viewpoint of both physicist and chemist. Steel he believed to result from iron by the addition, in the smelting or cementation processes, of particles of "sulfurs and salts," but these are definitely material particles and they could diffuse among the particles of iron under the influence of heat.

Réaumur clearly realized that "the arrangement and shape of the visible parts are usually the effect of certain natural tendencies and qualities of the invisible parts." He understood that behind the grains visible on the fracture there was an aggregation of smaller parts and beneath that perhaps others.

When I speak of the structure of wrought irons, I mean the shape, the size, and the arrangement of their molecules; and it is by means of their fracture, of the surface at the location where they have been broken, that one can judge how these molecules differ. When different kinds of stone are broken, the fractures show the difference in the grains of each one. When different kinds of wood are broken, fibers of different size, which are sometimes differently arranged, are seen in their fractures. When bars made of different irons are broken, such obvious diversity will be noticed in their fractures that to the unaided eye the fractures of these bars sometimes seem to differ more among themselves than they differ from other metals such as lead, tin, and silver. There is not only as much and more difference in color, but the difference is even greater in the shape and the arrangement of the parts. By paying only casual attention to the fractures of these different kinds of iron, it is seen at once that they can be divided into two groups: into irons with only grains or platelets and irons with fibers in their fracture. The fracture of irons of the first group resembles either the fracture of stones or that of bismuth; and the fracture of irons of the second group resembles that of wood. (The artisans say that the latter are fibrous.) But this classification is still too general. The detailed study we intend to make requires finer subdivisions.[29]

483

He goes on to list and illustrate seven classes of iron, starting with those having a fracture which "shows very brilliant and white platelets, which appear to be as many little mirrors, but of irregular shape and arrangement: they rather resemble those in the fracture of bismuth." Some of these grains were as much as one-sixth of an inch in diameter and the space between them was sometimes occupied by smaller parts. The next three classes contain increasing areas of finer and finer grained iron, the color progressively darkening. The softest irons, the last class and the best for making steel, have a fracture of fibers alone.

Réaumur assumes that the structure of iron is that shown by the fracture—i.e., that the iron of the first class, showing platelets, is itself composed of platelets, and similarly a granular or fibrous fracture indicates that the iron is composed of differently-shaped parts which he called grains or fibers. He realizes that a cutting tool may sometimes remove clumps of particles, but neither he nor any scientist for as much as a century and a half later realized that the difference between the fibrous and a platelet-like fracture could lie solely in the manner in which a crack propagates through the structure. As iron is converted into steel the platelets become smaller and smaller, due to the fact that "the shape and arrangement of the parts of iron are changed in proportion as the sulfurous and saline substances penetrate." The weight and volume of the iron increases, and consequently the parts have moved farther from each other, being separated by the substances that have diffused in. He uses the curious analogy of a tangled or felted mass of fibers and concludes that steel is better carded, by which he apparently means a more thoroughly subdivided rather than better arranged structure. He devised tests, nearly two hundred years before they came to be known as the Shepherd and Metcalf tests, to measure quantitatively the fracture grain size and its variation with heat treating temperature. He saw that different steels had different inherent grain sizes, and that the size of grain would increase with increasing temperatures above the point where any change occurred.

In addition to several good drawings of fractures, Réaumur introduces a sketch representing the microstructure of steel, which is remarkably similar to a modern photomicrograph of a medium-carbon steel but is apparently nothing but a hypothetical sketch to show how the parts and their interstices might be arranged. It does indeed bear some resemblance to the drawings in Descartes' *Principia* which symbolically illustrate the passage of light particles through matter. Réaumur says it represents a magnified view of a grain the size of a printed spot (which actually measures 1.17 mm. diameter). Another view, further enlarged, shows one of these particles to be itself composed of an aggregate of still smaller particles, similarly arranged. The model is a hierarchy of parts, each complex and with voids, apparently without regularity of shape or arrangement. On the hardening of steel Réaumur has this to say:

This grain, which is easily seen by the eye, is itself an accumulation of an infinite number of others grains, which we shall call the "molecules" of this grain. The microscope brings these molecules into the field of our vision. But these molecules are themselves composed of other parts. It is possible, if it seems desirable, to suppose that the latter are the elementary parts, although in reality we may have to continue the division vastly much farther before we reach them; however, we can stop here. We thus have to consider a grain, the molecules of which it is composed, and the elementary parts of the molecules. As the salts and sulfurs intimately penetrate the iron, we can at least assume that those by which steel outnumbers iron penetrate the molecules of the grain. If I expose to the fire a soft steel containing the grain on which we have concentrated our attention, the fire will melt the sulfurs and the salts of the molecules of this grain before it melts the molecules themselves. It will drive part of the sulfurs and salts by which steel outnumbers iron out of the molecules in which they were wedged. Whereas, before this, they had penetrated these molecules, they will now, as a first step, go into the gaps between them. This will be all the effect caused by a moderate fire. . . . Let us therefore not hesitate to admit that, when our grain has reached a certain degree of heat, the empty spaces between the molecules of which it is composed will be partly filled by a sulfurous matter which was not there previously and of which the molecules have been deprived; that part of this sulfurous matter, which the fire has started on its way out of the iron, has passed from the molecules themselves to the intervals left between them. In this state let us plunge the bar of steel with the grain we are studying into cold water. We shall instantaneously fix the sulfurs and the salts which float around together. We shall deprive them of their fluidity; they will no longer be in condition to reenter the molecules. However, the small intervals between these molecules of the grain will now be more completely filled, and filled by a substance which we can suppose to be almost as hard as we want it to be. The molecules of the grain will therefore be more firmly bound to each other. For this reason our grain of steel will be more difficult to divide or to break; in other words, our grain of steel has now become harder.[30]

This is a remarkably acute, purely structural explanation of the hardening of steel. Today we say that carbon diffuses at high temperatures into the interstices between the atoms of iron in the austenite crystal, that it is retained approximately in the same relation on rapid cooling (though the iron lattice changes its symmetry), while slow cooling allows the carbon to diffuse out of the iron grains and to collect into spaces between them to form a separate carbide. This picture, though reversed in detail, is of the same physical nature as Réaumur's. By refining Cartesian ideas on the general structure of matter and adjusting them to conform to his actual observations of the structure of metals as shown by their fractures, he had ended up with a picture which was an extremely satisfactory guide for subsequent experiment. Not until the microscopic technique permitted the true shapes of the various segregates and solutions to be seen could it be improved upon.

Réaumur also has interesting comments on the structure of gray and white cast irons. Gray iron, he reports, when viewed under the microscope, seems to be composed of "some sort of chemical vegetation made up with an infinite number of interlaced branches, each composed of little platelets arranged on top of one another." The latter is a delightful and precise description of dendrite of a kind that could be separated from the shrinkage cavity of a casting, but his illustrations lack the regularity of a real dendrite crystal, which was left for Grignon to depict fifty years later (see note 34).

In a memoir presented to the Academy two years after his book, Réaumur[31] describes a series of observations rather like those of Boyle (see note 7) on the fracture of cast metals which illustrate well the conceptual difficulty that is associated with the problem of microcrystallinity. Successively remelted and cast ingots of antimony sulphide had different arrangements of the needles on their fractures. Cooling rate was important, and the different needles interfered with each other, from which he concluded that it was impossible to deduce the configuration of the elementary parts or molecules. He showed that the direction of the fibers was that in which cooling had taken place, but he thought that extremely slow cooling (which actually must have given him a single crystal with a single flat facet as its fracture) left the molecules remaining fixed in the places where the fire had left them, for the ordering tendency of a previously solidified surface was lacking! Many metals could not be broken without deformation which disguised the arrangement and state of the parts, so he made use of the hot-shortness of copper, zinc, and particularly lead, at temperatures very near the melting point, to reveal the structure. In this case, however (as we now know), the fracture is intergranular rather than transgranular. Observing the difference between these rounded surfaces and the flat facets of normally brittle materials, he remarks "Perhaps it will be found that it is on the shape of the grains and their arrangement that the ductility of metals and of other materials depends."

In 1725,[32] the younger Geoffroy studied various alloys in the copper zinc system, noting by the appearance of the fracture under the microscope, and commenting on structures with a mixed yellow and white grain which supposedly correspond to our two-phase beta-gamma structures. William Lewis[33] observed two components in the fracture of gold platinum alloys which were incompletely melted, and noticed the tendency for this heterogeneity to disappear on remelting.

Macquer, Baumé, and others, early in the eighteenth century, not realizing that crystallinity was due to internal order rather than external shape, devised various menstrua for metals, and grew crystals from mercury solution. Later crystals were grown directly from the melt by appropriate separation of the crystal and the liquid before complete solidification. The best observations of this kind were those of Grignon,[34] Mongez,[35] and Guyton de Morveau,[36] the latter being the first to point out that the dendritic markings on the surface of an ingot of metal solidifying under a protective layer of slag (the age-old Star of Antimony) were related to three-dimensional crystals.

In 1775, Grignon published several memoirs on various aspects of the mineralogy and metallurgy of iron, most of which he claims to have presented to the Academy in 1761.[37] Grignon was involved in the commercial operation of blast furnaces and forges, and his observations came from observations on a much larger scale than those of his laboratory contemporaries. He sawed open the shrink-head of large castings, and even cooled an entire blast furnace with extreme slowness, by working the bellows at decreasing power, so as to produce very large crystals. "It would be interesting for the progress of physics," he remarks, "if all men who employ fire on a large scale as an instrument of their art would adopt an observant and analytical viewpoint; the extent of our knowledge would be greatly increased and the relative number of hydrogenous systems would diminish." He believed that "everything in nature is characterized by an essential form in which chance has no domain. Every substance has a characteristic predetermined figure which, in concert with the equally invariable quality of the substance, determines its properties. . . . Every molecule, similar to its neighbor, approaches and unites with it, and in the measure that the separating fluid is dissipated, they discard the obstacles and draw tighter their bonds with an invariable affinity. They attach themselves to each other in numbers related to their faces and to the opening of the angles of each molecule."[38] In seeing a body naturally cubic, he conceived that each molecule was in the form of a cube and so for other crystal forms. This idea became the basis of the crystallographic systems of Romé de l'Isle and Haüy, who named the units *molecules integrantes*. Grignon uses the word "molecule" for a smaller state of aggregation than Réaumur and many other writers. He supposes that they might be of the order of 1/1000 of a ligne, i.e. about two microns in size.

Grignon thought that the grains could be distorted by pressure against each other, but, though he comes very near to describing correctly the typical grain of a polycrystalline material, he seems to regard this not as a stable form in its own right, but as a distorted simpler polyhedral form. He is remarkably perspicacious in picking a 14-sided body as a typical grain, but unlike the ideal minimum-area tetrakaidecahedron he has apices at which more than three faces meet.

Grignon believed that iron would crystallize only from a solvent. In a paper of 1771,[39] he took issue with Romé de l'Isle who believed that water was the principal and perhaps only agent in nature for forming crystals and hence that it was useless to attempt to form metallic crystals in a laboratory by fire alone. To refute the belief that the roughly shaped figures produced by art were not real crystals in form he gives drawings of some beautifully formed crystals of iron. He approaches the idea that fire is a solvent, an idea that was clearly stated immediately afterwards by Guyton de Morveau (see note 36), who in 1776 pointed out that solidification was a true crystallization occurring by the evaporation of the excess of solvent: Fire is to metals what water is to salt, but metallic crystallizations have escaped observation for so long "because the operations of melting have been relegated to men animated more by the need to play than by the desire for knowledge." He believed in a regularity of form based on a secret mechanics despite the many accidents to which the products of fusion are subject, and he postulated that some day a kind of geometric progress would be seen in the composition of solids, which perhaps would serve as a microscope to reveal the figures of their elements. He illustrated his paper with some remarkable surface crystallization patterns. Two years later,[40] after examining a large plate of cast iron found in the bottom of one of Buffon's furnaces, he realized definitely that the surface crystallization and the separated crystals obtained by Grignon were essentially the same thing.

Mongez,[41] extending these solidification studies, developed a reliable method of growing metal crystals by draining away the liquid in a crucible when the metal partly solidified, thus obtaining a kind of metallic geode lined with crystals. He observed that metals generally formed trihedral or quadrangular pyramids joined at the base to give something like an octahedron, while the semi-metals were not so regular.

In the great flurry of chemical activity at the end of the eighteenth century, ideas on structure played relatively little part, yet there is an interesting cross connection with our topic. When the formation and reduction of metallic calces were explained on the basis of oxygen, it became necessary to provide an explanation also for the role of phlogiston in steel and cast iron, for the properties of these materials were definitely associated with the transfer of something or other to pure iron during heating with some phlogiston-rich material. The idea that it was the material element carbon in place of the vague phlogistonic principle arose at the end of a chain of events that began with experiments by the metallurgist Sven Rinman[42] on the etching and solution of different kinds of iron and steel, which arose from some Swedish experiments on the duplication of Damascus gun barrels. His observations were repeated precisely and elegantly by Bergman,[43] whence they were incorporated in the famous paper by Vandermonde, Berthollet, and Monge in 1786.[44]

⧉ THE NINETEENTH CENTURY:
METEORITES, PHYSICISTS, AND ENGINEERS

The nineteenth century inherited the background of theory and experimental observation described above and joined it to a developing knowledge in other fields to give modern metallurgy. We will consider, first, the developments of etching studies, then fracture, and finally the physical interpretation of the whole.

European attempts to duplicate the Wootz steel from which Damascus swords were made prompted many investigations into the structure of steel, even including some attempts by Michael Faraday to get an improved steel by alloying. Many pseudo-Damascus blades were made by a mixed forging technique akin to that used for textured gun barrels, but eventually Bréant in 1823,[45] under the influence of Berzelius' ideas on compounds and solutions, hit upon the correct explanation of the structure and made swords showing a true crystallization damask due to very high carbon content and slow cooling. He realized that carbon added to iron made steel, as a separate substance of fixed carbon content, and that beyond 1 per cent either graphite or a compound of iron with carbon forms. On slowly cooling an unsaturated steel, the more fusible "molecules of steel" tend to unite and separate from the iron. Such a material will form a poorly defined damask and is not very hard. If carbon is just right for steel the material will be uniform, but if the carbon is in excess, the iron will be all in the form of steel and the additional carbon will combine in a new proportion, i.e., cast iron. Slow cooling allows the molecules of the two compounds to arrange themselves in accordance with their respective affinities, the structure being larger the slower the rate of cooling.

A Russian factory for making the blades by Bréant's procedures was later set up by the Russian metallurgist Paul Anossoff.[46] The Duc de Luynes[47] made blades in 1844 both by crystallization and by a partial melting technique involving both wrought and cast iron that was a partial reversion to ancient Chinese steel-making methods. The finishing of all these swords, of course, required etching to reveal the structure that had been so laboriously produced.

Etching was being used quite widely in more immediately practical applications. First, following the observations of Grignon, Perret,[48] Rinman, and others, it had become a standard means of quickly distinguishing between steel and iron.[49] Next, in 1817, the English chemist Daniell[50] suggested etching for showing the heterogeneity of wrought iron and its dependence on the method of manufacture. The appearance of etched samples of wrought iron was related to their mechanical properties by Kirkaldy in 1862.[51] Tresca, in 1865 and later, used etching to study the flow of metals in punching, extruding, and other forming operations.

For a short time following its discovery in 1814 in France by one Alard, etched tin plate was used for decorative purposes under the name *moiré metallique,* and the pattern was correctly related to crystallization in the tin.

Meanwhile a most important observation had been made by the Austrian mineralogist Alois von Widmannstätten, who in 1808 etched a section of the Agram meteorite and showed the beautiful crystalline structure that now bears his name. A direct typographical imprint from such an etched surface was used to illustrate Schreibers' book on meteorites published in 1820.[52] Etching of iron meteorites thereafter became common, and it was natural that the English scientist Henry Sorby, who essentially founded the science of microscopical petrography, should proceed from the study of meteorites to that of the terrestrial iron and steel which abounded in his native Sheffield. In 1863 he developed for the first time really adequate methods of specimen preparation, and was able to show at the British Association Meeting in 1864 a number of micrographs of etched iron and steel in various forms and conditions. But despite enthusiastic comments on the part of the leading metallurgists of the day, he had no imitators and did not even publish his results in detail until 1885,[53] some years after Martens in Germany read a paper[54] on the same subject.

As soon as adequate methods of specimen preparation and microscope illumination were in existence, the structure of metals was easily seen and a true picture was rapidly assembled. Following Sorby the grains came to be regarded as separate though imperfectly developed crystals, which, although they are irregular in form, fit closely against one another in every direction. The bogey of crystallization under vibration was effectively laid when Sorby showed that metals were invariably crystalline, and that distorted crystals, being in an unstable state, would easily recover their normal crystalline polarity whenever the circumstances were such as to permit this arrangement. Grains "broken up, distorted and elongated" by cold deformation, recrystallized when heated to an essentially equiaxed structure, while on hot working the grains themselves were ductile and did not slide over each other. However, it was in the structure of steel that his observations were the most remarkable for he identified the "pearly constituent" with its duplex structure, and traced its origin to the decomposition of a structure that had been uniform at higher temperatures. Knowing that anomalies in the expansion and electrical properties of iron suggested an allotropic transformation, he related this to his structural changes.

It is necessary now to go back and follow studies of fracture. The increased use of iron in engineering structures soon gave rise to many failures of iron under repeated stresses ("fatigue"), with the fractures frequently brittle and showing crystalline facets. The problem became particularly acute with railway axles. The London Institution of Civil Engineers had a heated discussion on the subject in 1842 and several subsequent years in which, despite Robert Stevenson's disbelief, the predominating opinion was that percussion or vibration would produce crystallization in iron at room temperature, although there are no clear descriptions as to the state of the iron before it was crystallized. Though Kirkaldy[55] had actually shown in 1862 that the only difference was in the manner of fracture, not in the structure, revealed by etching, even as late as 1893 the members of the American Institute of Mining and Metallurgical Engineers could still indulge in a long and bitter discussion showing very little more understanding of the subject than their English counterparts half a century earlier. To this day the crystallization of steel under vibration is a popular myth, and, deplorably, constitutes a large fraction of the metallurgical "knowledge" of the man in the street.

It was common for nineteenth-century engineers performing tensile tests on materials to observe and record the appearance of the broken surfaces, though rare for them to comment significantly on the underlying structure. An exception is Robert Mallet[56] who examined the fracture of guns and enunciated certain rules of solidification which he said applied to all crystalline matter, and came to regard most of the differences of properties between different kinds of material as due to differences in crystalline aggregation.

It was long ago recognized that some materials could exist in more than one allotropic modification[57] and metals were brought into this class as a result of the important observation by Gore[58] in 1869 of the momentary elongation of iron at a critical temperature on cooling. Barrett[59] found this to be accompanied, in steel, by a sudden increase in temperature, which left no doubt that it was due to allotropy, though it was not until the work of Tschernoff[60] that the effect was related in any way to a structural change. Tschernoff's papers are remarkable for showing the amount of information that could be obtained on structure and structural changes without the aid of modern metallographic methods. He deduced that a profound change was taking place in the structure of steel from an observation of the change of tint on the machined surface of a gun forging that had been heated only at one end, and confirmed this by fracture and etching studies. He determined the increase in fracture grain size with increasing temperature of annealing, although he thought that steel in the high temperature form was amorphous, and that crystallization occurred during cooling. Carbon, he supposed, acted like water of crystallization. In a second paper[61] he discusses dendritic growth and crystal interference in relation to the structure and serviceability of ingots, and describes grains separated from an overannealed forging. His illustrations are the first ones to show the real shape of metal grains in three dimensions. He remarks that "It is evident that these grains have only a slight likeness to regular crystals; there is no regularity of formal angle and the edges are more or less crooked or curved." He gave a simple linear diagram showing the critical temperatures related to the various kinds of crystallization, and comments on their variation with carbon content. Translations of this work in English and French were published, and, together with Martens' paper of 1878, served to establish a wide metallurgical interest in structural ideas which has never since ceased. Shortly afterwards Sorby was persuaded to publish the results he had obtained in 1863–65, then to follow this with a second paper on higher magnification studies, and the science of metallography was well launched.

Although nineteenth century investigations by physicists of the elastic, electrical and other properties uncovered many a phenomenon which was later to be explained in structural terms, they showed on the whole remarkably little interest in the structure of metals and alloys. A most notable exception—though his paper remained almost unnoticed—was Félix Savart.[62] Savart begins by saying that "Hitherto cast metals have been regarded as solid substances which most closely approach conditions of homogeneity: they have been regarded as assemblies of an infinite number of tiny crystals joined together without order and at random, and there was no suspicion that in any mass of metal whatever there could be differences of elasticity or cohesion as great as, or perhaps greater than, those observed in a fibrous body such as wood."[63] Savart was interested in the elasticity of metals and studied the nodal vibration patterns on disks of various materials, particularly of lead. Cast disks of this metal showed quite irregular behavior, which he eventually traced to the presence of a number of crystals neither randomly nor perfectly regularly arranged. "These facts and many others of the same kind show clearly that metals do not possess a homogeneous structure, but only that they are not crystallized regularly. There remains thus only one supposition to be made, namely that they possess a semi-regular structure, as if at the moment of solidification there were formed many distinct crystals, of a

493

fairly considerable volume but whose homologous faces were not turned to the same points in space. According to this idea metals would be like certain grouped crystals, in which each one, considered by itself, presents a regular structure while the whole mass appears quite confused." Savart found that isotropic elastic properties could be obtained by vibrating the casting during solidification, or by cold working and annealing (both of which processes would give a fine random grain). Hammering not followed by annealing, or annealing not preceded by hammering, produced little effect. The kind of structure was not, he thought, peculiar to metals, but all solids were heterogeneous except, perhaps, fine sedimentary materials like chalk.

To return to the metallurgical aspects of the story; in the last two decades of the nineteenth century [64] there was a most active period of observation and speculation on the nature of steel precipitated by two important papers [65] by Floris Osmond, one advocating *"La théorie cellulaire des propriétés de l'acier"* (1885) and the other on thermal critical points (1887). The background of this was partly chemical, for Osmond began from the observation of Caron (1862) of the difference in chemical form of carbon in annealed and in quenched steels. This arose from a study of the residues left behind after dissolving the bulk of the metal in acid—the same technique that had previously led to important ideas when used by Rinman, Bergman, Berthollet, and Karsten, and which, as we have seen, had its roots in Oriental metalworking techniques.

Most of the observations on the range of existence of microconstituents as a function of temperature and composition were suddenly reduced to order in 1900, when Roozeboom [66] showed how Gibbs' phase rule could be applied to them, but arguments on the causes of the hardening of steel continued for a long time. It is interesting to note that, in this period as in earlier ones, studies of the highly complex material steel provoked the most meaningful experiments: Simpler substances which would have been much easier to understand failed to excite any interest. (Another interesting point is that the best work in both France and England was done under the aegis of national committees, which are not commonly supposed to inspire scientific research!) Even recrystallization and grain growth was observed metallographically first on steel, not brass. However, following the work of Charpy [67] in 1897 relating microstructure of alloys to their freezing point curves, the microscope became a standard companion of the pyrometer in constitution studies.

At about the same time the micro-mechanism of the deformation of crystalline materials was studied, particularly by Ewing and Rosenhain.[68] Thereafter, for a quarter of a century, classical metallography had its golden period of development, exploiting the standard methods of combined microscopy and thermal analysis. Metallurgists understood the principal ways in which crystals were formed and modified and used the results practically. They became accustomed to allotropic transformations and to the co-existence of phases of different structure, but they were slow to realize that there could be just as great changes of solubility in the solid state as in the liquid. Attempts to explain age hardening, which had been discovered in 1911 by an observant engineer without any help from physical or metallurgical theorists, eventually led to this. The excitement of metallography was beginning to decline as it became more a routine than a research tool, when an entirely new level of structure was opened for experimental study by the development of X-ray diffraction methods following von Laue's suggestion in 1912, widely applied to metallurgical problems in the 1920's. Although the frontier of research thereafter moved to the atomic scale with growing emphasis on lattice imperfections, important work such as that by Davenport and Bain[69] on isothermal transformations in steel and other work on the role of surface tension in determining microstructure showed that there was still much to learn even on the scale of structure accessible to the optical microscope.

References

Note: Unless English translations are listed below, quotations given in the paper from works in other languages have been translated by the author.

1 C. S. Smith, *Endeavour*, XVI (1957), 199–208. (Contains several illustrations of textured metal surfaces.)

2 A. France-Lanord, *Pays Gaumais*, X (1949), Nos. 1–3. A. France-Lanord, *Révue de Métallurgie*, XLIX (1952), 411–22. A. Liestöl, *Viking*, XV (1951), 71–96.

3 H. Williams, *Technical Studies in the Field of the Fine Arts*, III (1934), 16–18. J. G. Mann, *Proc. British Acad.*, XXVIII (1942), 17–44.

4 V. Biringuccio, *de la Pirotechnia* (Venice, 1540 [Eng. trans.; New York, 1942]).

5 L. Savot, *Discours sur les medailles antiques* (Paris, 1627).

6 M. Jousse, *Fidelle ouverture de l'art de Serrurerie*, (La Flêche, 1627). Partial English translation by Anneliese G. Sisco, 1956, unpublished.

7 R. Boyle, *Origine and Vertue of Gems* (London, 1672).

8 E. C. Millington, *Annals of Science*, V (1941–47), 253–69 and 352–67.

9 M. Boas, *Osiris*, XX (1952), 412–541.

10 D. Sennert, *Hypomnemata physica* (Frankfurt, 1636).
11 P. Gassendi, *Syntagmata philosophicum* (1649). Quotations are from English translation by W. Charleton, *Physiologia* (London, 1654), p. 337.
12 R. Descartes, *Principia philosophia* (Paris, 1644).
13 A. R. Hall, *The Scientific Revolution, 1500–1800* (London, 1954), p. 215.
14 R. Descartes, *Oeuvres,* ed. P. Tannery, II (1898), 486.
15 *Ibid.,* IX (1904), 276–78. Translated for this paper by A. G. Sisco.
16 A. LeGrande, *An Entire Body of Philosophy According to . . . Descartes,* tr. R. Blome (London, 1694).
17 C. and P. Perrault, *Essais de physique* (Paris, 1680).
18 N. Hartsoeker, *Principes de physique* (Paris, 1696).
19 J. Rohault, *Traité de physique* (Paris, 1671).
20 *Ibid.,* tr. by J. Clarke (London, 1723), II, 156–57.
21 *Ibid.,* I, 134.
22 I. Newton, *Opticks* (2nd ed.; London, 1718), pp. 243, 363–64 and 370.
23 H. Power, *Experimental Philosophy . . .* (London, 1664).
24 Robert Hooke, *Micrographia* (London, 1665).
25 Rohault, *Traité de physique,* p. 155.
26 A. Leeuwenhoek, *Phil. Trans.,* XXIII (1703), 1430.
27 *Ibid.,* XXVI (1709), 493.
28 R. A. F. de Réaumur, *L'Art de convertir le fer forgé en acier . . .* (Paris, 1722); English translation, Chicago, 1956.
29 *Ibid.,* (English translation), p. 111.
30 *Ibid.,* pp. 213–14.
31 R. A. F. de Réaumur, *Mem. Acad. Sci.* (1724), 307–16.
32 Geoffroy, *Mem. Acad. Sci.* (1725), 16–66.
33 W. Lewis, *Commercium Philisophico-Technicum* (London, 1763), p. 525.
34 P. C. Grignon, *Mémoires de physique, sur le fer. . . .* (Paris, 1775).
35 J. A. Mongez, *Journal de physique,* XVIII (1781), 74–76.
36 L. B. Guyton de Morveau, *Journal de physique,* VIII (1776), 348–53.
37 *Ibid.,* p. xi.
38 P. C. Grignon, *Mémoires,* p. 69.
39 *Ibid.,* p. 475.
40 L. B. Guyton de Morveau, *Journal de physique,* XIII (1779), 90–92.
41 Mongez, *Journal de physique,* XVIII (1781), 74–76.
42 S. Rinman, *Handlingar K. Vetenskaps Acad. Stockholm,* XXXV (1774), 1–14.
43 T. Bergman, "de Analysi Ferri," *Opuscula Physica et Chemica* (Upsala, 1783), III, 1–107. French translation by P. C. Grignon, Paris, 1783.
44 Vandermonde, Berthollet and Monge, *Mem. Acad. Sci.* (1786), 132–200.
45 Bréant, *Bull. Société d'Encouragement,* XXII, 1823, 222–27.
46 P. Anosoff, *Annuaire du journal des mines en Russie* (1841), 192–286.
47 C. d'A. de Luynes, *Mémoire sur la fabrication de l'acier fondu et damassé* (Paris, 1844); notes from German translation in *Polytechnische Centralblatt,* I (1845) (ii), 315–21.
48 J. J. Perret, *L'Art du Coutelier* (Paris, 1771), I, 219.

49 Comité de Salut Public, quoted by Vandermonde, Monge and Berthollet in *Avis aux ouvriers en fer sur la fabrication de l'acier,* reprinted in *Journal de physique,* XLIII (1793), 373–86.

50 J. F. Daniell, *Quarterly J. Sci.,* II (1817), 278–93.

51 D. Kirkaldy, *Experiments on Wrought Iron and Steel* (Edinburgh, 1862).

52 C. von Schreibers, *Beyträge zur Geschichte . . . der meteorischer stein- und metall-Massen* (Vienna, 1820), plate IX.

53 H. C. Sorby, *J. Iron and Steel Inst.,* i (1887), 255–88. (Paper read in 1885.)

54 A. Martens, *Zeit. des vereines Deutscher Ingr.,* XXI (1877), 11.

55 Kirkaldy, *Experiments on Wrought Iron and Steel.*

56 R. Mallett, *On the Physical Conditions Involved in the Construction of Artillery* (London, 1856).

57 It must be remembered that in the nineteenth century, allotropy meant to both chemists and physicists a *molecular* change, involving groupings of very few atoms without regard to the stacking of these into the crystal, which was energetically disregarded. Even Osmond, the prince of siderurgical allotropists, thought it quite natural for the A_2 point in iron to be unaccompanied by a crystallographic change. The chemical preoccupation with small units and simple combining proportions again confused the issue, and it was not until after X-ray diffraction studies that it was seen that for most metals and simple ionic substances there is no state of aggregation between the atom and the crystal. The crystal is the molecule, but it was not so thought of in the nineteenth century. Yet when the crystal was accepted, it brought in a new set of difficulties. Many well-established properties of solids (such as diffusion, elastic aftereffects, and even gross plastic deformation) would be impossible in a good crystal, though they had been easily explained on the basis of the concept of a variable distribution of molecular arrangements, postulated by Maxwell to explain elastic anomalies. Actually this differs only in geometry, not in principle, from the modern view of crystals with defects in them to allow noncrystalline behavior, but it took time for crystallized thinking to allow imperfections. These were reintroduced theoretically to explain anomalously intense X-ray reflections some years before their general physical necessity was again realized.

It seems to be generally true that the formulation of an improved fundamental concept can only be arrived at by ignoring some very real aspects of experimental knowledge; and only after a new scientific generation has arisen can the necessary modifications of the ideal picture be thought of. Perhaps it is impossible to make any completely true statement that will apply usefully to nature, and this not in the sense of the principle of indeterminancy but simply because of complexity. There is an essential fuzziness in nature that arises from the multitudinous factors affecting everything that can be observed. We understand phenomena only to the extent that we can isolate them in thought and experiment; but we should not forget the simplification, and we should not dogmatically extrapolate to situations wherein the once properly ignored factors may begin to have weight. If science is to grow in method as well as content it must develop stronger tools than statistical analysis to deal with complex systems.

58 G. Gore, *Proc. Roy. Soc.,* XVII (1869), 260–65.

59 W. F. Barrett, *Phil. Mag.,* XLVI (1873), 472–78.

60 D. Tschernoff, *Imp. Russian Tech. Soc.* (1868); English translation in *Proc. Instn. Mech. Engrs.* (1880), pp. 286–307.

61 D. Tschernoff, *Imp. Russ. Tech. Soc.* (1878); English translation in *Proc. Instn. Mech. Engrs.* (1880), pp. 152–83.

62 F. Savart, *Ann. Chim. Phys.,* XLI (1829), 61–75.

63 Although Savart refers to the *"infinité de petits crystaux"* as a common supposition, I cannot find any earlier writer who specifically describes the aggregates as composed of small *crystals,* although it is undoubtedly implicit in some of the discussions above. Even Coulomb, who considered elasticity and coherence as due to the *"parties intégrantes,"* does not presuppose any scale of order whatever. Structurally, his viewpoint is qualitative and essentially identical with that of the seventeenth century philosophers. Biot (*Traité de physique,* Paris, 1817) used a good corpuscular theory to explain the differences between solids, liquids and gases, and properly distinguishes between elasticity and cohesion in terms of the forces and arrangements of very small material particles held together by forces which operate only at very small distances, but even he seems to require no state of aggregation between that of the ultimate molecule and that of the entire body. His polarized light studies (1818), however, led him to suggest that many natural bodies which were not observed as good crystals were composed of innumerable invisible crystals.

64 R. F. Mehl, *A Brief History of the Science of Metals* (New York, 1948), offers an excellent discussion of developments since 1871.

65 F. Osmond and J. Werth, *Annales des Mines,* VIII (1885), 5–84, and F. Osmond, *Mém. Artillerie Marine,* XV (1887), 573–714; also published separately, Paris, 1888.

66 H. W. Bakuis Roozeboom, *Zeit. Phys. Chem.,* XXXIV (1900), 437.

67 G. Charpy, *Bull. Société d'Encouragement* (1897) 2 (v), 384–419.

68 J. A. Ewing and W. Rosenhain, *Phil. Trans. Royal Soc.,* CXCIII (1900), 353.

69 E. S. Davenport and E. C. Bain, *Trans. A.I.M.E.,* XC (1930), 117.

Acknowledgments

The writer gratefully acknowledges the support he has received from the Guggenheim Memorial Foundation and the U. S. National Science Foundation during the period of active research on which this paper is based.

STRUCTURE OF MATTER AND CHEMICAL

THEORY IN THE SEVENTEENTH

AND EIGHTEENTH CENTURIES

Marie Boas

🔹 It has always seemed puzzling to me that Lavoisier's work on the role of oxygen in combustion, dramatic and important as it was, could, together with a new system of chemical nomenclature, have permitted the complete reform of chemistry. As usually happens when one examines a dramatic event in the history of science, it turns out not to have been so simple. Seventeenth-century chemists had tried to have both a rational system of nomenclature and a rational theory of combustion; had tried, and had failed. Their failure was inevitable without a knowledge of pneumatic chemistry. But when one examines their attempts, it is plain that more than pneumatic chemistry was lacking. Much was required, including a different climate of opinion and an improved state of chemical knowledge. One further ingredient in the success of Lavoisier was a proper understanding of the nature of chemical elements; it is my contention here that this was a development of chemistry before Lavoisier, and that it was intimately connected with discussions of the particulate structure of matter though, paradoxically, a chemically satisfying atomic theory had to wait until Dalton's work in the years following Lavoisier's.

One of the great achievements of the scientific revolution was the universal acceptance of the mechanical philosophy. This demanded not only the acceptance of a particulate theory of matter, but also the use of such a theory of matter to explain the properties of matter. This was a problem for the physicists, but chemists were not backward, and, especially as chemistry became, increasingly, a branch of natural philosophy, they accepted one or another of the prevailing mechanical philosophies. It was slowly that chemists accepted the mechanical philosophy as a truly necessary ingredient of chemical theory, as distinct from natural philosophy. This was not just obscurantism; it was partly that until Dalton no con-

499

cept of atomic structure was really suitable for use in chemical theory, and partly that chemists tended, for good reason, to compartmentalize the various branches of chemistry. Since physical chemistry is a modern science it may seem anachronistic to say that when one considers the use chemists made in the seventeenth and eighteenth centuries of the theories of structure of matter which they adopted one must differentiate between physical chemistry and other sorts of chemistry. But, in fact, such a distinction always existed, and chemists primarily concerned with what we should now call physical chemistry were often referred to—and not always admiringly—as natural philosophers rather than as chemists. The same distinction appears in the textbook division between theoretical and practical chemistry. Eighteenth-century chemists nearly always distinguished between physical principles—ultimate particles—and chemical principles—elements. One should also remember that fire (and, *mutatis mutandis,* heat) was in the eighteenth century regarded as both a chemical instrument and a chemical substance: hence both its physical and its chemical structure had to be considered. So, too, air was far more a physical than a chemical substance, and the physical properties of air—atmospheric or factitious—were commonly discussed apart from their possible chemical properties.

The point I wish to emphasize is that throughout the period under discussion one must, frequently, distinguish between physical chemistry and ordinary chemistry. One could, that is, accept as physically true a particulate structure of matter—almost all chemists did from the mid-seventeenth century onwards—without necessarily making any use of it in discussing chemical composition or chemical reaction. For obvious reasons, I do not propose here to consider the work of such chemists in any detail. Here I shall only include the ideas of chemists who, in one way or another, actively made use of a particulate theory of matter. These chemists fall roughly into two groups. On the one hand are those chemists who combined a belief in the existence of small particles as the ultimate entities necessary for the structure of matter with a strong emphasis on chemical principles or elements as the real instruments of chemical change. Many late seventeenth-century chemists did this, as Stahl was to do in the early eighteenth century. Such chemists did discuss particles in connection with chemical theory, but the role they assigned to such corpuscles was, naturally, of no very interesting importance. Far more important—it is one of the real achievements of seventeenth-century chemistry—were the chemists who so firmly incorporated the notion of a particulate structure of matter into their thinking that they automatically spoke and thought of chemical composition and chemical reaction as involving the arrangement and rearrangement of chemical units in the form of particles. Whether these chemical units were simple substances or chemical elements such a point of view was an important ingredient in the shift from interpreting a reaction in terms of the saline or earthy or spirituous *part* of a substance to interpreting a reaction in terms of particles of real chemical entities, so that when one spoke of a particle of sulfur one meant just that and not a chunk of some doubtfully material "principle." This shift was an important one, only possible with increased chemical knowledge. But knowledge alone was not enough; one needed the changed point of view.

The almost universal acceptance in the seventeenth century of the mechanical philosophy does not, I am sure, need elaborating here.[1] Chemists were not backward in accepting the theory of matter this philosophy implied, though at first they held aloof from the true mechanical system. Before the end of the century most chemists had come to interpret and explain the more physical properties of chemical substances—solidity, fluidity, volatility, density, acidity—in terms of the size, shape, and motion of the particles. Even textbook writers, never anxious to embroil the student in contemporary and confusing controversy, began to include a short section on the structure of matter, usually beginning with particles and continuing with elements, presumed particulate. This theory was usually supported—as was, it should be remembered, any theory of elements—with elaborate experimental evidence. Yet many chemists felt that the prefatory sections of a book were the place for theories about the structure of matter, and disapproved of the tendency shown by the more advanced chemists to try to inject into chemical operations explanations in terms of particles. Thus Lemery, though he explained the physical properties of many substances, including acids, by means of the size and shape of their component particles, took the trouble to note his disapproval of the interpretation of chemical changes in such terms. He wrote:

When one throws corrosive sublimate into lime water, it takes on a yellow colour and loses so much of its corrosive force that one can safely take it by mouth. I do not aspire to explain this change of colour; I prefer to leave it to those with more leisure than I to explain the disposition of the pores that the acid and the lime must give the mercury so that it may modify or reflect the light in such a way that a material which formerly appeared very white is now yellow. I shall only say that the lime water softens or weakens the force of the sublimate because of the particles of lime which it contains, which particles, meeting and striking the sublimate, break part of the points in which the corrosiveness consists.[2]

Though Lemery here disapproves of a chemist's wasting his time on this kind of explanation, it is obvious that he believes that such an explanation is possible. He merely regards it as more suitable for exploration by a natural philosopher than by a chemist. A chemist should, apparently, merely observe changes, and not ask what happens to the small parts of matter during the changes he observes. No doubt Lemery agreed with other French chemists that Boyle, who tried to incorporate his theory of matter directly into his chemical theory, was a natural philosopher, not a chemist. This, I think, is an attitude which helps to explain why the eighteenth-century chemists so often preferred the writings of Beccher and Stahl to those of Boyle, who besides his tendency to physics never wrote the work on chemical philosophy which he once planned, which might have convinced chemists that he had a systematic theory.

For Boyle the ultimate structure of matter was too intimately a part of chemical theory ever to be completely ignored.[3] Not that he tried to do what Lemery so deplored, to explain all chemical phenomena in directly particulate terms, but rather that underlying all his chemical theory was a corpuscular theory of matter, a theory of matter developed by himself, and one which helped him to discard many outmoded concepts. As is well known, Boyle utterly rejected all occult qualities, whether chemical or physical, and he rejected at the same time all notion of chemical elements or principles as the ultimate building blocks of matter. This he did on both logical and experimental grounds. First, he rejected the contention that analysis by fire produced the same three—or four—or five elements from all substances; for what a green stick produced did not necessarily come up to the claims of chemists, and almost no other substance produced even as satisfactory a number of elements as a green stick. (It should be remembered that for the seventeenth century an element or principle was a substance universally present in all non-elementary substances.) Next, he showed that supposed elements like spirits and oils were in fact highly heterogeneous classes of substances; and that in any case these substances were not immutable. (It was the desire to prove this last point that led him to claim, against his own better experimental judgment, that he had converted water to earth.) Perhaps most important of all, he showed that the interpretation of chemical phenomena by means of a theory based on elements or principles was unsound. For example, chemists often insisted that salt (the element) was the cause of solidity, and hence the agent in causing chemical precipitation; Boyle showed by experiment that chemical precipitation was perfectly feasible when no salt was present. Again, some chemists made use of the acid-alkali theory, whereby all reactions were assumed to be the result of an acid-alkali combination; Boyle, by means of the tests he developed, using chemical indicators, demonstrated that some substances were neither acid nor alkali, and were yet reactive. The only elements he would accept on the basis of the then current definition of elements, substances universally present, and into which all substances could be resolved, were particles of matter.

He did, even, work out a scheme of corpuscular association.[4] The simplest corpuscles of matter he conceived as differing one from another in size, shape and motion, and these physical differences caused the different physical properties and ultimately chemical properties of various kinds of matter. These simple corpuscles, he thought, associated themselves into "minute masses or clusters . . . not easily dissipable into such particles as composed them." These masses could persist unchanged through reactions, even though they manifestly composed mixt bodies. Chemical elements, he thought, might be such mixt bodies. Sometimes Boyle spoke of the possibility of a more complex hierarchy of corpuscles: primary concretions, secondary mixts, and decompounded mixts, the first two roughly corresponding to the modern atom, the last to the modern molecule.

These ideas of Boyle were not necessarily as satisfying to a chemist as the more sweeping theories they were designed to replace. What he presented was really a chemical empiricism based upon a physical theory. But these ideas did help distract attention from elements and principles, and focus it where it belonged, on actual chemical entities. This is particularly true since Boyle endeavored constantly to explain composition, reaction and chemical properties primarily in terms of simple known substances and only secondarily in terms of corpuscles. Thereby he showed the way to a more profitable mode of thought in chemistry. It might here be of interest to compare Boyle's discussion of the reaction of mercury sublimate and oil of tartar (used by him as a test for the presence of an alkali in solution) with Lemery's discussion of the reaction of mercury sublimate and lime-water, quoted above. Boyle wrote that this reaction was very like those "of metalline bodies corroded by acid salts; so that the colour in our case results from the coalition of the mercurial particles with the saline ones, wherewith they were formerly associated, and with the alkalizate particles of the salt of tartar that swim up down in the oil."[5] That is to say, when mercury is joined to saline particles, the resulting compound is white; when it is joined to alkalizate particles, the resulting compound is yellow. This makes far more sense in chemical terms than does Lemery's explanation, and is, besides, much more likely to lead to profitable chemical investigations than any discussion in terms of either purely physical principles (like Lemery's particles) or purely chemical principles, (the salt, oils, and mercuries of the chemists). This is a sufficiently typical example of Boyle's method of chemical explanation to serve as illustration here. His attitude was, slowly, adopted by other chemists, until even when chemists continued to speak of elements or principles they modified their theory by accepting particles as the basis of simple, if nonelementary, real chemical entities. That is to say, they came to prefer discussing simple substances to discussing the simplest conceivable substances.

This is especially noticeable in those most directly influenced by Boyle. Thus Guillaume Homberg, a most influential figure in the *Académie des sciences,* where he seems to have been responsible for the introduction of theoretical and physical chemistry (which he presumably learned while in Boyle's laboratory), took a quite sceptical view of chemical elements. In his *Essais de chimie*[6] he differentiated carefully between the organic and the inorganic worlds, each of which he supposed to have a different set of chemical principles. Even within the inorganic world he found that the basic elements varied; for he thought that some minerals (metals and their ores) contained mercury, while fossils and salts did not. That is, elements were not the universal constituents of all matter, but merely the simplest substances from which various kinds of matter were built up, so that different kinds of matter contained different elements. Though Homberg did not indicate that he felt this to be a new concept it was not by any means the prevailing view in 1700 and implies a quite novel attitude toward chemical analysis and toward the structure of matter, an attitude to become more and more common as the eighteenth century proceeded.

Other chemists, though less clearly than Homberg, also and paradoxically spoke of elements and principles while regarding them as compounded of particles. Even Becher, founder of the theory of principles enlarged and publicized by Stahl, who made it even more reactionary and mystic than Becher, had a much more complex theory of the structure of matter than early seventeenth century chemists had had. Presumably he was at least partially influenced by his study with Boyle.[7] Becher differentiated between what he called mixts, compounds and aggregates; the distinction is not entirely clear, but essentially it involves a hierarchy of particles. Becher spoke of mixts as being composed of corpuscles, few in number, very small and not easily divisible, though themselves composed of lesser corpuscles. Compounds consisted of combinations of mixts, and aggregates of combinations of mixts or of mixts and compounds. Only aggregates were large enough to be, as he put it, "objects of our senses," that is experimentally detectable. This was the theory adopted by Stahl, who described earth and water as the "two material principles of mixts," and compounds and aggregates as the combination of mixts or of "several Atoms combined together."[8] In contrast Boerhaave, who completely accepted a particulate structure of matter, together with the concept of physical chemistry, described fire, water, air and earth as "instruments" of chemistry rather than as elements or principles, so that he did not explicitly discuss their possible elementary nature.[9]

Frequently eighteenth-century chemists, while accepting principles and elements, preferred to minimize their role in chemical reactions and to concentrate instead on the role of simple chemical substances in chemical composition and chemical reaction. This soon involved them in two problems: first, in the problem of affinity, which was inevitably bound up with the acceptance of the existence of an attractive force as an essential property of the particles of matter; and second, in the role of fire or heat, for fire tended to become the dominant chemical principle because of the stress laid upon calcination (rather than analysis by fire) as the key chemical reaction.

Boyle's great triumph had been the exclusion of occult qualities from chemistry and physics; there is no record of what he thought of Newton's addiction to what the latter variously denominated sociability and attraction. To be sure, most of Newton's specifically chemical papers were written after Boyle's death, but the term sociability and the thing itself were used chemically in both the letters on light and colors, and in the letter on ether addressed to Boyle.[10] Actually the clearest application of attraction to chemical theory was made by Newton in his *De natura acidorum,* first written in 1692, though published a dozen years later in John Harris' *Lexicon technicum.*[11] There Newton defined acids as being "endued with a great Attractive Force; in which Force their Activity consists," though at the same time he noted that their physical properties—volatility for example—depended upon the physical properties of the component particles, especially upon the size of the particles. This is one of the first statements of what came to be a characteristically eighteenth-century differentiation between chemical and physical characteristics. In the short paper on acids Newton hinted that there was a varying attractive force between particles; this is the aspect of the problem which he was to develop in more detail in the *Opticks,* in his discussion of metallic displacements. In the paper on acids he also sketched his notion of the hierarchy of particles; in this he was not particularly original, except for the idea that menstruums were incapable of transforming gold because they could only break down the bigger coalitions of particles, not the smaller coalitions which, when broken, would no longer constitute gold.

Even before Newton published these views, which foreshadow common notions of the eighteenth century, some chemists had reached similar conclusions on the basis of their own interpretation of Newtonian natural philosophy. Thus John Freind, making use of the brand of Newtonianism presented at Oxford in 1700 by John Keill, developed a Newtonian chemistry which he discussed publicly at Oxford in 1704.[12] He chose to consider the operations of chemistry as the result either of the dissociation of particles—calcination, distillation, sublimation—or of the coherence of particles—fermentation, digestion, extraction, precipitation, crystallization—in both cases under the influence of the universal attractive force between particles. Freind was so anxious to stress what he regarded as the purest form of the mechanical philosophy, a philosophy that included rigorously defined mathematical laws, that he minimized the importance of chemical combination and reaction; his operations seem little more than the mere coming together of various substances, and there is little real concept in his lectures of genuine chemical union. He did not even discuss chemical composition or chemical analysis, but spoke of the action of fire only in terms of calcination, distillation, and sublimation. His emphasis on the importance of attractive forces led him to conclude that crystal structure might in no way reflect the shape of the particles; for as the force of attraction will be different on different sides of the particle, the shape of the final crystal will be dependent on the attractive properties of the particle, not upon its size, shape and figure. It is altogether a very neat scheme of chemistry, though rather remote from actual chemical processes.

Few chemists, actually, accepted anything resembling Freind's elaborate structure. Yet Newtonian attraction entered firmly though not always overtly into chemistry. Affinity might be intended to be a neutral word, but it inevitably suggested attraction. Geoffroy, presenting his table of affinities to the *Académie des sciences* in 1718, carefully avoided all reference to attraction in Newtonian terms, though he did refer incidentally to the "rapports" between bodies, the disposition they had to unite.[13] It is only occasionally, indeed, that he even referred to the particles of bodies, merely talking of substances in general terms. Equally he avoided the use of the terms principles and elements. The substances with which Geoffroy was concerned were all simple chemical entities, and his descriptions of reactions were all in terms of these same chemical entities. Thus, when he described the reaction between sea salt and a solution of vitriol, he said "the vitriolic acid abandons its metal to join itself to the Earth of the sea salt. The acid of the salt lets go its hold and dissipates itself in air," unless there is some other substance present with which it can unite. Or again he referred to "the particles of iron abandoned by the acid of the vitriol" which were now free to unite with some other substance for which they had more affinity than they did for the vitriol. Throughout, Geoffroy indicates a very clear understanding of the way in which reactions proceed, together with a tacit acceptance of the particulate structure of matter. (It is amusing to find that while Geoffroy ignored the question of what affinity really is, Fontenelle, writing the history of the Academy for 1718, wistfully remarked that to explain affinity it would be useful to be able to speak of sympathies and attractions, "si elles étoient quelque chose.")[14]

Attraction was clearly a controversial force in 1718; how far the situation had changed by the end of the century is strikingly illustrated by comparing Geoffroy's discussion with that prefixed by Bergman to his *Dissertation on Elective Attractions,* first published in 1775.[15] Before he could give tables of affinity, Bergman felt it necessary to discuss the universal nature of Newtonian attraction, which, he pointed out, was a characteristic of particles that could be determined experimentally, as size and shape of the particles could not. Like most eighteenth-century chemists

Bergman concluded that the physical properties of matter—solidity, fluidity, and the like—depended on the physical properties of the component particles, while chemical properties rather depended upon the attraction between particles and, a necessary consequence, the atmosphere of heat which surrounded each individual particle. Chemists of the eighteenth century quite generally distinguished not only between chemical and physical properties but also between chemical and physical elements or principles. (The latter was the more popular word.) To most chemists fire—or heat or phlogiston—was a physical substance, to be explained in terms of the particles of matter; it was less a chemical substance than a chemical agent. It was usually thought to be a fluid by reason of its having very small and very swiftly moving particles, even though, commonly, other bodies were thought to be fluid because of the quantity of heat (or fire or phlogiston) contained in the pores. (For Macquer fire was the cause of fluidity, for Lavoisier caloric.) In general too it was held to be the presence and motion of fire (for Boerhaave, Stahl, and Macquer), or caloric (Lavoisier), or even phlogiston, which was responsible for exciting the particles of a substance and causing it to grow hot. Phlogiston was not necessarily identical with fire; for Macquer and others it was less simple than fire, for Stahl and others more nearly elementary than fire.[16] Very often, though regarded as a material elastic fluid, phlogiston came close to being a name for the "principle of fire," where "principle" was interpreted quite otherwise than it had been in the seventeenth century. Both Macquer and Lavoisier spoke of an "acidifying principle" and in the same way, though for Macquer the principle was unknown and perhaps unknowable, and for Lavoisier it was a simple substance, oxygen. Thus, in the eighteenth century, "principle" was no longer necessarily taken to mean the ultimate constituent; it might be an agent, more or less mechanical. This perhaps explains one reason why the eighteenth-century chemists, so logical and non-mystic, could treat Stahl with such respect and interest.

To see how far, in spite of outmoded nomenclature and confusion resulting from the widely diverse views held by various chemists, rational chemistry had yet accepted a generalized combination of on the one hand a particulate theory of matter and on the other Newtonian attraction, one can conveniently turn to Macquer's *Dictionary of Chemistry*, that pleasant and useful encyclopedia of chemical knowledge. Macquer evidently tried to make this a general dictionary rather than merely a vehicle for his own views; he is in the *Dictionary* much more cautious in his definitions than when writing a textbook. In the latter he accepted Earth, Water, Air, and Fire as principles or elements in the seventeenth century sense of ultimate substances ("at least they are really such with regard to us," he said) composed of material particles. In the *Dictionary* he went no further than to say that these were commonly regarded as elements or principles because never so far decomposed, though they are quite probably compound substances.[17] But on the other hand, in the *Dictionary* he described the structure of matter in far more detail than he felt was necessary in a textbook for beginners.[18] The real principles of bodies, he says, are particles, which are of two sorts: constituent particles (*parties constitutantes*) and integrant particles (*parties intégrantes*). Constituent particles unite in real union (that is, not as a mixture) to form compounds or mixts, and one can never separate such a substance into its constituent particles without decomposing the substance which, together, they constitute. (Thus, in the case of salt, the principles, or perhaps "proximate principles" are an acid and an alkali; if one were to decompose the salt and so separate the principles from one another, "there would no longer be common salt, but only the acid and alkali of the salt, which are very different substances from salt.") Integrant particles are less elementary; they are identical with the substance they compose, and are "the smallest molecule into which a body may be reduced without being decomposed." Such a molecule (Macquer's word) of salt would consist of "a single atom of acid and a single atom of alkali." When one splits up a molecule, one loses the original substance and acquires a new substance. Both kinds of particles enter into chemistry, as chemical properties depend on the manner into which the various kinds of particles are joined. The specific nature of the individual particles is less important, and is in any case unknown. Affinity is the property which makes both kinds of particles cohere; every substance probably has some affinity for every other substance, but the amount of affinity between substances varies. As chemistry is a uniquely experimental science, one can reasonably accept as constituent particles those which experiments have shown to be not further divisible.

That this represents what chemists in general believed in the mid-eighteenth century appears when one examines the views of other chemists. Thus Venel, the author of the article "Chymie" in Diderot's and D'Alembert's *Encyclopédie,* wrote, "The particles of an aggregation are by modern physicists called molecules or masses of the last composition or the last order, derived corpuscles, &c. and much more accurately by earlier physicists integrant particles or simple corpuscles." These corpuscles, he noted, were "inalterable," that is, the nature of the aggregation depends upon the integrant particles, which cannot be altered without changing the aggregation. Though he regarded this as a physicist's theory, it was pretty generally accepted by chemists as well. Guyton-Morveau, in the *Elements of Chemistry,* which was based upon lectures given at Dijon in the 1770's, defined "simple bodies" as those we cannot decompose; these could be of two sorts, either "natural elements"—fire, air, water, earth—or "chemical elements"—constituent particles, the unspecified principles which compose integrant particles, and the things with which chemists really deal. Though Guyton was inclined to think that there was only one kind of matter, whose structure determined the physical properties of substances, and even of the elements, he thought that was not a subject which the chemists should discuss in detail.[19]

Increasingly, elements and principles were, as in the examples already cited, regarded as those which chemists could not decompose; in the then state of analytical chemistry, this made for a certain caution. Bergman, writing an introduction to Scheele's work on air and fire, firmly expressed the belief that elements, though never as yet analyzed, were not homogeneous.[20] Scheele himself was a sceptic; he suspected that elements were not immutable even though he had shown that Boyle had not converted water into earth in a famous experiment. (This is the experiment Lavoisier disproved in his more famous pelican experiment.) Scheele was, moreover, quite sure that air was a compound body; so was fire, which contained air. Only phlogiston was "a true element and a simple principle," but phlogiston was rather outside ordinary analysis, since it was never found pure and by itself.[21]

These somewhat tentative statements of Guyton-Morveau, Bergman, and Scheele are very much like the more famous pronouncements of Lavoisier. It is well known that Lavoisier loftily claimed to be uncommitted on the subject of elements, though he had, in fact, sufficiently precise ideas. In the introductory memoir to the pleas for a reform of chemical nomenclature he wrote, "We should be contradicting everything we are about to set forth if we were to devote ourselves to a long discussion of the constituent principles of bodies and of their elementary molecules. We shall here simply consider those substances which we cannot decompose; those which we obtain as the last product of chemical analysis. No doubt someday those substances which we find simple will be decomposed in their turn."[22] This is by no means a revolutionary statement; nor, I feel sure, was it meant to be one. Like Macquer, Lavoisier was quite willing to speak of "principles" which were not elements, but rather the part of a compound which produced a particular characteristic; so he spoke of the "acidic principle" before he identified it with oxygen. Principles were not elements; and what elements might be was doubtful—especially as all substances were made up of particles. Most historians know the famous pronouncement in the Preliminary Discourse to the *Treatise on Chemistry* about the elements, in which he wrote,

All that one can say about the nature and number of elements is, as far as I am concerned, purely metaphysical. . . . I will only say that if by the word element we mean the simple and indivisible molecules which make up bodies, it is probable that we do not know what they are; if on the contrary we designate as elements or principles of bodies the idea of the last term to which one comes by analysis, all substances which we have not as yet been able to decompose by any means are, for us, elements.[23]

When Lavoisier added further that it is possible that the substances considered simple are themselves made up of other "principles" but that we can ignore these principles since they cannot be known experimentally, he was merely continuing the tradition of chemists like Macquer who held that chemistry as a uniquely experimental science should discuss only experimental discoveries.[24] None of this, however, should be taken to imply that Lavoisier did not consider matter to be particulate. In fact, he always took the particulate structure of matter for granted. Usually he spoke of molecules in the same sense as that in which Macquer spoke of integrant particles—the last subdivision of a substance. Similarly, Lavoisier's caloric was an elastic fluid composed of very small particles just as the fire or phlogiston of other chemists was an elastic fluid made up of particles.

Lavoisier, in fact, was only saying with the great weight of his scientific authority behind him what other chemists had been saying for some time: that it was the business of chemistry to concern itself with molecules, the smallest and simplest particles amenable to chemical examination, the last particles identical with the mass which they composed, particles of such properties that to subdivide them is to change the nature of the substance—a chemical atom. Smaller particles were presumed to exist, but to be the business of the physicist, not the chemist. Ironically, by this time the physicists were also growing sceptical of the utility of the old game of trying to explain properties directly in terms of the nature of the particles of matter. This had been new and exciting in the seventeenth century, but it was beginning to outlive its usefulness. One needed a radically new concept of matter. This was available in the atomic philosophies of Boscovich and Le Sage; but though both had interesting and novel things to say about the intimate structure of matter, neither was very constructive on the question of the cause of physical and chemical properties, nor were their particles detectable by experiment.[25] It is interesting to find that a mathematical physicist like D'Alembert agreed with the chemists in thinking that it was not much use talking about the nature of the ultimate particles of matter when they were experimentally undetectable.[26] It was no wonder then that chemists, actually making more and more use of molecules, the units of chemical union and reaction, disclaimed belief in the importance of studying the physicist's atoms.

This, I think, helps stress the important aspect of Dalton's atomic theory. Though weight was not, as far as Dalton was concerned, the only property of the atoms, it was the property which Dalton (and all subsequent chemists) most stressed, the property most useful to chemists, and the one property which was experimentally determinable. It was this experimentally determinable property, moreover, which made a substance what it was, chemically. Dalton's atoms were like the integrant particles of earlier chemists, the smallest consistent particles of a given substance. Dalton did not bother to define the elements which atoms composed; he accepted the current view summarized by Lavoisier, and spoke of simple substances as agreed upon chemical entities well known to all. (Though he did not include heat as a simple substance, he continued the common concept of an atmosphere of heat surrounding all atoms.) Dalton's atoms are the first envisaged by an eighteenth-century scientist which are, in the chemist's sense, both physical and chemical principles; that is, they are at once the last products of chemical analysis, and real physical units which account for the physical properties of bodies. This is what chemists had been striving for, consciously or unconsciously, throughout the eighteenth century, ever since they had first accepted the implications of the mechanical philosophies of the seventeenth century natural philosophers. Only the total acceptance of both the experimental discoveries of the age, and of the implications of this structure of matter for chemical theory, at last made a truly chemical atomism possible.

References

1 See my "Establishment of the Mechanical Philosophy", *Osiris*, X (1950), 413–541.

2 *Cours de chimie* (9th ed.; Paris, 1697), p. 195. Yet Lemery thought and said in his introductory material that chemistry was the only clue to the small particles of matter.

3 This argument is developed in some detail in my book *Robert Boyle and Seventeenth Century Chemistry* (Cambridge University Press, 1958).

4 See especially the "proposition" in the *Sceptical Chymist* (Everyman ed.), p. 31; *Origin of Forms*, in *Works of the Honourable Robert Boyle*, ed. Thomas Birch (5 vols.; London, 1744), II, 470; Royal Society Boyle Papers, Vol. 5, folio 41; *Mechanical Explication of Qualities*, in *Works* (1744), III, 610 (on volatility).

5 *History of Colours*, Part IV, expt. 40, in *Works* (1744), II, 64–66.

6 *Mémoires de l'académie royale des sciences pour 1702*, (2nd ed.; Amsterdam, 1737), 44–70. For Homberg's influence on the work of the Academy one must consult the Registres de physique in the archives of the *Académie des sciences*. His having worked with Boyle is mentioned by Fontenelle in his *Eloge*.

7 A convenient summary of Beccher's travels is to be found in chapter VIII of Thomas Thomson, *The History of Chemistry* (London, 1830), I, 246f.: his theory of matter is contained in *Physica subterranea* (Lipsiae, 1738); there is a convenient summary by Peter Shaw in *Philosophical Principles of Universal Chemistry* (London, 1730), footnote, p. 7.

8 G. E. Stahl, *Philosophical Principles of Universal Chemistry*, tr. Peter Shaw from the *Collegium Jenense* (much abridged), Part I, esp. 10–20.

9 *A New Method of Chemistry*, tr. Peter Shaw (2nd ed.; London, 1741), I, 155; also *Elements of Chemistry*, tr. Timothy Dallowe (London, 1735), I, 46, 78.

10 See Thomas Birch, *History of the Royal Society* (London, 1755–57), III, 247–61 and 262–69; and for the letter to Boyle, Birch, *Life of Boyle* (London, 1744), 234–47.

11 *Lexicon Technicum* (2nd ed.; London, 1723), II, B2 and B3.

12 *Chymical Lectures: in which almost all the Operations of Chymistry are Reduced to their True Principles, and the Laws of Nature* (London, 1712).

13 *Mémoires de l'académie royale des sciences* for 1718 (Paris, 1720), 256–69; and for 1720 (Paris, 1727), 200–19.

14 *Histoire de l'académie royale des sciences, année 1718* (Paris, 1727), "Sur les rapports de différents substances en Chymie," 45–47.

15 English ed., London, 1785; French ed. as *Traités des affinités chymiques* (Paris, 1788).

16 For a good summary, see P. J. Macquer, *Dictionnaire de chymie* (2nd ed.;

Paris, 1777), II, 474, s.v. "Phlogistique," and J. H. White, *The History of the Phlogiston Theory* (London, 1932); also Macquer, *Elémens de chymie théorique* (Paris, 1741), 12–15.

17 *Elements of the Theory and Practice of Chemistry* (2nd ed.; London, 1764), ch. I, 1–10 and *Dictionnaire de chymie,* s.v. "Elémens," 2nd ed., I, 527.

18 See especially s.v. "Aggrégation," I, 68–73; "Décomposition des Corps," I, 423–25; "Affinité," I, 53–68.

19 *Elémens de chymie théorique et pratique* (Dijon, 1777), I, 8–20, 50.

20 *Chemical Observations and Experiments on Air and Fire* (London, 1780); Bergman's introduction is found on pp. xv–xl; see esp., xxxii.

21 *Ibid.,* p. 3 and 95–103.

22 "Mémoire sur la Necessité de Réformer & de Perfectionner la Nomenclature de la Chimie" in Lavoisier, Guyton-Morveau, Fourcroy, Berthollet, *Méthode de nomenclature chimique* (Paris, 1787), esp. p. 17.

23 *Traité de chimie* (2nd ed.; Paris, 1793), "Discours Préliminaire," xvii–xviii.

24 *Dictionnaire de chymie,* s.v. "Décomposition des Corps."

25 R. J. Boscovich, *A Theory of Natural Philosophy,* tr. J. M. Child (Chicago, 1922), Part III, esp. art. 450–71, pp. 319–333, G. L. Le Sage, *Essai de chymie méchanique* (1758).

26 *Elémens de philosophie,* in *Oeuvres* (Paris, 1821). I, 331 (first published in 1759).

HENRY GUERLAC

on the Papers of Cyril Stanley Smith and Marie Boas

Something more than the common subject of chemistry links the two excellent papers of Professors Smith and Boas. Each study—one of them quite explicitly—contributes to our understanding of that great event of late eighteenth-century science, the Chemical Revolution.

Dr. Smith has taken up a central metallurgical problem and traced its gradual elucidation from the earliest times to our own day, but much of his paper deals with the work done in the seventeenth and eighteenth centuries. Like Miss Boas, he has a good deal to say about speculations on the nature of matter, showing the part they played in guiding the thought of scientists.

I am ill equipped to criticize Dr. Smith on his home ground, yet I should like to raise a question on one point. Was Descartes' contribution to metallurgical theory as important as Dr. Smith contends, even if—in the form Rohault's book gave to them—his ideas did help Réaumur interpret what he saw? It would seem to be the burden of the rest of Dr. Smith's paper that the real addition to our knowledge of metallic structure came not from theory, but from fracture experiments, the microscopic examination of specimens, and the careful observation of metallurgical practice. French scientists not at all beholden to Descartes, like the Chevalier de Grignon and Guyton de Morveau, were the first to show by experiment that the solidification of iron was in reality a crystallization. Nor can I quite accept Dr. Smith's suggestion that the victory of Newton over Descartes may have set back the progress of solid-state physics. Surely Newton's example, his respect for experiment, and above all his *Opticks*—from which Dr. Smith quotes with approval—served to stimulate fields that were little more susceptible than metallurgy to the elegant mathematical formulations of the *Principia*.

The metallurgical interests of eighteenth-century France, to which Dr. Smith has directed our attention in his fine paper, played an important, if often forgotten, role in preparing the Chemical Revolution. Réaumur was one of the first in France to urge that chemistry should escape from the prevailing thralldom to medicine and pharmacy and ally itself with the other useful arts. By mid-century the improvement of French mining and metallurgy had come to be recognized as a matter of central importance for the economic progress of the country. In the 1750's and 1760's Diderot's *Encyclopédie* was campaigning vigorously to re-orient chemical interests in this direction. The Baron d'Holbach wrote numerous articles on metallurgical and mineralogical subjects for the *Encyclopédie,* and translated into French important German and Scandinavian books on mineral chemistry and allied matters. It was this wave of translations, I believe, which brought Stahl's phlogiston theory into prominence in France during the youth of Lavoisier. This interest in metallurgy had a further result: It brought into contact with chemical problems, and with academic science, men of practical background, like the Chevalier de Grignon, an iron-master, expert on the design of blast-furnaces, and advisor to the Royal government. More important still, a new generation of chemists, among them Macquer, who had been trained as physicians or apothecaries, began to turn their attention to industrial problems. Without this significant transformation of the chemist's profession, without this broadening of interests, and the feverish activity of these years, the Chemical Revolution would remain quite inexplicable.

Miss Boas deals skillfully with the chemists' theory of matter in the period before Lavoisier. It is her contention that something akin to Lavoisier's pragmatic definition of a chemical element had been widely adopted by earlier eighteenth-century chemists, a fact not unconnected with the introduction into chemistry of a particulate theory of matter. A number of influential chemists, she reminds us, believed that the effective counters of chemical transformation—and it is with these the chemist is really concerned—were the "mixts" or "proximate principles." Despite the name, it is these which combine to form the familiar compounds with which the chemist deals, and to these that the compounds in turn can be broken down. It is these that Lavoisier retained as his elements. They are called "mixts" because, in theory at least, they can be reduced to "elements" or "principes primitifs," and the latter, in all likelihood, cannot be broken down further. And Miss Boas shows from the writings of Macquer in particular that this hierarchy of substances was thought to correspond to a hierarchy of particulate aggregation.

On the eve of Lavoisier's great discoveries, there was a surprising amount of agreement as to what the "principes primitifs," which Lavoisier was to banish from chemistry, actually were: They were the old peripatetic Four Elements. This should occasion some surprise, when we recall the disrepute into which the theory had fallen in the last years of the seventeenth century and the first decades of the eighteenth. Boyle and Homberg had doubted the existence of such chemical principles, and other chemists had advanced rival schemes, derived more or less from the Paracelsian *tria prima* with various additions and subtractions. Five-element theories were particularly common. But all these rival theories had one thing in common: They all denied that air could be one of the elements. I think one fact, and one fact chiefly, accounts for the resurrection of the Four Element theory in the second half of the eighteenth century. This was the discovery of Stephen Hales that air was a constituent of so many common substances. With the reinstatement of "air" his work permitted, it was not illogical (nor yet, as it proved, very helpful) to return to the Four Elements, since the later elementary theories had retained earth, water, and fire in various transparent disguises.

For the corpuscular or particulate theory of matter to be useful in chemistry, some explanation had to be forthcoming to account for the selectivity and specificity of chemical reactions, indeed to supply a motive force for chemical combinations. This was provided by the theory of *chemical attraction* which Miss Boas rightly shows to have been a necessary concomitant of the particulate theory of matter applied to chemistry. The more familiar facts of chemical affinity had been set forth in the famous table of E. F. Geoffroy in 1718. But, as Miss Boas says, Geoffroy neither adopted a corpuscular picture of what transpires, nor speculated on the nature of the forces at work. It was T. Bergman's *Dissertation on Elective Attractions* (1775) that taught Europe to interpret the phenomena of affinity in terms of Newtonian attraction. Yet Bergman was not the first to do so. Elsewhere I have called attention to the attractionist vocabulary in Joseph Black's paper on *magnesia alba* (1756). Even earlier, William Cullen had been treating the subject in his Glasgow chemical lectures. In 1759 we find Cullen referring in a letter to "single elective attractions" and "double elective attractions." We shall perhaps find that it was from Scotland that Bergman learned of the doctrine of chemical attraction which he elaborated in such detail.

Black used diagrams, which appear in a manuscript of his early lectures, to explain what happens in double displacement reactions. Because they illustrate rather well the view Miss Boas has been describing for us, and because they have never been published, I shall say something about them. Black represents what Macquer called the "molécule" or "partie intégrante" by a circle cut in two. In the two halves of the circle are the "atomes" or "parties constitutives," though Black does not use these terms.

Thus ammonium chloride is represented by ⊖ where the symbol

ȣ represents the volatile alkali, and the symbol ⋗ stands for any acid; and sodium carbonate by ⊕ . Mir is Black's symbol for Fixed Air (CO_2), and ȣ represents a fixed alkali. Black then goes on to say that "if a compound of a volatile alkali with any Acid is added to a Mild fixed Alkali, the fixed Air will unite with the volatile Alkali, while the fixed Alkali joins itself to the acid." This is illustrated by the following diagram:

AARON J. IHDE

on the Papers of Cyril Stanley Smith and Marie Boas

The very excellence of both Dr. Smith's and Dr. Boas' papers leaves one in a position of finding no real criticism to make. As a result of this, I am going to examine several points in connection with these two papers in the hope that it may be possible to extend the subjects to some extent.

Dr. Smith's paper illustrates the significance of the nature of the subject matter in connection with the pattern followed in the development of understanding of the subject. It is also a study of the inter-dependence of techniques of investigation. The successful investigation of metals had to await three developments: (1) a knowledge of crystallography; (2) microscopic techniques; and (3) the origin of phase rule. Any one of these three developments would have been insufficient for the rise of scientific metallurgy, but together they form a composite whose interaction made such scientific study of metals possible. The growth in crystallography from the beginning of the nineteenth century could not successfully stimulate the study of metals, since metals commonly form micro-crystals. These have no gross order such as is observed in crystals of chemicals such as the alums, phosphates, and tartrates. Since the normal growth of crystals in metals is hindered due to surface distortions, an internal order is present but this internal order resists systematic study without the aid of such techniques as x-ray crystallography, a development of the twentieth century. Internal order, of course, suggested itself to early students of crystallography who worked with large crystals where the systematic orientation of faces could only suggest an internal arrangement of atomic particles. Micro-crystals, however, require aids for study. Hence studies of metals could only become significant with the development of the second and third areas of investigation.

Microscopic techniques of crystallography became important with the perfection of the microscope and the development of suitable etching techniques. Dr. Smith has indicated clearly that etch patterns were fairly commonplace in metal working for a number of centuries before microscopic examination was introduced. I would like to suggest that the slowness in the perfection of the compound microscope was a factor in delaying the development of micro-techniques. Although metals were examined by Hooke and other early workers in the seventeenth century, the difficulty of working with the simple lens and the imperfections of the compound microscope made serious study difficult. As a result, all studies of materials under magnification became suspect. The difficulties in the compound microscope were not finally solved until well into the nineteenth century, by which time studies of optics made it possible to develop compound microscopes revealing a realistic image without the aberrations and optical ghosts characteristic of the images seen by earlier workers. The success of the microscope in biological studies may have been a factor in its introduction into the study of metals. It is interesting to note that the microscopic study of etch patterns in metals parallels the introduction of staining techniques in biological studies.

The interpretation of the information revealed by microscopic studies required the availability of the Phase Rule of J. Willard Gibbs. This was developed in the eighteen seventies but was not generally brought into successful application until more than a decade later. The classic work of Roozeboom on the phase pattern of iron pointed the way to successful application of the phase rule to all sorts of metallurgical problems.

Now I would like to turn my attention to a quotation from Dr. Smith's paper where he says, "Metals have always had a central place in chemical theory but until recently chemists have neither assumed nor explained any structure beyond the molecule. Moreover, since the time of Dalton they seem rather to have avoided the study of metals because so many intermetallic phases ignore the rules of simple combining proportions." This quotation points out the fact that at times the success of a concept in one area of science may have a retarding effect upon progress in other areas. The rule of simple combining proportions, or as it is frequently called, "the Law of Definite Proportions," was established at the end of the first decade of the nineteenth century only after the long battle between Berthollet and Proust. The success of the Proust position was so decisive that the matter received little critical study during the decades which followed. Possibly this was for the best as far as the progress of chemistry was concerned. Had chemists concerned themselves with the composition of solutions, glasses, and alloys the establishment of atomic theory might have been even slower than it was.

Anyone who has studied the vicissitudes of atomic theory during the period between 1810 and 1860 recognizes the tremendous problems which faced chemists of that day in connection with atomic weights, equivalent weights, reliable formulas, and matters of that sort. The Law of Definite Proportions was a useful concept in helping to bring order out of chaos.

Inconsistencies in chemical composition are much easier to deal with from the standpoint of a sound atomic hypothesis than otherwise. With the growth of a useful atomism, with an associated concept of valency, the composition of many inorganic and organic compounds became meaningful. Had chemists had to face the fact of variable composition in some of their common compounds it is doubtful if atomic theory might have been established as soon as it was.

It is in solid state chemistry that the Law of Definite Proportions has been found wanting. Not only in the case of metallic compounds are peculiar atomic ratios of the component elements to be found, but even in such solids as metallic oxides and sulfides. Ferrous oxide (FeO) presents a particularly fine example of this. Although the compound is frequently mentioned in freshman chemistry courses to illustrate the Law of Definite Proportions, the range of iron content is naturally somewhat variable. One never finds the ratio of one atom of iron to one of oxygen which is indicated by the formula, but observes, on accurate analysis, a somewhat lower ratio of iron to oxygen. Not only is the ratio of metal to non-metal lower than one to one, but the ratio is somewhat variable between a range of 0.84 to 0.95 atoms of iron per atom of oxygen. The reason for failure to show a simple whole number ratio or even a fixed ratio has become apparent only with the study of atom packing in the solid state which shows that such variation is not only possible but is to be expected. The existence of a considerable number of such compounds has led to the proposal that compounds be classified as *Berthollides* and *Daltonides;* the term *Berthollide* referring to such compounds as cuprous sulfide with a somewhat variable composition, and *Daltonide* referring to those with precisely fixed atomic ratios.

It occurs to me that it is sometimes desirable to have experimental data which are not completely precise. Had Berthollet been successful in convincing the chemical world that compounds do not have fixed proportions, the development of atomic theory would have been greatly hindered. The fact that the Proust position became the accepted one in view of the work of Dalton, meant that the trials and errors toward a successful formulation of chemical compounds would ultimately succeed on the basis of an atomic philosophy. Once the atomic philosophy was clearly developed, the existence of the *Berthollides* could still be incorporated into chemical philosophy on the basis of studies of solid state physics. This illustrates clearly, I think, the fact that science progresses from one state of approximation to another, and that progress may well be hindered when so much precise information is available that broad and useful concepts are overlooked.

Now in turning to the paper of Dr. Boas, I should like to point out that she has here, as in her earlier papers, done much to fill in the gap which existed in understanding the development of chemistry from the time of Agricola and Paracelsus. Her studies of the mechanical philosophy of Boyle and his contemporaries has done a great deal to show the influence of particulate hypotheses on the growth of chemical thought before the time of Lavoisier and Dalton.

I should like to further examine this period as one which was essential for the clearing out of a great deal of rubbish which had accumulated during the previous centuries and which was serving as an obstacle to progress in chemistry. Actually a study of this period is a confusing one because, as so often happens in the clearing out of rubbish, it is not destroyed but is tossed aside where it can again serve as an obstacle to progress. This, it seems to me, was one of the fundamental difficulties encountered in the development of seventeenth- and eighteenth-century chemistry.

Dr. Guerlac has outlined for us the various ideas about elements which were current in the chemical philosophies of the time. All of these elemental concepts had developed out of the growth of alchemy during a millennium and a half and out of the attempts to develop a medico-chemical philosophy during the sixteenth century. The philosophy of the four elements and the four qualities had made alchemy possible. This discipline had flourished on the basis of an apparently reasonable system but could not truly succeed because the system was based on false Aristotelian premises. In clearing out the accumulated rubbish a particulate philosophy could have a great deal of value, but there were perhaps other factors which were involved as well. It is these which I should like to examine.

First of all, I would like to suggest that the discard of the four elements of the Aristotelians may have been associated with the fall of Ptolemaic cosmology. We remember from the paper of Father Clark that the four elements held a very definite place in the universe of the middle ages. Earth, water, air, and fire represented the four sub-lunary spheres beyond which the spheres of the various planets were composed of the perfect element, the quintessence. However, the studies of Tycho Brahe on the nova of 1572 and the comet of 1577 revealed that these phenomena were occurring beyond the sphere of moon, in that perfect portion of the heavens where changes of this type should not be taking place. Fire was actually rejected as an element by Brahe and such philosophers as Peter Ramus and Jerome Cardan, thus discarding the sphere of fire between that of air and moon. Jean Péna actually denied the existence of celestial spheres and suggested that the sphere of air extended out to the fixed stars, thus eliminating the quintessence entirely.

Today the chemical elements appear to be unrelated to the motion of the heavens. This was not true of the elements of the middle ages and I suggest, although I am unable to document the suggestion, that the collapse of the Ptolemaic system may have been a factor in the decline of the alchemical concept of the elements.

Perhaps much more important were the rising doubts regarding the possibility of transmutation, together with an appreciation for identity of similar chemical substances arising from different sources. Doubts regarding transmutation of metals are seen as early as the works of Avicenna, who in his *Kitab al-Shifa* clearly expressed doubt about the possibility of preparing alchemical gold. Similar doubts were expressed by Albertus Magnus in his book on minerals. The Renaissance metallurgists, Biringuccio, Agricola, and slightly later, Ercker, expressed similar skepticism. Pallisy the Potter wrote that "by cupellation it is easy to show alchemical gold to be false." The physician, Nicholas Guibert, in 1603 and again in 1614, argued in opposition to Libavius that metals are species and therefore cannot be transmuted. Jean Riolan, an anti-Paracelcian of the same period, wrote that each metal is perfect in its own species. These comments all clearly reveal the beginning of a feeling that metals are unique in their characteristics and cannot be transformed one into another.

Along the same line was the recognition that substances coming from different sources might be identical in their properties and composition. Thus we find that Libavius reported that oil of vitriol (sulfuric acid) from the distillation of mineral vitriols was identical with that prepared by the combustion of sulfur with niter. Somewhat later Boyle, through his use of indicators, was able to test for acids and alkalies. He extended his testing techniques to the precipitation of salts with suitable reagents in order to identify various substances. Sala argued that the copper which was formed by the addition of iron to blue vitriol solutions was not a transmutation of iron into copper, but that rather the copper was present in a dormant state in the blue vitriol and was merely replaced from this substance by the iron. Sennert demonstrated that it was possible to recover gold which had been dissolved in aqua regia and indicated that the dissolved gold still represented a substance, gold. Van Helmont carried out certain experiments which approached in a primitive manner the law of conservation of matter when he dissolved sand in alkali and after a series of transformations recovered sand once more. Helmont also recognized that the substance *gas sylvestre,* identical with our carbon dioxide, was the same substance whether coming from fissures in caves, or from the action of acid on limestone, or from the heating of limestone. Glauber, in his numerous chemical investigations, came to understand the nature of double decomposition reactions in which the components of one substance may exchange partners with the components of another substance.

Thus it would appear that the progress in chemical thought which led to the great synthesis of Lavoisier was appearing in the seventeenth century with the development of a mechanistic philosophy applied to chemistry but along with this there was a clearing away of rubbish with the discard of Aristotelian cosmology and with a recognition of identity in chemical substances.

GLOSSARY | INDEX

GLOSSARY

MARIE BOAS is Associate Professor of History at the University of California, Los Angeles. She took her Ph.D. at Cornell and has taught at the University of Massachusetts and Brandeis University. Professor Boas' main areas of interest are the history of chemistry and of theories of matter in the seventeenth century. She is the author of *Robert Boyle and Seventeenth Century Chemistry*.

CARL B. BOYER is Professor of Mathematics at Brooklyn College. He took his Ph.D. at Columbia University and has also taught at Rutgers University and Yeshiva University. His principal areas of research are the history of mathematics and optics. Professor Boyer is the author of *The Concepts of the Calculus, History of Analytic Geometry,* and *The Rainbow: from Myth to Mathematics*.

MARSHALL CLAGETT is Professor of the History of Science at Wisconsin. He received his Ph.D. from Columbia University in 1941. His special field of interest is the history of medieval mathematics and mechanics. He is the author of *Giovanni Marliani and Late Medieval Physics, Greek Science in Antiquity, Science of Mechanics in the Middle Ages,* and co-author of *Medieval Science of Weights* and *Chapters in Western Civilization*.

JOSEPH T. CLARK, S.J., is Assistant Professor of Philosophy at Canisius College, Buffalo, New York. On the Board of Editorial Consultants of *Isis* for the area of Logic and Philosophy of Science, Father Clark is engaged in research on the work and influence of Pierre Gassendi (1592–1655). Father Clark is the author of *Conventional Logic and Modern Logic*.

I. BERNARD COHEN is Professor of the History of Science at Harvard University where he received his Ph.D. in 1947. His principal area of research has been the conceptual development of modern physical science with stress on the seventeenth and eighteenth centuries. Professor Cohen is the author, among other volumes, of *Benjamin Franklin's Experiments, Some Early Tools of American Science,* and *Franklin and Newton.*

A. C. CROMBIE is Senior Lecturer in the History of Science at the University of Oxford. He received his Ph.D. from the University of Cambridge. His principal areas of research have been in the history of medieval and early modern optics and conceptions of scientific method and modern biology. Professor Crombie is the author of *Augustine to Galileo: The Hitsory of Science A.D. 400–1650* and *Robert Grosseteste and the Origins of Experimental Science, 1100– 1700.*

E. J. DIJKSTERHUIS is Professor of Science at the Universities of Utrecht and Leyden, having previously taught at the Hogere Burgerschool Willem II at Tilburg. His major areas of research are Greek and seventeenth-century science. Professor Dijksterhuis is the author of many books, including *Val en Worp, De Mechanisering van het Wereldbeeld,* and *Archimedes.*

I. E. DRABKIN is Professor of Classical Languages at the City College of New York. He received his Ph.D. from Columbia University. Professor Drabkin is the editor and translator of Caelius Aurelianus, *On Acute and Chronic Diseases,* and co-editor of *A Source Book in Greek Science* and Caelius Aurelianus, *Gynaecia.*

CHARLES COULSTON GILLISPIE, Associate Professor of History at Princeton, received his Ph.D. from Harvard in 1949. His principal periods of research are the eighteenth and nineteenth centuries. Professor Gillispie is the author of *Genesis and Geology* and has edited *A Diderot Encyclopedia of Trades and Industry.*

JOHN C. GREENE is Associate Professor of History at Iowa State College, Ames, and received his Ph.D. at Harvard University. Professor

528

Greene's areas of major interest include the history of evolutionary concepts since the seventeenth century, and the role of science in American society in the eighteenth and nineteenth centuries. He is the author of numerous articles on biological thought in the eighteenth and nineteenth centuries, and of *The Death of Adam: Evolution and Its Impact on Western Thought.*

HENRY GUERLAC is Professor of History at Cornell University. He took his Ph.D. at Harvard University in 1941 and has taught at a number of institutions, including Yale, Harvard, and the University of Wisconsin. Professor Guerlac's major area of interest is the history of scientific thought in the seventeenth and eighteenth centuries. He is the author of *Science in Western Civilization, Selected Readings in the History of Science,* and numerous articles on the history of chemistry in the seventeenth and eighteenth centuries.

RUPERT HALL, Lecturer in the History of Science at Cambridge University, England, obtained his Ph.D. at the same institution. His major interest is in the development of physical science during the seventeenth century. Dr. Hall is the author of *Ballistics in the Seventeenth Century, The Scientific Revolution,* and one of the editors of *A History of Technology.*

ERWIN HIEBERT, Assistant Professor of the History of Science at the University of Wisconsin, took his Ph.D. there in 1954. Professor Hiebert has taught at San Francisco State College and Harvard. During 1954–55 he was a Fulbright Lecturer at the Max Planck Institute of Physics and the University of Göttingen, Germany. His principal area of research is nineteenth-century physical science with emphasis on the history of thermodynamics.

HENRY BERTRAM HILL is Professor of History at the University of Wisconsin, from which he received his Ph.D. in 1933. His field of special interest is French constitutional history, about which he has published numerous articles in journals both here and abroad. He is co-author of *Modern France* (Princeton, 1951), and co-editor of *Europe in Review* (Chicago, 1957).

AARON IHDE, Professor of Chemistry and the History of Science at the University of Wisconsin, took his Ph.D. there in 1941. He has also taught at Butler and Harvard Universities. His interests lie in the fields of pure-food legislation and the history of chemistry. Professor Ihde's publications in the latter field deal mainly with developments in nineteenth-century chemistry.

FRANCIS R. JOHNSON is Professor of English at Stanford University. He took his Ph.D. at Johns Hopkins University in 1935. His chief fields of interest are English literature of the Renaissance and the history of science during the period of the Renaissance. Professor Johnson is the author of *Astronomical Thought in Renaissance England* and of numerous articles in various scholarly journals.

THOMAS S. KUHN is Associate Professor of the History of Science at the University of California at Berkeley. He took his Ph.D. at Harvard in 1949 and taught there from 1950 to 1956. His principal area of research is the history of physical science and related technology since 1600. Professor Kuhn is the author of *The Copernican Revolution* and a number of technical and historical articles.

ROBERT K. MERTON is Professor of Sociology at Columbia University. He received his Ph.D. at Harvard in 1936. His major areas of research are sociology of science and social theory. Professor Merton has taught at Harvard and Tulane, and is the author of *Science, Technology and Society in Seventeenth-Century England, Social Theory and Social Structure,* and *Mass Persuasion;* and co-author of *The Student-Physician.*

ERNEST NAGEL is Professor of Philosophy at Columbia University where he took his Ph.D. in 1930. He has taught at the City College of New York and at Princeton University. Professor Nagel's principal areas of research are the philosophy of science, logic, and the history of logic. He is the author of a number of books including *Principles of the Theory of Probability, Sovereign Reason,* and *Logic without Metaphysics.*

DEREK J. DE S. PRICE received a Ph.D. in physics from the University of London and a Ph.D in the History of Science from Cambridge

University. Among his major areas of research are the history of scientific instruments and the history of astronomy. Dr. Price is the author of *The Equatorie of the Planetis* and of numerous articles on the history of scientific instruments including two chapters in volume three of the Oxford *History of Technology* (Singer, *et al.*).

DUANE H. D. ROLLER is Associate Professor of the History of Science at the University of Oklahoma. He received his Ph.D. from Harvard in 1954. His major area of research is the history of electricity and magnetism. Professor Roller is the author of *Development of the Concept of Electrical Charge* (with D. Roller) and *Foundations of Modern Physical Science* (with G. Holton).

GIORGIO DE SANTILLANA is Professor of History and Philosophy of Science at the Massachusetts Institute of Technology. He took his Ph.D. at Rome University in 1925. His major interests lie in the fields of antiquity and the Renaissance. Professor Santillana is the author of *Compendio di storia del pensiero scientifico* (with Federigo Enriques), Galileo's *Dialogue,* and *Crime of Galileo.*

RICHARD H. SHRYOCK is Professor of History at the University of Pennsylvania where he also received his Ph.D. He has taught at Ohio State University and Duke, and was Director of the Institute for the History of Medicine at Johns Hopkins University. His principal areas of research are American social and intellectual history and the history of modern medicine. Professor Shryock is the author of *American Medical Research Past and Present, Cotton Mather: First Significant Figure in American Medicine,* and *The Development of Modern Medicine.*

CYRIL STANLEY SMITH is Professor of Metallurgy at the University of Chicago. He received his D.Sc. in metallurgy at the Massachusetts Institute of Technology in 1926. Professor Smith came to Chicago after wartime service with the N.D.R.C. and the Manhattan District. His major interests lie in the fields of physical metallurgy and metal physics. He is the author of numerous articles and has published a series of annotated translations including *The Pirotechnia* of Vannoccio Biringuccio (with Martha T. Gnudi), Lazarus

Ercker's *Treatise on Ores and Assaying*, and *The Probierbüchlein* (the last two with Anneliese G. Sisco).

DOROTHY STIMSON is Emeritus Professor of History at Goucher College, Baltimore. Since her retirement she has taught at Sweet Briar, Vassar, Sarah Lawrence, and Mount Holyoke colleges. She took her Ph.D. at Columbia University. Professor Stimson's major areas of academic interest are the history of science and seventeenth-century English history. She is the author of *The Gradual Acceptance of the Copernican Theory of the Universe* and *Scientists and Amateurs: A History of the Royal Society*.

L. PEARCE WILLIAMS is Assistant Professor of History at the University of Delaware. He received his Ph.D. in 1952 from Cornell University and has taught at Yale University. His principal area of research has been science and education during the period of the French Revolution and First Empire. Professor Williams has contributed a number of articles dealing with this subject to *Isis*.

J. WALTER WILSON is Professor of Biology at Brown University where he took his Ph.D. in 1921. His principal areas of research are experimental embryology, cytology, and cancer biology. Aside from his major publications in anatomy, physiology, and pathology, Professor Wilson has written numerous articles on the history of cellular theory.

CONWAY ZIRKLE is Professor of Botany at the University of Pennsylvania. He received his Ph.D. from Johns Hopkins University in 1925. Professor Zirkle's major area of research is cytology and cytogenetics. He is the author of *The Beginnings of Plant Hybridization, Death of a Science in Russia,* and *Evolution, Marxian Biology and the Social Scene.*

INDEX

Galilean-Newtonian mechanics: and consequent research, 181

Galilei, Galileo (1564–1642): and university courses, 6; and academic strife, 7; and Archimedes, 12; on servility to ancient thought, 13; and the Florentine Arsenal, 28; and his philosophical status, 42; and natural philosophy, 42; and the anti-Copernican decree of 1616, 46; and science of dynamics, 47; third letter to Dini, 54; first integral of motion, 55; on sunspots, 56; and change of form, 56; on falling bodies, 86; and graphical presentation of Mertonian rule, 150; and *vis impressa,* 173; and classical dynamics, 180–81; and experimental method in physics, 181; and the macroscopic aspects of his work, 196; and relation between mathematical and non-mathematical parts of astronomy, 199; and the too-steady brightness of Venus, 213

Galton, Sir Francis (1822–1911), 464

Galvanism: theory of, and electric current, 323

Gassendi, Pierre (1592–1655): contra Descartes, 395; and metallurgy, 472; and the causes of brittleness in metals, 473

Gear-wheels: geometrical theory of, mentioned, 16

Gegenbaur, Karl (1826–1903), 416

Generation, Spontaneous: and experiments of Redi and Spallanzani, 410; and Virchow, 412

Geocentric. *See* geostatic

Geoffrey, Étienne François (1632–1731): and alloys in the copper zinc system, 487; and the table of affinities, 507, 517

Geology: development of, and the study of fossils, 407

Geometrical devices, principal: and Ptolemy's celestial kinematics, 117

Geometry: transformation of, and Archimedes, 11; and visual space, 52; and difference between projective geometry and elementary high school geometry, 110; Cartesian analytical geometry

and Oresme's *longitudo,* 127; tests in, and the *Académie des sciences,* 311

Geostatic hypothesis: and Ptolemy's celestial kinematics, 117

Geostatic systems: change to heliostatic, and mathematical techniques, 202

Gesner, Konrad von (1516–1565), 33

Ghiberti, Lorenzo (1378–1455): and optical theories, 36; and report on portals of Baptistry, 40; and the Guild of the Masters of Stone and Wood, 43; and the chain-and-girder system, 49

Ghirlandaio, Domenico (1449–1498), 43

Gibbs, Josiah Willard (1839–1903), 494, 520

Glauber, Johann Rudolf (1604–1668), 524

Goethe, Johann Wolfgang von (1749–1832), 255

Gold: transformation of, and menstruums, 506

Golden Age of Mathematics: definition, 186

Golden Age of Mechanics: definition, 186

Golden Rule of Mechanics: and elementary statics, 166

Golden Section: and Brunelleschi, 62

Gore, G., 492

Gothic architecture: and strict rules of proportion, 75

Gothic structure: "a denial of proportion," 59

Goucher College: and the teaching of the history of science, 224–34

Gozzoli, Benozzo (1420–1498), 43

Grain growth: observed metallographically on steel, 494

Graphical Methods (Oresme): and Galileo, 95

Gravesande, Willem Jakob (1688–1742), 395

Gravitation: synthesis of terrestrial and celestial mechanics (Newton), 182

Gravitational attraction: Newton and, 183–84

Gravity: and Galileo's comment on falling bodies, 86; and Scholasticism, 171–72

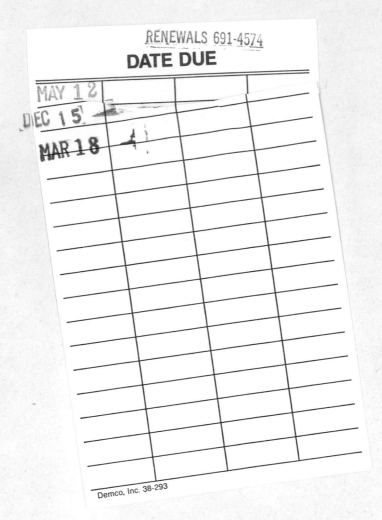